The Pineal Gland and Its Endocrine Role

NATO Advanced Science Institutes Series

A series of edited volumes comprising multifaceted studies of contemporary scientific issues by some of the best scientific minds in the world, assembled in cooperation with NATO Scientific Affairs Division.

This series is published by an international board of publishers in conjunction with NATO Scientific Affairs Division

A Life Sciences **B Physics**	Plenum Publishing Corporation New York and London
C Mathematical and Physical Sciences	D. Reidel Publishing Company Dordrecht, Boston, and London
D Behavioral and Social Sciences **E Applied Sciences**	Martinus Nijhoff Publishers The Hague, Boston, and London
F Computer and Systems Sciences **G Ecological Sciences**	Springer Verlag Heidelberg, Berlin, and New York

Recent Volumes in Series A: Life Sciences

Volume 60—The Use of Human Cells for the Evaluation of Risk from Physical and Chemical Agents
edited by Amleto Castellani

Volume 61—Genetic Engineering in Eukaryotes
edited by Paul F. Lurquin and Andris Kleinhofs

Volume 62—Heart Perfusion, Energetics, and Ischemia
edited by Leopold Dintenfass, Desmond G. Julian, and Geoffrey V. F. Seaman

Volume 63—Structure and Function of Plant Genomes
edited by Orio Ciferri and Leon Dure III

Volume 64—Gene Expression in Normal and Transformed Cells
edited by J. E. Celis and R. Bravo

Volume 65—The Pineal Gland and Its Endocrine Role
edited by J. Axelrod, F. Fraschini, and G. P. Velo

The Pineal Gland and its Endocrine Role

Edited by

J. Axelrod

National Institute of Mental Health
Bethesda, Maryland

F. Fraschini

University of Milan
Milan, Italy

and

G. P. Velo

University of Verona
Verona, Italy

Springer Science+Business Media, LLC

Proceedings of a NATO Advanced Study Institute on
the Pineal Gland and Its Endocrine Role,
held June 21–July 2, 1982,
at the Ettore Majorana Center,
in Erice, Sicily, Italy

Library of Congress Cataloging in Publication Data

NATO Advanced Study Institute on the Pineal Gland and Its Endocrine Role (1982: Ettore Majorana International Centre for Scientific Culture)
The pineal gland and its endocrine role.

(NATO advanced science institutes role. Series A, Life science; v. 65)
"Published in cooperation with NATO Scientific Affairs Division."
"Proceedings of a NATO Advanced Study Institute on the Pineal Gland and Its Endocrine Role, held June 21–July 2, 1982, at the Ettore Majorana Center, in Erice, Sicily, Italy" —T.p. verso.
Includes bibliographical references and index.
1. Pineal body—Congresses. 2. Melatonin—Physiological effect—Congresses. I. Axelrod, Julius, 1912– II. Fraschini, F. III. Velo, G. P. IV. North Atlantic Treaty Organization. Scientific Affairs Division. V. Title. VI. Series. [DNLM: 1. Pineal body—Congresses. 2. Melatonin—Congresses. WK 350 N279p 1982]
[QP188.P55N37 1982] 599′01′42 83-8134
ISBN 978-1-4757-1453-1 ISBN 978-1-4757-1451-7 (eBook)
DOI 10.1007/978-1-4757-1451-7

Originally published by Plenum Press, New York in 1983

PREFACE

The pineal gland has been a subject of interest and speculation for more than 2000 years. Greek anatomists were impressed by the observation that the pineal gland is an unpaired structure and they believed that it regulated the flow of thoughts. The philosopher Descartes proposed an important role for this organ in brain function. At the beginning of the 20th century experiments by several investigators indicated that the pineal influenced sexual function and skin pigmentation and was also responsive to light signals. With the isolation of melatonin from bovine pineal glands by Lerner and coworkers in 1958 the modern era of pineal research was initiated.

Within a few years the pathway for the biosynthesis of melatonin in the pineal was elucidated. Soon thereafter it was shown that the formation of melatonin was influenced by environmental lighting. Anatomists found that the pineal was innervated by sympathetic nerves and that the gland had photoreceptor elements. It was also shown that the gonads were influenced by light via the pineal gland. Research on the pineal gland became of increasing interest to anatomists, biochemists, pharmacologists and endocrinologists. With the expanding knowledge concerning the function of the pineal gland contributed by the wide variety of disciplines, it was thought that a study workshop would be timely.

For this study group, a limited number of investigators in anatomy, biochemistry, pharmacology, endocrinology and clinical medicine who were working on the pineal were asked to come together to tell about their investigations, exchange ideas and to suggest new directions in research. Also attending the meeting were graduate students and postdoctoral fellows. The Study Institute on "The Pineal and its Endocrine Role" was held at Erice, Sicily at the "Centro Ettore Majrana" and sponsored by NATO. The meeting fully satisfied our expectations. During 9 days of meeting, the pineal was discussed from many points of view. There was an informative and stimulating exchange by leading experts as well as junior investigators and graduate students. The lecture and discussion provided provocative hypotheses and suggestions for new research directions. Especially interesting was the potential role of the pineal in clinical medicine. The present volume includes the proceedings of this meeting as well as reviews on anatomy, biochemistry, physiology and endocrinology of the pineal gland and the most recent data concerning this gland.

The organizers and participants are grateful to NATO for their support. We also thank the "Centro Ettore Majorana" for their gracious hospitality. The lovely setting of Erice provided a peaceful place for a fruitful scientific meeting.

J. Axelrod, F. Fraschini and G.P. Velo

CONTENTS

CLINICAL ASPECTS

REGULATION OF CIRCADIAN RHYTHMS OF INDOLEAMINES IN THE PINEAL GLAND

Julius Axelrod

Section on Pharmacology, Laboratory of Clinical Science
National Institute of Mental Health, Bethesda, Maryland
20205

Although the pineal gland has been recognized for many centuries it was the discovery of melatonin that opened the modern era of research in this organ. McCord and Allen had observed in 1927 that an extract of the bovine pineal can blanch the skin of tadpoles. This prompted Lerner, a dermatologist and biochemist interested in pigmenation, to isolate the active blanching factor of the pineal (Lerner et al., 1978). The active blanching principle of the pineal organ was isolated and identified as 5-methoxy-N-acetyltryamine (which was named melatonin). Because of my interest in both indoles and transmethylation reaction, I initiated studies on the biosynthesis and regulation of melatonin metabolism in the pineal. Together with Wurtman we found that the pineal gland can act as a neuroendocrine transducer converting neuronal signals, which are controlled by environmental lighting, into endocrine messages. This led to the formulation of the "melatonin hypothesis" in which we proposed that the pineal gland is influenced by light-dark cycles to regulate the synthesis of melatonin, a compound that acts at distant target organs (Wurtman and Axelrod, 1965). Further work in our laboratory led to the exploration of this gland to study circadian rhythms and the beta-adrenergic receptors.

THE BIOSYNTHESIS AND METABOLISM OF MELATONIN IN THE PINEAL GLAND

Soon after the discovery of melatonin, Weissbach and I began studies on the biosynthesis of this indole. We soon isolated and purified an enzyme from bovine pineal which O-methylated several indoles (Axelrod and Weissbach, 1961) (Table 1). Although many hydroxy indoles could be O-methylated, N-acetylserotonin was by far

Table 1. Substrate Specificity of Hydroxyindole-O-Methyltransferase

Substrate	Relative Activity
N-acetylserotonin	100
Serotonin	7
5-Hydroxydimethyltryptamine	14
5-Hydroxyindole acetic acid	12
4-Hydroxy-N-acetyltryptamine	0
6-Hydroxydimethyltryptamine	0

Purified enzyme obtained from bovine pineal was incubated with [^{14}C]methyl-S-adenosylmethionine and substrate at pH 7.9. After one hour incubation the reaction mixture was examined for [^{14}C]-labelled methyl derivative after extraction into organic solvent (from Axelrod and Weissbach, 1961).

the best substrate for this enzyme which we named hydroxyindole-O-methyltransferase (HIOMT). The isolation of HIOMT was in great part due to the synthesis of the radioactive methyl donor [^{14}C]-methyl-S-adenosyl-L-methionine. This made it possible to label and isolate the transmethylated product [^{14}C]-methoxymelatonin from a bovine pineal gland preparation. HIOMT was found to be highly localized in the pineal gland of mammals, birds and amphibians. This observation convinced me that this organ is biologically active and worthy of further studies. We also found an enzyme in the pineal that converted serotonin to N-acetylserotonin (Weissbach et al., 1961). In view of these findings it was proposed that melatonin is synthesized almost exclusively in the pineal gland as follows: tryptophan ⟶ 5-hydroxytryptophan ⟶ serotonin ⟶ N-acetylserotonin ⟶ melatonin. To study the fate of melatonin in the body Kopin and I synthesized radioactive melatonin which was labelled on both the indole ring and acetyl group (Kopin et al, 1961). After the injection, melatonin was almost completely metabolized in the body. The major route of melatonin was via hydroxylation to form 6-hydroxymelatonin. This transformation was carried out by an enzyme in the liver microsomes that required NADPH and oxygen.

EFFECTS OF LIGHT ON INDOLEAMINE METABOLISM IN THE PINEAL

It has been long known that environmental lighting has an effect on the pineal. Holmgren (1918) had found that the amphibian pineal has photoreceptors and later it was reported that these receptors responded to light. Fiske et al. (1960) observed that continuous light reduced the weight of the rat pineal and increased that of the ovaries. Pineal extracts were then found to decrease light induced hypertrophy of the ovaries and persistent estrus in the rat (Wurtman et al., 1961). These findings indicated a relationship between light, the pineal and ovaries. When Wurtman joined my laboratory we found that injecting melatonin into rats had the same effect on the rat gonads as bovine pineal extracts. These results suggested that melatonin could be a putative gonadal hormone secreted by the pineal and that light influences its formation. Since we had just described the route of biosynthesis of melatonin in the pineal, the effect of light on the intermediate metabolism of this indole was studied.

We first examined the effect of continuous light and darkness on the melatonin forming enzyme, HIOMT (Wurtman et al., 1963). Rats kept in constant light showed a decrease in enzyme activity as compared to those in darkness. This indicated that constant light suppressed the activity of HIOMT and this would reduce the production of the melatonin, a gonad inhibiting compound. How did environmental lighting affect the enzymes in the pineal gland which lies deep between two cerebral hemispheres? Kappers (1960) had, a few years previously, found that the mammalian pineal gland is innervated by sympathetic neurons arising in the superior cervical ganglia. This provided an opportunity to trace that pathway by which light signals relay their messages to the pineal gland (Wurtman et al., 1964). Blinded rats prevented the reduction of pineal HIOMT in rats kept in constant light. This indicated that the retina is necessary for environmental light signals to reach the pineal gland. The sympathetic nerve terminals innervating the pineal gland were then destroyed by the bilateral removal of the superior cervical ganglia. Upon denervation there was no longer a difference in HIOMT in rats kept in light or darkness. In another experiment the preganglionic nerves between the brain and the superior cervical ganglia were cut. This also abolished the effect of light and darkness on the melatonin forming enzyme. All these experiments indicated that information about environmental lighting that affects HIOMT reaches the brain via the retina, brain, superior cervical glanglia and sympathetic noradrenergic nerves. The above observations led to the proposal that the pineal gland is a neuroendocrine transducer transforming neuronal signals concerning environmental lighting from noradrenergic sympathetic nerves to endocrine signals (melatonin) acting on distant organs (gonads) (Wurtman and Axelrod, 1965).

CIRCADIAN RHYTHMS OF INDOLEAMINES IN THE PINEAL GLAND

In 1963 Quay reported that the levels of serotonin in the pineal were high during the daytime and low during the night. Since acetylated serotonin is the precursor of melatonin, this result was consistent with our findings that the melatonin-forming enzyme was higher at night than during the daytime. Together with Snyder, a postdoctoral fellow in my laboratory, a extremely sensitive assay for serotonin was developed (Snyder et al., 1965) which enabled us to explore the controlling mechanisms for the day-night changes in pineal serotonin. We observed that if rats were held in continuous darkness the diurnal rhythm in pineal serotonin persisted (Snyder et al., 1965) indicating that it was truly circadian driven by an internal clock. In rats kept in continuous light the day-night differences in serotonin was suppressed. Prompted by our previous findings that light and sympathetic nerves affect HIOMT, the role of the retina and the sympathetic nervous system on the circadian rhythm of pineal serotonin was examined. In blinded rats the circadian rhythm in pineal serotonin persisted, again supporting its circadian nature. The indoleamine rhythm was abolished when the superior cervical ganglia was removed. Depleting the brain and sympathetic nerves of noradrenaline by the administration of reserpine suppressed the serotonin rhythm. Cutting the nerves preganglionically and interrupting nerve connections from the brain to the superior cervical ganglia also abolished the pineal serotonin rhythm. These findings indicated that the circadian rhythm of serotonin in the pineal is regulated by sympathetic nerves innervating pineal cells presumably by the changes in the release of noradrenaline. Later we showed that there were day-night differences in the turnover and presumably the release of noradrenaline in the pineal (Brownstein and Axelrod, 1974). The rhythm in pineal noradrenaline turnover persisted in blinded rats and was suppressed by continuous light. These results suggested that more of the neurotransmitter was released during the nighttime. Decentralization experiments indicated that the circadian rhythm of serotonin was generated by a clock in the brain (Snyder et al., 1965). Later, work by Moore (1974) showed that the site of this clock in the brain is the suprachiasmatic nucleus.

In 1970, Klein and Weller found a circadian rhythm in N-acetyltransferase, in the rat pineal, which was 180° out of phase with the serotonin rhythm (Fig. 1). Soon after the onset of darkness there is a 30 to 50-fold increase in enzyme activity. Day-night rhythms in the pineal melatonin was found which like N-acetyltransferase was highest during the night and lowest during the day (Wurtman and Moskowitz, 1977). As was reported with serotonin, pineal N-acetylserotonin rhythm was suppressed by removal of the superior cervical ganglia or by decentralization. Bilateral lesions in the suprachiasmatic nucleus also abolished the circadian rhythm of N-acetyltransferase.

SYNTHESIS OF MELATONIN IN PINEAL ORGAN CULTURE

In collaboration with the late Harvey Shein and Wurtman (Axelrod et al., 1968), it was observed that the rat pineal gland in organ culture can synthesize [^{14}C]melatonin from [^{14}C]tryptophan. The pathway of biosynthesis of melatonin in pineal organ culture was similar to that observed in the intact rat pineal. This preparation provided an opportunity to examine the role of the sympathetic nerve neurotransmitter noradrenaline at a cellular level. The addition of noradrenaline to the rat pineal gland in organ culture markedly stimulated the formation of melatonin from tryptophan. The addition of serotonin had little effect. To examine whether noradrenaline stimulates melatonin synthesis via adrenergic receptors, the effect of alpha and beta-adrenergic blocking agents were examined. The beta-adrenergic blocking agent propranolol prevented the stimulation of melatonin synthesis in organ culture by noradrenaline while the alpha-blocking compound phenoxybenzamine had no such effect (Wurtman et al., 1971). Dibutyryl cyclic AMP also stimulated the synthesis of melatonin indicating that the beta-adrenergic receptor involved in melatonin synthesis is coupled to adenylate cyclase.

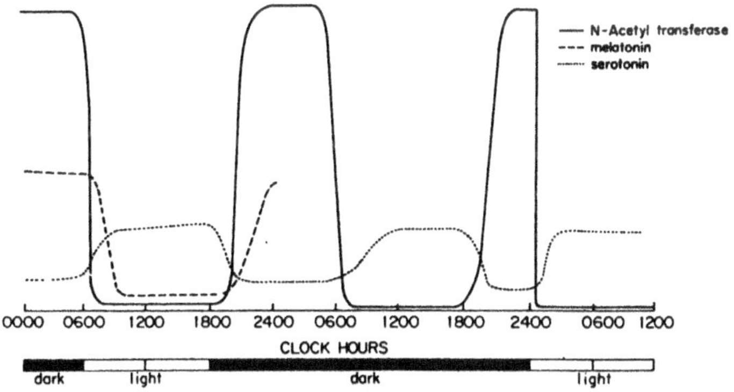

Fig. 1. Circadian rhythms of serotonin, N-acetyltransferase and melatonin in the rat pineal gland (from Axelrod, 1974).

At about the same time Klein and Berg (1970) demonstrated a marked stimulation of N-acetylserotonin transferase (NAT) by noradrenaline in organ cultures. They also found an elevation of cyclic AMP in rat pineal organ culture after the addition of noradrenaline. Because of the large amplitude of N-acetylserotonin transferase rhythm in vivo and its capacity to be stimulated by a beta-adrenergic receptor, work on the pineal in our laboratory and others were directed mainly towards this enzyme. The development of a simple assay for measuring N-acetyltransferase (Deguchi and Axelrod, 1972a) made it possible for rapid advances in uncovering the cellular and subcellular mechanisms driving the rhythms of melatonin and other indole amines in the pineal gland.

ADRENERGIC REGULATION OF THE PINEAL CIRCADIAN RHYTHM

As mentioned above pineal N-acetyltransferase showed a marked circadian rhythm in the rat. To determine the role of the adrenergic system in regulating this rhythm a number of pharmacological manipulations were tried (Deguchi and Axelrod, 1972b) (Fig. 2). The administration of the beta-adrenergic blocking agent propranolol blocked the nighttime rise in N-acetyltransferase. Depletion of noradrenaline from nerves by reserpine also prevented the marked elevation of N-acetyltransferase at night. When cycloheximide, a protein synthesis inhibitor, was injected before the onset of darkness the circadian rhythm of N-aceytltransferase was abolished. Denervation of the superior cervical ganglia, decentralization or lesions in the suprachiasmatic nucleus also abolished the nighttime

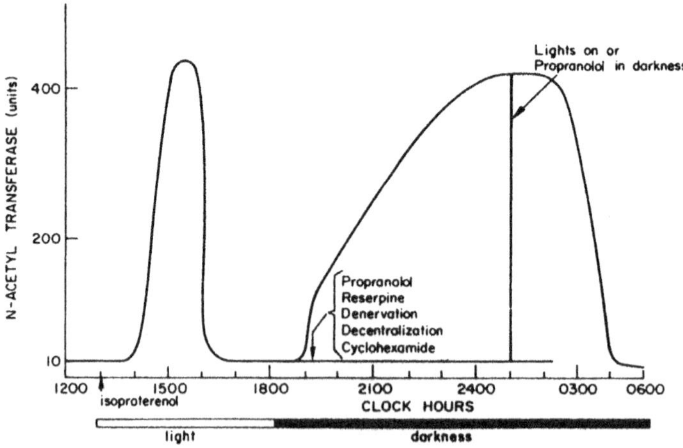

Fig. 2. Effect of adrenergic drugs and other manipulations on the circadian rhythms of N-acetyltransferase in rat pineal.

elevation N-acetyltransferase. Thus indicating that the pacemaker for circadian rhythm arises in the suprachiasmatic nucleus. The circadian messages are then transmitted to preganglic fibers of the superior cervical ganglia and on to the beta-adrenergic recepors on the pineal cell surface. These experiments and others indicated that with the onset of darkness there is an increased release of noradrenaline onto the beta-adrenergic receptor of the the pineal gland. This then initiates a series of events leading to the synthesis of N-acetyltransferase intracellularly (Fig. 3). Denervation of the superior ganglia, decentralization or lesions in the suprachiasmatic nucleus also blocked the nighttime elevation of N-acetyltransferase.

When rats are exposed to a brief period of light during the night when the pineal N-acetyltransferase levels are highest there is a rapid fall of this enzyme to the low daytime levels (Deguchi and Axelrod, 1972c) (Fig. 2). Injection of isoproterenol blocked the light induced fall of the enzyme. The administration of a beta-blocking drug at night caused a precipitous fall in the

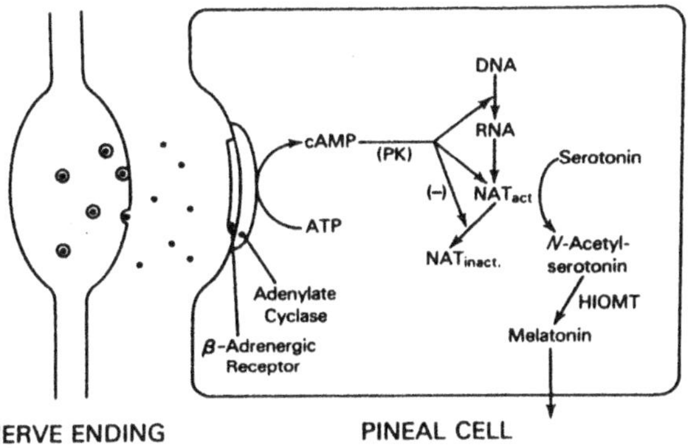

Fig. 3. Intracellular events in the stimulation of melatonin synthesis by the beta-adrenergic receptor. NAT_{act} is active form N-acetyltransferase, NAT_{inact} is inactive form. HIOMT is hydroxyindole-O-methyltransferase. • is noradrenaline. PK is protein kinase.

high levels of N-acetyltransferase. Thus the maintenance of the elevated levels of N-acetyltransferase at night requires the continuous stimulation of the pineal beta-adrenergic receptor.

SUPER- AND SUBSENSITIVITY OF THE PINEAL BETA-ADRENERGIC RECEPTOR AND THE AMPLIFICATION OF CIRCADIAN RHYTHMS

Increased responsiveness (supersensitivity) after beta-adrenergic stimulation by sympathetic nerve denervation is a well-known phenomena in many biological systems. The rat pineal gland was found to be a useful model to study changes in responsiveness in the beta-adrenergic receptor. Depriving the pineal gland of beta-adrenergic stimulation by surgical denervation, chemical sympathectomy with 6-hydroxydopamine or exposure of rats for long periods of light caused a marked increase of N-acetyltransferase synthesis, adenylate cyclase, accumulation of cyclic AMP after the administration of the beta-adrenergic agonist isoproterenol (Deguchi and Axelrod, 1973). These biochemical changes in beta-adrenergic activity occurred rapidly and appeared to be a function of the daily light-dark cycle. Pineal gland from rats housed under diurnal lighting conditions are supersensitive to beta-adrenergic stimulation as measured by induction of N-acetyltransferase by beta-adrenergic agonists at the end of the 12 hour light period and subsensitive at the end of the dark period. These rapid changes of sensitivity of the beta-adrenergic receptor are presumably due to the decreased release of noradrenaline from nerves innervating the pineal gland during the day and increased release at night. Subsensitivity can occur very rapidly, a single injection of isoproterenol causing a markedly decrease response in the pineal to a challenging dose of the drug given five hours later.

The role of catecholamines in producing super- and subsensitivity was examined in pineal organ culture (Axelrod, 1974) (Table 2) When the pineal gland was deprived of its neurotransmitter by denervation it was five times more responsive to the induction of N-acetyltransferase by a beta-adrenergic agonist in organ culture than the intact gland. When pineal glands of rats were exposed to excessive amounts of catecholamines by repeated injections of isoproterenol they were much less responsive to N-acetyltransferase induction as the gland of the untreated rats. These experiments indicated the activity of the post-synaptic beta-adrenergic receptor and the consequent formation of the pineal hormone depend on the previous exposure of the receptor to the neurotransmitter.

The beta-adrenergic receptor can be measured by its ability to specifically bind [^{3}H]dihydroalprenolol, a potent beta-adrenergic antagonist (Zatz et al., 1976). Binding of [^{3}H]dihydroalprenolol

Table 2. Supersensitivity and Subsensitivity in Rat Pineals.

l-Isoproterenol (M)	N-Acetyltransferase (Units)		
	Intact	Denervated	l-Isoproterenol-treated
1×10^{-9}	13	330	--
5×10^{-9}	330	1330	26
2×10^{-9}	680	2190	70
1×10^{-7}	940	1180	320
1×10^{-6}	1720	1490	1380

Rats were denervated by bilateral removal of the superior cervical ganglia 7 days before they were killed. Isoproterenol-treated rats received the drug (2.0 mg/kg) 8, 16, and 24 hours before they were killed. Pineals were cultured for 10 hours with indicated concentrations of isoproterenol, and, N-acetyltransferase activity was measured (from Deguchi and Axelrod, 1973).

to pineal membrane was found. This binding was found to be rapid, reversable, saturable and stereospecific. A marked variation in the number of available specific binding sites of [^{3}H]dihydroalprenolol in the rat pineal was observed (Romero et al., 1975) (Fig. 4). During the daytime with the decrease of sympathetic nerve activity the number of receptor sites, as measured by binding of the radioactive ligand, increased. During the night the number of receptors fell and reached a minimum at the end of the dark period.

The beta-adrenergic receptor is coupled to adenylate cyclase (Fig. 3). Diurnal changes in the activity of the adenylate cyclase as measured by the accumulation of cyclic AMP in the pineal was found (Axelrod and Zatz, 1977). Pineals taken from light exposed rats showed a greater accumulation of cyclic AMP in response to isoproterenol than did glands taken from dark exposed animals. Cholera toxin which bypasses the beta-adrenergic receptor and acts directly on the adenylate cyclase produced a greater accumulation of cyclic AMP in supersensitive glands. These differences in cyclic AMP accumulation were reflected in the induction of N-acetyltransferase activity.

Cyclic AMP acts to relay its biologic message (formation of melatonin) via a protein kinase (Zatz and Axelrod, 1978) (Fig. 3). As in the case of the beta-adrenergic receptor and adenylate cyclase the activity of pineal protein kinase varied with the activity of

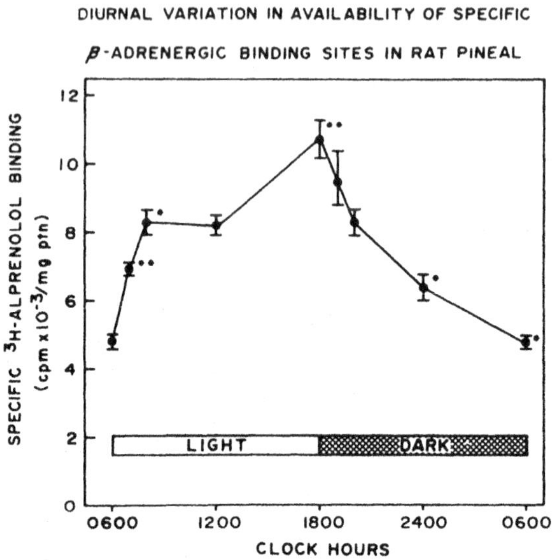

Fig. 4. Circadian rhythm in beta-adrenergic binding sites in rat pineal (from Romero et al., 1975).

light-dependent sympathetic nerve activity. There was a greater activity of protein kinase during the daytime than during the night. Further experiments showed that cyclic AMP acts at three sites in the biosynthesis of the protein, N-acetyltransferase; at the level of transcription via messenger RNA at the level of translation of the RNA message and at the level of degradation of NAT.

All these results indicate that there is a complex cascade of cellular reactions that amplifies and dampens the neurotransmitter signal to affect the activity of N-acetyltransferase, the enzyme that generates the pineal hormone melatonin. During the daytime there is a reduced release of noradrenaline but the amount of neurotransmitter release is not sufficient to trigger the synthesis of N-acetyltransferase. Because of the lower level of noradrenaline interacting with the pineal cell surface there is a gradual increase in the number of beta-adrenergic receptors, adenylate cyclase and cyclic AMP dependent protein kinase. As night falls there is an increased discharge of noradrenaline onto a supersensitive receptor and because there is also an increased amount of adenylate cyclase, cyclic AMP and protein kinase the neurotransmitter signal is amplified about 30- to 50-fold to stimulation in the synthesis of N-acetyltransferase. As night progresses the beta-adrenergic receptor number decreases and so does the cyclic AMP cascade and the pineal becomes gradually desensitized and this is reflected in the decreasing synthesis of N-acetyltransferase.

In the past few years there has been rapid advances in the ability to directly measure melatonin in body fluids and tissues. This makes it now possible to examine the role of pineal melatonin in normal body functions and in disease in man. Recent evidence indicates that the melatonin hypothesis that Wurtman and I proposed in 1965 in animal studies also applies to man. Several investigators have found that there is a circadian rhythm in melatonin synthesis controlled by sympathetic nerves and regulated by light. The secreted pineal hormone melatonin then acts at distant organs probably the gonads.

REFERENCES

Axelrod, J., 1974, The pineal gland: A neurochemical transducer. Science, 184:1341.

Axelrod, J., Shein, H.M. and Wurtman, R.J., 1969, Stimulation of C^{14}-tryptophan by noradrenaline in rat pineal in organ culture. Proc. Natl. Acad. Sci. USA, 62:544, 1969.

Axelrod, J. and Weissbach, H, 1961, Purification and properties of hydroxyindole-O-methyl transferase. J. Biol. Chem., 236: 211.

Axelrod, J. and Zatz, M., 1977, The beta-adrenergic receptor and the regulation of circadian rhythms in the pineal gland. in: "Biochemical Actions of Hormones," Vol. IV, Chapter 5, G. Litwack, ed., Academic Press, New York.

Brownstein, M.J. and Axelrod, J., 1974, Pineal gland: 24-Hour rhythm in norepinephrine turnover. Science, 184:163.

Deguchi, T. and Axelrod, J., 1972a, Sensitive assay for serotonin N-acetyltransferase activity in rat pineal. Anal. Biochem., 50: 174.

Deguchi, T. and Axelrod, J., 1972b, Induction and super-induction of serotonin N-acetyltransferase by adrenergic drugs and denervation in the rat pineal. Proc. Natl. Acad. Sci. USA, 69::2208.

Deguchi, T. and Axelrod, J., 1972c, Control of circadian change of serotonin N-acetyltransferase activity in the pineal organ by the beta-adrenergic receptor. Proc. Natl. Acad. Sci. USA, 69:2547.

Deguchi, T. and Axelrod, J., 1973, Supersensitivity and subsensitivity of the beta-adrenergic receptor in pineal gland regulated by catecholamine transmitter. Proc. Natl. Acad. Sci. USA, 70:2411.

Fiske, V.M., Bryant, K., Putnam, J., 1960, Effect of light on the weight of the pineal in the rat. Endocrinology, 66:489.

Holmgren, N., 1918, Zur Kenntnis der Parietalogran von Rana temporari. Ark. Zool., 11:1.

Kappers, J.A., 1960, The development, topographical relations and innervation of the epiphysis cerebri in the albino rat. Z. Zellforsch. Mikrosk. Anat. 5:163.

Klein, D.C. and Berg, G.R., 1970, Pineal gland: Stimulation of melatonin production by norepinephrine involves cyclic AMP mediated stimulation of N-acetyltransferase. Adv. Biochem. Psychopharmacol., 3:241.

Klein, D.C. and Weller, J.L., 1970, Indole metabolism in the pineal gland: A circadian rhythm in N-acetyltransferase. Science, 169:1093.

Kopin, I.J., Pare, C.M.B., Axelrod, J. and Weissbach, H., 1961, The fate of melatonin in animals. J. Biol. Chem., 236:3072.

Lerner, A.B., Case, J.D., Takahashi, Y., Lee, T.H. and Mori, W., 1958, Isolation of melatonin, the pineal gland factor that lightens melanocytes. J. Amer. Chem. Soc., 80:2587.

McCord, C.P. and Allen, F.P., 1917, Evidences associating pineal gland function with alterations in pigmentation. J. Exp. Zool., 23:207.

Moore, R.Y., 1979, Visual pathways and the central neural control of diurnal rhythms. in: "The Neurosciences, Third Study Program," F.O. Schmitt, F.G. Worden, eds, MIT Press, Cambridge, Mass.

Quay, W.B., 1963, Circadian rhythm in rat pineal serotonin and its modifications by estrous cycle and photoperiod. Gen. Comp. Endocrinol., 3:473.

Romero, J.A., Zatz, M., Kebabian, J.W and Axelrod, J., 1975, Circadian cycles in binding of ^{3}H-alprenolol to beta-adrenergic receptor cites in rat pineal. Nature, 258:435.

Snyder, S.H., Axelrod, J. and Zweig, M., 1965, A sensitive and specific fluorescence assay for tissue serotonin. Biochem. Pharmacol., 14:831.

Snyder, S.H., Zweig, M., Axelrod, J. and Fischer, J.E., 1965, Control of the circadian rhythm in serotonin content of the rat pineal gland. Proc. Natl. Acad. Sci. USA, 53:301.

Weissbach, H., Redfield, B.G. and Axelrod, J., 1961, Enzymic acetylation of serotonin and other naturally occurring amines. Biochim. Biophys. Acta, 54:190.

Wurtman, R.J. and Axelrod, J., 1965, The pineal gland. Sci. Amer. 213:50.

Wurtman, R.J. Axelrod, J. and Fischer, E., 1964, Melatonin synthesis in the pineal gland: Effect on light mediated by the sympathetic nervous system. Science 143:1328.

Wurtman, R.J., Axelrod, J. and Phillips, L.S., 1963, Melatonin synthesis in the pineal gland: Control by light. Science, 142: 1071.

Wurtman, R.J. and Moskowitz, M.A., 1977, The pineal organ. New Eng. J. Med., 296:1329 and 1383.

Wurtman, R.J., Roth, W., Altschule, M.D. and Wurtman, J.J., 1961, Interactions of the pineal and exposure to continuous light on the organ weights of female rats. Acta Endocrinol. 36:617.

Wurtman, R.J., Shein, H.M. and Larin, F., 1971, Mediation by beta-adrenergic receptors on the effect of norepinephrine in pineal synthesis of [^{14}C]serotonin and [^{14}C]melatonin. J. Neurochem. 18:1683.

Zatz, M. and Axelrod, J., 1978, Regulation of sensitivity to beta-adrenergic stimulation in the rat pineal. in: "Neuronal Information Transfer," A. Karlin, M. Tennyson and H.J. Vogel, eds., Academic Press, Inc., New York.

Zatz, M., Kebabian, J.W., Romero, J.A., Lefkowitz, R.J. and Axelrod, J, 1976, Pineal beta-adrenergic receptor: Correlation of binding of ^{3}H-l-alprenolol with stimulation of adenylate cyclase. J. Pharmacol. Exp. Ther. 196:714.

ASPECTS OF EVOLUTION OF THE PINEAL ORGAN

Andreas Oksche

Department of Anatomy and Cytobiology
University of Giessen, 6300 Giessen
Federal Republic of Germany

INTRODUCTION

The pineal organ is an ancient derivative and integral component of the brain. (Recent comprehensive reviews have been published by Kappers and Pévet, 1979; Leonhardt, 1980; Vollrath, 1981; Oksche and Pévet, 1981.) Phylogenetically it has changed from a 'third eye' endowed with photoreceptor cells to an endocrine gland influenced by visual stimuli from the retina (Fig. 1). The pineal organ has the capacity to translate photic (and apparently also other sensory and neural) information into a neuroendocrine response (cf. Wurtman et al., 1968. In a subsequent step this neuroendocrine message acts on various target organs. The unusual anatomical and physiological characteristics of the pineal organ make it an interesting subject of analysis both in brain research and endocrinology. In evolutionary terms the pineal organ displays features of a photoreceptor, biological clock, and endocrine gland (cf. Oksche and Pévet, 1981). It is a member of the group of circumventricular organs (Vigh and Vigh-Teichmann, 1981) and a component of the photoneuroendocrine systems (Oksche and Hartwig, 1979). The pinealocyte can be considered as a paraneuron (Ueck and Wake, 1979) originating from a special type of cerebrospinal fluid (CSF)-contacting neuron (Vigh and Vigh-Teichmann, 1981; see for references) displaying both sensory and secretory properties (i.e. a sensoneuroendocrine cell).

The diversity of structure and function of the vertebrate pineal complex reflects a surprisingly high degree of evolutionary and adaptive capacity (cf. Ralph, 1970; Quay,

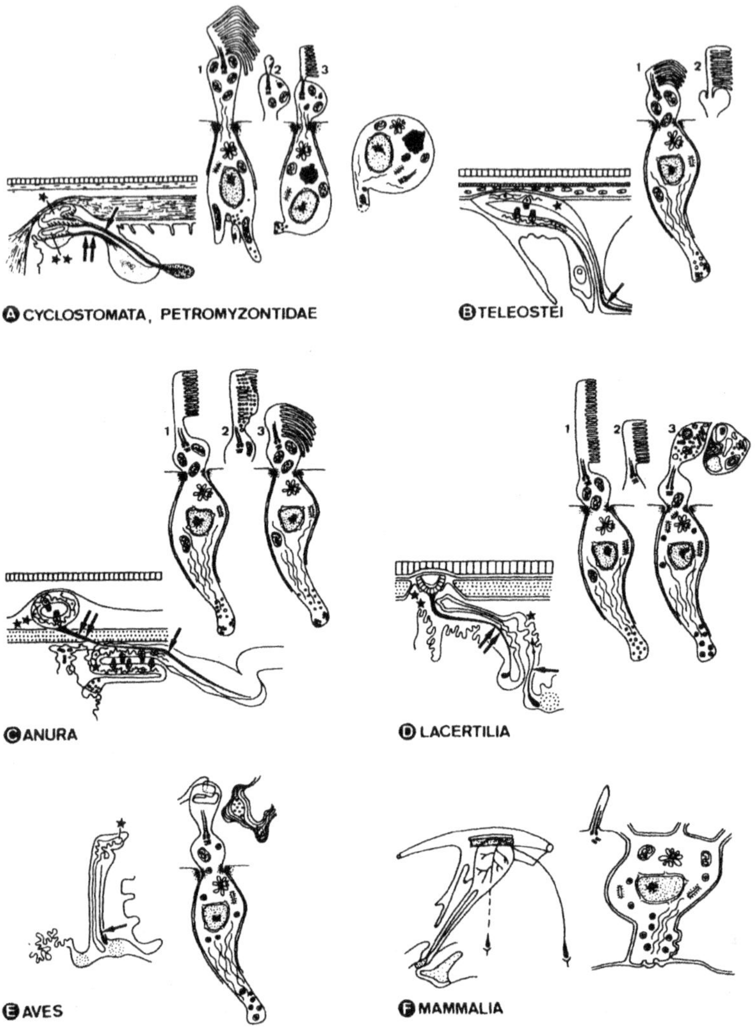

Fig. 1. Comparative representation of pineal complexes. Digrammatic midsagittal views (after Studniĉka, 1905) in relation to the respective features of the pinealocytes of the receptor line (modified after Oksche, 1980). Single star, pineal organ (epiphysis cerebri); double star, parapineal organ (cyclostomes, lacertilians) or frontal organ (anurans); single arrow, pineal tract; double arrow, parietal tract (cyclostomes, lacertilians) or frontal-organ tract (anurans). Note diversity in ultrastructural details of the pinealocytes (1-3) indicating transmission of outer segment structures and increasing secretory capacity. For details, see text (cf. Meiniel, 1980).

1974). Profound evolutionary changes have produced remarkable variations of the sensory and secretory apparatus of the pineal organ; at the ultrastructural and molecular levels new manifestations of these alterations have been discovered.

During the last two decades a high priority has been given to the elucidation of the function of the mammalian pineal gland (cf. Wurtman et al., 1968; Vollrath, 1981). However, there is no doubt that the pineal organs and complexes of submammalian vertebrates provide a most important key for the morphological and physiological analysis of the pineal gland of mammals including man. The present comparative report is based on two decades of intense interdisciplinary research cooperation with Professor Dodt and associates (cf. Dodt, 1973) and on continuous exchange of thoughts and information with Professor Collin culminating in a recent joint report (Collin and Oksche, 1981). Consideration is given to a number of previous surveys by Oksche (1965, 1970, 1971, 1980) and Collin (1969, 1971, 1977, 1979). In the present review a replica of these comprehensive treatises will be avoided; thus emphasis will be placed on general trends and concepts in comparative pineal research.

PROGRESS IN COMPARATIVE ANALYSIS OF SUBMAMMALIAN PINEAL SYSTEMS

State of Knowledge in 1970

The CIBA Symposium held in London in 1970 (cf. Wolstenholm and Knight, 1971) marked the end of a decade of innovative and expanding comparative research in the field. In the late fifties the structural analysis of the brain was much helped by the introduction of electron microscopy. Using this new technique, Steyn and Eakin, both in 1959, discovered that the parietal eye of lizards is endowed with true photoreceptor cells resembling the cones of the lateral eye (cf. Eakin, 1973). This progress was further enhanced by the first recordings of action potentials from the pineal complex of anurans, teleosts and lacertilians. In 1961, Dodt and his associates showed that the pineal 'eye' (frontal organ) of the frog is a true photoreceptor organ capable of chromatic and achromatic types of response (cf. Dodt, 1973). In 1962, collaborative work of Dodt and Oksche initiated a series of correlated neurophysiological and ultrastructural studies focussed on the pineal organs of different teleosts, amphibians, lacertilians and birds (for references, see Oksche, 1971; Oksche et al., 1971; Dodt, 1973). This work was followed by extensive comparative investigations of Collin, 1969; see also Collin, 1971). By the use of Nauta-type methods Paul et al. (1971) succeeded in tracing the pineal tract to the pretectal region and the tegmentum (the reticular system) of the frog. By means of fluorescence microscopy (according to Falck-Hillarp) the sympathetic input into the pineal organ was investigated in a number of species belonging to different classes of vertebrates (see Wolstenholme and Knight, 1971; Quay, 1974). Secretory granules were shown to occur in pinealocytes bearing a rudimentary outer segment (Collin 1969, 1971; Oksche, 1970, 1971); at the same time a monoamine fluorophore was demonstrated in pinealocytes of the sensory type (Owman and Rüdeberg, 1970). From these observations Oksche (1971) concluded that in pineal photoreceptor organs the input of photic information may be converted into a neurohormonal mechanism in one and the same sensory cell. This led to the definition of sensory pinealocytes as photo-neuroendocrine cells of neuronal origin, a conceptual extension of the photoneuroendocrine systems as defined by Scharrer (1964).

Recent Developments (1970 to 1982)

In 1974, three reports contributed substantial new information toward an understanding of the basic organization of the pineal sense organs. By means of microspectophotometry Hartwig and Baumann (1974) showed that the outer segments in the pineal organ (epiphysis cerebri) of *Rana esculenta* contain a rhodopsin-like photopigment 502_1 (specific optical density 0.011-0.016 µm), whereas the outer segments of the photoreceptor cells in the frontal organ display a photolabile substance with an absorption maximum between 560-580 µm. Wake et al. (1974) presented a diagram adding more precise details to the organization of the intrinsic neuronal circuitry in the frontal organ and the epiphysis of the frog. In addition to the basically bineuronal chain consisting of receptor cells and neurons participating in the formation of the pinealofugal pathways, the pineal complex of anurans contains stellate acetylcholinesterase-positive interneurons. The latter may play an essential role in the generation of the achromatic or the chromatic types of pineal response. Vigh and Vigh-Teichmann (1974) came to the conclusion that sensory pinealocytes of submammalian vertebrates, and even certain cilium-bearing forms of mammalian pinealocytes, resemble the cerebrospinal fluid-contacting neurons. This conclusion was further underlined by the observations (1) that the lumen of most pineal sense organs is in open communication with the third ventricle, and (2) that photoneuroendocrine systems of the brain are concentrated around the third ventricle (Oksche and Hartwig, 1975).

From the comparative point of view two concepts published in 1977 deserve particular attention. Collin (1977) gave a detailed account on the rudimentation of the photoreceptor cells in vertebrate pineal organs. He emphasized the gradual transformation of the photoreceptor cell during phylogeny and established homologies between pineal photoreceptors, rudimentary photoreceptor cells and secretory pinealocytes. These cells were claimed to form a single cell line, i.e. 'cells of the receptor line' (CRL), in contrast to the line of supportive ependymal elements and secondary neurons. In principle, the CRL of non-mammalian vertebrates are secretory in function. However, the rudimentation of the outer segment appears to be a necessary prerequisite for enhancing full secretory capacity. Ueck and Wake (1977) classified the pinealocyte as a paraneuron (Fujita 1977). They emphasized that all forms of pinealocytes possess a receptive input region, a conductive region and an output region; thus, the pinealocyte is a recepto-secretory functional unit. The neuroectodermal origin of pinealocytes has been well established. All criteria characteristic of a paraneuron appear to hold true for pinealocytes. The boundary between paraneurons and neurons is, however, a matter for individual interpretation. Further, it is noteworthy that about 1977 Collin and Pévet were the first to show that mammalian pinealocytes and the rudimentary photoreceptor cells of sauropsids elaborate, in addition to the well-known indoleamines, a proteinaceous secretion. These pioneering studies led to a series of further investigations (see Collin and Pevet in Oksche and Pévet, 1981). All three above-mentioned concepts suggest a relationship of pinealocytes with paraneurons and also with secretory neurons.

After extensive fluorescence- and electron-microscopic studies of the pineal organ in Lampetra planeri (Petromyzontidae) Meiniel (1980) came to the conclusion that the peculiar serotonin-containing pinealocytes of this lamprey are members of a second sensory cell line leading from photoreceptor cells to secretory pinealocytes. Subsequently, this concept was extended to other vertebrates (Meiniel, see Oksche and Pevet 1981). In the opinion of Meiniel, only the less conspicuous sensory elements of the second cell line are endowed with a secretory capacity thus becoming the stem cells of the secretory pinealocytes.

To prove the existence of two different lines of pineal receptors in other classes of vertebrates, an improved indoleamine-fluorescence technique (used without pharmacological pretreatment) and the light-dependent uptake of tritiated deoxyglucose were applied in the analysis of the pineal organ (epiphysis cerebri) in the frog (Hartwig and Oksche, 1981). An indoleamine fluorophore was found in the apical and supra-nuclear positions of sensory pinealocytes; however, cells displaying very long, birefringent outer segments were free of this fluorescence. No amine fluorescence was found in the frontal organ, i.e. the distal component of the pineal complex. After incubation with deoxyglucose in constant darkness (high spontaneous activity of photoreceptor cells) strongly labeled pinealocytes bearing long outer segments could be observed. The pattern of the silver grains in the pineal organ changed depending on the lighting conditions. Since cyclic changes in the structure and function of the receptors cannot be excluded, the existence of two independent receptor-cell lines remains a possibility. By the use of light- and electron-microscopic cytochemical techniques melatonin-like immunoreactivity was demonstrated in cells of the receptor line in the pineal organ of a teleost, the pike (Falcon et al., 1981). The intensity of the immunoreaction depended on the phase of the photoperiod. Remarkably, at the electron-microscopic level, not all cells of the photoreceptor type appeared to be immunoreactive. The existence of the fluorophore of 5-HT as well as of melatonin-immunoreactive material in pineal photoreceptor cells supports the concept of a photoneuroendocrine capacity of these elements (Hartwig and Oksche, 1981; Falcon et al., 1981).

Vigh and Vigh-Teichmann (1981) applied an antiserum against bovine retinal rods to different pineal systems and were able to show that the outer segments of the pineal sensory cells of teleosts, anurans and chelonians give a positive immuno-reaction. Very recently this reaction was successfully tested on the pineal organ of Phoxinus phoxinus (Fig. 2), the classical experimental animal of von Frisch (1911) and Scharrer (1928), pioneers in the field of photoneuroendocrine research Vigh-Teichmann et al., 1982). Curiously, the receptors in the parietal eye and the pineal organ of several lizards were negative to this antiserum; however, the negative reaction of the lacertilian retinal receptors with the antiserum used may indicate a basic difference in the chemical composition of opsins. On the other hand, the irregular lamellar complexes in the pineal of the pigeon displayed a positive opsin immuno-reaction (Vigh and Vigh-Teichmann, 1981). The latter result is important in suggesting new lines of research into the function of the avian pineal organ (see below). It is important to note that the CSF-contacting neurons in different vertebrates, which had been claimed as potential candidates for the 'deep diencephalic photoreceptors', were completely opsin-negative; also the occasional 9+0 cilia of mammalian pinealocytes (opossum, hedgehog) did not display a positive opsin immunoreaction. In addition, the areas of the ventricular wall described by Hartwig (see Oksche and Hartwig, 1979) to contain a microspectrophotometrically detectable photolabile compound were completely negative to the opsin antiserum. Since micro-spectrophotometry fails to detect photopigments in pinealocytes of the receptor line displaying geometrically irregular outer segments, the opsin immunoreaction will become a very important

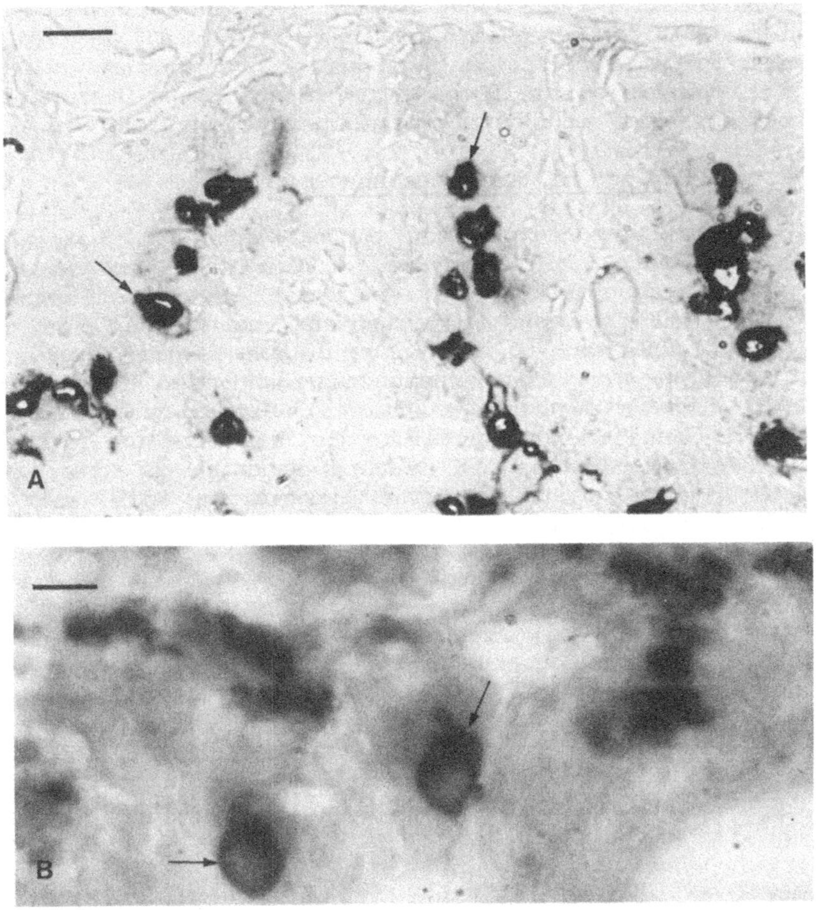

Fig. 2A,B. Pineal organ of Phoxinus phoxinus (Teleostei) A. Opsin-immunoreactive outer segments (arrow). B. Acetylcholinesterase-positive neurons (arrow). The other reactive (dark) structures represent inner segments of the photoreceptor cells. Bars = 20 µm. For details, see text. Unpublished microphotographs; see Vigh-Teichmann et al. (1982) for methods.

technique in comparative studies of the pineal (as well as other extraretinal) photoneuroendocrine systems.

Further, progress in demonstrating and mapping central pinealofugal and pinealopetal pathways by the use of tracers such as horseradish peroxidase (HRP) has been of great value in pineal research. In this context, only some of the findings from our laboratory will be mentioned (Korf and Wagner, 1980,

1981; Korf et al., 1982; Korf and Møller, unpublished results). They have established clear patterns of pineal afferents and efferents in a lacertilian and an avian species and shown that, in addition to the sympathetic input, central pinealopetal fiber systems exist in several mammalian species (see below).

Comparative Cytology of the Pinealocytes

In comparative terms the pineal organ is composed of (i) pinealocytes (photoreceptors, rudimentary photoreceptors, secretory pinealocytes), (ii) ependymal or glial supportive cells, and (iii) neurons (interneurons, neurons of the pineal tract). In addition, in the lumen of the pineal organ, which in sub-mammalian vertebrates communicates with the third ventricle, free cells, mostly macrophages but probably also supraependymal elements of neuronal or glial origin, can be observed. In an open pineal lumen the outer segments of the sensory pinealocytes are bathed in the circulating CSF, and there a release and also an uptake of substances may occur (Oksche and Hartwig, 1975). This aspect is of importance in connection with the conspicuous discharge from and renewal of lamellar systems in the outer segment of pineal photoreceptor cells. In a number of lower vertebrates, e.g. the frog, indoleamine fluorophores have been observed within the pineal lumen, seemingly in connection with a net-like mucopolysaccharide or mucoprotein matrix.

The following review will concentrate exclusively on cells belonging to the receptor line (CRL) as defined by Collin and Oksche (1981).

The cells of the receptor line (CRL) are derived from sensory, photoreceptive pinealocytes. The photoreceptor cells (PC) are gradually transformed into rudimentary photoreceptor cells (RPC) and, subsequently, into secretory pinealocytes (Pi). The process of rudimentation began already in anamniotes; and intermediate forms exist.

The achromatic and chromatic types of electric response of pineal sense organs are well established (Dodt, 1973; Hamasaki and Eder, 1977). However, the biological significance of this message conducted to different areas of the brainstem (see below) is still not understood.

The CRL in non-mammalian vertebrates are secretory in function. (1) Gradual rudimentation and subsequent disappearance of the photoreceptor pole, and (2) transformation of the inner cellular pole from an area involved in neurotransmission into a vascular contact are important prerequisites of increasing secretory activity. It is, as yet, unsolved whether a fully differentiated PC already possesses a full-range of secretory capacity. The discrete secretory activity of these cells is fully manifested in the RPC parallel to the regression of photosensory structures. However, the PC may display a highly active granular endoplasmic reticulum indicating protein synthesis (Owman and Rüdeberg, 1970; Oksche, unpublished results). Dense-core granules do not belong to the typical picture of the PC; however, after maintenance in constant darkness, Omura and Ali (1981) were able to accumulate a conspicuous number of dense-core vesicles in the PC of a teleost species. Dense-core vesicles are very abundant in the RPC of different lacertilians and birds. HIOMT and melatonin occur in the pineal sense organs (cf. Oksche and Pévet, 1981). In addition, a secretion of proteinaceous material takes place in the CRL, even in the PC (see above). With respect to sensory and secretory functions, it is impossible to draw an exact borderline between the PC and the RPC. A direct photoreceptive capacity of the RPC cannot be excluded. The problem of the presence and nature of membrane receptors in the PC and the RPC is completely open to discussion. In analogy to the PC, the RPC may contain synaptic ribbons. However, in the RPC synaptic ribbons do not mark true synaptic contacts (cf. Collin and Oksche, 1981); this pattern of synaptic ribbons already resembles the arrangement of synaptic ribbons in mammalian pinealocytes (cf. Vollrath, 1981).

On the basis of the above-mentioned evidence the existence of photoneuroendocrine cells can be postulated. Gradual evolution of cellular mechanisms leads from the perception of appropriate stimuli to the delivery of nervous, neurohormonal and non-neurohormonal humoral messages (cf. Collin and Oksche, 1981). This process includes, at the cellular level, a translation of photic stimuli into a neuroendocrine reaction.

Intrinsic Neuronal Circuitry

In contrast to the retina, the pineal sense organs possess generally a bineuronal chain. The second neuron displays a positive acetylcholinesterase reaction and is synaptically coupled with photoreceptor cells; it is endowed mostly with a single dichotomously branching process. The axon of these acetylcholinesterase-positive neurons joins the pineal tract. Stellate acetylcholinesterase-positive nerve cells, especially conspicuous in the pineal complex of anurans, can be considered as interneurons. Unfortunately, the synaptic connections of these cells, the corresponding neuropil areas, and the respective neurotransmitters have not been elucidated in detail, although they appear to be essential for the generation of the achromatic or the chromatic types of pineal response (cf. Oksche and Hartwig, 1979). According to Meissl (for references see Collin and Oksche, 1981), taurine, glutamic acid and glycine are present in significant concentrations in the pineal organ of the trout. However, the neurotransmitters of the individual neurons have as yet not been determined. Serotonin, added in vitro to the incubation medium, effected a decrease in the electrical activity of the pineal nerve in the frog, whereas the application of melatonin was without effect on the activity of the frontal organ (Donley and Meissl; for references see Oksche and Hartwig, 1979).

The acetylcholinesterase-positive neurons of the avian pineal organ are apparently homologues to the similarly reacting elements of the pineal sense organs in anamniotes. The intrinsic circuitry of the nerve cells is, however, less conspicuous than, e.g., in anurans and has not yet been deciphered in greater detail. There is no convincing evidence suggesting that such acetylcholinesterase-positive nerve cells occur in the pineal organ of mammals; the acetylcholinesterase-reactive neurons in the pineal organ of the ferret Herbert, cf. Wolstenholme and Knight, 1971) are apparently of habenular origin. The rather scarce, scattered neurons of most mammalian pineal organs have been claimed to belong to the autonomic nervous system; the origin of these cells should, however, be re-examined.

Pinealofugal and Pinealopetal Neuronal Connections

Pineal sense organs communicate with the brain via neural pathways consisting of myelinated and non-myelinated nerve fibers (cf. Oksche, 1971; Collin and Oksche, 1981); they transmit a nervous message to the brain. In phylogenetic terms the pinealofugal projection shows signs of sequential regression; at the same time the neuroendocrine communication gains increasing importance. Pinealopetal nerve fibers of central origin have been clearly demonstrated in pineal sense organs by neurophysiological and neuroanatomical methods (cf. Dodt, 1973; Oksche and Hartwig, 1979). It should be mentioned that efferent systems also occur in the lateral eye and other sense organs. Reciprocal neural connections appear to be characteristic of the pineal sense organs.

In degeneration studies, using methods of the Nauta type, Paul et al. (1971) had already shown (see Introduction) that the pineal tract in Rana esculenta projects to the pretectal area and to the reticular formation of the mesencephalic tegmentum. The existence of a pineal projection to the deep mesencephalic gray in the frog was confirmed by neurophysiological methods (Gaillard et al., 1977; Cadusseau et al., 1979).

In the rainbow trout, in addition to the above-mentioned

areas, the existence of a habenular projection of the pineal organ has been shown by means of cobalt chloride iontophoresis (Hafeez and Zerihun, 1974). Our knowledge of the patterns of pineal innervation has been greatly extended by tissue analysis following iontophoretic application of horseradish peroxidase (HRP). Korf and Wagner (1981) succeeded in showing that the parietal eye of the lizard, Lacerta sicula projects not only to the left habenular complex, the pretectal area and the dorsolateral nucleus of the thalamus, but also to the paraventricular nucleus. On the other hand, they showed that the parietal eye receives an input from the same paraventricular area. A reciprocal innervation is also characteristic of the pineal organ of the house sparrow, Passer domesticus (Korf et al., 1982). The pineal tract of this passerine bird (consisting of approximately 1,000 mostly unmyelinated nerve fibers) arises from intrinsic acetylcholinesterase-positive neurons, which seem to be homologues of similar neurons in pineal sense organs. Projections of this tract were observed in both habenular nuclei and in the periventricular gray of the hypothalamus. On the other hand, distinct pinealopetal projections originate in the habenular complex and in the area of the paraventricular nucleus. Further, a distinct pinealopetal projection of central origin also exists in mammals as has been shown in the guinea pig (Korf and Wagner, 1980) (Fig. 3A,B) and the Mongolian gerbil (Korf and Møller, unpublished results). These pinealopetal fibers originate either in the habenular complex or in the periventricular zone adjacent to the paraventricular nucleus. Only very occasionally, labeled neurons are found in the suprachiasmatic nucleus; thus, the existence of a direct connection between the suprachiasmatic nucleus and the pineal organ has not been confirmed. The pinealopetal projections of central nerve fibers in the rat are not very conspicuous; in this species the sympathetic input is clearly predominant (cf. Kappers, 1982, this volume). Finally, in the primitive pineal gland of the hedgehog oxytocin- and vasopressin-immunoreactive nerve fibers have been demonstrated (Nürnberger and Korf, 1981). (For other references the quoted papers should be consulted).

What do the pineal sense organs tell to the brain? It should be kept in mind that the achromatic pineal systems may be capable of measuring solar radiation energy and the chromatic systems - the changing spectral composition of daylight (cf. Dodt and Meissl in Oksche and Pévet, 1981). The neural connection of pineal sense organs with the reticular formation of the brainstem, especially via the tegmental projection, may be involved in phototactic reactions (cf. Oksche and Hartwig, 1979), the control of thermoregulatory behavior (cf. Ralph et al., 1979), and some unknown activating mechanisms. A direct connection with the neuroendocrine apparatus of the brainstem is not evident but cannot be excluded. The role of the neural connections of the pineal organ in the establishment and maintenance of circadian mechanisms is open to discussion (cf. Korf et al., 1982). The suprachiasmatic nucleus of birds and mammals is a powerful circadian oscillator; significant variation exists in the coupling of the avian pineal organ with the secondary self-sustained oscillators (Gwinner, 1978). The central innervation of the mammalian pineal organ is by no means rudimentary. The central projection to the pineal organ might not be involved in circadian mechanisms; it may serve other functions, e.g. as yet unknown types of neural feedback, and thus modulate the glandular activity of the pineal organ. It should not be overlooked that the pineal organ is a derivative and integral component of the brain.

Vascular Pattern

In contrast to innervation, vascular patterns of the pineal organ have not been examined in great detail. In our experience intravital studies of pineal circulation, whenever technically possible, are from a functional point of view superior to all injections of dyes and vascular casts. The knowledge of the vasculature and microcirculation in pineal sense organs is very limited. If the pineal organ of lower vertebrates played an essential role in humoral color-change mechanisms, the active agent (melatonin) should be released into the bloodstream either directly or via the CSF. In the frog, Mautner (1965) studied the arterial supply and venous drainage of the pineal complex. In addition to the major lateral connections with venous sinuses, vascular collaterals extend from the epiphysis cerebri to the choroid plexus of the third ventricle; however, this link is very discrete and probably without great functional significance. There is no indication for a portal vascular system between the epiphysis cerebri and other neuroendocrine centers in the frog brain. The pathways of release of indoleamines in the pineal organ of anurans and other pineal sense organs known to contain indoleamines are enigmatic. It is an open question whether certain types of PC, apparently precursors of the RPC, may terminate on blood vessels instead of forming synapses with secondary neurons. A direct release of indoleamines into the CSF appears possible although it is not proved.

In the present context the complex problem of the blood-brain-barrier cannot be discussed in detail. It is known, however, that in mammals the pineal parenchyma is outside the blood-brain-barrier (cf. Vollrath, 1981). However, in the frog, injection of trypan blue does not lead to a staining of the pineal complex, in contrast to the heavily stained

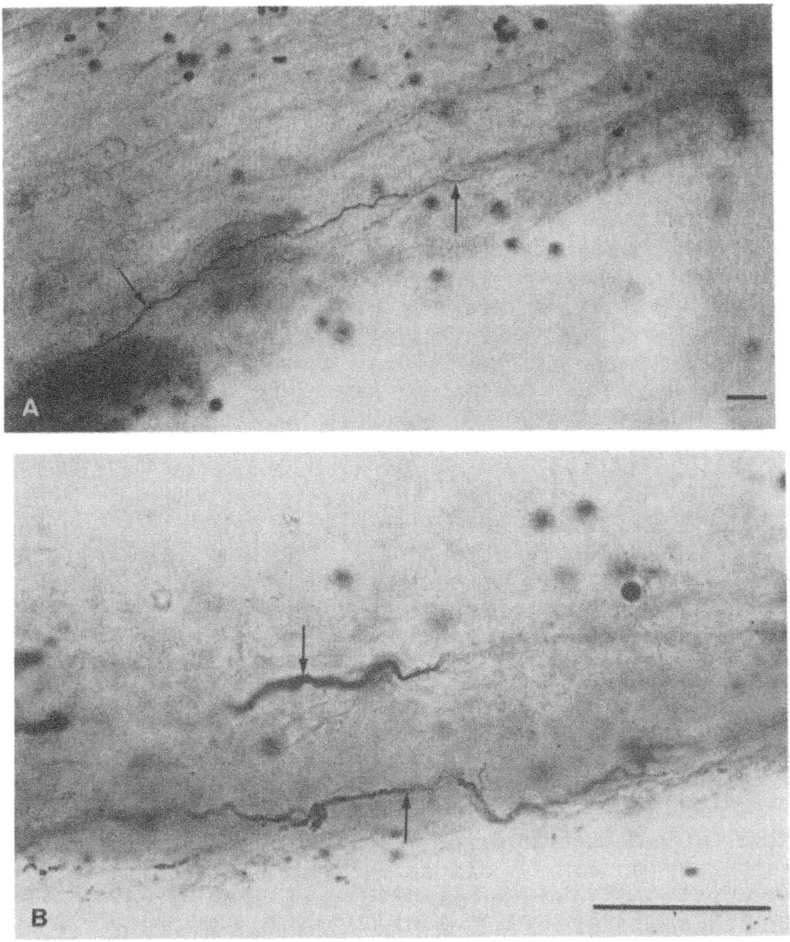

Fig. 3A,B. Labeled nerve fibers of central origin (arrow) in the pineal stalk of the guinea pig. Iontophoretic application of horseradish peroxidase in the distal portion of the pineal organ. Bars = 20 µm. For details, see text. Unpublished microphotographs, courtesy Dr. H.W. Korf. See Korf and Wagner (1980) for methods.

choroid plexus (Oksche, unpublished results). Further, there is no evidence showing that capillaries of the fenestrated type occur in the pineal organ of anurans. If this can be proved to be true also for other pineal sense organs, it may indicate a principal difference in the organization of sensory and glandular pineal organs.

The Avian Pineal Organ

The peculiar comparative position of the avian pineal organ has been discussed in detail by Oksche and Hartwig (1979), Collin and Oksche (1981), and Korf et al. (1982). The outer segments of avian pinealocytes exhibit distinct signs of structural regression; this regression coincides with a spectacular increase in the secretory capacity of the pinealocytes. On the other hand, the isolated avian pineal organ in vitro displays rhodopsin-like photosensitivity (Deguchi, 1981) and oscillatory activity of enzymes involved in melatonin synthesis (Binkley et al., 1978; Deguchi, 1980; Takahashi et al., 1980). According to the usual ultrastructural organization of the retina, the irregular lamellar whorls of the avian pinealocytes can hardly be regarded as functional structures. Retinae displaying a comparable degree of genetic or experimentally induced alteration of receptor ultrastructure are incapable of producing an electroretinogram. However, the lamellar complexes of avian pinealocytes may display a positive opsin immunoreaction, e.g. in the pigeon (Vigh and Vigh-Teichmann, 1981). Strangely enough, this immunoreaction is negative in the morphologically far-better developed outer segment structures of Passer domesticus (I. Vigh-Teichmann, personal communication). In electrophysiological studies conducted in different laboratories the pineal organ of P. domesticus has not shown a direct response to illumination (cf. Dodt, 1973). The photic response of the pineal organ of the pigeon to light stimuli (Semm, 1982, this volume) differs from the reactivity pattern (on- and off- effect) of typical pineal sense organs (Semm, personal communication). The possibility should be discussed that light stimuli perceived by the avian pineal organ may initiate some kind of chemical reaction instead of leading to the generation of action potentials.

To date no structural correlate has been found that would help to interpret the response of the avian pineal organ to the forces of the magnetic field (Semm, 1982, this volume).

The avian pineal organ is a self-sustained circadian pacemaker that acts on a population of secondary self-sustained oscillators located outside the pineal body, e.g. in the hypothalamus (Gwinner, 1978). The coupling within this system shows distinct interspecific differences. The oscillators of birds can be synchronized by light. In photoperiodic responses of birds, however, two functions must be considered, time and photosensitivity, the latter being a periodic (circadian) function. The problem of the avian circadian and photoneuroendocrine systems is very complex due to the fact that they encompass the deep encephalic photoreceptor(s), the retino-hypothalamic connection and the pineal organ (cf.

Yokoyama et al., 1978).). Thus the effects of pinealectomy must be viewed in context with other components of the oscillator system. In principle, the avian pineal organ is an independent oscillator; the implantation of a denervated pineal gland in a pinealectomized bird (P. domesticus) not only reinstates the circadian locomotor rhythm but also the donor phase of this rhythm (Zimmerman and Menaker, 1975, 1979).

CONCLUSIONS

The pineal organ of vertebrates has undergone profound evolutionary changes which have produced remarkable variations in its sensory and secretory apparatus. The pineal sense organs of lower vertebrates are endowed with indoleamine-containing sensory cells (sensoneuroendocrine cells), which may be capable of the translation of photic information into a neuroendocrine response. The concentration of sensory and neuroendocrine properties in one cell line or even in one cell may offer functional advantages. The photoneuroendocrine cells of primitive pineal organs were possibly involved in the proper timing of biological events. Achromatic pineal organs might have served (and still serve) as a dosimeter for solar radiation energy, and the chromatic eye-like components of pineal complexes as a sensor for discrimination of changes in the spectral composition of daylight. The target areas of the pinealofugal neural projections are scattered over a wide area of the central nervous system including reticulomotor and neuroendocrine areas of the brainstem. The pineal projection to the pretectal area displays some overlap with the input from the lateral eyes. Thus, a fine equilibrium in the retinal and pineal control of rhythmic motor and endocrine events seems to have arisen. In parallel with the regression of their outer segments, pinealocytes of the receptor line appear to have evolved further a capacity for generating humoral signals. Humoral agents may be mediated via the bloodstream and/or the circulating cerebrospinal fluid, and then may be taken up by specialized membrane receptors of target organ cells. The above-mentioned mechanisms appear to be involved in color change, endocrine control of reproduction, phototactic and locomotor reactions, thermoregulatory behavior and, in a wider sense, in the mechanisms of biological clocks. This remarkable range of biological effects becomes understandable when the divergent pattern of the neural projection of pineal sense organs to extended areas of the brainstem is taken into account; this type of connectivity may result in triggering a broad spectrum of different functions. In addition, cells endowed with melatonin receptors may also show a scattered pattern of distribution.

It is possible that during an early phase of phylogeny pineal organs directly or indirectly participated in complex reactions serving adaptation to the environment, and thus they were enhancing survival. The functional significance of the pineal parenchyma is emphasized by the discovery, using advanced techniques, of pineal tissue in several mammalian species where it was previously thought to be missing (cf. Hofer et al., 1976). It is also present in several species of anamniotes claimed to be devoid of a pineal organ; however, in crocodiles all efforts to demonstrate pineal tissue have been negative. A possible correlation between the geographical latitude and the extent of development of the pineal organ needs further examination. In general, the pineal organs tend to be prominent in vertebrates at higher latitudes and smaller at lower latitudes (Ralph, 1975); for example, the pineals of Antarctic penguins are very large and highly glandular Piezzi and Gutiérrez, 1975). Further, differences between day- and night-active animals and the special features of the pineal organ in blind cave-dwelling or burrowing species deserve particular attention (Pévet, cf. Oksche and Pévet, 1981).

The photoneuroendocrine capacity is a fundamental factor which must be considered when developing concepts dealing with the evolution of the pineal organ. The results and concepts reported in this paper emphasize the importance of the comparative structural, physiological and biochemical analysis in pineal research.

SUMMARY

The pineal organ is a derivative and integral component of the brain. Phylogenetically it has changed from a 'third eye' capable of direct achromatic and/or chromatic responses to light to an endocrine gland influenced by visual stimuli from the retina. In its phylogenetically ancient form the pinealocyte resembles a cerebrospinal fluid-contacting neuron endowed with sensory and secretory properties. Such pinealocytes have been considered to be photoneuroendocrine cells. Pinealocytes of the photoreceptor type display in their outer segments a microspectrophotometrically identifiable photopigment and an opsin-immunoreactive protein. On the other hand, certain pinealocytes bearing an outer segment contain microspectrofluorimetrically detectable levels 5-HT/5-HTP.

Synthesis, storage and release of indoleamines, and probably also of peptides and/or proteins, appear phylogenetically at an early stage and are fundamental properties of pineal systems. The capacity of formation of elementary granules parallels the regression of the geometrically regular outer segment structures and illustrates that the secretory function is becoming of increasing importance. The secretory pinealocyte has been considered to be a paraneuron.

In addition to their neuroendocrine messages acting on various target organs and systems, the pineal sense organs are capable of sending neural messages to the brain, e.g. to the habenular ganglion, pretectal region, tegmental reticular gray and hypothalamus, via pinealofugal tracts. Pinealopetal projections have been traced from the brain to the pineal organ in sub-mammalian and mammalian vertebrates. The avian pineal organ has retained certain neural elements characteristic of pineal sense organs.

In evolutionary terms, the pineal organ displays the capacities of a photoreceptor, biological clock and endocrine gland. It is a member of the group of circumventricular organs and a component of the photoneuroendocrine systems. The special anatomical and physiological features of the pineal organ enable it to translate photic information into a neuroendocrine response.

ACKNOWLEDGEMENTS

The investigations of the author cited in this report were supported by research grants from the Deutsche Forschungsgemeinschaft (SPP 'Mechanismen biologischer Uhren: Ok 1/24; 1/25). The author is greatly indebted to Professor J. Priedkalns (Adelaide) and Dr. H.W. Korf (Giessen) for their help in preparing the manuscript. Figure 1 was prepared by Miss K. Michael (Giessen). The secretarial assistance of Mrs. Julia Brazier (Adelaide) is gratefully acknowledged.

REFERENCES

Binkley, S., Riebman, J.B., and Reilly, K.B., 1978, The pineal gland: a biological clock in vitro; Science, 202: 1198.

Cadusseau, J., Gaillard, F., and Galand, G., 1979, Pineal response types in the frog's brain under white light exposure, Exp. Brain Res., 36:41.

Collin, J.-P., 1969, "Contribution à l'étude de l'organe pinéal. De l'épiphyse sensorielle à la glande pinéale: modalités de transformation et implications fonctionelles", Ann. St. Biol. Besse Chandesse, Suppl. 1.

Collin, J.-P., 1971, Differentiation and regression of the cells of the sensory line in the epiphysis cerebri, in: "The Pineal Gland", G.E.W. Wolstenholme, and J. Knight, eds., Churchill Livingstone, London-Edinburgh.

Collin, J.-P., 1977, La rudimentation des photorécepteurs dans l'organe pinéal des Vértebrés, in: "Mécanismes de la Rudimentation des Organes chez les Embryons de Vertébrés", Vol. 266, A. Raynaud, ed., Centre National de la Recherche Scientifique, Paris

Collin, J.P., 1979, Recent advances in pineal cytochemistry. Evidence of the production of indoleamines and proteinaceous substances by rudimentary photoreceptor cells and pinealocytes of Amniota, in: "The Pineal Gland of Vertebrates including Man", (Progress in Brain Research, Vol. 52), J.A. Kappers, and P. Pévet, eds., Elsevier/ North-Holland Biomedical Press, Amsterdam-New York.

Collin, J.-P., and Oksche, A., 1981, Structural and functional relationships in the nonmammalian pineal gland, in: R.J. Reiter, ed., "The Pineal Gland", Vol. I: Anatomy and Biochemistry, CRC Press, Boca Raton.

Deguchi, T., 1980, A circadian oscillator in cultured cells of chicken pineal gland, Nature, 282:94.

Deguchi, T., 1981, Rhodopsin-like photosensitivity of the isolated chicken pineal gland, Nature, 290:706.

Dodt, E., 1973, The parietal eye (pineal and parietal organs) of lower vertebrates, in: "Handbook of Sensory Physiology", Vol. VII, Part 3, R. Jung, ed., Springer, Berlin.

Eakin, R.M., 1973, "The Third Eye", University of California Press, Berkeley.

Falcon, J., Geffard, M., Juillard, M.-T., Delaage, M., and Collin, J.-P., 1981, Melatonin-like immunoreactivity in photoreceptor cells. A study in the teleost pineal organ and the concept of photoneuroendocrine cells, Biol. Cell, 42:65.

Frisch, K.v., 1911, Beiträge zur Physiologie der Pigmentzellen in der Fischhaut, Pflüger's Arch. ges. Physiol., 138-319.

Fujita, T., 1977, Concept of paraneurons. Arch. histol. jap., 40 (Suppl.):1.

Gaillard, F., Mironneau, C., and Galand, G., 1977, Light stimulation and pineal unitary responses in frog's central nervous system, Vision Res., 17:667.

Gwinner, E., 1978, Effects of pinealectomy on circadian locomotor activity rhythms in European starlings, Sturnus vulgaris, J. Comp. Physiol., 126:123.

Hafeez, M.A., and Zerihun, L., 1974, Studies on central projections of the pineal nerve tract in rainbow trout, Salmo gairneri, Richardson, using cobalt chloride iontophoresis, Cell Tissue Res., 154:485.

Hamasaki, D.I., and Eder, D.J., 1977, Adaptive radiation of the pineal system, in: "Handbook of Sensory Physiology", Vol. VII, Part 3, R. Jung, ed., Springer, Berlin-Heidelberg-New York.

Hartwig, H.G., and U. Baumann, 1974, Evidence for photosensitive pigments in the pineal complex of the frog, Vision Res., 14:597.

Hartwig, H.G., and Oksche, A., 1981, Photoneuroendocrine cells and systems: a concept revisited, in: "The Pineal Organ: Photobiology-Biochronometry-Endocrinology", A. Oksche, and

P. Pévet, eds., Elsevier/North-Holland Biomedical Press, Amsterdam-New York-Oxford.

Hofer, H.O., Merker, G. and Oksche, A., 1976, Atypische Formen des Pinealorgans der Säugetiere, Verh. Anat Ges. 70:97.

Kappers, J.A., 1982, this volume.

Kappers, J.A., and Pévet, P., eds., 1979, "The Pineal Gland of Vertebrates including Man", (Progress in Brain Research Vol. 52), Elsevier/North-Holland Biomedical Press, Amsterdam-New York.

Korf, H.-W., and M. Møller, unpublished results.

Korf, H.-W., and Wagner, U., 1980, Evidence for a nervous connection between the brain and the pineal organ in the guinea pig, Cell Tissue Res., 209:505.

Korf, H.-W., and Wagner, U., 1981, Nervous connections of the parietal eye in adult Lacerta s. sicula. Rafinesque as demonstrated by anterograde and retrograde transport of horseradish peroxidase, Cell Tissue Res., 219:567.

Korf, H.-W., Zimmerman, N.H., and Oksche, A., 1982, Intrinsic neurons and neural connections of the pineal organ of the house sparrow, Passer domesticus, as revealed by anterograde and retrograde transport of horseradish peroxidase, Cell Tissue Res., 222:243.

Leonhardt, H., 1980, Ependym und Circumventriculäre Organe, in "Neuroglia I", A. Oksche, ed., in: "Handbuch der Mikroskopischen Anatomie des Menschen", Vol. IV/10, A. Oksche, and L. Vollrath, eds., Springer, Berlin-Heidelberg-New York.

Nürnberger, F., and Korf, H.-W., 1981, Oxytocin- and vasopressin-immunoreactive nerve fibers in the pineal gland of the hedgehog, Erinaceus europaeus L., Cell Tissue Res., 220:87.

Mautner, W., 1965, Studien an der Epiphysis cerebri und am Subcommissuralorgan der Frösche, Z. Zellforsch. Mikrosk. Anat., 67:234.

Meiniel, A., 1980, Ultrastructure of serotonin-containing cells in the pineal organ of Lampetra planeri (Petromyzontidae). A second sensory cell line from photoreceptor cell to pinealocyte, Cell Tissue Res., 207:407.

Oksche, A., 1965, Survey of the development and comparative morphology of the pineal organ, in: "Progress in Brain Research", Vol. 10, J.A. Kappers, and P. Pévet, eds., Elsevier, Amsterdam.

Oksche, A., 1970, Zur Differenzierung sensorischer und sekretorischer Strukturelemente im Zentralnervensystem, Verh. Dtsch. Zool. Ges., 64:72.

Oksche, A., 1971, Sensory and glandular elements of the pineal organ, in: "The Pineal Gland", G.E.W. Wolstenholme, and J. Knight, eds., Churchill Livingstone, London-Edinburgh.

Oksche, A., 1980, The pineal organ - a component of photoneuroendocrine systems: evolution, structure, function, in: "Hormones, Adaptation and Evolution", S. Ishii et al.,

eds., Japan Sci. Soc. Press, Tokyo/Springer-Verlag, Berlin.
Oksche, A., unpublished results.
Oksche, A., and Hartwig, H.G., 1975, Photoneuroendocrine systems and the third ventricle, in: "Brain-Endocrine Interaction", Vol. II., K.M. Knigge et al., eds., Karger, Basel.
Oksche, A., and Hartwig, H.G., 1979, Pineal sense organs - components of photoneuroendocrine systems, in "The Pineal Gland of Vertebrates including Man", (Progress in Brain Research, Vol. 52), J.A. Kappers, and P. Pévet, eds., Elsevier/North-Holland Biomedical Press, Amsterdam-New York.
Okache, A., and Pévet, P., eds., 1981, "The Pineal Organ. Photobiology - Biochronometry - Endocrinology", (Developments in Endocrinology, Vol. 14), Elsevier/North-Holland Biochemical Press, Amsterdam-New York-Oxford.
Oksche, A., Ueck, M., and Rüdeberg, C., 1971, Comparative ultrastructural studies of sensory and secretory elements in pineal organs, in: "Subcellular organization and function in endocrine tissues", (Memoirs of the Society for Endocrinology, No. 19), H. Heller, and K. Lederis, eds., Cambridge University Press, London-New York.
Omura, Y., and Ali, M.A., 1981, Ultrastructure of the pineal organ of the Killifish, Fundulus heteroclitus, with special reference to the secretory function. Cell Tissue Res., 219:355.
Owman, C., and Rüdeberg, C., 1970, Light, fluorescence, and electron microscopic studies on the pineal organ of the pike, Esox lucius L., with special regard to 5-hydroxytryptamine, Z. Zellforsch. Mikrosk. Anat., 107:522.
Paul, E., Hartwig, H.-G., and A. Oksche, 1971, Neurone und zentralnervöse Verbindungen des Pinealorgans der Anuren, Z. Zellforsch. Mikrosk. Anat., 112:466.
Piezzi, R.S., and Gutiérrez, L.S., 1975, Electron microscopic studies on the pineal organ of the Antarctic penguin (Pygoscelis papua), Cell Tissue Res, 164:559.
Quay, W.B., 1974, "Pineal Chemistry in Cellular and Physiological Mechanisms", Charles C. Thomas, Springfield, Ill.
Ralph, C.L., 1970, Structure and alleged functions of avian pineals. Am. Zool., 10:217.
Ralph, C.L., 1975. The pineal organ and geographical distribution of animals. Int. J. Biometeorol., 19:289.
Ralph, C.L., Firth, B.T., and Turner, J.S., 1979, The role of the pineal body in ectotherm thermoregulation, Am. Zool., 19:273.
Scharrer, E., 1928, Die Lichtempfindlichkeit blinder Elritzen (Untersuchungen über das Zwischenhirn der Fische I.), Z. Vgl. Physiol. 7:1.
Scharrer, E., 1964, Photo-neuro-endocrine systems: general concepts, Ann. N.Y. Acad. Sci., 117:13.
Semm, P., 1982, this volume.

Semm, P., personal communication.

Studnička, F.K., 1905, "Die Parietalorgane", in: "Lehrbuch der vergleichenden mikroskroskopischen Anatomie der Wirbeltiere", Part 5, A. Oppel, ed., Gustav Fischer, Jena.

Takahashi, J., Hamm, H., and Menaker, M., 1980, Circadian rhythms of melatonin release from individual superfused chicken pineal glands in vitro. Proc. Natl. Acad. Sci. USA, 77:2319.

Ueck, M., and Wake, K., 1977, The pinealocyte-a paraneuron? A review. Arch. histol. jap., 40 (Suppl.):261.

Ueck, M., and Wake, K., 1979, The pinealocyte-a paraneuron, in: "The Pineal Gland of Vertebrates including Man", (Progress in Brain Research, Vol. 52), J.A. Kappers, and P. Pévet, eds., Elsevier/North-Holland Biomedical Press, Amsterdam-New York.

Vigh-Teichmann, I., personal communication.

Vigh, B., and Vigh-Teichmann, I., 1974, Vergleich der Ultrastruktur der Liquorkontaktneurone und Pinealozyten, Verh. Anat. Ges., 68:433.

Vigh, B., and Vigh-Teichmann, I., 1981, Light- and electron-microscopic demonstration of immunoreactive opsin in the pinealocytes of various vertebrates, Cell Tissue Res., 221:451.

Vigh - Teichmann, I., Korf, H.W., Oksche, A., and Vigh, B.,1982, Opsin-immunoreactive outer segments and acetylcholin-esterase-positive neurons in the pineal organ of Phoxinus phoxinus (Teleostei, Cyprinidae), Cell Tissue Res. (in press).

Vollrath, L., 1981. "The Pineal Organ", in "Handbuch der Mikroskopischen Anatomie des Menschen", Vol. VI/7, A. Oksche, and L. Vollrath, eds. Springer, Berlin-Heidelberg-New York.

Wake, K., Ueck, M., and Oksche, A., 1974, Acetylcholinesterase-containing nerve cells in the pineal complex and sub-commissural area of the frogs, Rana ridibunda and Rana esculenta, Cell Tissue Res., 154:423.

Wolstenholme, G.E.W., and Knight, J., eds., 1971, "The Pineal Gland", Churchill Livingstone, London-Edinburgh.

Wurtman, R.J., Axelrod, J., and Kelly, D.E., 1968, "The Pineal", Academic Press, New York.

Yokoyama, K., Oksche, A., Darden, Th. R., and Farner, D.S., 1978, The sites of encephalic photoreception in photoperiodic induction of the growth of the testes in the White-crowned Sparrow, Zonotrichia leucophrys gambelii, Cell Tissue Res., 189:441.

Zimmerman, N.H., and Menaker, M., 1975, Neural connections of sparrow pineal: role in circadian control of activity. Science, 190:477.

Zimmerman, N.H.. and Menaker, M., 1979, The pineal: a pacemaker within the circadian system of the house sparrow. Proc. Natl. Acad. Sci. USA, 76:999.

COMPARATIVE GROSS AND FINE MORPHOLOGY OF THE MAMMALIAN PINEAL GLAND

J. Ariëns Kappers

Houtweg 13
1251 CS Laren N.H.
The Netherlands

Some knowledge of the structure of an organ is indispensable for the investigation and understanding of its function. This is also very evident in case of the mammalian pineal organ the structure of which clearly points to its endocrine function.

Its classical name, and also its official one now, is "corpus pineale". This term was given to the structure by Galen (ca 130 - ca 200 A.D.) who observed that, in man, the organ is shaped like the cone of a pine tree. In general, however, the shape, size and also the location of the mammalian pineal greatly vary among mammals. Before dealing with this matter, a short survey of the principles of pineal embryology is necessary because the ontogenetic development of the organ elucidates much about its adult morphology and its relationship to neighbouring structures.

Just like in other vertebrates, the mammalian pineal organ develops from the neuroepithelial layer which forms the roof of the dorsal part of the diencephalon, the epithalamus. In fetal stages, this layer, which is situated between the habenular commissure rostrally and the posterior commissure caudally, evaginates forming a sac-like structure. In the rat, this first indication of a pineal anlage is observed between the 12th and the 14½th day of fetal life (Gardner, 1949; Ariëns Kappers, 1960; Vollrath, 1981, also for literature). Next, the pineal evagination deepens and its wall becomes thicker by cell proliferation. At days 16 and 17, the wall forms infoldings so that follicles appear. Between the 17th and 18th days ectodermal mesenchyma grows into the pineal anlage between the follicles. From this mesenchyma, connective tissue strands and blood vessels originate. The blood vessels are situated close to the parenchymal follicles the cells of which form perivascular palisades. On day 19 of fetal life, the organ has practically lost its follicular structure now consisting of a dense and solid cell mass. Between this day and birth the organ grows rapidly. Perivascular connective tissue forms septa which separate cell groups, cell cords or islands which show regional cytological differences. In the marginal lobated islands, for instance, the nuclei are pale, while in the central cell islands the nuclei contain more chromatin. Already on the 18th day of fetal life, commissural nerve fibres penetrate the pineal anlage for about three-quarters of its length (see Vollrath, 1981, for literature). Sympathetic nerve fibres reach the pineal only postnatally. The rat pineal obtains its adult pattern of green fluorescence intensity pointing to the presence, in sympathetic nerve fibres, of an aminergic catecholamine at postnatal days 10-11, while the final number of sympathetic varicose nerve terminals is probably established only at day 21 after birth.

In rat, the bulk of the pineal tissue, rapidly growing in size, gradually shifts in a dorsocaudal direction finally reaching a superficial position in regard to the brain. It, however, maintains its connection with the ependyma of the intercommissural region and some pineal cells which remain situated at this site by means of a slender pineal stalk consisting of pineal cells. In this stalk, nerve fibres are present with which we will deal in our paper on pineal innervation. In adult rats of most strains, the bulk of pineal tissue is situated just ventral to the confluens sinuum. The great cerebral vein and the prosencephalic median vein drain into the rostral part of this large venous receptacle. As we will see later, this topographical relationship of the main superficial pineal part to the large intracranial veins is of importance for understanding the drainage of the rat pineal veins.

As has been mentioned, the superficial part of the pineal gland is connected by a pineal stalk with a much smaller deep part of pineal tissue which keeps its conservative position against the ependyma of the roof of the third ventricle between the habenular and the posterior commissure. Along with the formation of a pineal stalk, the pineal recess, which is the original recess of the third ventricle into the primordial saccular pineal anlage, practically disappears in rat.

In man, the ontogenetic development of the pineal is somewhat more complicated. Its very first anlage appears in the beginning of the second month of gestation. From the neuroepithelial sac, evaginating from the roof of the third ventricle in the epithalamus, a rostral and a caudal lobe develop by cell proliferation. Vascular connective tissue grows in from the periphery of the anlage causing some lobulation of the pineal parenchyma. Between the cords of pineal cells, vascularized strands of connective tissue are now present. A somewhat larger connective tissue septum, originally separating the rostral from the caudal lobe, completely disappears at day 150 of fetal life and the two lobes fuse (Hülsemann, 1971). Some structural differences between the two lobes have been described, but not studied in detail (see Vollrath, 1981, for literature). Before the 60th day of pineal embryonic development, commissural fibres already penetrate into the organ, while sympathetic nerve fibres arrive at the caudal pole of the pineal between day 60 and 82 (Hülsemann, 1971), or even somewhat earlier (Møller, 1976). Varicosities and nerve terminals of such fibres, frequently observed at day 150 of gestation, contain clear as well as dense-core or granulated vesicles.

During its ontogenetic development, the human pineal keeps its original position close to the ependyma between the habenular and the posterior commissure, the lamina intercalaris. The adult organ occupies the depression between the mesencephalic rostral colliculi. Its stalk is only very short. Rostrally it is divided into two lamina which enclose the pineal recess which is still present in the adult human pineal. The ventral lamina comprises the posterior commissure and the dorsal lamina the habenular commissure. Most of the organ is covered by tela chorioidea. The pineal lies in the subarachnoid space and is, therefore, bathed in subarachnoidal cerebrospinal fluid. We will deal later with the drainage of the veins of the human pineal.

As has been mentioned before, the shape, size and location of the pineal gland greatly vary among mammals. This has already become clear from the above comparison of the rat pineal with the human pineal. Many types of classification have been offered. The most recent one is that by Vollrath (1981). It is based on the relationship of the organ to the third ventricle and the shape and topographical arrangement of the pineal tissue. According to this classification, a pineal belongs to the proximal type or type A when most of its tissue is situated close to the third ventricle, that is, as we have seen, in its embryologically conservative position. When the pineal is of an elongated type, its length being about more than twice its largest width, it is designated as belonging to the proximo-intermedio type, or type AB. Pineal glands of which the most substantial amount of glandular tissue lies either superficially or topographically very closely related to the cerebellum, due to a considerable ontogenetic shift of the bulk of the structure in a dorsocaudal direction, belong to the intermediate-distal type or type ABC. For further details of this very useful classification in which also a more or less greater reduction of the three parts, *i.e.*, the proximal, the intermediate and the distal part of the gland, is taken into account, we refer to Vollrath (1981). In his book, Vollrath also published tables of the shape, the dimensions and the weights of the pineal in a great number of mammals.

Some words may still be said about pineal size. Relating pineal weight either to body weight or to brain weight (Legait et al., 1975, 1976a,b) it has been shown that large interspecies variations exist in such relationships as well as in regard to the so-called epiphysial index (see Legait et al., 1976b; Quay, 1980, for definition and details).

In general, pineal activity is known to be correlated with pineal volume. Variations in pineal volume have been studied in three species of rodents (Quay, 1956, 1978; Legait et al., 1975). Although pineal weight is known to show, *inter alia*, circadian and seasonal variations, interspecies variations are much greater. Apparently, therefore, interspecies differences in pineal weight and volume are also based on genetic differences. Although the larger size of a pineal gland does not indicate *per se* that the organ is of greater functional importance it is nevertheless acceptable to assume that very large pineals play a more important physiological role than extremely small ones. As has been stressed several times (Ariëns Kappers, 1981; Pévet, 1982), the pineal can be considered to be principally involved in the integration of information based on environmental conditions. As was already demonstrated in rodents (Reiter, 1973, 1978; Hoffmann, 1979), this holds especially for daylength in temperate zones. However, also other environmental conditions such as temperature, rainfall, food supply, etc., depending on the ecological situation in which the species concerned is living (Pévet, 1979, 1981; Pévet et al., 1981; Vivien-Roels, 1981), may play an important role. Evidently, variations in pineal size exist which are due to the adaptation of the species to geographical, climatological and meteorological conditions (Pévet, 1982). Strikingly, animals having a very large pineal gland such as the sea lion, seal, walrus (Tilney and Warren, 1919), northern fur seal (Elden et al., 1971), Weddell seal (Cuello and Tramezzani, 1969), elephant seal and some Cervidae (Ralph, 1975) are all living under arctic or holarctic conditions ranging to high altitudes and high latitudes. In contrast, animals such as the anteater, sloth, armadillo and pangolin (Oksche, 1965; Ralph, 1975; Quay, 1965) in which pineal cells have been found to be quite absent or to be present in an extremely small amount only, and animals such as the rhinoceros, the elephant dugong (Oksche, 1965; Ralph, 1975; Hill, 1945) and the red kangaroo which have a small pineal, are all living under tropical/equatorial conditions. An apparent latitudinal relationship to pineal size has also been demonstrated in some rodents.

Next to this it should be remarked that, in general, small pineals appear to be correlated with a nocturnal life and larger pineals with a diurnal life of the animals concerned (Ralph, 1975). Moreover, a correlation seems to exist between pineal size and the degree of homeothermia. Many heterothermic endotherm animals have a small pineal, a fact which should perhaps be correlated with the capacity of pineal function for adaptation to the thermal environment. Endothermic mammals which live in a rather harsh thermal environment, such as is the case in arctic zones, but which nevertheless maintain a constant and warm body temperature and which have also a necessarily seasonal-conditioned reproductive pattern, mostly show a well-developed pineal organ (see Ralph, 1975, for details).

The great variability in pineal size among mammals is probably associated with the functional capability of the pineal to adapt them to several very different environmental conditions. If external conditions are not very rough as is true for seasonal changes in temperature in tropical/equatorial zones (DeLany and Happolid, 1979), the pineal will be relatively small. However, desert life is difficult for mammals to endure. This could, perhaps, explain why in some tropical/equatorial species the pineal has been reported to be large and most active. In the opinion of Pévet (1982, see also for details) it would be of special importance to investigate the correlation between pineal size, not with latitude, but with latitude-related biotopes such as equatorial wet forest, savanna, desert zones, temperate, arctic and holarctic zones in combination with zonation in altitude (see also DeLany and Happolid, 1979, for details), all representing biotopes in which the influence of environmental conditions on the mammals concerned differs.

In the experimental animal most used, the rat, the pineal complex shows rather considerable variations depending on the various strains. Basically, it consists of a proximal or deep part which is situated close to the third ventricle, a pineal stalk and a distal or superficial part. Especially the pineal stalk may show variations which are also individually conditioned (Boeckmann, 1980). In Sprague-Dawley rats, for instance, the stalk may be more or less reduced. If total reduction occurs, the pineal complex exclusively consists of a deep and a superficial part. In the albino rat used in the investigations of the present author, the relatively very small deep part of the pineal was constantly connected by a slender stalk containing differentiated pinealocytes with a very well-developed distal or superficial part. In the hamster, a small but distinct amount of pineal tissue is situated close to the habenular commissure forming the "deep pineal", while the bulk of pineal tissue is situated distally and superficially. Occasionally, clusters of pinealocytes could be observed between these two main parts of the pineal complex (Sheridan and Reiter, 1970), but often this connection exclusively consisted of myelinated and non-myelinated nerve fibres and blood vessels, a true pineal stalk entirely lacking. Removing for experimental purposes the superficial part of the pineal complex is relatively easy. It should, however, be realized that removing this part in animals having also a "deep pineal" does not lead to a total pinealectomy. However, according to the results obtained after such an operation it appears that, at least in the hamster, removal of the superficial part of the pineal complex only can practically be considered to be a total removal of the gland.

In the rabbit pinealectomy is especially difficult because the long, club-shaped organ consists of a distinct proximal part, a long stalk and a major distal or superficial part the terminal end of which is invaginated into the confluens sinuum and either one of the two transverse sinuses (Smith, 1971; Romijn, 1973).

In many mammals, a conspicuous *suprapineal or supra-*

habenular recess is present. This is formed by an evagination of the roof of the third ventricle ontogenetically arising just rostral to the habenular commissure and extending in a dorsocaudal direction. The ventral wall of this recess may cover the deep part of the pineal complex and part of the pineal stalk, such as in rat and guinea-pig. If the pineal is exclusively of the deep type, this recess may cover most of the total gland. Because subarachnoidal tissue is present between the ventral ependymal wall of the recess and the pineal there is no direct contact between this wall and the pineal tissue proper which is, moreover, covered by a pial envelope and sometimes by a sheath of glial tissue. Therefore, it can practically be excluded that, via the ependymal wall of the suprapineal recess, a direct transport of pineal secretory compounds is possible to the cerebrospinal fluid present in this recess.

Some findings have, however, provided evidence that, in mammals, the cerebrospinal fluid may at least serve as a transport medium for pineal compounds since it has been found to contain melatonin. The question is how such compounds may reach the cerebrospinal fluid, either direcly or indirectly. As we have seen it is not very reasonable to suppose that pineal compounds can reach the cerebrospinal fluid via the ependyma of the suprapineal recess. Interestingly, however, in the pineal recess of hamster, vole, guinea-pig and Rhesus monkey, an area is present in which the ependymal lining of this recess shows interruptions through which the processes of pinealocytes penetrate so that the surface of the protruding cells is in immediate contact with the cerebrospinal fluid (Hewing, 1978, 1980). This situation reminds of earlier phylogenetic stages of the pineal and suggests the possibility of a direct release of pineal compounds into the cerebrospinal fluid. It is, nevertheless, quite sure that by far most pinealocytes present in the pineal complex release their secretory products into the blood by way of the pineal venous drainage as we will see later.

By all authors studying this subject it has been stressed that the *blood supply* of the mammalian pineal gland is very large. The minimum rate of rat pineal blood flow per gram exceeds that of most endocrine organs, equals that of the neurohypophysis and is surpassed only by that of the kidney (Goldman and Wurtman, 1964). The oxygen consumption of pineal tissue is, however, rather low (Quay, 1962, 1963) and decreases with age. Interestingly, at the onset of light a pronounced vasoconstriction has been observed (Quay, 1972). Removal of the pineal sympathetic innervation, known to influence positively the production of melatonin by the pinealocytes, especially during the night, reduces the pineal blood flow by one third (Goldman, 1967). These facts, shortly mentioned here, already point to the special role of the pineal blood flow as a vehicle for the transport of pineal secretory products.

Arterial blood reaches the organ either by branches of the posterior choroidal arteries, which originate from the posterior cerebral arteries, or by branches arising directly from the posterior cerebral arteries. Before entering the pineal, these very thin arteries run in the pineal envelope of the organ. In the human, pineal arterioles penetrating the organ follow the connective tissue septa then giving off capillaries. In many other mammals, however, occasional arterioles are present only, the pineal being directly penetrated by capillaries which originate from arterioles in the pial pineal covering. In the organ, the capillaries form an extensive capillary network which is drained by pineal venules which form one or more larger veins. In man, having a proximal pineal gland only, a single pineal vein drains into the great cerebral vein which terminates in the sinus rectus (Gladstone and Wakeley, 1940), while, according to another author (Bargmann, 1943) venous pineal blood is also drained into the bifurcation of the two internal cerebral veins which, more caudally, give rise to the great cerebral vein. In rodents, in which the distal or superficial part of the pineal complex is very well-developed, pineal veins drain directly into the great cerebral vein. From here the blood is transported via the confluens sinuum and the transverse sinuses into the general circulation. In the Wistar rat, about 12-16 peripheral collecting pineal veins were counted which all drain into the great cerebral vein. Usually, one central vein was observed to be present in every pineal gland (Hodde, 1979, 1981).

Because the ultrastructure of pineal capillaries and their relationship to the pineal parenchyma is of great importance for understanding pineal function we will shortly deal with these subjects here although, otherwise, we will not deal with pineal ultrastructure, a few topics excepted. In the rat, mouse and ground squirrel the junctions between adjacent endothelial cells of the capillaries are variable in shape. Typical tight junctions are absent, but zonulae occludentes have been observed in the chinchilla, while maculae adherentes have been demonstrated in the cow, sheep and dormouse (see Vollrath, 1981). Not in all mammalian pineals fenestrations or pores in the capillary endothelium could be observed. In some mammals they are rather scarce, but in the pineal capillary endothelium of the mouse and the ground squirrel they occur frequently. The pores are mostly bridged by a very delicate membranous diaphragm of about 5 nm in thickness. The pores measure between 50 and 70 nm in diameter.

The cytoplasm of the endothelial cells contains the organelles usually present such as mitochondria, a Golgi complex, ribosomes and cisterns of rough endoplasmic reticulum. Occasionally, filaments, microtubules and multivesicular bodies are present. A large number of pinocytotic vesicles is a regular feature of the endothelial cells. Their size varies according to the species concerned. The basal lamina of the capillary endothelium forms the inner lamina of the perivascular spaces which are a very prominent characteristic in most mammalian pineals. The outer lamina of the perivascular space is, in principle, formed by the basal membrane covering the pineal parenchyma proper. This outer lamina, however, can be more or less reduced being, therefore, discontinuous. Extensions of perivascular spaces, which do not contain a capillary, may penetrate more or less deeply into the pineal parenchyma separating clusters or cords of pinealocytes. Such spaces have even been supposed to be continuous with intercellular spaces within the pineal parenchyma. In this way, a complicated network of communicating spaces is distributed all over the gland.

The width of the perivascular spaces varies among mammals and so does their content. When the space is narrow it contains only a few nerve fibres of sympathetic origin such as has been shown in the squirrel monkey, cow, sheep and rabbit. In contrast, the perivascular spaces in cat are wide and contain many nerve fibres. In rat they are even extremely wide. In them, nerve fibres, terminals of such non-myelinated nerve fibres, collagenous fibres, fibrocytes and the processes of glial cells have been demonstrated.

The autonomic nerve fibres present in the perivascular spaces may, varying among mammals, enter these spaces along with the vessels penetrating the pineal from its outer surface. In that case they may either terminate within the perivascular spaces or enter, from here, the pineal parenchyma to end between the pinealocytes and their processes. A number of fibres, however, directly penetrates into the pineal parenchyma. Some of such fibres reach, from here, the perivascular spaces to terminate in these spaces, while most of them end within the pineal parenchyma proper. Neurotransmitter substance released from the sympathetic nerve terminals present in the perivascular spaces can reach the pinealocytes by way of the extensions of these spaces and the intercellular spaces, which, together, form an extensive and continuous intrapineal canalicular network as mentioned above. Probably, neurotransmitter released in the perivascular spaces can also reach parenchymal cells more directly via the outer lamina of these spaces which borders the pineal parenchyma, especially if the outer lamina is discontinuous. In some mammals, the pineal parenchyma is separated from the perivascular space not only by a basal lamina, but also by a thin, often likewise discontinuous layer of astrocytic glial processes.

From the foregoing it will be clear that, in many mammals, a very firm tissue layer separating the perivascular space from the pineal parenchyma which, here,for the most part consists of the terminals of pinealocyte processes, is absent. As mentioned, the separation between the pineal parenchyma and the perivascular space is formed only by an often discontinuous basal lamina and/or a thin glial sheath which may likewise be discontinuous. It can, therefore, not wonder that in some mammals the terminal endings of the pinealocyte processes occasionally herniate or bulge directly into the perivascular spaces. In the cat such bulging endings are covered by an extremely thin glial barrier and the basal lamina. In the Rhesus monkey, the terminals of the pinealocyte processes penetrating deeply into the perivascular spaces are exclusively covered by the basal lamina, while, in the rat, the terminals of such processes bulging into the perivascular spaces are often completely devoid of either a glial sheath or a basal lamina (Wartenberg, 1968). Pinealocyte processes directly terminating in perivascular spaces have also been observed in the dog, hamster, guinea-pig, mouse and gopher (see Vollrath, 1981, for literature). Evidently, a close topographical relationship exists between the terminals of pinealocyte processes – in which at least part of the pineal secretory products accumulate and from which these are released – and the perivascular spaces. This situation in combination with the fact that in many mammals the endothelium of the capillary wall is fenestrated points morphologically already to the possibility that pineal secretory products can easily be taken up by the blood and illustrates the endocrine function of the mammalian pineal organ.

A feature much discussed especially in clinical circles and shortly to be mentioned here is the presence, in the pineal, of *brain sand, corpora arenacea* or *acervuli*, all synonyms for the same type of concretions. As such concretions are opaque in X-ray photographs of the human brain they form a point of orientation for the radiologist. It has long been maintained that the occurrence of brain sand would in some way be negatively related with the secretory activity of the organ, especially since, in man, they are more numerous in old age. Such concretions are not exclusively present in the pineal of Primates including man as was thought in earlier days, but they may also occur in other mammals such as, for instance, in the rat. Their number generally greatly varies as does their shape, histological appearance and topography. In principle, corpora arenacea have been observed in glial and stromal compartments. In them, the presence of hydroxyhepatite or carbonate apatite, iron, magnesium and phosphorus have been demonstrated. It should be noted that such concretions have not only been found in the pineal gland, but also in the habenulae and in the leptomeningeal tissue adjacent to the pineal. From the rather extensive literature (see Vollrath, 1981) it can be concluded that both the origin and the formation of these acervuli and their functional meaning, if any, are far from clear. Formerly it has often been held that the presence of brain sand would point to an atrophy and functional degeneration of the gland. It is, however, now known that this is not the case. It was, for instance, demonstrated that the metabolic activity of the human pineal does not alter in old age in which, generally, more acervuli are present (Wurtman et al., 1964) and that there is no correlation between the degree of pineal calcification and either the overall histological structure of the gland (Cassano et al., 1961) or pineal enzyme levels including those enzymes which are of crucial importance for the production of pineal indoleamines which are generally considered to play a most important role in pineal function (Otani et al., 1968).

Pineal cysts are also a rather regular feature. They may already be found at birth. In some cases it is quite apparent that such cysts can be considered pinched off parts of the normally occurring pineal recess of the third ventricle because, in such cases, the wall of the cysts consists of ependymal and pineal parenchymal cells. In other cases, however, cysts are clearly present in glial patches which may point to a more or less pathological process such as likewise occurs in other parts of the central nervous system. Apparently, such cysts are of no consequence for a normal pineal function. As this paper deals with the normal mammalian pineal gland, pineal pathology will not be dealt with.

In some mammals, such as the rat, the presence of *striated muscle fibres* has been observed in the pineal. Ultramicroscopically, such fibres show embryonic features and either or not a close topographical relationship to blood vessels. Their origin and functional significance, if any, is quite unknown. Most probably they originate from the mesenchyma invading, in embryonic stages, the neuroepithelial anlage of the organ.

In most mammals, pineal *connective tissue* is scarce. In man, connective tissue septa and strands may separate islands, cords, rosettes or follicles of pinealocytes. The septa often contain blood vessels and nerve fibre bundles. Mast cells have also been observed to be present in small numbers, while pigment-containing melanophores could be demonstrated in the pineal stroma in a number of mammalian species. It is most questionable whether melanophores are involved in pineal function.

As will more elaborately be dealt with in the paper on pineal innervation, in some few mammals *nerve cells* were noted to be present within the pineal parenchyma. They are certainly not a constant feature of the mammalian pineal. Abundant nerve fibre bundles and single nerve fibres are, however, constantly present. They are of different origin as will be mentioned in our second lecture.

Proceeding now to the histology and cytology of the pineal parenchyma proper it can be stated that its major component is the *pinealocyte*, sometimes termed pineocyte. In the rat, pinealocytes make up about 82% of the total pineal cell content (Wallace et al., 1969). The pinealocyte is a polymorphous cell element especially in the human pineal in which unipolar, bipolar and multipolar pinealocytes have been demonstrated by means of the silver carbonate staining technique. In most mammals, however, the pinealocyte consists of a cell body and a variable number of cell processes which vary in length and width. The terminal endings of the pinealocyte processes are often club-shaped. Ultramicroscopically, they are sometimes difficult to distinguish from nerve terminals, the more so because they may likewise contain granular vesicles. The terminal endings of pinealocyte processes are situated either within the parenchyma between adjacent pinealocyte perikarya or processes, or in close vicinity to perivascular spaces from which they are then separated by a thin basal lamina, a thin glial sheath or both, the glial sheath and/or the basal lamina being often discontinuous as has been mentioned before. As has also been described, in some mammals the endings of pinealocyte processes bulge into the perivascular spaces.

As to the *phylogenetic and ontogenetic origin of the mammalian pinealocyte* the following may be remarked. It has been maintained that, phylogenetically, the mammalian pinealocyte is the final stage of a so-called sensory cell line. Earlier stages in this line are the true photosensory cells or photoreceptor cells and the secretory rudimentary photoreceptor cells which are present in the pineal organ of non-mammalian vertebrates (Collin, 1971). Rather recently it could be demonstrated that, also ontogenetically, pinealocytes show signs of such an origin. In this regard, the presence of *centrioles* and *cilia* in mammalian pinealocytes is of special interest.

If at all present, centrioles may show a number of rather unusual features. Apparently, they may transform into microtubular sheaves or bouquets (Wolfe, 1965; Lin, 1969, 1970, 1972). Probably, such a transformation does not point to their degeneration, but to the formation of structures which recall outer segments or cilia as present in the above-mentioned phylogenetic precursor cells of pinealocytes, such as cilia of the 9 + 0 pattern (Vollrath, 1981). No centrioles are present in the adult rat pinealocyte, but they still occur in the pinealocytes of fetal (Clabough, 1973) and of newborn (Zimmerman and Tso, 1975) rats. Here they display the usual fine structure becoming elongated between 15 days and 6 weeks of age (see Vollrath, 1981, for further details). Abortive remnants of a ciliary apparatus, such as a ciliary rootlet in continuity with a centriole, have even been described in pinealocytes of the adult guinea-pig (Lin, 1972) and of the adult hedgehog (Pévet and Saboureau, 1973).

True cilia occur in a number of very different cells of the adult mammalian organism, such as in some ganglion cells. Their presence in adult mammalian pinealocytes attracted special attention because they have also been considered to point to the phylogenetic as well as the ontogenetic origin of pinealocytes from their precursor cells of the sensory cell line. In rat pinealocytes, the presence of cilia, like that of centrioles, has been noted in fetal (Clabough, 1970, 1973) and in early postnatal stages (Zimmerman and Tso, 1975) of development.

In pinealocytes of the adult rat cilia are, however, very rare. In the pinealocytes of most other mammals the presence of clilia shows great variability. In the pinealocytes of the noctule bat (Pévet et al., 1977), mole-rat (Pévet et al., 1976) and mole (Pévet, 1974) cilia are, however, frequently observed. Like in the phylogenetic precursor cells of pinealocytes they are of the 9 + 0 type. In the adult mole each pinealocyte has been reported to contain a cilium (Pévet and Collin, 1976). Structures reminiscent of rudimentary outer segments such as are present in the secretory rudimentary photoreceptor cells in the pineal organ of lizards and birds, have been demonstrated in pinealocytes of the noctule bat (Pévet et al., 1977), while even concentric lamellar structures resembling outer segmental discs were observed in close vicinity to cilia (Pévet and Collin, 1976; Pévet et al., 1977). The findings mentioned clearly indicate that, indeed, the mammalian pinealocyte is derived not only phylogenetically, but also ontogenetically from precursor cells of the so-called sensory cell line. At least in some mammals such an origin has been demonstrated during the ontogenetic development of the pinealocyte.

Mammalian pinealocytes not seldomly contain more than one *nucleus* showing one or more nucleoli. The nuclei are polymorphic. During the times light microscopy was exclusively available for pineal investigation, much has been speculated about the true nature of the so-called "Kernkugeln" or nuclear pellets which present themselves as nuclear inclusions. Electron microscopical research has, however, clearly demonstrated that these are, in fact, parts of the cell cytoplasm which are invaginated into the lobulated nuclei. Pinealocyte nucleoli are prominent, but contain little RNA. Nuclear size shows small variations according to the site of the pinealocytes either in proximal or distal regions of the gland, while differences in size have also been described in the so-called cortical and medullar region of the organ. Nuclear size also varies with the age of the animal, the seasons, the functional state of the gonads and, in general, with the activity of the pineal gland. Such changes are, however, small and do not offer very reliable indications as the functional state of the pinealocyte is concerned, at least not without very stringent statistical investigations. The circadian changes in size shown by pinealocytes can be expected to be related with the well-known circadian changes in pineal gland secretory activity as stated by cytochemical and pharmacological research.

In the fully adult mammalian pineal the pinealocytes show very little mitotic activity. In the rat it has, however, been demonstrated that the mean number of mitoses follows a circadian rhythmicity showing a peak at midday (Renzoni and Quay, 1964; Quay and Renzoni, 1966).

On morphological grounds, different types of pinealocytes have been distinguished. The most frequent distinction made is that between so-called light and dark pinealocytes. By some authors it has, however, been claimed that such a distinction is merely based on different functional states of the same cell type (see Vollrath, 1981). Recently, however, two different populations of pinealocytes have been described on the ground of their different ultrastructure and possible function (Pévet, 1977a, 1979). Pinealocytes of population I, representing the majority of pinealocytes, contain a well-developed Golgi apparatus producing granular vesicles, while pinealocytes of population II are characterized by the presence of accumulations of material or by the formation of vacuoles containing flocculent material of moderate electron density, both in the cisterns of the endoplasmic reticulum which is more abundant in these cells. In the mole, this type of pinealocyte shows a rather rough endoplasmic reticulum comprising lozenge-shaped paracrystalline structures (Pévet, 1974, 1976, 1977a, 1979; Pévet and Smith, 1975) which are of a proteic nature disappearing after pronase treatment (Pévet, 1977b). Formation of vacuoles by the endoplasmic reticulum has, under natural conditions, been observed in the hedgehog (Pévet and Saboureau, 1973; Pevet, 1976) and the noctule bat (Pévet et al., 1977) and,under experimental conditions, in the hedgehog (Pévet and Saboureau, 1974) and rat (Karasek et al., 1976; Karasek and Marek, 1978). Such material has been considered a special type of secretory product, in some way related to the antigonadotropic activity of the pineal gland (Pévet, 1977b; Pévet and Ariëns Kappers, 1977a,b). In the mole, the paracrystalline structures have been demonstrated to increase in number in males during the period of high sexual activity and in females during oestrus, gestation and lactation (Pévet and Smith, 1975). These findings are, at least, suggestive of some relationship between their presence, the functional state of this type of pinealocyte and the functional state of the sex organs. In castrated rats, the endoplasmic reticulum in such pinealocytes has been shown to produce vacuoles containing moderately dense material which is released into intercellular spaces (Karasek et al., 1976). Similar vacuoles the origin of which could, however, not be established exactly, have been demonstrated in hedgehog pinealocytes at the beginning of the physiological phase of quiescence (Pévet and Saboureau, 1973) and also when such a quiescence was experimentally induced (Pévet and Saboureau, 1974). Not in all mammalian pineals two populations of pinealocytes showing clearly the different characteristics mentioned can be distinguished. In such cases, in which the organ shows one population of pinealocytes only, a single pinealocyte may combine the cytological and functional properties of cells belonging to the two different populations to be found in other mammals. Further investigations on this subject are certainly necessary and could prove to be most fruitful for the undertanding of pineal function.

In this necessarily short survey of the comparative gross anatomy of the mammalian pineal organ, its development, microscopic anatomy and cytology we cannot deal with all of the structures present in pinealocytes. We will, however, shortly mention the presence of *lipid droplets* in such cells. Pinealocytes contain varying amounts of such droplets and their size, location in the cell and shape are also variable. Much has been speculated about their possible functional significance. The chemistry and histochemistry of pineal lipids has been thoroughly reviewed by Quay (1974). No lipid compounds have been identified which are characteristic of pinealocytes only. Probably, pineal lipids are primarily involved in the general cell metabolism and in membrane structure and function of pinealocytes, not playing any role as hormones or hormone precursors. It is, however, of some interest that pineal lipid content in the rat demonstrates a circadian rhythmicity showing a peak at midday and being lowest at midnight (Quay, 1965, 1974). Moreover, pineal lipid content is in some way correlated with the activity of the gonadal axis. In female rat it was observed to be lowest just before the start of ovulation and maximal at the time of dioestrus (Zweens, 1965). Pineal lipid content proved also to be age-dependent. In the human it increases with age (Quay, 1957), but it was shown to decrease in female rats between 8 and 15½ weeks of age (Zweens, 1965). Under different experimental conditions either an increase or a decrease in the mammalian pineal lipid content has been observed (see Table 16 in Vollrath, 1981). From the above findings the conclusion can be drawn that the lipid content of pinealocytes is somehow correlated with their function, but that the exact role played by these lipids is still enigmatic.

Finally we will turn to another constituent of the

mammalian pineal parenchyma, the *glial cell* which has also been termed interstitial cell. The number of pineal glial cells greatly varies among mammals. In the pineal parenchyma of the rat their number amounts to about 12% of the total number of pineal cell elements (Wallace et al., 1969), while in cat and monkey the ratio of glial cell perikarya to pinealocyte perikarya is less than 1 : 10 (Wartenberg, 1968). In the rat pineal gland the glial cells differ in some aspects from the classical astrocytes, more especially because they have some ultrastructural features in common with the so-called dark pinealocytes. This is the reason why, by some authors, they have been designated by the more neutral term interstitial cells. Recently, however, light microscopical studies in which antibodies against characteristic glial marker proteins were used, suggested that the "interstitial cells" in the rat pineal are, indeed, astrocytes (Møller et al., 1978).

In general, mammalian pineals which are superficially situated contain but few glial cells and glial processes, whereas deep pineals mostly show an abundance of astrocytic perikarya and processes (Hülsemann, 1967; Sheridan and Reiter, 1973). Astrocytes in the pineal may be of either the fibrous or the cytoplasmic type, just like in the brain. In the pineal, however, fibrous astrocytes apparently outnumber protoplasmic astrocytes. In the human pineal, which is deeply situated, glial fibres being the processes of fibrous astrocytes are very abundant (Hülsemann, 1967). The glial fibres form an intricate network which surrounds groups of pinealocytes. This network is specially dense in the periphery of the organ where glial patches have occasionally been observed. In other primates, regional differences in the amount of glial fibres and their arrangement have also been described (Hülsemann, 1967; Eckhardt, 1972; Merker, 1968). Not all glial fibres necessarily originate from within the pineal organ proper (see Vollrath, 1981, for literature). In the rhesus monkey, glial fibres reach the pineal via its short stalk. Here such fibres evidently originate from the subependymal layer of the third ventricle (Eckhardt, 1972).

In non-primates the amount of glial fibres shows great interspecies variations. In the pineal of some mammals even no glial fibres could be demonstrated at all, while in that of others they are relatively scarce (see Vollrath, 1981, for literature). The localization of glial fibres within the pineal also varies. They may be distributed in the marginal zone of the organ, in its centre, close to blood vessels and in connective tissue septa.

As protoplasmic astrocytes are concerned, little information is available from light microscopic investigations. As has, however, been mentioned before a glial barrier may separate pinealocytes, especially the terminals of their processes, from the perivascular spaces as is the case in cat and monkey (Wartenberg, 1968). This barrier is formed by cytoplasmic glial cells and their processes. About the exact function of pineal glial cells nothing is known that fundamentally differs from what we know about the function of astrocytic glia in general.

REFERENCES

Ariëns Kappers, J., 1960, The development, topographical relations and innervation of the epiphysis cerebri in the albino rat, Z.Zellforsch., 52:163.

Ariëns Kappers, J., 1981, Evolution of pineal concepts, in: "The Pineal Organ – Photobiology – Biochronometry – Endocrinology", A.Oksche and P.Pévet, eds., Elsevier, Amsterdam.

Bargmann, W., 1943, Die Epiphysis cerebri, in: "Handbuch der mikroskopischen Anatomie des Menschen", Vol. VI/4, W.von Möllendorff, ed., Springer, Berlin.

Boeckmann, D., 1980, Morphological investigation of the deep pineal of the rat, Cell Tiss. Res., 210:283.

Cassano, C., Torsoli, A., Peruzy, A.D., and De Martino, C., 1961, Studi su l'epifisi. Indagini nell'animale da esperimento e nell'uomo, Folia endocrinol. (Pisa), 14:755.

Clabough, J.W., 1970, Ultrastructural study of pineal cytogenesis in fetal rats and hamsters, Anat. Rec., 166:291.

Clabough, J.W., 1973, Cytological aspects of pineal development in rats and hamsters, Amer. J. Anat., 137: 215.

Collin, J.-P., 1971, Differentiation and regression of the cells of the sensory line in the epiphysis cerebri, in: "The Pineal Gland", G.E.W.Wolstenholme and J.Knight, eds., Churchill Livingstone, Edinburgh-London.

Cuello, A.C., and Tramezzani, J.H., 1969, The epiphysis cerebri of the Weddell seal: its remarkable size and glandular pattern, Gen. comp. Endocrinol., 12: 153.

DeLany, M.J., and Happolid, D.C.D., 1979, "Ecology of African Mammals", Longman, London.

Eckhart, W., 1972, Über den Verlauf der Gliafasern in der Epiphysis cerebri des Rhesusaffen *(Macaca mulatta)*, Acta anat. (Basel), 81:237.
Elden, C.A., Keyes, M.C., and Marshall, C.E., 1971, Pineal body of the northern fur seal *(Callorhinus ursinus)*. A model for studying the probable function of the mammalian pineal body, Amer. J. vet. Res., 32:639.
Gardner, J.H., 1949, Development of the pineal body in the hooded rat, Anat. Rec., 103:538.
Gladstone, R.J., and Wakeley, C.P.G., 1940, "The Pineal Gland", Baillière, Tindall and Cox, London.
Goldman, H., 1967, The nervous control of blood flow in the pineal body, Life Sci., 6:2071.
Goldman, H., and Wurtman, R.J., 1964, Flow of blood in the pineal body of the rat, Nature (London), 203:87.
Hewing, M., 1978, A liquor containing area in the pineal recess of the golden hamster *(Mesocricetus auratus)*, Anat. Embryol., 153:295.
Hewing, M.,1980, Cerebrospinal fluid-contacting area in the pineal recess of the vole *(Microtus agrestis)*, guinea-pig *(Cavia cobaya)* and Rhesus monkey *(Macaca mulatta)*, Cell Tiss. Res., 209:473.
Hill, W.C.O., 1945, Notes on the dissection of two dugongs, J. Mammal., 26:153.
Hodde, K.C., 1979, The vascularization of the rat pineal organ, Progr. Brain Res., 52:39.
Hodde, K.C., 1981, "Cephalic Vascular Patterns in the Rat", Thesis, Univ. Amsterdam.
Hoffmann, K., 1979, Photoperiod, pineal, melatonin and reproduction in hamsters. Progr. Brain Res., 52:397.
Hülsemann, M., 1967, Vergleichende histologische Untersuchungen über das Vorkommen von Gliafasern in der Epiphysis cerebri von Säugetieren, Acta anat. (Basel), 66:249.
Hülsemann, M., 1971, Development of the innervation in the human pineal organ. Light and electron microscopic investigations, Z.Zellforsch., 115:396.
Karasek, M., and Marek, K., 1978, Influence of gonadotropic hormones on the ultrastructure of rat pinealocytes, Cell Tiss. Res., 188:133.
Karasek, M., Pawlikowski, M., Ariëns Kappers, J., and Stepién, H., 1976, Influence of castration followed by administration of LH-RH on the ultrastructure of rat pinealocytes, Cell Tiss. Res., 167:325.
Legait, H., Bauchot, R., and Contet-Audonneau, J.L., 1976a, Etude des corrélations liant les volumes des lobes hypophysaires et de l'épiphyse au poids somatique et au poids encéphalique chez les Chiroptères, Bull. Assoc. Anat. (Nancy), 60, no. 168:175.

Legait, H., Bauchot, R., Stephan, H., and Contet-Audonneau, J.L., 1976b, Etude des corrélations liant le volume de l'épiphyse au poids somatique et encéphalique chez les rongeurs, les insectivores, les chiroptères, les prosimiens et les simiens, Mammalia 40:327.
Legait, H., Legait, E., Contet-Audonneau, J.L., Mur, J. M., Roux, M., Richoux, J.P., Sirjean, D., and Dussart, G., 1975, Etude des corrélations liant les volumes de l'hypophyse, de l'épiphyse et de l'OSH au poids somatique et au volume de l'hypothalamus chez les Rongeurs, Bull., Assoc. Anat. (Nancy), 59, no. 164:185.
Lin, H.-S., 1969, A special morphological expression of centriolar formation. Anat. Rec., 163:313.
Lin, H.-S., 1970, The fine structure and transformation of centrioles in the rat pinealocyte, Cytobios, 2: 128.
Lin, H.-S., 1972, Transformation of centrioles in pinealocytes of adult guinea-pigs. J. Neurocytol., 1:61.
Merker, G., 1968, Licht- und elektronenmikroskopische Studien über die Fasergliastruktur der Epiphysen-Subcommissuralregion der Primaten, Z. Zellforsch., 92:232.
Møller, M., 1976, The ultrastructure of the human fetal pineal gland. II. Innervation and cell junctions, Cell Tiss. Res., 169:7.
Møller, M., Ingild, A., and Bock, E., 1978, Immunohistochemical demonstration of S-100 protein and GFA protein in interstitial cells of the rat pineal gland, Brain Res., 140:1.
Oksche, A., 1965, Survey of the development and comparative morphology of the pineal organ, Progr. Brain Res., 10:3.
Otani, T., Györkey, F., and Farrell, G., 1968, Enzymes of the human pineal body, J. clin. Endocrinol., 28: 349.
Pévet, P., 1974, The pineal gland of the mole (*Talpa europaea* L.). The fine structure of pinealocytes, Cell Tiss. Res., 153:277.
Pévet, P., 1976, "Correlations between Pineal Gland and Sexual Cycle. An Electron Microscopical and Histochemical Investigation on the Pineal Gland of the Hedgehog, Mole, Mole-rat and White Rat", Thesis, Univ. Amsterdam.
Pévet, P., 1977a, On the presence of different populations of pinealocytes in the mammalian pineal gland, J. Neural Transm., 40:289.
Pévet, P., 1977b, The pineal gland of the mole (*Talpa europaea* L.). IV. Effect of pronase on material

present in cisternae of the granular endoplasmic reticulum of pinealocytes, Cell Tiss. Res., 182:215.

Pévet, P., 1979, Secretory processes in the mammalian pinealocyte under natural and experimental conditions, Prog. Brain Res., 52:149.

Pévet, P., 1981, Ultrastructure of the mammalian pinealocytes, in: "The Pineal, Anatomy and Biochemistry", R.J.Reiter, ed., CRC Press, Boca Raton (Fl.), USA.

Pévet, P., 1982, The anatomy of the pineal gland of mammals, in: "The Pineal Gland", R.Relkin, ed., Elsevier, New York (in press).

Pévet, P., and Ariëns Kappers, J., 1977a, Secretory processes in the mammalian pineal gland. An ultrastructural identification, Acta endocrinol., (Kjöbenhavn), Suppl. 212:157.

Pévet, P., and Ariëns Kappers, J., 1977b, Secretory processes differing from that producing granular vesicles in the mammalian pinealocytes. An ultrastructural identification. XXVIIe Intern. Congr. Physiol. Sci., Paris, July 18-23, Abstract no. 1760.

Pévet, P., and Collin, J.-P., 1976, Les pinéalocytes de Mammifère: Diversité, homologies, origine. Etude chez la Taupe adulte (*Talpa europaea* L.), J. ultrastruct. Res., 57:22.

Pévet, P., and Saboureau, M., 1973, L'épiphyse du Hérisson (*Erinaceus europaeus* L.) mâle. I. Les pinéalocytes et leur variations ultrastructurales considérees au cours du cycle sexuel, Z. Zellforsch., 143:367.

Pévet, P., and Saboureau, M., 1974, Effect of serotonin administration on the ultrastructure of pinealocytes during the period of maximal sexual activity of the male hedgehog (*Erinaceus europaeus* L.), Experientia (Basel), 30:1069.

Pévet, P., and Smith, A.R., 1975, The pineal gland of the mole (*Talpa europaea* L.), II. Ultrastructural variations observed in the pinealocytes during different parts of the sexual cycle, J. Neural Transm., 36:227.

Pévet, P., Ariëns Kappers, J., and Nevo, E., 1976, The pineal gland of the mole rat (*Spalax ehrenbergi* Nehring). I. The fine structure of pinealocytes, Cell Tiss. Res., 174:1.

Pévet, P., Ariëns Kappers, J., and Voûte, A.M., 1977, Morphologic evidence for differentiation of pinealocytes from photoreceptor cells in the adult noctule bat (*Nyctalus noctula* Schreber), Cell Tiss. Res., 182:99.

Quay, W.B., 1956, Volumetric and cytologic variation in the pineal body of *Peromyscus leucopus* (Rodentia)

with respect to sex, captivity and day-length, J. Morphol., 98:471.
Quay, W.B., 1957, Cytochemistry of pineal lipids in rat and man, J. Histochem. Cytochem., 5:145.
Quay, W.B., 1962, The respiration of ependyma and pineal cells in vitro with diverse substrates and concentrations of inorganic ions, Amer. Zool., 2:439.
Quay, W.B., 1963, Pineal and ependymal respiration with diverse substrates and inorganic ions, Amer. J. Physiol., 204:245.
Quay, W.B., 1965, Histological structure and cytology of the pineal organ in birds and mammals, Progr. Brain Res., 10:49.
Quay, W.B., 1972, Pineal vasoconstriction at daily onset of light: its physiological correlates and control, Physiologist, 15:241.
Quay, W.B., 1974, "Pineal Chemistry", Charles Thomas, Springfield (Ill.), USA.
Quay, W.B., 1978, Quantitative morphology and environmental responses of the pineal gland in the collared lemming *(Dicrostonyx groenlandicus)*, Amer. J. Anat., 153:545.
Quay, W.B., 1980, Greater pineal volume at higher latitudes in Rodentia: Exponential relationship and its biological interpretation, Gen. comp. Endocrinol., 41:340.
Quay, W.B., and Renzoni, A., 1966, Twenty-four-hour rhythms in the pineal mitotic activity and nuclear and nucleolar dimensions, Growth, 30:315.
Ralph, C.L., 1975, The pineal and geographical distribution of animals, Intern. J. Biometr., 19:289.
Reiter, R.J., 1973, Pineal control of a seasonally reproductive rhythm in male golden hamsters exposed to natural daylight and temperature. Endocrinology, 92: 423.
Reiter, R.J., 1978, Interaction of photoperiod, pineal and seasonal reproduction as exemplified by findings in the hamster, in:"Progress in Reproductive Biology" R.J.Reiter, ed., Karger, Basel.
Renzoni, A., and Quay, W.B., 1964, Daily karyometric and mitotic rhythms of pineal parenchymal cells in the rat, Amer. Zoologist, 4:416.
Romijn, H.J., 1973, Structure and innervation of the pineal gland of the rabbit, *Oryctolagus cuniculus* (L.). I. A light microscopic investigation, Z. Zellforsch., 139:473.
Sheridan, M.N., and Reiter, R.J., 1970, Observations on the pineal system in the hamster. I. Relations of the superficial and deep pineal to the epithalamus, J. Morphol., 131:153.

Sheridan, M.N., and Reiter, R.J., 1973, The fine structure of the pineal gland in the pocket gopher, *Geomys bursarius*, Amer. J. Anat., 136:363.

Smith, A.R., 1971, The topographical relations of the rabbit pineal to the large intracranial veins, Brain Res., 30:339.

Tilney, F., and Warren, L.F., The morphology and evolutional significance of the pineal body, Amer. Anat. Memoirs, no. 9, The Wistar Institute, Philadelphia.

Vivien-Roels, B., 1981, Activité sexuelle et glande pinéale, La Recherche, 113:833.

Vollrath, L., 1981, The pineal organ, in: "Handbuch der mikroskopischen Anatomie des Menschen", Vol. VI/7, A.Oksche and L.Vollrath, eds., Springer, Berlin-Heidelberg-New York.

Wallace, R.B., Altman, J., and Das, G.D., 1969, An autoradiographic and morphological investigation of the postnatal development of the pineal body, Amer. J. Anat., 126:175.

Wartenberg, H., 1968, The mammalian pineal organ: Electron microscopic studies on the fine structure of pinealocytes, glial cells and on the perivascular department, Z. Zellforsch., 86:74.

Wolfe, D.E., 1965, The epiphysial cell: an electron-microscopical study of its intercellular relationships and intracellular morphology in the pineal body of the albino rat, Progr. Brain Res., 10:332.

Wurtman, R.J., Axelrod, J., and Barchas, J.D., 1964, Age and enzyme activity in the human pineal, J. clin. Endocrinol., 24:299.

Zimmerman, B.L., and Tso, M.O.M., 1975, Morphologic evidence of photoreceptor differentiation of pinealocytes in the neonatal rat, J. Cell Biol., 66:60.

Zweens, J., 1965, Alterations of the pineal lipid content in the rat under hormonal influences, Progr. Brain Res., 10:540.

PERSPECTIVES OF COMPARATIVE ANATOMY OF THE MAMMALIAN PINEAL GLAND

Lutz Vollrath

Department of Anatomy
Johannes Gutenberg-University
Mainz
Federal Republic of Germany

INTRODUCTION

As the precise role of the pineal gland has been established in a few mammalian species only, every effort should be made to elucidate the function of this potentially important neuroendocrine organ in as many species as possible. One way in which the morphologist can contribute to reach this goal is to carry out comparative studies. Comparative anatomical studies with respect to the pineal gland can be, and should be, carried out at different structural levels: (a) At the gross anatomical level; (b) at the tissue level; (c) at the cellular level, and (d) at the level of cell organelles.

GROSS ANATOMY

From a comparative point of view, the pineal gland is a most interesting organ. The morphologist is particularly impressed by the great structural diversity of the organ at the gross anatomical level, though it should be noted that due to the small size microscopic methods may have to be used to examine the organ and its topographic relationships. The present contribution restricts itself to the situation in mammals, which alone give enough food for thought.

An often discussed point is whether a pineal gland is regularly present or not (Bargmann, 1943; Oksche, 1965; Ralph, 1975). From a recent survey (Vollrath, 1981), it can be seen that the organ has been clearly demonstrated in a large number of mammalian species. Therefore, the species are of special interest which lack a pineal organ or have very small, atrophic pineals. One Order of mammals in which the pineal is apparently not regularly present is the Xenarthra. Whereas Bargmann (1943) mentioned an absence of the pineal in *Dasypus novemcinctus*, *D. villosus* and *Bradypus* and Ralph (1981) in *D. novemcinctus*, Hofer et al. (1976) found diffuse pineal tissue arranged around the intercommissural recess in the latter species. The absence of pineal tissue has been established also for some marsupials, i.e. *Petaurus breviceps* and *P. norfolcensis* (Johnson, 1977).

An atrophic pineal gland has recently been described for the equatorial African species *Heterocephalus glaber*, measuring as little as 0.002135 mm^3 and representing the smallest rodent pineal gland in absolute size reported so far (Quay, 1981). This finding is of particular importance as it conforms to the trend noted by Ralph (1975) and Quay (1980a) that rodents whose geographic distributions are centred in lower latitudes have relatively small pineal glands, whereas rodents at higher latitudes are characterized by greater pineal volume. In view of this situation it would appear wise to carry out comparative anatomical and comparative physiological studies, concentrating on species inhabiting lower latitudes, in particular as their pineal glands reveal ultrastructural peculiarities (Pévet and Kuyper, 1978; Pévet and Yadav, 1980).

Moreover, attempts should be made to breed small equatorial rodents in the laboratory to have a sufficient amount of experimental animals available for systematic studies. In my view, it would be highly desirable to select rodents characterized by very small, yet non-atrophic pineal glands which could be easily quantitatively surveyed with respect to a perhaps mosaic-like structure of the parenchyma (Vollrath, 1979), direction and course of pinealocyte processes, distribution of peripheral sympathetic as well as central commissural fibres and ultrastructural peculiarities.

Apart from the size of the pineal gland, comparative anatomical studies have to take into consideration the form and the position of the pineal complex. A classification has been proposed (Vollrath, 1979, 1981) which enables researchers to indicate with a few letters the type of pineal gland present. According to this classification, an AßC pineal gland is a dumbbell-shaped organ in which the proximal part (area A) lies in the immediate vicinity of the third ventricle and the distal part (area C) just beneath the skull, in close relation to the cerebellum; the ß indicates that the intermediate area of the pineal complex (area B) is distinctly reduced in size when compared with areas A and C. The respective letter is omitted when the area it stands for is missing.

The usefulness of this classification has yet to be demonstrated, but one advantage is that researchers are beginning to look very carefully at the arrangement of pineal tissue and are beginning to realize the distinct interstrain differences that exist. These are particularly noticeable in rats. In the albino rats studied by Gregorek et al. (1977), a distal pineal was regularly present whereas a proximal or deep pineal was always lacking. By contrast, in the 106 Sprague-Dawley rats studied by Boeckmann (1980) a deep pineal gland was always present. The Wistar rats investigated by Korf and Møller (unpublished observations) resembled those described by Gregorek et al. (1977). In view of these diverse findings it may be wise to breed a specific strain of albino rats characterized by a uniform pineal complex. Another interesting approach might be to select extreme forms and to examine what effect cross-breeding has on the pineal complex. In this context it is interesting to note that according to Gardner (1953) the superficial pineal of the hooded rat is not connected to the epithalamus. This poses the question as to how the central innervation (Buijs and Pévet, 1980; Korf and Wagner, 1980; Schneider et al., 1981) of the pineal gland is effected.

The classification referred to above is also a useful basis for studying whether or not closely related mammalian species have similar types of pineal complexes. It is premature to answer this question as an insufficient number of mammalian species has been investigated in this respect. Nevertheless, some interesting facts emerge. The Primates, Chiroptera, Tupaiodea, Insectivora, Marsupialia and Monotremata studied so far are mainly characterized by proximal or Type A pineals; in Artiodactyla and Perissodactyla there appears to be the tendency of the pineal complex to extend into a posterior direction, forming what can be classified as a Type AB pineal; a similar trend is noticeable in fissipede carnivores and especially the pinnipede carnivores, which have Type ABC pineals. The greatest variability is found in rodents, ranging from a Type A pineal (Glis glis) to an αC pineal (Mesocricetus auratus) (for further details, see Vollrath, 1981). This structural diversity of the pineal complex clearly calls for an explanation, but before this goal can be achieved the pineal complexes of many more mammalian species have to be analysed. In this context it is interesting to note that the latitudinal trend in relative pineal size observed in rodents (Quay, 1980a) is not seen in Insectivora, Microchiroptera, Artiodactyla, Prosimii and Anthropoidea (Quay, 1980b).

HISTOLOGY

Recent systematic comparative studies of the pineal gland at the tissue level are lacking. Among the many interesting points to be studied, three aspects are particularly important. One is the arrangement of pinealocytes within the parenchyma. Often it has been observed that the pinealocytes are not randomly arranged, but form cords, clusters or follicles (cf. Vollrath, 1981). A follicular arrangement of pinealocytes is of interest as these cells, in certain respects, appear to resemble photoreceptors more closely than randomly arranged pinealocytes. Fig. 1 of the detailed paper by Wolfe (1965) on the rat pineal gland shows an abundance of mitochondria in the juxta-luminal areas, resembling the ellipsoid-regions of inner segments of photoreceptors. Whether these photoreceptor-like pinealocytes differ in function from "ordinary" pinealocytes remains to be elucidated. An interesting point is that pinealocytes arranged in follicles, in some species, appear to be restricted to certain areas of the pineal gland. In sheep, they were found to be restricted to the dorsal tip of the gland (Lang, 1959). In Pirbright White guinea pigs follicles are conspicuous, however, the proximal area and in the distal area a small zone facing the colliculi are regularly devoid of these structures (Jung and Vollrath, 1982). Interestingly, the follicles were more prominent in continuously illuminated guinea-pigs than in animals kept under LD 12:12.

A second important point is the characterization and the investigation of the distribution of the cells immunoreactive for arginine vasotocin-, α-MSH-, LHRH- and somatostatin-like compounds (Pévet et al., 1980) as well as their relationships to interstitial or glial cells. Should these cells turn out to be a special category of pinealocytes, then ultrastructural studies would be particularly important in order to identify the different types of pinealocytes.

Thirdly, the innervation invites comparative studies, in particular as the central innervation reaching the pineal gland via the commissures is becoming more and more important. Recent studies using the horse-radish peroxidase technique have shown in a variety of mammalian species that central fibres from diverse sources innervate the gland and, particularly important, not just the proximal part of the pineal complex but also the distal parts (Korf and Wagner, 1980; Nürnberger and Korf, 1981; Korf an Møller, 1982). Further studies have to elucidate the main centres of the brain that regularly send nerve fibres to the pineal gland via the commissures and to clarify their functional significance. An interesting question is whether or not the central fibres and the peripheral sympathetic nerves supply different areas of the parenchyma. Electron microscopic studies have revealed well developed neuropil areas within the pineal parenchyma characterized by synapses between axonic and dendritic profiles (Schneider et al., 1981). As it is not yet clear whether the perikarya of these neurons lie outside the pineal, a search for small intrapineal nerve cells is necessary in order to assess the functional significance of the nerve fibres leaving, or reaching, the pineal gland and to understand the intrapineal neuronal circuitry in mammals.

Finally quantitative studies on both types of innervation, central commissural and peripheral sympathetic, are of interest. An important first step in this direction has been taken by Karasek et al. (1982a) who assessed the number of nerve endings per unit area in a variety of mammalian species and found highly significant interspecies differences. Moreover, an interesting inverse correlation between adrenergic nerve fibre endings and pineal noradrenaline concentrations on one hand and synaptic ribbons on the other was noted. Although we still do not know the function of the synaptic ribbons in the mammalian pineal gland, these studies provide an impetus for carrying out further studies on possible functional interrelationships between synaptic ribbons and sympathetic nerve fibres.

CYTOLOGY

Comparative studies at the cellular level are also needed, as such basic questions as to whether pinealocytes are monopolar, bipolar and/or multipolar cells and whether particular forms of pinealocytes can be attributed to a particular type of pineal organ have not been answered. However, this kind of study should be preceeded by detailed investigations of individual species, as differences may exist in the different regions of the pineal complex. An important problem is to assess the length of the neurite-like pinealocyte processes and to examine to what extent they form bundles and run in a particular direction within the parenchyma.

The uneven distribution of cell organelles, especially mitochondria, in some pinealocytes invites studies as to the presence or absence of polarized pinealocytes in different species, a question that is particularly relevant with respect to pinealocytes arranged in follicles. As reviewed in detail (Vollrath, 1981), pinealocyte mitochondria exhibit a number of unusual features that appear to make comparative studies worthwhile Besides giant mitochondria which have been found in some species, the mitochondrial crests show great variation: In addition to shelf-like cristae, tubular and tubulovesicular cristae have been described. In view of the fact that mitochondria with tubular cristae are typical of steroid-producing cells a systematic search for such mitochondria on a comparative basis may be useful. Finally, it should be recalled that Clabough (1971) noted in the hamster that small mitochondria with few cristae were exclusively present in what was termed P_1 pinealocytes, while larger mitochondria with a plexiform array of crests in a dense matrix were typical of P_2 pinealocytes. As the problem of different types of pinealocytes is still unsettled, every attempt should be made to solve it.

Comparative studies are also needed with respect to the synaptic ribbons and spherules, especially on a quantitative basis and under experimental conditions. As clearly shown by Karasek et al. (1982a) synaptic ribbons exhibit prominent interspecies differences in amount, being especially abundant in ground squirrel and chipmunk. According to personal experience, the ribbons are least frequent in the pineal gland of the laboratory mouse. The inverse correlation between ribbons and noradrenergic nerve terminals described by Karasek et al. (1982a) indicates that stimulating results can be obtained when comparative studies are carried out on a quantitative basis. Experimentally the question could be answered whether the striking increase in number of synaptic ribbons in animals exposed to continuous illumination (Lues, 1971; Vollrath and Huss, 1973; Roux et al., 1977) which is in contrast to the low number of ribbons during the photophase of the 24 hour cycle, is generally present or not.

Comparative studies may also provide useful information with respect to the topographical relationships of synaptic ribbons. Whereas in the guinea-pig pineal 93.1 % of the ribbon fields bordering the plasmalemma faced adjacent pinealocytes, in the vole, Microtus agrestis, the majority of ribbons lay opposite glial cells and only about 15-20 % faced clearly identified pinealocytes (Hewing, 1981). These results clearly indicate that the concept that the synaptic ribbons may be involved in a functional coupling of adjacent pinealocytes as proposed for the guinea-pig (Vollrath and Huss, 1973; Vollrath, 1973) does not appear to be generally valid.

Occasionally "paired" ribbon fields (Vollrath and Huss, 1973) are present in the pineal gland, especially after exposure to continuous illumination (Vollrath and Huss, 1973; Vollrath, 1973). These structures in which ribbons lie directly opposite one another, yet in different pinealocytes, are functionally even more enigmatic than the ordinary synaptic ribbons. Comparative studies by Karasek et al. (1982a) have clearly indicated that the number of "paired" ribbon fields differs in various species.

Finally, comparative studies are required with respect to the synaptic spherules (see Vollrath, this volume). As in the guinea-pig pineal these structures may be peculiar to diurnal pinealocytes of the proximal pineal gland, it would be interesting to know whether the spherules are regularly present in the different species and whether they are confined to a particular type of pinealocyte. Currently, studies are in progress in my laboratory to answer this question.

Another component of the pinealocyte that will greatly benefit from quantitative comparative studies is the dense-core vesicles (DCV's). Their precise functional significance is not known, though quantitative studies have shown that they are more numerous during photophase than scotophase, which speaks against an assumption that they contain melatonin. Comparative studies in a number of mammalian species have supported the visual impression that the DCV's are quite variable in amount and are predominantly present in pinealocyte processes in some species (Karasek et al., 1982b). Again, what this means in terms of function remains to be elucidated.

PINEALOCYTE-SPECIFIC (?) ORGANELLES

A potentially important component of the mammalian pinealocyte is what has been called microtubular sheaves (Wolfe, 1965; Lin, 1969, 1970, 1972). These structures are transformed centrioles. The transformation involves elongation of the centriole, apposition of fine-granular, electron-dense material at the outer surface of the centriole and lateral tilting of their microtubules, giving rise to microtubular sheaves. The sheaves are interesting from a comparative point of view, as they are present in rat (Wolfe, 1965; Arstila, 1967), guinea-pig (Lin, 1972), gerbil (Welsh and Reiter, 1978) and hamster (Bucana et al., 1974) but are apparently absent in mouse and monkey (Lin, 1970). In the rat apparently all the centrioles transform into microtubular sheaves between 15 days and 6 weeks of age (Lin, 1969, 1970), as in adult rats typical centrioles are entirely lacking (Wolfe, 1965; Lin, 1970). Comparative studies should be carried out using morphometric techniques to assess (i) the relative amounts of centrioles and microtubular sheaves in different species, (ii) the length of the microtubular sheaves under different physiological and experimental conditions, (iii) their topographical relationships and (iv) to find out whether their frequency can be correlated with other structural features of the pineal gland in a meaningful manner. Such a comparative approach may provide useful hints as to the functional significance of the microtubular sheaves, in particular as it has been suggested that the formation of the sheaves may represent an abortive attempt of the pinealocyte to produce outer segments (Vollrath, 1981) which in photoreceptors are linked to the inner segment by 9 x 2 + 0 cilia.

CONCLUSIONS

The present survey has shown that comparative studies are an important tool to study specific aspects of the mammalian pineal gland. These studies may not solve the pineal problem, but would at least provide incentives for further investigations which may eventually lead to a fuller understanding of some of the inherent problems and especially of the structural and functional complexity of the pineal system.

REFERENCES

Arstila, A. U., Electron microscopic studies on the structure and histochemistry of the pineal gland of the rat. Neuroendocrinology, 2 Suppl.: 1 (1967)

Bargmann, W., 1943, Die Epiphysis cerebri. In: Hdb. mikrosk. Anat. Mensch . W. v. Möllendorff, ed., Springer, Berlin. VI,4: 309

Bucana, C. D., Nadakavukaren, M. J. and Frehn, J. L., 1974, Novel features of hamster pinealocyte ultrastructure. Tissue Cell 6: 85

Clabough, J. W., 1971, Ultrastructural features of the pineal gland in normal and light deprived golden hamsters. Z. Zellforsch 114: 151.

Boeckmann, D., 1980, Morphological investigation of the deep pineal of the rat. Cell Tiss. Res. 210: 283.

Buijs, R. M. and Pévet, P., 1980, Vasopressin- and oxytocin-containing fibres in the pineal gland and subcommissural organ of the rat. Cell Tiss. Res. 205: 11.

Gardner, J. H., 1953, Innervation of pineal gland in hooded rat. J. comp. Neurol. 99: 319.

Gregorek, J. C., Seibel, H. R. and Reiter, R. J., 1977, The pineal complex and its relationship to other epithalamic structures. Acta Anat. 99: 425.

Hewing, M., 1981, Topographical relationships of synaptic ribbons in the pineal system of the vole (Microtus agrestis). Anat. Embryol. 162: 313.

Hofer, H. O., Merker, G. and Oksche, A., 1976, Atypische Formen des Pinealorgans der Säugetiere. Verhandl. Anat. Ges. 70: 97.

Johnson, J. I. Jr., 1977,Central Nervous System of Marsupials. In: The Biology of Marsupials. D. Hunsaker II, ed., Academic Press, New York, San Francisco, London, 157.

Jung, D. and Vollrath, L., 1982, Structural dissimilarities in different regions of the pineal gland of Pirbright White guinea-pigs. J. Neural Transmiss.(in press).

Karasek, M., King, T. S., Brokaw, J., Hansen, J. T., Petterborg, L.J. and Reiter, R. J., 1982a, Inverse correlation between "synaptic" ribbon number and the density of adrenergic nerve endings in the pineal gland of various mammals. J. Comp. Neurol. (in press)

Karasek, M., King, T. S., Petterborg, L. J., Hansen, J. T., Bartke, A. and Reiter, R. J., 1982b, Dense-core vesicles in the mammalian pinealocyte and their relation to secretory processes. TSEM Journal 13: 13.

Korf, H.-W. and Møller, M., 1982, Zentrale Innervation der Epiphyse bei verschiedenen Säugetierarten. 77. Vers. Anat. Ges. 1982: 28. Abstract.

Korf, H.-W. and Wagner, U., 1980, Evidence for a nervous connection between the brain and the pineal organ in the guinea-pig. Cell Tiss. Res. 209: 505.

Lang, K., 1959, Anatomische und histologische Untersuchungen der Epiphysis cerebri von Rind und Schaf. Thesis, München

Lin, H.-S., 1969, A special morphological expression of centriolar formation.
Anat. Rec. 163: 313.

Lin, H.-S., 1970, The fine structure and transformation of centrioles in the rat pinealocyte.
Cytobios 2: 128.

Lin, H.-S., 1972, Transformation of centrioles in pinealocytes of adult guinea-pigs.
J. Neurocytol. 1: 61.

Lues, G., 1971, Die Feinstruktur der Zirbeldrüse normaler, trächtiger und experimentell beeinflußter Meerschweinchen.
Z.Zellforsch. Mikrosk. Anat. 114: 38.

Nürnberger, F., Korf, H.-W., 1981, Oxytocin- and vasopressin-immunoreactive nerve fibers in the pineal gland of the hedgehog, Erinaceus europaeus L.
Cell Tiss. Res. 220: 87.

Oksche, A., 1965, Survey of the development and comparative morphology of the pineal organ.
Progr. Brain Res. 10: 3.

Pévet, P. and Kuyper, M.A., 1978, The ultrastructure of pinealocytes in the golden mole (Amblysomus hottentotus) with special reference to the granular vesicles.
Cell Tiss. Res. 191: 39

Pévet, P. and Yadav, M., 1980, The pineal gland of equatorial mammals. I. The pinealocytes of the Malaysian rat (Rattus sabanus).
Cell Tiss. Res. 210: 417.

Pévet, P., Ebels, I., Swaab, D. F., Mud, M. T. and Arimura, A., 1980, Presence of AVT-, α-MSH-, LHRH- and somatostatin-like compounds in the rat pineal gland and their relationship with the UMO5R pineal fraction. An immunocytochemical study.
Cell Tiss. Res. 206: 341.

Quay, W. B., 1980a, Greater pineal volume at higher latitudes in rodentia: exponential relationship and its biological interpretation.
Gen. Comp. Endocrinol. 41: 340.

Quay, W. B., 1980b, Pineal volume and latitudinal and chronobiological characteristics: Comparison of five mammalian orders.
Am. Zool. 20: 898.

Quay, W. B., 1981, Pineal atrophy and other neuroendocrine and circumventricular features of the naked mole-rat, Heterocephalus glaber (Rüppel), a fossorial, equatorial rodent.
J. Neural Transm. 52: 107.

Ralph, C. L., 1975, The pineal organ and geographical distribution of animals.
Int. J. Biometeorol. 19: 289.

Ralph, C. L., 1981, Melatonin production by extra-pineal tissue. In: Melatonin: Current Status and Perspectives. N. Birau, W. Schloot, eds., Pergamon Press, Oxford. Advances in the Biosciences 29: 35.

Roux, M., Richoux, J. P. and Cordonnier, J. L., 1977, Influence de la photopériode sur l'ultrastructure de l'épiphyse avant et pendant la phase génitale saisonnière chez la femelle du lérot (Eliomys quercinus). J. Neural Transm. 41: 209.

Schneider, T., Semm, P. and Vollrath, L., 1981, Ultrastructural observations on the central innervation of the guinea-pig pineal gland. Cell Tiss. Res. 220: 41

Vollrath, L., 1973, Synaptic ribbons of a mammalian pineal gland. Circadian changes. Z. Zellforsch. Mikrosk. Anat. 145: 171.

Vollrath, L. and Huss, H., 1973, The synaptic ribbons of the guinea-pig pineal gland under normal and experimental conditions. Z. Zellforsch. Mikrosk. Anat. 139: 417.

Vollrath, L., 1979, Comparative morphology of the vertebrate pineal complex. Progr. Brain Res. 52: 25.

Vollrath, L., 1981, The Pineal Organ. In: Hdb. mikr. Anat. Mensch. A. Oksche, L. Vollrath, eds. VI/7. Springer, Berlin-Heidelberg-New York.

Welsh, M. G. and Reiter, R. J., 1978, The pineal gland of the gerbil, Meriones unguiculatus. I. An ultrastructural study. Cell Tiss. Res. 193: 323.

Wolfe, D. E., 1965, The epiphyseal cell: an electron-microscopic study of its intercellular relationships and intracellular morphology in the pineal body of the albino rat. Progr. Brain Res. 10: 332.

THE USE OF ELECTRON MICROSCOPY AND STEREOLOGY IN THE STUDY OF THE MAMMALIAN PINEAL GLAND

Lutz Vollrath

Department of Anatomy
Johannes Gutenberg-University
Mainz
Federal Republic of Germany

INTRODUCTION

Now that transmission electron microscopy (TEM) has been used for many years to study the pineal gland it is timely to evaluate what progress has been achieved by using this technique and to speculate which aspects of pineal research may benefit most by its application in the future.

THE BASIC ULTRASTRUCTURE OF THE PINEALOCYTE

One of the main achievements of TEM is the clear demonstration that the functionally important intrinsic cells of the pineal organ are photoreceptor cells in lower vertebrates, closely resembling the cones of the retina. The intrinsic cells of the mammalian pineal gland, the pinealocytes, have some structural features in common with photoreceptor cells, i.e. presence of synaptic ribbons, ellipsoid-like aggregations of mitochondria, but on the whole it would appear that they represent a cell type sui generis, the ultrastructure of which reveals little about its "profession". As extensively reviewed (Vollrath, 1981a), the mammalian pinealocyte is a neuron-like cell consisting of a cell body, or perikaryon, and cytoplasmic processes.

Little is known from TEM studies about the number and the length of the processes that emanate from the perikaryon. This kind of information could be very useful as it seems possible that the pinealocytes, by way of special structures ("synaptic" ribbons and spherules, vesicle-crowned rodlets), are functionally interrelated (Hopsu and Arstila, 1965; Vollrath and Huss, 1973; Vollrath, 1973) and form intricate neuronal circuits. Pictures obtained in human pineals (Hortega, 1922) showing pinealocyte perikarya with long or short cytoplasmic processes suggest that different types of pinealocytes may be present, resembling the Golgi Type I and Golgy Type II neurons. Interestingly, silver impregnation studies in sheep, cow (Lang, 1959), dog, cat (Zach, 1960) and horse (Faßbender, 1962) have have revealed monopolar pinealocytes in the periphery of the pineal gland, whereas pinealocytes with 2-4 processes occupy the centre of the gland. Systematic ultrastructural studies as to the number of processes leaving a pinealocyte perikaryon are lacking. In one study, in which serial sections were used, it was found that cat pinealocytes have only one (or perhaps two) processes (Wartenberg, 1968). In view of the structural complexity of mammalian pinealocytes, reconstructions of a number of cells from serial sections would be highly desirable.

The ultrastructure of the pinealocytic cytoplasm has been extensively reviewed (Pévet, 1979, 1981a; Vollrath, 1981a). Here it is relevant to note that in my opinion the pinealocytes of different mammalian species appear basically similar and resemble neurons in many respects. Clearly defined sub-types of pinealocytes such as observed in the mole, Talpa europaea (Pévet and Collin, 1976), have not yet been convincingly described in detail for other mammalian species. The distinction made by Pévet (1977, 1979, 1981a) based mainly on the presence or absence of small granular vesicles is debatable, as in my experience these granules are so rare in many species, and in some strains of animals even lacking, as to use them as distinct markers of a specific sub-type of pinealocyte. Nevertheless, the search for different types of pinealocytes should continue, for a number of reasons (see below).

Among the various cytoplasmic components of mammalian pinealocytes, mitochondria, microtubules and microfilaments often stand out due to their concentrations in certain cell regions. Peculiar cell organelles of the mammalian pinealocyte, the so-called "synaptic" ribbons and the microtubular sheaves, the latter representing centriole derived structures, have attracted, and no doubt will continue to attract, special attention.

QUANTITATIVE ELECTRON MICROSCOPY

Now that the basic ultrastructure of the mammalian pinealocyte and the pineal gland as a whole is known, the question arises as to how further progress can be achieved. As the ultrastructure of the pinealocyte is such that its function cannot be deduced at first sight, if at all, special approaches have to be used to unravel its mysteries. One such approach is quantitative electron microscopy. What has been achieved with this technique in the past and what can be attained in the future?

QUANTITATIVE CHANGES OF SYNAPTIC RIBBONS AND SPHERULES

In what was probably the first quantitative ultrastructural approach to the mammalian pineal gland the functionally obscure "synaptic" ribbons were investigated. As synaptic ribbons are present also in the photoreceptor cells of the pineal in lower vertebrates, they were regarded as "possible reminders of the original neurosensory function" of the pinealocyte (Kappers, 1971). Experimental studies in guinea-pigs, involving exposure to continuous illumination or continuous darkness for several weeks revealed that the ribbons cannot be regarded as functionless, phylogenetic relics (Vollrath and Huss, 1973). Under constant light, the ribbons showed a striking increase in size and number as well as changes in shape. These results then led to the question as to whether synaptic ribbons show numerical changes also under physiological conditions. In an attempt to answer this question the number of ribbons was counted in guinea-pigs over a period of 24 h and it was found that the amount of ribbons decreases during the day and increases during the night (Vollrath, 1973). These day/night differences have been confirmed for other species such as rat (Kurumado and Mori, 1977; King and Dougherty, 1980; Karasek and Vollrath, 1982), baboon (Theron et al., 1979, 1981) and goldfish (McNulty, 1981). As these results are in complete agreement with the day/night differences in melatonin formation, one may question whether this kind of quantitative approach to the pineal problem leads to new insight. In my opinion - it does. First, the numerical changes of synaptic ribbons under physiological and experimental conditions led to the concept that the structures in question may functionally couple adjacent pinealocytes and that the pinealocytes are clearly neuronal or neuroendocrine in nature (Vollrath and Huss, 1973; Vollrath, 1973). This concept in turn triggered studies into the electrophysiological nature of mammalian pinealocytes carried out in conjunction with Dr. Semm, now at the Zoological Institute in Frankfurt/M., which have led to a better understanding of mammalian pinealocyte function (see below). Finally, right now we are witnessing that quantitative studies dealing with pineal "synaptic" structures may lead to an interesting new concept.

The synaptic structures I am referring to are the so-called synaptic spherules (cf. Vollrath, 1981a), also known under the terms rosette-like structures (rabbit, Leonhardt, 1967; Romijn, 1973, 1975) and vesicle-crowned balls (guinea-pig, Lues, 1971) in the pineal, and synaptic body in the organ of Corti (Ades and Engström, 1974; Baird, 1974) and the vestibular organ (Wersäll and Bagger-Sjöbäck, 1974).

That these structures may be functionally different from synaptic ribbons has been suggested by the findings that in pregnant guinea-pigs spherules show a strong increase in number whereas ribbons increase less conspicuously (Lues, 1971). In rabbits, sympathectomy of the pineal gland resulted first in an increase in the number of spherules, followed by that of the ribbons (Romijn, 1975). Recently, the day/night changes of ribbons and spherules were investigated in the rat pineal gland in vivo and in vitro (Karasek and Vollrath, 1982). It was found that in vivo both ribbons and spherules show corresponding changes in number, i.e. low numbers at daytime and high numbers at night. In vitro, the day/night changes did not occur, but interestingly after 3 h in culture the spherules had increased considerably in number, again suggesting that spherules and ribbons may differ in their response.

That the spherules are of great potential importance occurred to us when we studied them in the guinea-pig. In this species the pineal gland is a dumbbell-shaped, Type ABC or AßC organ (Vollrath, 1979, 1981a). Quantitative assessment revealed that ribbons and spherules may differ in their distribution within the parenchyma and may differ in their rhythmicity (Vollrath, Schultz and McMillan, unpublished observations). In all three pineal regions investigated (proximal = A, intermediate = B, distal = C) the ribbons were equally abundant and showed the expected behaviour, i.e. small numbers during daytime and high numbers at night. However, when looking at the spherules it is striking to note that during day time they are more abundant in the proximal area than in the intermediate and distal regions and that at night time they exhibit low numbers in all three pineal regions. This indicates that ribbons and spherules, in the proximal region, show an inverse behaviour. What could this mean in terms of function? Three aspects have to be considered. First, as in the majority of instances, ribbons and spherules are not found in one and the same pinealocyte profile, different populations of pinealocytes could exist, characterized by ribbons and spherules respectively. Second, it may mean that there are differences in the circadian rhythmicity of the pinealocytes, one group of pinealocytes being diurnal and the other nocturnal. In this context it is interesting that electrophysiological studies carried out in guinea-pigs also point into the direction of different rhythmicity (Semm and Vollrath, 1980): One group of cells exhibited an electrical activity corresponding to the melatonin forming activity of the pineal gland, i.e. low activity during daytime and high activity at night. A second category of pineal cells showed a inverse electrical behaviour: High activity during daytime and low activity at night. In view of the morphological results obtained it is tempting to speculate that the day-active pinealocytes correspond to the cells that contain synaptic spherules whereas the night-active cells are identical with the ribbon-containing pinealocytes. Pinealocyte profiles exhibiting both spherules and ribbons could correspond to those cells which did not exhibit circadian rhythmicity electrophysiologically (Semm and Vollrath, 1980). More detailed studies have to be carried out to prove or to disprove this concept. The third implication of the morphological findings is that the different regions of the guinea-pig pineal complex contain different types of pinealocytes in different amounts and hence may exhibit functional heterogeneity. In this context the question arises as to which role the central commissural fibres play, that are prominent in the proximal pineal region (cf. Schneider et al., 1981).

Finally, morphometric data are of value also with respect to the length of synaptic ribbons. In the pineal gland of the goldfish it was found that synaptic ribbons are short during the day and almost twice as long during night (McNulty, 1981).

QUANTITATIVE DATA ON DENSE-CORE VESICLES

In view of the fact that the mammalian pinealocyte is an endocrine cell it is logical to search for morphological correlates of its secretory products. According to Pévet (1977, 1979, 1981a,b) flocculent material in cisterns of the endoplasmic reticulum or in vacuoles represents one type of secretory product and dense-core vesicles (DCV's) another. It should be noted though that in commonly used experimental animals the two types of secretory processes are not so conspicuous as to solve the inherent problems at first sight. It is therefore relevant to enquire what kind of information can be gained fron quantitative studies. Quantitative studies so far concern only the DCV's, which are known to be highly variable in number in pinealocytes of different mammalian species (cf. Pévet, 1981b). Recently, hard quantitative data have been obtained with respect to the number of DCV's present within a clearly defined unit area of pinealocyte parenchyma and equally important, with respect to their location within the pinealocyte (Karasek et al., 1982a). The following sequence (from lowest to highest numbers) was obtained: Fox, cat, Richardson's ground squirrel, Sprague- Dawley rat, Eastern chipmunk, Cotton rat, Djungarian hamster, white-footed mouse and laboratory mouse, a clear tendency being present for the DCV's in some species to accumulate in pinealocytic processes. No systematic studies have been carried out yet to determine whether the amount of DCV's present can be correlated with the serotonin and melatonin levels of the pineal gland.

In different species data are available with respect to the 24 h rhythmicity of DCV's. In the rabbit it has been shown that in the vicinity of the Golgi apparatus DCV's exhibit a distinct increase in number between 7 a.m. and noon, followed by a prominent decline coming to a halt at midnight; in the terminals of pinealocytic processes DCV's show a peak 7 hrs later suggesting a transport of the DCV's from the perikaryon to the processes (Romijn et al., 1976). In the laboratory mouse (Benson and Krasovich, 1977; Kachi, 1979), Mongolian gerbil (Welsh et al., 1979a) and Syrian hamster (Krasovich and Benson, 1982) the numbers of DCV's also increase during the day and decrease at night. In view of the consistency of the results obtained in different mammalian species it can be concluded that the DCV's appear to be intimately related to circadian events in the pineal gland. Their numerical increase during the day speaks against the supposition that they contain melatonin; they may contain serotonin, as their qualitative behaviour parallels the circadian rhythmicity of serotonin (cf. Vollrath, 1981b). This conclusion is in agreement with the cytochemical findings by Lu and Lin (1979) and Juillard and Collin (1980). According to another view (Collin, 1981), they main contain a specific protein termed pinealin. Whatever the DCV's contain, it is clear that it is a substance that is present in larger amounts during the day than at night.

STEREOLOGICAL METHODS

In the above studies the quantitative methods applied were mostly rather simple. Pineal tissue covering a specified number of holes of electron microscopic supporting grids was scanned and all the structures in question were counted. This rather crude method is applicable only when the structures to be quantitated are relatively rare, as is the case with synaptic ribbons and DCV's. Among the different morphometric techniques available, stereological methods appear to have the greatest scope for future quantitative investigations of the pineal gland.

Stereology is a special type of morphometric analysis based on "a geometric analysis of structures and textures; it includes methods that allow direct derivation of metric properties of structures from two-dimensional sections on the basis of geometric-statistical reasoning" (Weibel, 1969). It is an important tool to correlate structure and function. Technical details may be obtained from the publications by Weibel (1969, 1979, 1980) and others.

In the field of pineal research, stereological methods have so far been applied with three aims: (i) to obtain data on the volume fractions of the different pineal components; (ii) to monitor circadian changes, and (iii) to study seasonal rhythmicity.

COMPOSITION OF PINEAL TISSUE

In one stereological study in which the data were obtained by means of the point counting method, the pineal glands of 10 adult male Wistar rats, 14 weeks old, were analysed (Krstić, 1977). In a first step the material was studied at the light microscopic level. The data obtained are given in Table 1.It should be noted that the data stem from immersion fixed material. Hence, the volume fraction of capillaries may differ when perfusion-fixed pineals are investigated. The presence of two types of cell, dark and light, is also most likely due to immersion fixation.

Table 1

Morphometric analysis at the light microscopic level. Rat pineal gland.
From Krstić, 1977.

Capillaries	5.8 ± 0.15 %[a]
Pericapillary spaces and pineal canaliculi	13.1 ± 0.22 %
Dark cells	2.2 ± 0.09 %
Light cells	78.8 ± 0.27 %

[a]expressed as % volume fraction mean ± SEM

The stereological analysis of individual rat pinealocytes at the electron microscopic level is shown in Table 2. These kinds of data are of value for comparative studies only.

Table 2

Stereological analysis of "light" pinealocytes of rats.
From Krstić, 1977.

Nucleus	22.3 ± 0.1 %[a]
Mitochondria	6.1 ± 0.06 %
Golgi apparatus	0.56 ± 0.02 %
Endoplasmic reticulum, rough and smooth	5.37 ± 0.07 %
Lipid droplets	1.19 ± 0.03 %
Lamellar bodies	0.1 ± 0.008%
Cytoplasmic matrix	64.38 ± 0.13 %

[a]expressed as % volume fraction, mean ± SEM

When a structure-function correlation at the organ level is intended, a more sophisticated approach is necessary, taking into consideration all the structural components of the pineal gland, i.e., blood vessels, perivascular spaces, nerve fibres, connective tissue if present, pinealocytes, interstitial cells and last, but not least, the intercellular canaliculi. These 0.2 - 0.4 µm wide canaliculi which extend from the perivascular spaces and appear to enmesh individual pinealocytes (Quay, 1973, 1974; Krstić, 1975) are especially important as their width changes over a period of 24 h, being wide during the day and small at night. Moreover, as pinealocytes may have a follicle-like arrangement, the central lumen or remnants of it have to be taken into consideration.

Finally, it should be noted that a stereological analysis of the pineal gland would be incomplete without dealing with pinealocyte perikarya and processes separately.

STEREOLOGICAL STUDIES WITH RESPECT TO CIRCADIAN RHYTHMICITY

One promising application of stereology is the study of circadian rhythmicity. In one such investigation a careful analysis of the various components of gerbil pinealocytes was carried out (Welsh et al., 1979a). It was found that most of the parameters studied were smallest between 7 a.m. and 1 p.m., followed by an increase continuing into the dark phase. From these results and those on melatonin synthesis it looks as though the morphological parameters increase in anticipation of the increased melatonin formation in scotophase. Further studies have to show whether the changes observed in the gerbil principally apply to other species as well. In view of the well established melatonin rhythmicity of the pineal gland, the pineal gland appears to be an ideal object to correlate structure and function.

One morphological parameter, the volume fraction of microtubules, has been studied in two species. In the gerbil, microtubule volume fraction increased from 7 a.m. to 7 p.m., followed by a slight decrease at night (Welsh et al., 1979a). In the baboon, Papio ursinus, the cytoplasmic content of microtubules was small during photophase and high during scotophase (Theron et al., 1981). These divergent results show how important it is to carry out detailed stereological studies in a variety of species before generalizations with respect to circadian rhythmicity are made.

Recently, the conspicuous nucleoli of rat pinealocytes have been investigated with respect to 24-hour-rhythmicity (Lew et al., 1982). Four parameters were studied: (i) Total nucleolar volume, (ii) granular and (iii) fibrillar components as well as (iv) clear areas. According to these studies the nucleolus is largest 1 h after the onset of light and smallest towards the end of the dark period (LD 12:12), the amplitude of the daily change being ± 15 %. The granular component exhibited a daily amplitude of ± 29 %, the maxima and minima occurring at the same time as the peaks and troughs of total nucleolar volume. In view of the many unresolved questions with respect to the structure and function of the nucleolus, studies dealing with the prominent nucleoli of the pinealocytes may yield important information.

STEREOLOGICAL STUDIES WITH RESPECT TO CIRCANNUAL RHYTHMS

Stereological analyses are also an important tool to study circannual rhythmicity of the pineal gland. At the light microscopic level seasonal changes of the pineal gland have been clearly established (cf. Vollrath, 1981a,b). In one ultrastructural study using stereology it was found that in the ground squirrel the volume densities of mitochondria and lipid droplets increased from a minimum in winter to a maximum in the fall, whereas the surface density of endoplasmic reticulum showed an inverse behaviour. The means for number of Golgi bodies per test area of cytoplasm were small in winter and consistently high in spring, summer and fall. DCV's were highly variable in amount, showing a distinct depression in spring only (McNulty and Dombrowski, 1980).

Before systematic stereological studies on an annual basis are carried out, in my view a number of points should be taken into consideration. First, it is desirable to perform a 24-hour study in the species under question, with two aims: One is to establish that 24-hour rhythmicity of a particular parameter is clearly demonstrable and to clarify the order of magnitude of the daily changes; only when these data are available, the annual data can be seen in the right perspective. Moreover, proper sampling techniques and appropriate statistical analyses are important prerequisites to exclude, or to confirm, a mosaic-like structure of the pineal parenchyma (cf. Vollrath, 1979) and to cope with this situation accordingly. As usually the sampling is done in different sections and in different regions of the gland, an appropriate statistical test should be used, such as a nested or hierarchic analysis of variance. Secondly, it should be taken into consideration that infradian rhythms, i.e., rhythms with periods longer than 28 h, may be present in the pineal gland (Vollrath et al., 1975, 1981; Welker et al., 1982), which may give tha false impression of annual rhythmicity. Moreover, ultradian rhythms may complicate the picture, the superimposition of ultradian and circadian rhythms resulting in beat frequencies and an attenuation of the daily amplitudes (Wallen and Yochim, 1974). Thirdly, as seasonal studies are by nature long-term investigations it should be made sure that the animal supply from the dealer is carefully controlled, i.e.,that the dealers do not buy extra animals from other dealers, change the diet etc. As a last point it should be mentioned that the stereological methods applied give valid results only when the proper methods of randomization are observed.

APPLICATION OF STEREOLOGY IN EXPERIMENTALLY-AFFECTED ANIMALS

It should be noted that stereological studies, especially when carried out on a 24-hour- or an annual basis, are very laborious, time-consuming, and expensive. These disadvantages are of lesser concern in investigations whose aim it is to study the effects of a special treatment on the pineal gland.

In one study in which the function of the pineal gland was severely impaired by sympathectomy (Welsh et al., 1979b, gerbils) a clear decrease was noted with respect to total pinealocyte size, smooth and rough endoplasmic reticulum, free cytoplasm, mitochondria and presumptive secretory vesicles, whereas synaptic ribbons increased in volume. This study has also revealed the important point that sham-operation is not without effect on the pinealocytes. Hence, in addition to the experimental and sham-operated animals, untreated controls should be included in the studies to obtain a complete picture.

CONCLUSIONS

The present survey has shown that quantitative studies at the ultrastructural level are an important tool to investigate the structure and function of the pineal gland. In contrast to other organs, stereological analyses have been surprisingly little used to unravel the structural and functional complexity of the mammalian pineal gland.

In my view, undoubtedly a new era of pineal ultrastructural research will begin when the approach so often used in the past is abandoned, viz. to carry out painstaking experiments and subsequently to be content with purely descriptive and often meaningless and debatable semiquantitative results which cannot be subjected to rigorous statistical analyses.

The application of stereological techniques to all the components of the mammalian gland will be of great value, especially when they are carried out on a comparative basis. Then we will be able to make definite statements about a number of important points: The nerve fibre density in different species and in different regions of the pineal complex. The volume fraction of the blood vessels in the pineal gland can be compared with other organs. The length, width, course and volume fraction of pinealocyte processes can be objectively assessed and attempts can be made to relate these findings to the structure of the pineal complex. Is it just a coincidence that the small, roughly sperical rat pineal gland contains pinealocytes with few and short processes whereas the elongate, dumbbell-shaped guinea-pig pineal gland has relatively long processes forming bundles which not rarely run in preferred directions?

Last, but not least, hard quantitative morphological data provide an impetus for correlations which may eventually lead to new concepts. One such study is that by Karasek et al. (1982b) in which an inverse correlation between synaptic ribbon number and the density of adrenergic nerve endings was found in the pineal gland of various mammals.

REFERENCES

Ades, H. W. and Engström, H., 1974, Anatomy of the inner ear. In: Handbook of Sensory Physiology, V/1. Auditory System. Anatomy, Physiology (Ear). W.D. Keidel, W.D. Neff, eds. p. 125. Springer, Berlin-Heidelberg-New York

Baird, I. L., 1974, Anatomical features of the inner ear in submammalian vertebrates. In: Handbook of Sensory Physiology, V/1: 159. Auditory System. Anatomy, Physiology (Ear). W.D. Keidel, W.D. Neff, eds. Springer, Berlin-Heidelberg-New York

Benson, B. and Krasovich, M., 1977, Circadian rhythm in the number of granulated vesicles in the pinealocytes of mice. Effects of sympathectomy and melatonin treatment. Cell Tiss. Res. 184: 499.

Collin, J.-P., 1981, New data and vistas on the mechanisms of secretion of proteins and indoles in the mammalian pinealocyte and its phylogenetic precursors; the pinealin hypothesis and preliminary comments on membrane traffic. In: The Pineal Organ: Photobiology-Biochronometry-Endocrinology. A. Oksche, P. Pévet, eds., p. 187. Elsevier/North Holland Biomedical Press, Amsterdam,New York, Oxford.

Fassbender, E., 1962, Topographie und mikroskopisch-anatomischer Feinbau der Epiphysis cerebri des Pferdes. Gegenbaurs morphol. Jb. 103: 457.

Hopsu, V. K. and Arstila, A. V., 1965, An apparent somato-somatic synaptic structure in the pineal gland of the rat. Exp. Cell. Res. 37: 484.

Hortega, P. del Rio, 1922, Contitución histológica de la glándula pineal. Arch. de Neurobiol. 3: 359.

Juillard, M.-Th. and Collin, J.-P., 1980, Pools of serotonin in the pineal gland of the mouse: The mammalian pinealocyte as a component of the diffuse neuroendocrine system. Cell Tiss. Res. 213: 273.

Kachi, T., 1979, Demonstration of circadian rhythm in granular vesicle number in pinealocytes of mice and the effect of light: semiquantitative electron microscopic study. J. Anat. 129: 603

Kappers, J. A., 1971, Discussion remark. In: The pineal gland. G.E. Wolstenholme, J. Knight, eds. p. 121 Churchill Livingstone, Edinburgh-London

Karasek, M. and Vollrath, L., 1982, "Synaptic" ribbons and spherules of the rat pineal gland: Day/night changes in vitro? Exp. Brain Res. 46: 205.

Karasek, M., King, T. S. and Petterborg, L. J., 1982a, Dense-core vesicles in the mammalian pinealocyte and their relation to secretory processes. TSEM Journal 13: 13.

Karasek, M., King, T. S., Brokaw, J., Hansen, J. T., Petterborg, L.J. and Reiter, R. J., 1982b, Inverse correlation between "synaptic" ribbon number and the density of adrenergic nerve endings in the pineal gland of various mammals. J. Comp. Neurol. (in press)

King, T. S. and Dougherty, W. J., 1980, Neonatal development of circadian rhythm in "synaptic" ribbon numbers in the rat pinealocyte. Amer. J. Anat. 157: 335.

Krasovich, M. and Benson, B., 1982, A study of the relationship between photoperiod and pinealocyte granulated vesicles in the golden Syrian hamster. Cell Tiss. Res. 223: 155.

Krstić, R., 1975, Scanning electron microscope observations in the canaliculi in the rat pineal gland. Experientia 31: 1072

Krstić, R., 1977, Glande pinéale de Rat. Analyse morphométrique aux microscopes photonique et électronique. Labirint, str.: 121.

Kurumado, K. and Mori, W., 1977, A morphological study of the circadian cycle of the pineal gland of the rat. Cell Tiss. Res. 182: 565.

Lang, K., 1959, Anatomische und histologische Untersuchungen der Epiphysis cerebri von Rind und Schaf. Thesis, München

Leonhardt, H., 1967, Über axonähnliche Fortsätze, Sekretbildung und Extrusion der hellen Pinealozyten des Kaninchens. Z. Zellforsch. Mikrosk. Anat. 82: 307.

Lew, G. M., Payer, A. and Quay, W. B., 1982, The pinealocyte nucleolus. Ultrastructural and stereological analysis of twenty-four-hour changes. Cell Tiss. Res. 224: 195.

Lu, K.-S. and Lin, H.-S., 1979, Cytochemical studies on cytoplasmic granular elements in the hamster pineal gland. Histochem. J. 61: 177.

Lues, G., 1971, Die Feinstruktur der Zirbeldrüse normaler, trächtiger und experimentell beeinflußter Meerschweinchen. Z. Zellforsch. Mikrosk. Anat. 114: 38.

McNulty, J. A., 1981, Synaptic ribbons in the pineal organ of the goldfish: Circadian rhythmicity and the effects of constant light and constant darkness. Cell Tiss. Res. 215: 491.

McNulty, J. A. and Dombrowski, T. A., 1980, Ultrastructural evidence for seasonal changes in pinealocytes of the 13-lined ground squirrel, Spermophilus tridecemlineatus: A qualitative and quantitative study. Anat. Rec. 196: 387.

Pévet, P., 1977, On the presence of different populations of pinealocytes in the mammalian pineal gland. J. Neural Transm. 40: 289.

Pévet, P., 1979, Secretory processes in the mammalian pinealocyte under natural and experimental conditions. Progr. Brain Res. 52: 149.

Pévet, P., 1981a, Ultrastructure of the mammalian pinealocyte. In: The Pineal Gland. Vol I. Anatomy and Biochemistry. R.J. Reiter, ed. p. 121. CRC Press, Boca Raton, Florida.

Pévet, P., 1981b, Peptides in the pineal gland of vertebrates. Ultrastructural, histochemical, immunocytochemical and radioimmunological aspects. In: The Pineal Organ: Photobiology-Biochronometry-Endocrinology. A. Oksche, P. Pévet, eds., p. 211. Elsevier/North-Holland Biomedical Press. Amsterdam-New York-Oxford.

Pévet, P. and Collin, J. P., 1976, Les pinéalocytes de Mammifère: Diversité, homologies, origine. Etude chez la Taupe adulte (Talpa europaea L.) J. Ultrastr. Res. 57: 22.

Quay, W. B., 1973, Twenty-four-hour rhythmicity of pineal canaliculi and evidence for their intrinsic humoral regulation. The Physiologist 16: 427.

Quay, W. B., 1974, Pineal canaliculi: demonstration, twenty-four-hour rhythmicity and experimental modification. Amer. J. Anat. 139: 81.

Romijn, H. J., 1973, Structure and innervation of the pineal gland of the rabbit, Oryctolagus cuniculus (L.) II. An electron microscopic investigation of the pinealocytes. Z. Zellforsch. Mikrosk. Anat. 141: 545.

Romijn, H.J., 1975, The ultrastructure of the rabbit pineal gland after sympathectomy, parasympathectomy, continuous illumination, and continuous darkness. J. Neural Transm. 36: 183.

Romijn, H. J., Mud, M. T. and Wolters, P. S., 1976, Diurnal variations in number of Golgi-dense core vesicles in light pinealocytes of the rabbit. J. Neural Transm. 38: 231.

Schneider, T., Semm, P. and Vollrath, L., 1981, Ultrastructural observations on the central innervation of the guinea-pig pineal gland.
Cell Tiss. Res. 220: 41.

Semm, P. and Vollrath, L., 1980, Electrophysiological evidence for circadian rhythmicity in a mammalian pineal organ.
J. Neural Transm. 47: 181.

Theron, J. J., Biagio, R., Meyer, A. C. and Boekkooi, S., 1979, Microfilaments, the smooth endoplasmic reticulum and synaptic ribbon fields in the pinealocytes of the baboon (Papio ursinus).
Am. J. Anat. 154: 151.

Theron, J. J., Biagio, R. and Meyer, A. C., 1981, Circadian changes in microtubules, synaptic ribbons and synaptic ribbons fields in the pinealocytes of the baboon (Papio ursinus).
Cell Tiss. Res. 217: 405

Vollrath, L., 1973, Synaptic ribbons of a mammalian pineal gland. Circadian changes.
Z. Zellforsch. Mikrosk. Anat. 145: 171.

Vollrath, L., 1979, Comparative morphology of the vertebrate pineal complex.
Progr. Brain Res. 52: 25.

Vollrath, L., 1981a, The Pineal Organ. In: Hdb. mikr. Anat. Mensch. A. Oksche, L. Vollrath, eds. Vol. VI/7.
Springer, Berlin-Heidelberg-New York.

Vollrath, L., 1981b, Cytology of rhythmic phenomena in the pineal organ. In: The Pineal Organ: Photobiology- Biochronometry-Endocrinology.
A. Oksche, P. Pévet, eds. p.139.
Elsevier/North-Holland Biomedical Press, Amsterdam-New York-Oxford.

Vollrath and Huss, 1973, The synaptic ribbons of the guinea-pig pineal gland under normal and experimental conditions.
Z. Zellforsch. Mikrosk. Anat. 139: 417.

Vollrath, L., Kantarjian, A. and Howe, C., 1975, Mammalian pineal gland: 7-day rhythmic activity?
Experientia 31: 458.

Vollrath, L., Becker, U., Diehl, B. J. M., Schühle, A. and Welker, H., 1981, Rhythmic changes in the rat pineal gland.
In: Pineal Function. C.D. Matthews, R.F. Seamark, eds. p. 217.
Elsevier/North-Holland Biomedical Press, Amsterdam-New York-Oxford

Wallen, E. P. and Yochim, J. M., 1974), Rhythmic function of pineal hydroxyindole-O-methyl transferase during the estrous cycle: An analysis.
Biol. Reprod. 10: 461

Wartenberg, H., 1968, The mammalian pineal organ: Electron microscopic studies on the fine structure of pinealocytes, glial cells and on the perivascular compartment. Z. Zellforsch. Mikrosk. Anat. 86: 74.

Weibel, E. R., 1969, Stereological principles for morphometry in electron microscopic cytology. Int. Rev. Cytol. 26: 235.

Weibel, E. R., 1979, Stereological methods. Vol. 1: Practical methods for biological morphometry. Academic Press, London- New York-Toronto.

Weibel, E. R., 1980, Stereological methods. Vol. 2: Theoretical foundations. Academic Press, London-New York-Toronto

Welker, H. A., Schühle, A. and Vollrath, L., 1982, Infradian rhythms of serotonin and serotonin-N-acetyltransferase in the pineal gland of male rats. J. interdiscipl. Cycle Res. (in press).

Welsh, M. G., Cameron, I. L. and Reiter, R. J., 1979a, The pineal gland of the gerbil, Meriones unguiculatus. II. Morphometric analysis over a 24-hour period. Cell Tiss. Res. 204: 95.

Welsh, M. G., Hansen, J. T. and Reiter, R. J., 1979b, The pineal gland of the gerbil, Meriones unguiculatus. III. Morphometric analysis and fluorescence histochemistry in the intact and sympathetically denervated pineal gland. Cell Tiss. Res. 204: 111.

Wersäll, J. and Bagger-Sjöbäck, D., 1974, Morphology of the vestibular sense organ. In: H. H. Kornhuber, ed. Handbook of Sensory Physiology, Vol. VI,1, p. 123. Springer, Berlin-Heidelberg-New York.

Zach, B., 1960, Topographie und mikroskopisch-anatomischer Feinbau der Epiphysis cerebri von Hund und Katze. Zentralbl. Veterinärmed. 7: 273.

INNERVATION OF THE VERTEBRATE PINEAL ORGAN

J. Ariëns Kappers

Houtweg 13
1251 CS Laren N.H.
The Netherlands

In the present survey we will deal with the innervation of the pineal organ proper only, not with that of other parietal organs such as the parapineal organ, the frontal organ of anuran amphibians, and the parietal eye of lizards which have been termed "accessory pineal organs".

Phylogenetically, the pineal develops from a directly photoreceptive, mainly sensory organ into an indirectly photosensitive secretory organ by which also other external as well as internal stimuli are integrated. The latter is often termed a neuroendocrine gland, not because its secretory cells which release their products into the blood would be true nerve cells, but because these cells are of neural origin. Evidently, the functional changes which the organ phylogenetically shows will not only be reflected by changes in the structure of the cells by which it is composed, but also by changes in its innervation.

INNERVATION OF THE NON-MAMMALIAN PINEAL

Before going into more detail, some fundamental structural and functional principles should be mentioned first. Especially in more primitive non-mammalian vertebrates such as fish and amphibians, the pineal can primarily be considered a directly photoreceptive organ conveying photic stimuli to brain centres along a neural pathway. In principle, this pathway consists of two types of cells, photoreceptive neurosensory cells and nerve cells. Embryologically, both cell types originate from the same neuroepithelial anlage. In the adult stage both are situated in the pineal epithelium or pineal parenchyma. Sensory nerve cells may, in part, also be situated in the pineal stalk which connects the pineal organ proper with the brain.

Neurosensory cells in general combine two different functions. In the first place they are directly sensory in being susceptible to stimuli originating in the environment of the organism. In addition they are capable of "translating" or transducing such stimuli into a neural impulse. This means that the external stimulus causes an action potential in the very same cell by which this stimulus is received. Such neurosensory cells show a polar surface which is structurally specialized for the reception of specific external stimuli and a basal process which is comparable to an axon. The polar surface of photosensitive neurosensory cells or photoreceptor cells is characterized by an outer segment which, as has been shown in a previous lecture, consists of a more or less complicated lamellar structure. The terminal ending of the basal process of any neurosensory cell conveyes the neural impulse, generated in it by the external stimulus, by means of a synaptic contact to the dendrites, or more directly to the soma of a true nerve cell changing the spontaneous activity of this nerve cell. The axon of the sensory nerve cell then conducts the effect of the neural impulse, originally generated in the neurosensory cell by external stimuli, to brain centres. In case of the lateral eyes of all vertebrates as well as in the pineal organ of many non-mammals, such sensory nerve cells are situated in the sensory organ proper, while, for instance, the neurosensory cells present in the olfactory epithelium convey their action potential to nerve cells which are situated in the brain, *i.e.*, in the olfactory bulbs.

As has been mentioned in our earlier lecture, the pineal organ of all vertebrates develops from the epithalamic neuroepithelial layer which forms the roof of the third ventricle between the habenular and the posterior commissure. During its ontogenetic development this layer bulges out in a dorsal direction forming the primary pineal anlage. In the lowest vertebrates this pineal anlage then grows along a shorter or longer distance in a rostral direction forming at its most rostral end a pineal vesicle which remains in connection with the epithalamus by means of a primary hollow pineal stalk, differing in length. The lumen of this stalk is in open communication with the third ventricle. In some lower vertebrates the pineal end-vesicle keeps its lumen, while the walls of the pineal stalk may fuse so that its lumen disappears and the stalk becomes solid and the open communication between the pineal lumen and the third ventricle does not exist any longer.

Part of the cells of the neuroepithelium from which the pineal organ proper originates differentiates into photoreceptor cells, sensory cells and, in some lower vertebrates, intercalated nerve cells, while other cell elements differentiate into supporting or interstitial cells. Being non-neural in function, this latter cell type will not be discussed here. As has been mentioned before, sensory nerve cells may also be present in the pineal stalk, not exclusively in the pineal organ proper. The axons of the sensory nerve cells grow in the direction of the brain via the walls of the pineal stalk forming together the pineal tract. From the foregoing it is evident that the direction of the neural impulse, generated in the neurosensory photoreceptor cells by photic stimuli and conveyed by sensory nerve cells is to the brain. Therefore, this type of pineal innervation is termed afferent or pinealofugal innervation.

By a number of different investigations it has been demonstrated that, most probably, secretory cells are already present in the pineal organ of fishes and amphibians. In principle, in such lower vertebrates three cell types could be secretory in function: *(1)* photoreceptor cells having a well-developed fully functional outer segment and conveying a neural impulse to sensory pineal neurons may show, next to their neurosensory function also a secretory function, *(2)* part of the photoreceptor cells which show degeneration of their outer segment have lost their neural function if their basal process is not in synaptic contact with pineal sensory neurons, now exclusively exerting a secretory function, *(3)* ontogenetically prospective photoreceptor cells do not develop an outer segment at all, but differentiate directly into purely secretory elements. In the pineal organ of the cyclostome *Petromyzon planeri* pinealocyte-like secretory cells have even been described (Meiniel, 1980). It cannot be entirely excluded that all of these three possibilities are, indeed, realized, even in rather primitive vertebrates. More investigations are, however, necessary, but many studies confirmed that the pineal organ of fish and amphibians is already involved in the production of secretory compounds, indoleamines and probably also peptides (see Ariëns Kappers, 1981, for literature and discussion). It might be possible that, in such lower vertebrates, photic stimuli are not exclusively transduced into a neural, but also into a secretory response. It should, however, be stressed that some problems are still to be solved, especially as the virtual secretory response to photic stimuli of pinealocyte-like cells is concerned, which, next to the so-called secretory rudimentary photoreceptor cells (Collin, 1971) are the most likely candidates for a secretory function.

In analogy to the conditions present in mammalian pineals the question can be raised whether the secretory production in such primitive non-mammalian vertebrates is neurally regulated at least in so far as such a production would not be exclusively regulated by direct photic stimuli. As we will mention later, the presence of such a pinealopetal or efferent pineal regulating innervation is doubtful in fish, while it is still not very well proved to exist in amphibians.

In *fish* the histochemical technique for demonstrating acetylcholinesterase has given the best results for observing the presence of pineal sensory nerve cells in the pineal epithelium. Their number, which varies greatly among the different species, is smaller than that of the photoreceptor cells. This means that a number of photoreceptive neurosensory cells convey their neural impulse to the same sensory nerve cell. In some fish, nerve cells are more densely packed in the stalk region than in the pineal end-vesicle in which such cells are also not equally distributed. Multipolar as well as pseudo-unipolar sensory nerve cells have been distinguished, the latter type being the most common. Synaptic contacts of a special type between the basal process of photoreceptor cells and the dendrites of the intrapineal sensory nerve cells have been described, but very little is known about the intrapineal connectivity perhaps existing between such nerve cells and, when present, intercalated nerve cells.

As has been mentioned before, the axons of the sensory nerve cells form the afferent or pinealofugal pineal tract. This tract contains mostly non-myelinated nerve fibres. In fish their number may vary from 120 to more than 2.000. Their diameter is also variable. In some pineal tract fibres dense-core or granular vesicles measuring 50 - 140 nm in diameter have been observed. Their functional significance is, so far, unknown. In two species of fish indications for the presence of axo-

axonal contacts between pineal tract fibres have been demonstrated.

With the cobalt chloride iontophoresis technique, the central projections in the brain of pineal tract fibres have been studied in detail (Hafeez and Zerihun, 1974). In *Salmo* such fibres could be traced to the parapineal organ, lateral habenular nucleus, pretectal area, di- and mesencephalic periventricular grey, dorsomedial and dorsolateral thalamic nuclei, nucleus of Darkschewitsch, cells of origin of the medial longitudinal fasciculus and the dorsal tegmentum which suggests that the afferent neural pathway from the pineal influences the function of many important brain centres. Interestingly, no fibres running to the subcommissural organ, optic tectum and interpeduncular nucleus could be observed in *Salmo*. According to the experience of a number of authors using less sophisticated methods, in most fish pineal tract fibres could not be followed beyond the level of the posterior commissure. In *Phoxinus*, the pineal tract was found to be connected with the right habenular ganglion (Bhargava, 1973). Recently, some pineal tract fibres of the teleost *Puntius sophore* were observed to bifurcate after reaching the brain. Some branches enter the optic tectum and the subcommissural organ, while others enter the ependymal lining of the third ventricle just ventral to the posterior commissure, a few fibres even projecting into the ventricle (Sathyanesan and Sastry, 1982).

Electrophysiological recordings from the fish pineal end-vesicle and the pineal stalk provided clear evidence that the pineal is light-receptive (see Dodt, 1973, for a review). In cyclostomes and fish, light inhibits the spontaneous electrical activity of the organ. Two different responses to light could be distinguished. Usually, an achromatic response is obtained, the effect of stimulation solely depending on luminance and not on wavelength. A few cells, however, show a chromatic response being inhibited by light of short wavelength and excited by exposure to longer wavelengths.

In fish no clear evidence exists about the presence of pinealopetal or efferent nerve fibres. Using the Falck-Hillarp technique, some catecholaminergic green fluorescent fibres were observed in the peripheral meningeal tissue of the pike *(Esox lucius)*, but not within the pineal parenchyma proper (Owman and Rüdeberg, 1970). As to the granular vesicles observed in some pineal tract fibres it is not known whether these are present in either afferent or efferent nerve fibres. The exact nature of their content is also still enigmatic.

In the pineal of *amphibians*, well-developed true photoreceptor cells as well as secretory rudimentary photoreceptor cells have been described in detail by many authors (see Vollrath, 1981, for a review). In the rudimentary photoreceptor cells, the outer segments show signs of degeneration. It is, however, questionable whether such cells represent true and permanently rudimentary photoreceptor cells or true photosensory elements the outer segments of which are involved in a process of more or less cyclic break-down and renewal. It has, however, been supposed that a true phylogenetically progressive degenerative process of the directly photoreceptive capability of pineal photoreceptor cells in favour of their secretory capability sets in clearly in amphibians. The basal process of secretory rudimentary photoreceptor cells is no longer in synaptic contact with pineal sensory neurons, but ends on the limiting membrane of the pineal epithelium.

The basal process of true amphibian photoreceptor cells is in axo-dendritic or axo-somatic synaptic contact with intrapineal sensory nerve cells. Three types of such synaptic connections have been distinguished: *(1)* ribbon synapses, the ribbons having also been termed vesicle-crowned rodlets, *(2)* convential synapses, and *(3)* synapses characterized by the presence of subsurface cisterns. Ribbon synapses are rather typical of vertebrate photoreceptor cells in general. The axo-dendritic synapses have been subdivided in two main types (Bayrhuber, 1972). In type I, 2-5 dendrites lie opposite a synaptic ribbon, while, in type II, 1-6 ribbons or vesicle-crowned rodlets lie opposite a single dendrite. Usually, the dendrites are not invaginated into the photoreceptor cells.

Conventional synapses lacking ribbons have been observed less frequently. They may connect intercalated neurons, also termed interneurons (Bayrhuber, 1972; Korf, 1976). Synapses showing subsurface cisterns have been observed in the pineal of *Rana pipiens*. They have been held to be involved in efferent transmission of feedback mechanisms (Kelly and Smith, 1964).

For the investigation of different types of ampbibian pineal nerve cells and their relationships best results have been obtained by applying the acetylcholinesterase reaction after Karnovsky and Roots (Wake et al., 1974; Ueck, 1971). In *Rana ridibunda* and *Rana esculenta* the pineal organ contains large and small multipolar nerve cells, pseudo-unipolar cells and a few bipolar ones. It has been suggested that the pseudo-unipolar cells give rise to the pineal tract fibres, whereas the multipolar nerve cells would be interneurons. However, the wiring of the pineal nervous circuitry and its functional significance is a most complicated matter which has still to be elucidated.

A system of axons running through the pineal lumen has also been observed (Vigh-Teichmann et al., 1973). The fibres are synaptically connected with the plasmolemma of supporting pineal cells. They show some resemblance with cerebrospinal fluid-contacting axons in other parts of the central nervous system present in other classes of vertebrates. In the amphibian pineal, these non-myelinated axons contain dense-core vesicles, but their origin and function are unknown.

The amphibian pineal tract consists of both myelinated and non-myelinated nerve fibres, the number of the latter prevailing. Part of the fibres in this tract originate in the frontal organ, present in anuran amphibians, which will not be dealt with here. In addition to undoubtedly pinealofugal nerve fibres, the pineal tract of anurans also contains some possibly aminergic fibres showing dense-core vesicles measuring 85 - 95 nm in diameter (Oksche and Vaupel-von Harnack, 1965). Similar fibres were observed in the peripheral regions of the organ (Ueck, 1968). Green fluorescent nerve fibres being evidently aminergic have, however, been mostly observed in the peripineal leptomeningeal tissue. According to some authors (Ueck, 1968) such fibres would be exclusively confined to blood vessels, but others (Iturizza, 1967) have pointed out that they do not show special relationships with vessels and are particularly abundant in the ventral region of the pineal. Their pinealopetal nature would be indicated by a complete disappearance of fluorescence in the pineal and an accumulation of fluorescent material in the proximal stump of the nervous pathway after transsection of the posterior commissure. Although observations by other authors also indicate the possible presence of an efferent pinealopetal innervation (Kelly and Van de Kamer, 1960; Kelly and Smith, 1964; Paul, 1972) the origin of the fibres involved and their exact course and function still need to be established. Theoretically the regulation of secretory processes in the amphibian pineal by pinealopetal nerve fibres is attractive because it is improbable that the secretory function of the secretory rudimentary photoreceptor cells is directly regulated by photic stimuli.

The amphibian pineal tract leaves the organ at its caudal border and comes in close contact with the base of the commissural organ giving off some fibres which, however, do not originate in the pineal organ but in the frontal organ. The majority of the tract fibres runs further caudalward and penetrates the posterior commissure. Fibres originating in the pineal organ proper reach the pretectal region and the periventricular grey, while both the pineal and the frontal organ may also influence the function of the habenular nuclei and the dorsal thalamus.

Electrophysiological studies of the amphibian pineal organ show an exclusively achromatic response to light proving that the pineal is, indeed, directly photosensitive (see Vollrath, 1981, for the extensive literature on this subject).

In *reptiles* the pineal organ is characterized by a progressive reduction of its photoreceptive sensory function and a progressive development of its secretory capability. This is also reflected by the afferent innervation of the reptilian pineal. Considering the scarcely present fully differentiated true photoreceptor cells in most reptiles, it is obvious that the number of pineal sensory nerve cells is also very rare (*Lacerta:* Ariëns Kappers, 1976). The relatively few afferent or pinealofugal nerve fibres originating from these nerve cells and forming the pineal tract are, for the most part, myelinated. They run to the habenular and posterior commissure, and, more caudally, to the region of the subcommissural organ and the periventricular fibre layer. Their exact function is, so far, unknown.

The basal process of the rare well-developed photoreceptor cells forms synaptic connections with a number of dendrites of a sensory nerve cell, while a number of terminals of basal processes of several photosensory cells may also form synaptic connections with a single dendrite of a sensory nerve cell. The synapses are of the ribbon or vesicle-crowned rodlet type.

Efferent or pinealopetal non-myelinated monoaminergic nerve fibres containing dense-core vesicles have often been described in reptilian pineal organs. They are noradrenergic as is shown by their green fluorescence and the increase of their granular vesicular content after the administration of false sympathetic transmitters such as 5-hydroxydopamine and 5-hydroxydopa. Moreover, the administration of tritiated hydroxytryptamine followed by electron microscopic autoradiography and a number of pharmacological tests point to the presence of both noradrenaline and serotonin in these fibres. Like in the mammalian pineal, the presence of serotonin in such noradrenergic fibres is very likely due to their uptake of this compound which is produced by the pineal secretory parenchymal cells.

The exact origin of the autonomic pineal efferent fibres is not known. In the snake *Natrix* they largely derive from a caudal and dorsal meningeal nerve bundle reminding of the conary nerves in mammals. Most probably, these peripheral noradrenergic nerve fibres originate in the superior part of the sympathetic trunks to regulate the function of the pineal organ which, in snakes, is exclusively secretory. In the pineal, the fluorescent fibres are primarily situated in the perivascular spaces accompanying the blood vessels which penetrate into the organ from the periphery. Some fibres, however, apparently innervate parenchymal lobules (Quay et al., 1968). In the lizard, such fibres are not in true synaptic contact with the secretory rudimentary photoreceptor cells (Wartenberg and Baumgarten, 1969). Some fibres lie freely in the pineal lumen sometimes in close topographical association with the degenerated outer segments of such cells. The functional significance of the peripheral pinealopetal autonomic innervation has, in reptiles, not been experimentally established. However, as has already been mentioned, it can be assumed that it regulates the secretory function of the organ.

In *birds*, the afferent pineal innervation is still more reduced than in reptiles, snakes excepted in which no such innervation is present at all, due to the phylogenetically continuing regression of the neural directly photoreceptive function of the organ. Rather large interspecies differences do, however, exist regarding the development of pinealofugal innervation. Even by means of the Karnovsky-Roots histochemical technique for the demonstration of acetylcholinesterase, sensory nerve cells cannot easily be demonstrated (Ueck and Kobayashi, 1972a,b). In the Japanese quail, nerve cell bodies were not observed at all, while they are rare in the dove, but more abundant in the chicken. In the house sparrow and the canary, sensory nerve cell bodies and their processes are also present.

Avian afferent pineal innervation has been most extensively studied in the house sparrow. Mostly, the sensory nerve cells are bipolar, occasionally, however, pseudo-unipolar (Ueck and Kobayashi, 1972a). Their axons form small nerve bundles which run in the pineal stalk. The presence of such a pineal tract has also been demonstrated in other birds. Most often its fibres are non-myelinated and always non-fluorescent suggesting their afferent function.

In the basal process of the photoreceptor cells which, for the most part, are secretory rudimentary photoreceptor cells, synaptic ribbons have been demonstrated (Bischoff, 1969; Ueck, 1970a,b), but there is little evidence for the presence of functional synapses between such photoreceptor cells and the sensory neurons. It is also typical of the much reduced afferent pineal innervation that, for instance in the sparrow, several hundreds of rudimentary photoreceptor cells face a single nerve cell only. In accordance with the evident preponderant secretory function of the avian pineal, its pinealofugal or afferent innervation is far from well-developed and does not appear to play an important functional role.

Practically nothing is known about the sites of ending of the afferent nerve fibres. In some birds they could be followed into the posterior commissure, while some reach the habenular commissure or seemingly disappear in the intercommissural region.

In agreement with the more important secretory function of the avian pineal, the pinealopetal innervation by peripheral efferent autonomic nerve fibres is by far better developed than the pinealofugal afferent one. As has been demonstrated using several techniques, most efferent fibres are noradrenergic. It has been supposed that, in the chicken, a noradrenergic as well as a cholinergic efferent pineal innervation may be present (Wight and MacKenzie, 1970). Most probably, however, just like in mammals some sympathetic noradrenergic nerve fibres also demonstrate a cholinergic reactivity (Ueck and Kobayashi, 1972a). In some species, the sympathetic fibres show, next to a green fluorescence, also a yellow fluorescence due to the presence of serotonin which has evidently been taken up by such fibres from the secretory parenchymal cells as is also the case in reptiles and mammals.

The sympathetic origin of the avian pinealopetal

peripheral nerve fibres has been clearly established. The green fluorescence disappears after sympathetic ganglionectomy (Hedlund and Nalbandov, 1969). In the pineal, the number as well as the location of such fibres vary considerably among birds. Their location may be exclusively perivascular, both perivascular and parenchymal, or entirely parenchymal (see Vollrath, 1981, for literature). The terminals of the noradrenergic fibres are not in direct contact with pineal parenchymal cells. As usual, such fibres which are non-myelinated contain dense-core as well as electron-lucent vesicles.

In a large number of avian species, cell staining with chrome-alum haematoxylin phloxine or aldehyde-fuchsin are embedded in the pineal parenchyma close to the habenular and posterior commissure (Oksche et al., 1972). Supposedly they are extrahypothalamic neurosecretory elements (Quay and Renzoni, 1963, 1966). Futher investigation of such cells and their processes would be of much interest because of the recent observations on the presence of extrahypothalamic neurosecretory nerve fibres in the mammalian pineal gland (see later).

In respect to the afferent innervation morphologically observed in some avian pineals it cannot completely be ruled out that, in such cases, the organ is directly photosensitive and influences the function of brain centres. In most species, however, the number of true photoreceptor cells is small, the synaptic connections between the basal process of photoreceptor cells, if present, and of the secretory rudimentary photoreceptor cells have not been thoroughly studied, and the number of pinealofugal nerve fibres is relatively small. Pertaining to this afferent pineal innervation, electrophysiological investigations did not give unequivocal results (see Vollrath, 1981). In some studies, light stimuli did not exert any influence on spontaneous electrical pineal activity, while, in others, light flashes resulted in a marked inhibition of electrical discharges. This was, however, only the case if the optic nerves were intact which would speak against a direct pineal photoreceptivity, at least in the species concerned, the quail.

Curiously, light has been shown to directly effect the chemical composition of explanted avian pineal cells. In cultured chick pineal, darkness stimulated the production of N-acetyl transferase (Wainwright and Wainwright, 1978), the crucial enzyme involved in the production of melatonin. In birds, the problem whether the pineal is directly photosensitive or not is complicated by the fact that there is evidence of the presence of extraretinal as well as of extrapineal photoreceptivity (see Menaker and Oksche, 1974).

INNERVATION OF THE MAMMALIAN PINEAL

In mammals, the pineal is a purely secretory endocrine organ. The pineal-specific cells, the secretory pinealocytes which are phylogenetically as well as, at least clearly in some mammals, ontogenetically derived from precursor cells of the sensory cell-line, do not transduce direct photic stimuli into a neural impulse which is conveyed to brain centres via sensory pineal neurons. In the adult mammalian pineal, sensory neurons are lacking with a possible exception of one species only (see later). However, light or its absence, darkness, still plays a most important role in the function of the mammalian pineal. The secretory function of the pinealocytes is, besides a number of other factors (Ariëns Kappers, 1979), regulated by efferent pinealopetal nerve fibre systems which convey photic, but also external stimuli of other types such as olfactory and acoustic stimuli, to the organ. We will now deal with the systems of efferent mammalian pineal innervation.

An *orthosympathetic innervation* of the pineal gland was suggested to be present in mice as early as 1904 by the father of neuroanatomy, Ramon y Cajal, and subsequently by some few other authors. It was then described in man and monkey by Le Gros Clark in 1940 and finally demonstrated and experimentally confirmed in rat by superior cervical ganglionectomy by the present author (Ariëns Kappers, 1960, also for earlier literature) who erroneously held this efferent type of pineal innervation to be the exclusive one. In the sixties, mammalian pineal orthosympathetic innervation has drawn special attention because mainly American authors proved that this neural pathway conveys indirect photic stimuli to the organ regulating the secretory function of pinealocytes, especially as the metabolism of indoleamines is concerned. As has been demonstrated by autoradiography, administration of false transmitters, chemical sympathectomy, fluorescence histochemistry, electron microscopy and pharmacological experiments, the neurotransmitter substance in the sympathetic fibres is noradrenaline. Noradrenaline, when released from the fibre terminals and probably also from preterminal varicosities, is taken up by β-adrenergic receptors present in the cell membrane of pinealocytes. Via the adenyl cyclase-cAMP system, noradrenaline then stimulates the production of melatonin in these cells.

As was observed in some reptiles and birds, the terminal part of the sympathetic nerve fibres in the pineal of a number of mammals contains serotonin which is produced by pinealocytes and taken up and stored by the nerve fibres. The presence of octopamine has also been demonstrated (see Vollrath, 1981, for literature). In the pineal parenchyma, the sympathetic fibre terminal endings do not show true synaptic connections with pinealocytes, ending freely. In general, the fibres are incompletely surrounded by lemmocyte cytoplasm (Bondareff, 1965; Arstila, 1967; Ariëns Kappers, 1969), while the basal lamina usually covering nerve fibre bundles is also discontinuous (Ariëns Kappers, 1969). Most probably, the noradrenaline released from intraparenchymal sympathetic nerve fibres affects the function of the pinealocytes by way of diffusion. The same may be true for noradrenaline released from the terminals of perivascular sympathetic nerve fibres, the more so as the perivascular spaces show outpocketings penetrating into the parenchyma. As has been mentioned in our first lecture, it has, moreover, been demonstrated by light- (Quay, 1974) and by electron microscopy (Krstić, 1975) that the pinealocytes are surrounded by a system of continuous fine channels which enhance the diffusion of compounds. Together with the extensions of the perivascular spaces, these channels form an extensive canalicular network.

The orthosympathetic nerve fibres are postganglionic, originating from the perikarya of the superior cervical ganglia. Having entered the skull with other fibres of the same origin they reach the pineal gland along two somewhat different pathways. In rat, most fibres run to the organ in two bilateral symmetrically nervi conarii which course to the dorsocaudal pole of the pineal. A number of fibres also enter the organ by way of the perivascular spaces of pial vessels which penetrate into the gland (Ariëns Kappers, 1960). After superior cervical ganglionectomy all sympathetic pineal fibres disappear as has been shown by silver staining (Ariëns Kappers, 1960), while, after transection of the nervi conarii by far most of the green fluorescent noradrenergic autonomic fibres in the organ are absent (Owman, 1964). In some species, such as in the cat and bat, the two symmetrical conary nerves fuse before reaching the pineal (Kenny, 1965). This is also the case in man (Le Gros Clark, 1940) and monkey (Kenny, 1961). The fibres in the conary nerves distribute directly in the pineal parenchyma.

The way in which sympathetic nerve fibres enter the organ varies among mammals as does their intrapineal distribution. In some mammals, sympathetic fibres are practically exclusively situated in the perivascular spaces, while in others they are found exclusively in the pineal parenchyma proper. As has been mentioned, a perivascular and a parenchymal distribution also occur (see Vollrath, 1981, for literature). Moreover, perivascular nerve fibres sometimes enter the pineal parenchyma, while parenchymal fibres reach the perivascular spaces. In species in which the pineal complex is constituted of a superficial and a deep part, autonomic nerve fibres were observed to leave the superficial part to run uninterruptedly through the pineal stalk to the deep part of the gland and, in some mammals, even to the habenular nuclear complex.

In the terminals of the sympathetic fibres as well as in their preterminal varicosities small and large dense-core vesicles measuring 40 - 60 and 60 - 120 nm, respectively, are present. Spherical and flat electron-lucent vesicles also oocur. By a number of different techniques it has been shown that the more abundant small granular vesicles and the flat clear vesicles are involved in the storage and release of noradrenaline, the flat clear vesicles representing depleted granular vesicles. The spherical electron-lucent vesicles may be cholinergic in nature constituting a cholinergic link in postganglionic transmission (Eränkö et al., 1970; Machado and Lemos, 1971; Vollrath, 1981, also for the extensive literature on this subject). It has been suggested that, functionally, the small dense-core or granular vesicles are basically similar to the larger ones.

After the onset of darkness, the number of granular vesicles strikingly decreases, while the number of flat electron-lucent vesicles increases (Matsushima and Ito, 1972; Matsushima et al., 1979). This agrees with the concept that noradrenaline, the amount of which increases during darkness (Wurtman et al., 1967), is soon released at that time of the day. As has been mentioned, noradrenaline release induces an increased production of melatonin by the pinealocytes during darkness (Wurtmann et al., 1968).

Many experiments have shown that information about environmental conditions, especially changes in photic input in the lateral eyes, is conveyed to the pineal gland by way of its peripheral orthosympathetic innervation. The central neural pathway involved has been thoroughly studied by Moore (1978a,b) who will speak himself about this subject during this course. Therefore, I will only shortly indicate this pathway here. The ganglion cell layer of the retina conveys photic stimuli bilaterally to the suprachiasmatic nuclei. These hypothalamic nuclei project caudalward into the periventricular hypothalamus and the ventral tuberal area, while neurons of this latter area project to the lateral hypothalamic nucleus. Neurons of this nucleus have efferent connections, running in the medial forebrain bundle, with the sympathetic intermediolateral cell column in the upper thoracic part of the spinal cord. From this part of the cell column, preganglionic orthosympathetic fibres run to the superior cervical ganglion. The central neural pathway from the lateral eyes to the cervical sympathetic fibres has been confirmed by electrophysiological experiments (Brooks et al., 1975; Dafny and McClung, 1975; McClung and Dafny, 1975). Furthermore, it has been shown that effects of light on the pineal are not mediated by any of the endocrine organs (Fiske et al., 1962) and that light does not exert any direct influence on the function of cultured pinealocytes.

As was first shown in the rat (Ariëns Kappers, 1960) and later confirmed in rat and other mammals by several authors, nerve fibres originating from both the habenular and posterior commissure course in the pineal stalk. However, most of them form hairpin loops joining again their commissure of origin, especially the habenular, at the contralateral side. Some fibres, however, were found to run not only in the pineal stalk, but also to reach the pineal organ proper not returning to their commissure of origin. Because of a number of reasons, such fibres were first regarded to be aberrant commissural fibres going astray (Ariëns Kappers, 1960, 1965). However, recent findings fully confirmed the concept held by much earlier authors (see Ariëns Kappers, 1960) that the habenulo-pineal complex is a functional neural entity. As was earlier observed (Schapiro and Salas, 1971), even after bilateral superior cervical ganglionectomy light stimuli are still able to influence rat pineal function while, more recently, a number of neuroanatomical and electrophysiological investigations proved the presence of a *central epithalamo-pineal neural connection*, especially between the habenular region and the mammalian pineal gland (see Ueck, 1979; Guérillot et al., 1979; Semm et al., 1981; Vollrath, 1981). Some of the pathways involved are pinealofugal and either catecholaminergic or acetylcholinergic, while others are clearly pinealopetal (see Ariëns Kappers, 1981, and Vollrath, 1981, for literature). The most important results obtained can be summarized as follows. In rat, nerve fibres from the habenular nuclei, but probably also stria medullaris fibres, run via the pineal stalk to the pineal gland. By recording average photic responses from the rat pineal in light and dark adaptation, it was observed that in the dark-adapted state responses exhibited a higher amplitude (Dafny, 1980). It has been assumed that photic stimuli are transmitted to the pineal along two separate routes: *(1)* a fast one via the central habenulo-pineal pathway, and *(2)* a slower one via the peripheral sympathetic innervation of the pineal. It was also found (Semm and Vollrath, 1979) that the guinea-pig pineal contains two different populations of spontaneously active pinealocytes. The cells of the first population respond to olfactory, acoustic and short-term photic stimuli which evidently reach the gland via the central habenulo-pineal neural pathway. After sympathectomy the spontaneous activity of such cells in the pineal is diminished, but not quite suppressed. The cells of the second category do not respond to short-term sensory stimulation. Assumedly, melatonin production happens in pinealocytes showing an increased electrical activity during the night and a decreased activity during the day. These cells cease firing after removal of the peripheral sympathetic pineal innervation which is known to enhance melatonin production during the night. Because, moreover, the activity of some pinealocytes is proved to be stimulated by light and inhibited by darkness it was, furthermore, hypothesized that a positive feedback mechanism regulates melatonin synthesis which is induced by darkness and terminated by light. The activity of the two different populations of pinealocytes has been supposed to depend on the central and the peripheral innervation of the pineal, respectively.

In the guinea-pig, nerve fibres coursing from the habenular nuclei to the pineal have also been demonstrated, while fibres joining the striae medullares and running in the habenula reach the gland somewhat more dorsally (Semm and Vollrath, 1979). On the base of electrophysiological findings it was suggested that the habenular nuclei, probably the lateral ones, modify pineal gland activity, while, vice versa, the pineal influences single unit activity in the habenular nuclei. In exactly which way the latter possibility is realized is still open to further research. It is known that, in the ferret, intramural pineal cholinergic nerve cells send fibres to the habenular region, while from this region nerve fibres run to this pineal ganglion (Trueman and

Herbert, 1970; Herbert, 1971; David and Herbert, 1973; David et al., 1973). To our knowledge in no other adult mammal such a cholinergic pinealo-habenular connection has, so far, been observed.

In some mammalian fetuses, however, morphological research suggested the presence of a pinealofugal nerve fibre bundle closely associated with clusters of nerve cells (Møller, 1979, also for literature). It has been held that this bundle may represent a vestigial homologue of the non-mammalian pineal tract which, perhaps, persists in adult mammals (Ueck, 1979) having been incorporated in the pineal parenchyma and the pineal stalk (Møller et al., 1975). As has been mentioned, however, the ferret is the only mammal in which, in the adult state, pineal sensory neurons have at present been observed.

The investigations shortly surveyed clearly established the presence of a central pineaopetal innervation via the habenular region. They, furthermore, suggest that *(1)* other sensory stimuli than just photic ones may regulate pinealocyte activity via this pathway, and that *(2)* photic stimuli of a different duration influence the activity of different pinealocyte populations along two different pinealopetal neural pathways, the peripheral sympathetic and the central habenulo-pineal pathways.

Irregularly occurring intrapineal nerve cells have been observed in several mammals including primates. The distinct intramural pineal ganglion in the ferret has already been mentioned. Its cells are acetylcholinergic and probably of a sensory nature. Intramural cholinergic nerve cells of a quite different nature are present in the rabbit pineal. They have been considered postganglionic parasympathetic neurons. The *parasympathetic pineal innervation* has, so far, only been investigated in the macaque monkey (Kenny, 1961) and, more extensively, in the rabbit (Romijn, 1972, 1973a,b, 1975a,b). In the rabbit, preganglionic parasympathetic nerve fibres probably originate from the superior salivatory nuclei in the medulla oblongata. Some of such fibres run along with the facial nerves, the greater petrosal nerves and the conary nerves to the pineal gland in which they establish axo-somatic and axo-dendritic synapses with intramural pineal postganglionic parasympathetic neurons, the number of which is about 40. Postganglionic cholinergic parasympathetic nerve fibres originating from these cells innervate the pinealocytes. Other parasympathetic preganglionic fibres, likewise originating from the superior salivatory nuclei, do, however, synapse with postganglionic parasympathetic cells present in the greater petrosal nerves. The axons of these cells run to the pineal via th nervi conarii to distribute directly in the pineal parenchyma. In the rat pineal, cholinergic nerve terminals exclusively situated in the organ and presumably making tru synaptic contacts with pinealocytes have also been described (Wood, 1973). However, bilateral stereotaxic coagulation of rabbit facial nerves caused only minor changes in the ultrastructure of pinealocytes (Romijn, 1975a), while the administration of parasympatholytic drugs showe comparable results (Romijn, 1976). By some authors such a parasympathetic pineal innervation has, however, not been accepted on biochemical grounds, choline acetyltransferase being barely or not at all detectable in the pineal. The very little activity measured could, in fact, be due to the presence of carnitine acetyltransferase (LaBella and Shin, 1968; Schrier and Klein, 1974). So far it would appear that there is little evidence that the parasympathetic autonomic system exerts any important effect on mammalian pineal function.

Extramural pineal nerve cells forming distinct ganglia which are probably functionally associated with the organ have occasionally been observed (see Bargmann, 1943). Two of such ganglia showing a somewhat different topography have been distinguished. After their discoverers they have been named parietal or Warburg's ganglion and Pastori's ganglion. By one author (Kenny, 1965) a small Warburg's ganglion was observed in a young adult human between the suprapineal recess and the dorsal surface of the gland. In older individuals it could, however not be found. By other authors (Møllgaard and Møller, 1973) a Warburg's ganglion was demonstrated in human fetuses on the dorsal surface of the great cerebral vein which lies in between the ganglion and the pineal, while a ganglion of Pastori was present at the caudal pole of the gland, ventral to the great cerebral vein. It has been suggested that both extramural pineal ganglia are parasympathetic (Kenny, 1965).

Neurosecretory nerve fibres are also present in the mammalian pineal gland. They stain with chrome-alum haematoxylin aldehyde fuchsin or alcian blue. Such extrahypothalamic neurosecretory fibres have been demonstrated to occur in small amounts in the pineal of the hedgehog (Bargmann, 1964; Hartmann, 1957; Suomalainen, 1960; Oksche, 1965), the Rhesus monkey (Lukaszyk and Reiter, 1974, 1975) and the Mongolian gerbil (Japha et al., 1974). Recently, vasopressin- and oxytocin-containing nerve fibres were demonstrated in the area of the subcommissural organ and in the pineal of the rat (Buijs et al., 1978; Buijs and Pévet, 1980). By radioimmunoassay, vasopressin, oxytocin, neurophysin I and neurophysin II were also quantitatively determined in the bovine pineal (Pévet et al., 1980). The ratio of vasopressin to oxytocin was similar to that of neurophysin I to neurophysin II. This suggests that the neurophysins present in the pineal are, indeed, the carrier proteins of the two peptidergic hormones rather than of arginine-vasotocine, the pineal presence of which was earlier supposed (see Dogterom et al., 1979).

Such extrahypothalamic neurosecretory fibres were also specifically demonstrated in the rat pineal by the unlabelled antibody-enzyme technique (Buijs and Pévet, 1980). They run via the subcommissural organ or the habenular commissure into the pineal stalk to end in the rostral part of the organ proper. It has been speculated that they are possibly involved in the regulation of water balance. In the guinea-pig pineal, probably identical nerve fibres have been observed using the horseradish peroxydase method (Korf and Wagner, 1980). Their origin from the region of the neurosecretory paraventricular hypothalamic nucleus suggests their peptidergic nature. From here, the fibres course via the habenular or the posterior commissure and the pineal stalk to the caudal part of the pineal gland. Further investigations concerning the functional significance of such neurosecretory nerve fibres in the pineal organ are necessary.

In concluding this survey it appeared that, in parallel with the change in function which can be observed in the pineal organ during its phylogenetic development, pineal innervation also changes from a practically purely pinealofugal or afferent innervation into a pinealopetal or efferent innervation, a more or less gradual process which is fully realized in mammals. The peripheral sympathetic pineal innervation is certainly of the utmost importance for pineal function. Besides, however, many investigations have now shown that central efferent nerve fibres of very different origins convey sensory input and also special chemical input to the organ influencing its function.

REFERENCES

Ariëns Kappers, J., 1960, The development, topographical relations and innervation of the epiphysis cerebri, in the albino rat, Z. Zellforsch., 52:215.

Ariëns Kappers, J., 1965, Survey of the innervation of the epiphysis cerebri and the accessory pineal organs in vertebrates, Progr. Brain Res., 10:87.

Ariëns Kappers, J., 1976, The sensory innervation of the pineal organ in the lizard, *Lacerta viridis*, with remarks on its position in the trend of pineal phylogenetic structural and functional evolution, Z. Zellforsch., 81:581.

Ariëns Kappers, J., 1969, The mammalian pineal organ, J. neuro-visc. Rel., Suppl. 9:140.

Ariëns Kappers, J., 1979, Short history of pineal discovery and research, Progr. Brain Res., 52:3.

Ariëns Kappers, J., 1981, Evolution of pineal concepts, in: "The Pineal Organ, Photobiology — Biochronometry — Endocrinology", A.Oksche and P.Pévet, eds., Elsevier, Amsterdam.

Arstila, A.U., 1967, Electron microscopic studies on the structure and histochemistry of the pineal gland of the rat, Neuroendocrinology, 2, Suppl.:1.

Bargmann, W., 1943, Die Epiphysis cerebri, in: "Handbuch der mikroskopischen Anatomie des Menschen", Vol. VI/4, W.von Möllendorff, ed., Springer, Berlin.

Bargmann, W., 1954, Neurosekretion und hypothalamisch-hypophysäres System, Verh. anat. Gesellsch., 51:30.

Bayrhuber, H., 1972, Über die Synapsformen und das Vorkommen von Acetylcholinesterase in der Epiphyse von *Bombina variegata* (L.), (Anura), Z. Zellforsch., 126:278.

Bhargava, H.N., 1973, The pineal organ of the minnow, *Phoxinus phoxinus* L., and a note on the effects of temperature, light, background responses, hypophysectomy and gonadectomy, Zool. Jhrb. (Abt. Anat. Ontogen. d. Tiere), 91:478.

Bischoff, M.B., 1969, Photoreceptoral secretory structures in the avian pineal organ, J. ultrastruct. Res., 28:16.

Bondareff, B., 1965, Submicroscopic morphology of granular vesicles in sympathetic nerves of rat pineal body, Z. Zellforsch., 67:211.

Brooks, C. McC., Ishikawa, T., and Koizumi, K., 1975, Autonomic system control of the pineal gland and the role of this complex in the integration of body function, Brain Res., 87:181.

Buijs, R.M., and Pévet, P., 1980, Vasopressin- and oxytocin-containing fibres in the pineal gland and subcommissural organ of the rat, Cell Tiss. Res., 205:11.

Buijs, R.M., Swaab, D.F., Dogterom, J., and Van Leeuwen, F.W., 1978, Intra- and extra-hypothalamic vasopressin and oxytocin pathways in the rat, Cell Tiss. Res., 186:423.

Cajal, Ramon y, 1904, "Textura del Sistema nervioso del Hombre y de los Vertebrados", II, 2de parte, Moya, Madrid.
Collin, J.-P., 1971, Differentiation and regression of the cells of the sensory line in the epiphysis cerebri, in: "The Pineal Gland", G.E.W.Wolstenholme and J.Knight, eds., Churchill Livingstone, Edinburgh-London.
Dafny, N., 1980, Photic input to rat pineal gland conveyed by both sympathetic and central afferents, J. neural Transm., 48:203.
Dafny, N., and McClung, R., 1975, Pineal body: Neural recording, Experientia (Basel), 31:321.
David, G.F.X., and Herbert, J., 1973, Experimental evidence for a synaptic connection between habenula and pineal ganglion in the ferret, Brain Res., 64:327.
David, G.F.X., Herbert, J., and Wright, G.D.S., 1973, The ultrastructure of the pineal ganglion in the ferret, J.Anat. (London), 115:79.
Dodt, E., 1973, The parietal eye (pineal and parietal organs) of lower vertebrates, in: "Handbook of Sensory Physiology", Vol. VII/3B, E.Jung, ed., Springer, Berlin-Heidelberg-New York.
Dogterom, J., Snijdewindt, F.G.M., Pévet, P., and Buijs, R.M., 1979, On the presence of neuropeptides in the mammalian pineal gland and subcommissural organ, Progr. Brain Res., 52:465.
Eränkö, O., Rechardt, L., Eränkö, L., and Cunningham, A., 1970, Light and electron microscopic histochemical observations on cholinesterase-containing sympathetic nerve fibres in the pineal body of the rat, Histochem. J., 2:479.
Fiske, V.M., Pound, J., and Putman, J., 1962, Effect of light on the weight of the pineal organ in hypophysectomized, gonadectomized, adrenalectomized or thiouracil-fed rats, Endocrinology, 71:130.
Guérillot, C., Lefray, P., Pfister, A., and Da Lage, C., 1979, Contribution to the study of the pineal stalk fibres in the rat, Progr. Brain Res., 52:97.
Hafeez, M.A., and Zerihun, L., 1974, Studies on central projections of the pineal tract in rainbow trout, *Salmo gairdneri* Richardson, using cobalt chloride iontophoresis, Cell Tiss. Res., 154:485.
Hartmann, F., 1957, Über die Innervation der Epiphysis cerebri einiger Säugetiere, Z. Zellforsch., 46:416.
Hedlund, L., and Nalbandov, A.V., 1969, Innervation of the avian pineal body, Amer. Zoologist, 9:1090.
Herbert, J., 1971, The role of the pineal gland by light of the reproductive cycle of the ferret, in: "The Pineal Gland", G.E.W.Wolstenholme and J.Knight,

eds., Churchill Livingstone, Edinburgh-London.

Iturizza, F.C., 1967, Histochemical demonstration of biogenic amines in the pineal gland of the toad, *Bufo arenarum*, J. Histochem. Cytochem., 15:301.

Japha, J.L., Eder, T.J., and Goldsmith, E.D., 1974, Morphological and histochemical features of the gerbil pineal system, Anat. Rec., 178:381.

Kelly, D.E., and Van de Kamer, J.C., 1960, Cytological and histochemical investigations on the pineal organ of the adult frog *(Rana esculenta)*, Z. Zellforsch., 52:618.

Kelly, D.E., and Smith, S.W., 1964, Fine structure of the pineal organs of the adult frog, *Rana pipiens*, J. Cell Biol., 22:565.

Kenny, G.C.T., 1961, The "nervus conarii" of the monkey. (An experimental study), J. Neuropathol. exp. Neurol., 20:563.

Kenny, G.C.T., 1961, The innervation of the mammalian pineal body. (A comparative study), Proc. Austr. Assoc. Neurol., 3:133.

Korf, H.-W., 1976, Histological, histochemical and electron microscopical studies on the nervous apparatus of the pineal organ in the tiger salamander *(Ambystoma tigrinum)*, Cell Tiss. Res., 174:475.

Korf, W.-H., and Wagner, U., 1980, Evidence for a nervous connection between the brain and the pineal organ in the guinea-pig, Cell Tiss. Res., 209:505.

Krstić, R., 1975, Scanning electron microscope observations in the canaliculi in the rat pineal gland, Experientia (Basel), 31:1072.

LaBella, F.S., and Shin, E., 1968, Estimation of cholinesterase and cholin acetyltransferase in bovine anterior pituitary, posterior pituitary, and pineal body, J. Neurochem., 15:335.

Le Gros Clark, W.E., 1940, The nervous and vascular relations of the pineal gland, J. Anat. (London), 74:471.

Lukaszyk, A., and Reiter, R.J., 1974, Neurosecretion in the pineal gland of *Macaca rhesus*, Experientia (Basel), 30:654.

Lukaszyk, A., and Reiter, R.J., 1975, Histological evidence for the secretion of polypeptides by the pineal gland, Amer. J. Anat., 143:451.

Machado, A.B.M., and Lemos, V.P.J., 1971, Histochemical evidence for a cholinergic sympathetic innervation of the rat pineal body, J. neuro-visc. Rel., 32:104.

Matsushima, S., and Ito, T.: Diurnal changes in sympathetic nerve endings in the mouse pineal: semiquantitative electron microscopic observations, J. neural Transm., 33:275.

Matsushima, S., Morisawa, Y., and Mukai, S., 1979, Diurnal variation in large granulated vesicles in sympathetic nerve fibres of the mouse pineal – Quantitative electron microscope observations, J. neural Transm., 45:63.

McClung, R., and Dafny, N., 1975, Neurophysiological properties of the pineal body. II. Single unit recording, Life Sci., 16:621.

Meiniel, A., Ultrastructure of serotonin-containing cells in the pineal organ of *Lampetra planeri* (Petromyzontidae). A second sensory cell line from photoreceptor cell to pinealocyte, Cell Tiss. Res., 207:407.

Menaker, M., and Oksche, A., 1974, The avian pineal, in: "Avian Biology", Vol. 4, D.S.Farner and J.R.King, eds., Academic Press, New York-London.

Møller, M., 1979, Presence of a pineal nerve (nervus pinealis) in fetal mammals, Prog. Brain Res., 52:103.

Møller, M., Møllgaard, K., and Kimble, J.E., 1975, Presence of a pineal nerve in sheep and rabbit fetuses, Cell Tiss. Res., 158:451.

Møllgaard, K., and Møller, M., 1973, On the innervation of the human fetal pineal gland, Brain Res., 52:428.

Moore, R.Y., 1978a, The innervation of the mammalian pineal gland, Progr. reprod. Biol., 4:1.

Moore, R.Y., 1978b, Neural control of pineal function in mammals and birds, J. neural Transm., Suppl. 13:47.

Oksche, A., 1965, Survey of the development and comparative morphology of the pineal organ, Progr. Brain Res., 10:3.

Oksche, A., and Vaupel-Von Harnack, M., 1965, Elektronenmikroskopische Untersuchungen an den Nervenbahnen des Pinealkomplexes von *Rana esculenta*, Z.Zellforsch., 68:389.

Oksche, A., Kirschstein, H., Kobayashi, H., and Farner, D.S., 1972, Electron microscopic and experimental studies of the pineal organ in the white-crowned sparrow, *Zonotrichia leucophrys gambelli*, Z.Zellforsch., 124:247.

Owman, Ch., 1964, Sympathetic nerves probably storing two types of monoamines in the rat pineal gland, Intern. J. Neuropharmacol., 3:105.

Owman, Ch., and Rüdeberg, C., 1970, Light, fluorescence, and electron microscopic studies on the pineal organ of the pike, *Esox lucius* L., with special regard to 5-hydroxytryptamine, Z. Zellforsch., 107:522.

Paul, E., 1972, Innervation und zentralnervöse Verbindungen des Frontalorgans von *Rana temporaria* und *Rana esculenta*. Faserdegenration nach operativer Unterbrechung des Nervus pinealis, Z. Zellforsch., 128:504.

Pévet, P., Reinharz, A.C., and Dogterom, J., 1980, Neurophysins, vasopressin and oxytocin in the bovine pineal gland, Neurosci.Lett., 16:301.
Quay, W.B., 1974, Pineal canaliculi: demonstration, twenty-four-hour rhythmicity and experimental modification, Amer. J. Anat., 139:81.
Quay, W.B., and Renzoni, A., 1963, Comparative and experimental pineal structure and cytology in passeriform birds, Riv. Biol., 56:363.
Quay, W.B., and Renzoni, A., 1966, Studies on the commissuro-pineal "neurosecretory cells" of birds, Riv. Biol., 59:231.
Quay, W.B., Ariëns Kappers, J., and Jongkind, J.F., 1968, Innervation and fluorescence histochemistry of monoamines in the pineal organ of a snake *(Natrix natrix)*, J.neurovisc. Rel., 31:11.
Romijn, H.J., 1972, "Structure and Innervation of the Pineal Gland of the Rabbit, *Oryctolagus cuniculus* (L.), with some functional Considerations. A light and electron microscopic Investigation", Thesis, Free Univ., Amsterdam.
Romijn, H.J., 1973a, Structure and innervation of the pineal gland of the rabbit, *Oryctolagus cuniculus* (L.). I. A light microscopic investigation, Z. Zellforsch., 139:473.
Romijn, H.J., 1973b, Parasympathetic innervation of the rabbit pineal gland, Brain Res., 55:431.
Romijn, H.J., 1975a, The ultrastructure of the rabbit pineal gland after sympathectomy, parasympathectomy, continuous illumination, and continuous darkness, J. neural Transm., 36:183.
Romijn, H.J., 1975b, Structure and innervation of the pineal gland of the rabbit, *Oryctolagus cuniculus* (L.). III. An electron microscopic investigation of the innervation, Cell Tiss. Res., 157:25.
Romijn, H.J., 1976, The influence of some sympatholytic, parasympatholytic and serotonin-synthesis-inhibiting agents on the ultrastructure of the rabbit pineal organ, Cell Tiss. Res., 167:167.
Sathyanesan, A.G., and Sastry, V.K.S., 1982, Pineal innervation of the third ventricular ependyma in the Teleost, *Puntius sophore* (Ham.), J. neural Transm., 53:187.
Schapiro, S., and Salas, M., 1971, Effect of age, light and sympathetic innervation on electrical activity of the rat pineal gland, Brain Res., 28:47.
Schrier, B.K., and Klein, D.C., 1974, Absence of choline acetyltransferase in rat and rabbit pineal gland, Brain Res., 79:347.
Semm, P., and Vollrath, L., 1979, Electrophysiology of

the guinea-pig pineal organ: Sympathetically influenced cells responding differently to light and darkness, Neurosci. Lett., 12:93.

Semm, P., Schneider, T., and Vollrath, L., 1981, Morphological and electrophysiological evidence for habenular influence on the guinea-pig pineal gland, J. neural Transm., 50:247.

Suomalainen, P., Stress and neurosecretion in the hibernating hedgehog, Bull. Museum Comp. Zool., Harvard College, 124:271.

Trueman, T., and Herbert, J., 1970, Monoamines and acetylcholinesterase in the pineal gland and habenula of the ferret, Z. Zellforsch., 109:83.

Ueck, M., 1968, Granulierte marklose Nervenfasern in der Epiphysenregion von Anuren, Z. Zellforsch., 90:389.

Ueck, M., 1970a, Zur Ultrastruktur der Epiphysis cerebri der Vögel, Verh. zool. Gesellsch., 1969, Zool. Anz., Suppl., 33:509.

Ueck, M., 1970b, Weitere Untersunchungen zur Feinstruktur und Innervation des Pinealorgans von *Passer domesticus* L., Z. Zellforsch., 105:276.

Ueck, M., 1971, Strukturbesonderheiten der Anurenepiphyse nach prolongierter Osmierung und Anwendung der Acetylcholinesterase-Reaktion, Z. Zellforsch., 112:526.

Ueck, M., 1979, Innervation of the vertebrate pineal, Progr. Brain Res., 52:45.

Ueck, M., and Kobayashi, H., 1972a, Vergleichende Untersuchungen über acetylcholinesterasehaltige Neurone im Pinealorgan der Vögel, Z. Zellforsch., 129:140.

Ueck, M., and Kobayashi, H., 1972b, Secretory and neuronal elements of the avian pineal organ, Gen. comp. Endocrinol., 18:624.

Vigh-Teichmann, L., Vigh, B., and Aros, B., 1973, CSF contacting axons and synapses in the lumen of the pineal organ, Z. Zellforsch., 144:139.

Vollrath, L., 1981, The pineal organ, in: "Handbuch der mikroskopischen Anatomie des Menschen", Vol. VI/7, A.Oksche and L.Vollrath, eds., Springer, Berlin-Heidelberg-New York.

Wainwright, S.D., and Wainwright, L.K., 1978, Regulation of the diurnal cycle in activity of serotonin acetyl-transferase in the chick pineal gland, Canad. J. Biochem., 56:685.

Wake, K., Ueck, M., and Oksche, A., 1974, Acetylcholinesterase-containing nerve cells in the pineal complex and subcommissural area of the frogs, *Rana ridibunda* and *Rana esculenta*, Cell Tiss. Res., 154:423.

Wartenberg, H., and Baumgarten, H.G., 1969, Über die elektronenmikroskopische Identifizierung von noradrenergic Nervenfasern durch 5-Hydroxydopamin und 5-Hydroxydopa im Pinealorgan der Eidechse *(Lacerta muralis)*, Z.Zellforsch., 94:252.

Wight, P.A.L., and MacKenzie, G.M., 1970, Dual innervatio of the pineal of the fowl, Gallus domesticus, Nature (London), 228:474.

Wood, J.G., 1973, The effects of niamid and reserpine on the nerve endings of the pineal gland, Z.Zellforsch. 145:166.

Wurtman, R.J., Axelrod, J., Sedvall, G., and Moore, R.Y., 1967, Photic and neural control of the 24-hour norepinephrine rhythm in the rat pineal gland, J. Pharmacol. exp. Ther., 157:487.

Wurtman, R.J., Axelrod, J., and Kelly, D.E., 1968, "The Pineal", Academic Press, New York-London.

THE DIFFERENT CLASSES OF PROTEIC AND PEPTIDIC SUBSTANCES PRESENT IN THE PINEAL GLAND

Paul Pévet[1,2]

1) The Netherlands Institute for Brain Research Amsterdam, and 2) Dept. of Anatomy and Embryology, University of Amsterdam, The Netherlands

It is now well known that the pineal gland is principally involved in long-term adaptation of functions such as reproduction to environmental conditions. The mechanism by which the pineal acts on the gonadal axis is, however, not yet known. Recently it has been hypothesized that the 5-methoxyindoles (melatonin, 5-methoxytryptamine, 5-methoxytryptophan, 5-methoxytryptophol and 5-methoxyindole-3-acetic acid) synthesized by the pineal gland as well as by the retina, Harderian gland and intestine would be implicated in a system by which the pineal and some other organs such as the brain would be able to perceive, to differentiate and to integrate environmental information such as photoperiod, temperature, food, etc. In response, the pineal, then, would synthesize and release proteic/peptidic hormone(s) which would act on the reproductive axis (Pévet et al.,1981b; Pévet and Haldar-Misra, 1982). Considering this hypothesis, the study of the pineal active proteic/peptidergic fractions – the discovery of which dates back as far as the 1920's – appears to be of fundamental interest. According to the literature, the pineal appears to contain a large number of proteic or peptidic compounds, some of them being identified or partially characterized, others not (see table 1). However, although present in the pineal, these various compounds can be of one or more of the following origins:

1) synthesis in organs other than the pineal, but found there because they are present in cell elements such as nerve axons which connect the gland to different brain structures;

2) synthesis in organs other than the pineal, with a

Table 1. Proteins, peptides and proteic/peptidic fractions reported to be present in the pineal of mammals

Arg: arginine, Gly: glycine, Glu: glutamine, Leu: leucine, LHRH: luteinizing hormone releasing hormone, α-MSH: α-melanocyte stimulating hormone, MIF: α-MSH-release inhibitor factor, MW: molecular weight, PAG: pineal antigonadotrophin, PPIF: pineal prolactin inhibiting factor, PPRF: pineal prolactin releasing factor, Pro: proline, TRH: thyrotropin releasing hormone, TSL: threonylserinyllysine, Tyr: tyrosine, VIP: vasoactive intestinal peptide.

Name/terminology	Species	References
I IDENTIFIED PEPTIDES AND PROTEINS		
A) *Neurohypophysial principles*		
Neurophysins	cow, hamster, mouse, man, pigeon, rat, sheep	Reinharz et al.,1981*, Weindl & Sofroniew,1982
Oxytocin	cow, hamster, hedgehog, mouse, man, rat, sheep	Pévet et al., 1981a*, Geelen et al., 1981, Nürnberger & Korf,1981
Vasopressin	cow, hamster, hedgehog, mouse, man, rat, rabbit, sheep	Pévet et al., 1981a*, Geelen et al., 1981 Nürnberger & Korf, 1981
Vasotocin**	cat, chicken, cow, eel, gerbil, guinea pig, hamster, lizard, man, mouse, quail, rat, sheep, snake, trout	Pavel, 1979, Vaughan, 1981,* Vivien-Roels et al., 1979, 1981*
B) *Other hypothalamic hormones*		
LHRH ***	cow, pig, rat, sheep	Pévet, 1982a*, Pévet et al., 1981a, King & Millar, 1981
MIF	rat	Kastin et al., 1980

Table 1 (continued)

Name/terminology	Species	References
TRH***	cow, frog, pig, sheep, rat	Pévet, 1982a*
Somatostatin***	rat	Pelletier et al., 1975
C) *Various other peptides*		
ACTH (1-39)	rat	Vuolteenaho & Leppäl-uoto, 1981
Angiotensin I	rat	Coculescu et al., 1977
Angiotensin II	rat	Changaris et al., 1979
β-Endorphins	rat	Rossier et al., 1977; Vuolteenaho et al., 1980
Enkephalin	rat	Rossier et al., 1979
α-MSH	rat	Pévet et al., 1982a*
Substance P	cat, human	Powel et al., 1977
VIP	cat, pig, rabbit	Uddman et al., 1980;
TSL	cow	Orts et al., 1980
II PARTIALLY IDENTIFIED PEPTIDES AND/OR PROTEINS		
E5 (MW ≃ 2000, peptide of 14 amino-acids. Ring with an S-S bridge. Lateral chain of 5 amino-acids with Pro-Arg-Gly(NH2) for the N-terminal amino-acids)	cow	Neacşu, 1971, 1972
p-Glu-χ-Leu-Arg-Pro-Gly(NH2)	sheep	Millar et al., 1981
Pinealin (MW ≃ 10.000, 14 amino acids)	cow	Milcou et al., 1963a
χ-Pro-Arg-Gly(NH2)	cow, sheep, rat	Pévet et al., 1981c
χ-Tyr-Pro-Leu-Gly(NH2)	rat	Kastin et al., 1981

(continued)

Table 1 (continued)

Name/terminology	Species	References
III NON IDENTIFIED PROTEIC OR PEPTIDIC FRACTIONS ****		
Anestrine	cow	Bianchi & Osima, 1960
Altschule's pineal extract	cow	Altschule, 1957
Anovulin	cow	Chazov et al., 1972
Crinofizin	cow	Milcu et al., 1976
Epiglandol	cow	Hofstätter, 1914
Epiphysan	cow	Engel, 1935; Hofstätter, 1938
Epifizhormon/epiphysormone	cow	Parhon et al., 1940
Epithalmin	cow	Dil'man & Anisimov, 1975
Extracts IIa, IIIa	cow	Ota et al., 1975
Fraction F_2 (MW < 1500)	sheep	Ebels et al., 1965; Moszkowska et al., 1974
Fraction F_3 (MW < 700 when purified)	sheep	Moszkowska et al., 1973
Fractions A_1 and A_3	cow	Vaughan et al., 1974; Blask et al., 1976
Fraction III (MW ≃ 500-1000)	cow	Thieblot et al., 1979
PAG (MW ≃ 500-1000)	cow, human	Benson et al., 1972; Matthews et al., 1971
Peptide II F (MW ≃ 3000)	cow	Rudman et al., 1970
PPRF	cow	Blask et al., 1976
PPIF	cow	Blask et al., 1976
PP7,2	sheep	Noteborn et al., 1982
UMO5R (MW ≃ 500-100)	cow, sheep	Ebels et al., 1973
XM300R PP7,2	sheep	Noteborn et al., 1982

* Review article to which we refer for details and references.
** Its presence is doubtful in mammalian species (see text).
*** The large amount found by some authors is doubtful (see text).
**** Many fractions isolated from the pineal and termed 'pineal extracts' have not been quoted in this table.

secondary uptake by the gland from the general circulation;
3) synthesis in the pineal by specific pineal cells.
Such a classification of the proteic/peptidergic compounds into three different classes is very important to consider. Indeed, although all of them can be isolated from the gland, only the proteic/peptidic compounds of class three can be considered as true pineal hormones.

The present review deals with the results obtained the last few years in studies concerned with this problem, using different techniques such as electron microscopy, ultracytochemistry, immunocytochemistry, radioimmunoassay and bioassays.

I. PEPTIDES AND PROTEINS OF THE FIRST CLASS

Complex nervous connections between the central nervous system and the pineal gland are known to exist in vertebrates (see the articles by Ariëns Kappers and Moore, this volume). Thus, neurotransmitters such as for example noradrenaline and acetylcholine (at least in some species) are present in nerve endings located in the pineal and consequently can be isolated from pineal extracts. However, they are not and have never been considered as pineal factors. Similarly, since it is known that in the brain neuropeptides may act as neurotransmitters (Buijs et al., 1982), the existence of peptidergic nerve fibres might explain the presence of neuropeptides in the pineal. Vasopressin (AVP) and oxytocin (OT) have been detected in the pineal gland of numerous mammalian species (see table 2) but at a relatively low concentration. Their presence has been demonstrated – by the use of the unlabeled antibody enzyme method – to be due, at least in the rat (Buijs and Pévet, 1980) and the cow (Pévet, unpublished data), to the presence in the pineal of AVP- or OT-containing fibres. In the rat these fibres are extrahypothalamic fibres, originating – at least partly – from the magnocellular AVP- and OT-producing nuclei which run via the subcommissural organ and the deep pineal (if present) or via the habenular commissure into the pineal stalk, terminating in the rostral part of the pineal (Buijs and Pévet, 1980). This work has recently been confirmed in the hedgehog by Nürnberger and Korf (1981). These authors indeed have observed that the perikarya giving rise to these AVP- or OT-containing fibres were located in the dorsal part of the paraventricular nucleus, the fibres entering the pineal also via the habenular commissure. The extrahypothalamic AVP- or OT-containing fibres explain also the presence of the two neurophysins described in the pineal

Table 2. Neurohypophysial Hormone Concentrations in the Mammalian Pineal
G = gland; n.d. = not detectable; Pr = protein, WT = wet tissue

Species	AVP	OT	References
Cow	≃ 1.6 pg/mg WT	≃ 0.75 pg/mg WT	Fisher and Fernstrom, 1981
	3.7 $\pm$ 0.7 pg/mg WT	3.7 $\pm$ 1.7 pg/mg WT	Dogterom et al., 1980
	407 $\pm$ 66 pg/G	238 $\pm$ 41 pg/G	Fernstrom and Fisher, 1980
Hamster	Trace	Trace	Pévet et al., 1981a
Human	306.4 $\pm$ 27.7 pg/G	386 $\pm$ 42.1 pg/G	Geelen et al., 1981
Mouse	Trace	Trace	Pévet et al., 1981a
Rabbit	0.032 $\pm$ 0.008 pg/µg Pr	-	Negro-Vilar et al., 1980
Rat	0.186 $\pm$ 0.019 pg/µg Pr	-	Negro-Vilar et al., 1980
(Sprague-Dawley	≃ 1.99 pg/mg WT	≃ 2.46 pg/mg WT	Fisher and Fernstrom, 1981
(Wistar)	2.7 $\pm$ 0.8 pg/mg WT	1.3 $\pm$ 0.8 pg/mg WT	Dogterom et al., 1980
(Brattleboro)	n.d.	8.7 $\pm$ 3.7 pg/mg WT	Dogterom et al., 1980
Sheep	19.74 $\pm$ 2.3 pg/mg WT	35.3 $\pm$ 7.1 pg/mg WT	Reinharz et al., 1981

of different mammalian species (details in Reinharz et al., 1981). Neurophysins, indeed, are cystein-rich proteins which in the pineal, like in the neurohypophysis, are associated with the neurohypophysial hormones (Pévet et al., 1980c).

Vasotocin (AVT) is one of the neurohypophysial hormones in the nonmammalian vertebrate species. For some authors (review in Pavel, 1979), AVT would be also a hormone present in and synthesized by the pineal gland of mammals. This conclusion now appears to be very improbable (see details in Pévet, 1981b, 1982a,b; Pévet et al., 1981a, and in the next subheading), but AVT is really present in the pineal of some nonmammalian vertebrate species (table 3). Vivien-Roels et al. (1981) have suggested that this presence could be due — in analogy to mammals — to the presence in the pineal of extrahypothalamic fibres containing the neurohypophysial hormone (here AVT). Such an interpretation which, to date, has not yet been proved, seems very likely because Weindl and Sofroniew (1982) have detected by means of immunocytochemistry neurophysin-containing nervous fibres in the pineal of the pigeon.

The existence of other peptidergic fibres in the pineal explains also the presence of numerous other neuropeptides. For example, Uddman et al. (1980) and Møller et al. (1981) have demonstrated that the VIP immunocytochemically detected in the pineal of the cat, pig and rabbit, was exclusively found in nerve fibres running in the pineal and especially in the pineal stalk. Following Hansen and Karasek (personal communication) the substance-P detected by radioimmunoassay, for example in the human pineal (300-500 pg/mg tissue; Powel et al., 1977), would be also located exclusively in nerve fibres and endings, at least in the cat.

However, these results cannot be generalized to all neuropeptides. For some of them indeed, the problem appears to be more complicated. For example, as will be explained under the next subheadings, the determination of the exact origin of the LHRH-immunoreactivity found in the pineal is a difficult problem. However, the presence in the pineal of some extrahypothalamic LHRH-containing fibres, although not yet clearly demonstrated, is very probable. Indeed, Barry (1979), in the female squirrel monkey, has described an LHRH-reactive perikaryon in the habenular commissure or in the rostral part of the pineal, a perikaryon which might send fibres into the pineal. Similarly, although not any immunocyto-

Table 3. Vasotocin Concentration in the Pineal of Nonmammalian Vertebrates

G = gland; Pr = protein

Species	AVT	References
Birds		
Chicken	0.180 ± 0.04 pg/μg Pr	Negro-Vilar et al., 1980
	≃ 9.1 pg/G	Fisher and Fernstrom, 1981
	31 ± 5 pg/G	Vivien-Roels et al., 1981
	312 ± 39 pg/G	Fernstrom et al., 1980
Quail	44 ± 21 pg/G	Vivien-Roels et al., 1981
Reptiles		
Lizard (Lacerta agilis)	210 ± 74 pg/G	Vivien-Roels et al., 1981
Snake (Natrix tessala)	245 ± 89 pg/G	Vivien-Roels et al., 1981
Tortoise (Testudo hermanni)	181 ± 118 pg/G	Vivien-Roels et al., 1981
Amphibians		
Frog (Rana esculenta)	324 ± 27 pg/G	Vivien-Roels et al., 1981
	≃ 8.5 pg/G	Fisher and Fernstrom, 1981
Fishes		
Salmo gairdneri	58 ± 15.7 pg/G	Vivien-Roels et al., 1979; Holder et al., 1979
	124 ± 17 pg/G	Vivien-Roels et al., 1981
Salmo gairdneri Richardson	30.7 ± 7.2 pg/G	Holder et al., 1979; Vivien-Roels et al., 1979
Salmo trutta morpha fario	34.8 ± 9.1 pg/G	Holder et al., 1979; Vivien-Roels et al., 1979
Salvelinus frontalis	24.0 ± 6.7 pg/G	Holder et al., 1979; Vivien-Roels et al., 1979
Anguilla anguilla	32.8 ± 6.9 pg/G	Holder et al., 1979; Vivien-Roels et al., 1979

chemical study has confirmed the radioimmunological observations of White et al. (1974) and of Jackson et al. (1977) and though, at the contrary, Youngblood et al. (1979) have chemically demonstrated that the immunoreactivity observed in bovine and sheep pineal was due to an unknown compound different from the genuine peptide, it is not possible to exclude completely the presence of some TRH-containing fibres in the pineal.

Concerning somatostatin, Pelletier et al. (1975) have noted that in the rat the reaction was only present close to the capillaries. Thus, here also, although it is known that some of the observations are due to cross reaction of the antibody used with an unknown compound (see after), the presence of some somatostatin-containing fibres is a possibility.

α-MSH is known as a pituitary hormone but is also produced by the nervous system (see review in Swaab, 1982). Consequently we have also to examine the possibility that the α-MSH found in the pineal would be also a neurotransmitter present in nerve fibres or endings. To date, however, not any information permits to support this idea (see after for the problem of the presence of α-MSH in the pineal).

II. PEPTIDES AND PROTEINS OF THE SECOND CLASS

The pineal gland is an endocrine organ capable of synthesizing and releasing material which can affect numerous organs and systems and which, in turn, is influenced by a variety of stimuli and signals. Especially the pineal, like all endocrine organs, appears to be under feed-back control by target tissues (see review in Ariëns Kappers, 1978). This peculiar aspect of pineal physiology is important to consider here. Indeed, a system feedback operating between the hypothalamus or the hypophysis and the pineal would implicate that hypothalamic or hypophysial hormones are transported via the general circulation into the pineal. Looking now at the list of identified proteins and peptides found in the pineal (table 1), it appears that some of them are hypothalamic and hypophysial principles. Consequently, the possibility that the presence of such hormones within the pineal would be due to an accumulation of circulating hormones has to be taken into consideration.

This aspect of the pineal peptide/protein problem has not been deeply investigated. Clear results, however, have been obtained with two hypothalamic hormones.

After intravenous injection of labelled MIF, Redding et al. (1973) have observed that the radioactive hormone was especially concentrated in the pituitary and the pineal. A similar observation was made after injection of labelled LHRH by Redding and Schally (1973), Admundson and Wheatons (1979) and Trentini et al. (1980). The last mentioned group noted also that the pineal gland accumulated ^{3}H-LHRH in an amount significantly higher than the pituitary gland. These experiments which demonstrated that the pineal, at least in rat, is able to concentrate in a notable amount circulating MIF and LHRH, can easily explain the presence in the pineal of these hypothalamic hormones. Such a phenomenon of accumulation, however, does not permit to explain all results published in the literature and especially the finding that the pineal would contain and synthesize a very large amount of 'hypothalamic hormones' (White et al., 1974). Cross reactions of the antibodies used with unidentified compounds could be also responsible of some of these observations (see next sub-heading). Moreover, as explained before, LHRH for example is probably also present in extrahypothalamic nervous LHRH-containing fibres.

The physiological significance of such a phenomenon of accumulation of circulating hormones is not known but at least it demonstrates that a feed back system operates between the hypothalamus and the pineal. Moreover, this raises the question of the possibility of a similar mechanism for other hormones such as hypophysial ones, a phenomenon which would explain the high amount of α-MSH detected in the pineal by some authors (Olivier and Porter, 1978; Vaudry et al., 1978; O'Donohue et al., 1980). As after hypophysectomy the amount of immunoreactive α-MSH does not change (Vaudry et al., 1978; O'Donohue et al., 1980), this presence of α-MSH in the pineal cannot be attributed to an accumulation of circulating hypophysial hormone. It is very probable that these results are due to a cross reaction of the used antibodies with an unknown compound (see below). However, as in intact as well as pinealectomized rat, exogenously administered H^{3}-α-MSH was demonstrated to be also accumulated in the pineal (Kastin et al., 1976), the presence of a small amount of accumulated α-MSH, TRH or other peptidic of proteic hormones is probable and at least cannot be excluded. Interestingly, Vuolteenaho and Leppäluoto (1981) who have detected ACTH (1-39) in the rat pineal (42-166 fmol/gland), concluded also that this ACTH in fact originated from the pituitary.

III. PEPTIDES AND PROTEINS OF THE THIRD CLASS

Processes involved in the synthesis and release of proteic/peptidic substances have been identified at the ultrastructural level (see below). The existence of proteins or peptides of the third group thus cannot be discarded. To date, however, none of these substances have been structurally identified and by consequence their studies are always indirect ones.

A. *Fluorescent Histochemical Studies*

The pinealocytes of the hedgehog (Pévet, 1976), the mole (Pévet, 1976; Pévet et al., 1976), the rabbit (Smith, 1972; Smith et al., 1972a,b) and the rat (Pévet et al., 1975; Smith et al., 1975) contain a yellow autofluorescent substance which was demonstrated to be, in the rabbit, a protein containing much tryptophan (Smith, 1972). A comparison of the fluorescent histochemical observations with the ultrastructural ones in the mole (Pévet, 1974, 1976; Pévet et al., 1976) suggests that this autofluorescent material could be identical with the proteinaceous material (or with part of it) which is located in the cisterns of the rough endoplasmic reticulum (the 'ependymal-like secretory process', see below). This opinion is supported by the fact that in rat after castration or injection of gonadotropins, both experiments which provoke an increase in number of pinealocytes containing autofluorescent material (Pévet, 1976; Pévet et al., 1975; Smith et al., 1975), an increase in number of vacuoles characteristic of the ependymal-like secretory process has been described (Karasek and Marek, 1978; Karasek et al., 1976a,b). This, however, remains to be demonstrated.

B. *Immunocytochemical studies*

(1) *LHRH Immunoreactivity*

In 1974 White et al., using a radioimmunoassay, demonstrated the presence of a very large amount of LHRH in the pineal gland of the cow, pig and sheep and concluded that this gland might be considered as a supplemental source of hypothalamic releasing hormone. This result was the center of a large controversy. Some authors denied its presence in the ovine pineal (Carson et al., 1977) while others (Duraiswani et al., 1976; Gradwell et al., 1976; Joseph, 1976; Millar and Tobler,

1981; Wheaton, 1980), although confirming the presence of detectable amounts of LHRH, noted that the LHRH-immunoreactivity found in the pineal of cow, rat and sheep was considerably lower than that in the hypothalamus. New results obtained in the last few years by different teams, however, now permit to have a better idea on this LHRH problem. The low LHRH-immunoreactivity observed in the pineal is in fact due, as was demonstrated in the sheep by a combination of immunological and chemical techniques by King and Millar (1981), to the presence of the genuine peptide. This presence is probably due, as explained before (subheadings I and II), to the existence in the pineal of LHRH-containing extrahypothalamic fibres and/or to a phenomenon of accumulation in the pineal of circulating LHRH. Immunocytochemically, however, an intense positive reaction has also been obtained in the rat pineal gland (Pévet et al., 1980b). In this study, three different antibodies against LHRH were used. All stained equally well both rat hypothalamic tissue and synthetic LHRH-containing agarose beads. In the pineal, however, only one gives positive results (see details in Pévet et al., 1980b). This difference between the antibody potencies in the rat pineal, the rat hypothalamus and the synthetic LHRH-containing agarose beads suggests strongly that the intense reaction in the pineal is due to the presence of a compound different from LHRH but recognized by the antibody. This cross reaction of some anti-LHRH with an unknown compound, a LHRH-like compound, could explain the high LHRH immunoreactivity found in the pineal by White et al. (1974). Indeed, if we use the same antibody which has given us a positive reaction in immunocytochemical studies, in a radioimmunoassay system (Arimura, personal communication), a large amount of immunoreactive LHRH similar to that described by White et al. (1974) is detected in the pineal. The presence of such a LHRH-like compound has recently been confirmed immunocytochemically by Piekut and King (1981) and chemically by Miller et al. (1981), who isolated from bovine pineal a substance recognized by different antibodies against LHRH but structurally different from the hypothalamic LHRH.

However, to date the nature of the immunocytochemically stained LHRH-like compound which, as demonstrated by means of organ culture (Pévet et al., 1980a), is synthesized in the pineal itself, has not been determined and more specially, we don't know whether this LHRH-like compound is identical to the peptide immunologically related to LHRH detected by Millar et al. (1981) in the bovine pineal.

(2) *α-MSH, Somatostatin and TRH Immunoreactivity*

α-MSH is a hormone synthesized primarily in the intermediate lobe of the pituitary and also in the central nervous system (for details see Swaab, 1982). A large α-MSH immuno- and bioactivity has also been detected in the rat pineal by different authors who assume that this immuno- and bioreactivity is caused by the presence of α-MSH (O'Donohue et al., 1979, 1980; Oliver and Porter, 1978; Vaudry et al., 1978). Moreover, as hypophysectomy does not deplete pineal α-MSH content, O'Donohue et al. (1980) concluded that very likely this α-MSH is synthesized by the pineal itself. Using antibodies against α-MSH, cells have been immunocytochemically stained in the rat pineal (Pévet and Swaab, 1979; Pévet et al., 1980b). After 6 days culture in an *in vitro* system with synthetic medium, some cells in the pineal explant are immunocytochemically stained (Pévet et al., 1980b). This means that the immunocytochemical result is not due to the accumulation of circulating α-MSH (see subheading II) but to the synthesis by the pineal of an α-MSH immunoreactive compound. This immunocytochemical observation which confirms the observation made by the earlier cited authors does not permit to conclude with them that α-MSH is present in the pineal gland. Indeed, the three antibodies against α-MSH used in this study, although they stained equally well both α-MSH containing agarose beads and cells of the intermediate lobe of the rat pituitary, have quite different staining properties in the pineal (see details in Pévet and Swaab, 1979). These differences in antibody potencies permit the conclusion that the α-MSH immunoreactivity observed in the rat pineal is due to the presence of an α-MSH-like compound different from pituitary α-MSH, a conclusion which had already been made on the basis of bioassays by Rudman and Scott (1975).

Somatostatin has been immunologically detected in the rat pineal gland (Pelletier et al., 1975). However, this result has been later contested by Berolowitz et al. (1978). Probably, as suggested by the immunocytochemical study of Pévet et al. (1980b) in which only one of the seven antibodies against somatostatin tested gives a positive reaction in the pineal while all of them react in the different control tests, this discrepancy is due to the fact that it is a somatostatin-like compound different from the genuine peptide which is present in the pineal. This is probably also the case for the TRH immunoreactivity detected by White et al. (1974) in the pineal of cow, pig and sheep. Indeed, Youngblood et al. (1979) have chemically demonstrated that the immunoreactivity observed in sheep bovine pineal was due to an unknown compound different from TRH.

(3) *The Vasotocin Controversy*

In 1963 Milcou, Pavel and Neacsu discovered in extracts of bovine pineal glands a peptide with pressor and oxytocic activities. Although "unfortunately no synthetic arginine vasotocin was available as standard", they indicated that the chromatographic characteristics of this peptide were similar to those of vasotocin and concluded that the pineal peptide was "arginine vasotocin, or a hormone with similar structure". This result was immediately checked by Ebels et al. (1965) who were not able to confirm the presence of AVT in the pineal of sheep and cow. One year later, however, Pavel and Petrescu (1966), after a comparative study of the antigonadotrophic effects of the purified peptide and of synthetic AVT, concluded again that the peptide was identical with AVT. This result once more was contested by Neacşu (1972) who, interestingly, was co-author with Pavel of the original 1963 paper. This author, after a chemical analysis, concluded that the pineal peptide (termed by him "fraction E5") was different from AVT and was composed of 14 amino-acids. The last-mentioned study was never considered, probably because two years earlier Cheesman (1970) and Cheesman and Farris (1970) had isolated and structurally identified AVT from bovine pineal extracts. This work of Cheesman – which to date has not yet been confirmed – was considered as the definite proof of the existence of pineal AVT. The vasotocin controversy was thus provisionally stopped and in the next few years Pavel and his group, still using the bioassay approach, "demonstrated" that the pineal gland of all mammals so far examined, including man, contained AVT (or lysine vasotocin, in pigs); that AVT was synthesized by the ependymal cells located in the pineal recess and by the cells of the subcommissural organ (SCO), and that AVT was released into the cerebrospinal fluid (CSF) (see details and references in Pavel, 1978). At the same time, the physiological properties of AVT were extensively investigated by the physiologists (see review in Vaughan, 1981, and Vaughan and Blask, 1978) and the results obtained were such that at the end of the seventies – and although in the same period AVT was also considered as a neurohormone of the fetus implicated in the maintenance of its own intra-uterine water balance (Vizsolyi and Perks, 1976) – practically no pinealogist questioned the presence of AVT in the mammalian pineal gland. It is interesting to note here that it is probably this general consensus which is at the origin of the new revival of the vasotocin controversy. Indeed, the acceptation of the concept that AVT is a pineal antigonadotrophic agent has led to intensified efforts to measure the AVT content in tissues and body fluids. The bioassays being particularly tedious techniques, especially for large numbers of samples, radioimmunoassays were developed. Different assays had been made already in the mid-seventies, but none of them was really specific for AVT, the antibodies used being antibodies against AVP or OT (Rosenbloom and Fisher, 1975; Legros et al., 1976).The first radioimmunoassay for AVT using an AVT-antiserum was developed by Dogterom et al. (1978, 1979, 1980). Using this radioimmunoassay, these authors were not able to find an AVT immunoreactivity in the pineal, subcommissural organ or fetal neurohypophysis of any of the mammalian species studied and they concluded that the AVT biological activity found by some authors (see above) was probably due to a compound different from AVT itself. Negro-Villar et al. (1979) who had also developed a specific radioimmunoassay for AVT, were also unable to detect AVT immunoreactivity in the pineal of the rat and rabbit and they strongly questioned the role of AVT as a pineal hormone in mammals. These observations were quickly contested by Fernstrom et al. (1980) who, having also developed a radioimmunoassay specific for AVT, concluded that AVT is present in the mammalian pineal. One year later, however, the last-mentioned authors, having refined their technique, reevaluated their results and concluded that AVT is probably not present in the mammalian pineal (Fisher and Fernstrom, 1981). At the same time Bowie also reconsidered her original immunocytochemical work which had led her to accept the presence of AVT in the rat pineal (Bowie and Herbert, 1976) and concluded that this result was due to a cross reaction of the antibody used with an albumin-like substance (Bowie, 1980).

It thus appears that the long standing vasotocin controversy has now, at least apparently, been solved, the more so since when the physiological effects of AVT are considered, it is possible to note, with Vaughan, 1981), that "... where tested, AVP (which is really present in the pineal) has many of the same effects as AVT". It should be realized, however, that these negative results obtained with radioimmunoassay do not contest the reality of the presence of an AVT-biological activity, a presence which might be due to either vasopressin and oxytocin, two peptides present in the pineal of mammals (see subheading I) and which, taken together, would react in different bioassays or to a compound different from AVT itself but related to it, an idea already suggested by Neacșu (1972) and which could eventually explain the present "inexplicable" chemical results of Cheesman and Farris (1970).

Using the antibody against AVT raised by Fernstrom, a positive staining similar to that described by Bowie and Herbert (1976) has been obtained immunocytochemically in the rat pineal (Pévet et al., 1980b). Moreover, the relative number of immunostained cells in the pineal of 3 week old rats appears to be larger than in that of adult rats. These immunocytochemical results which have to be related to the observation that, as compared to adult rats, a higher AVT biological activity is found in the pineal of young rats (Pavel et al., 1975), and to the positive radioimmunological results obtained in the rat pineal by Fernstrom et al. (1980), indicate clearly the presence of an AVT immuno- and bioactivity. During this immunocytochemical study four antibodies which all stained equally well tissue of avian neurohypophysis, of a turtle hypothalamus and synthetic AVT-containing agarose beads, were used. Of these antibodies only that raised by Fernstrom gave positive results. This difference between the antibody potencies in the rat pineal, the avian neurohypophysis, the tortoise hypothalamus and in the synthetic AVT-containing agarose beads has permitted the conclusion that the reaction obtained in the pineal was due to the presence of an AVT-like compound different from AVT itself to which the antibody of Fernstrom would cross-react. Interestingly, this interpretation is now confirmed by Fisher and Fernstrom (1981) who, after reconsidering their first observation (Fernstrom et al., 1980), concluded that their results were probably due to antiserum nonspecificity. This interpretation is also supported by the fact that using our radioimmunoassay for AVT, vasotocin has never been detected in the mammalian pineal while, using the bioassay developed by Holder and Guerné (1976), a biological activity similar to that of AVT has been found in all species studied (Pévet, 1982a; Holder et al., 1982). It agrees also with the conclusion raised by Neacșu in 1972 that the biological activity of AVT in the mammalian pineal gland is due to a compound different from AVT itself.

Some preliminary results have permitted us to suggest that this compound could be a peptide possessing the same Pro-Arg-Gly(NH_2) tripeptidic carboxyterminal end as vasotocin (Pévet et al., 1979, 1980a,c; Pévet, 1981b, 1982a).

(4) *Perspectives*

Looking at the results published in the literature it appears that very often a close correlation exists between results obtained with the bioassays, the radioimmunoassays and the immunocytochemical studies. For example, as explained before, using either the bioassay technique or immumocytochemistry, a higher pineal "AVT-like" content is found in the young rat than in the adult one. Positive immunocytochemical results were obtained with antibodies which give also positive result in a radioimmunoassay system. It would thus be tempting to "logically" conclude that the compound responsible for a given biological activity is the compound which is recognized by the corresponding antibody and consequently to decide to use these antibodies for further isolation and purification of the compounds concerned.

The results obtained in immunocytochemistry, however, oblige us to be careful with such a conclusion. Indeed, surprisingly, similar staining patterns were obtained with all antibodies tested, anti-angiotensin-I (Changaris et al., 1979), anti-AVT (Bowie and Herbert, 1976; Pévet et al., 1980b), anti-α-MSH (Pévet et al., 1980b), anti-LHRH (Pévet et al., 1980b; Piekut and Knigge, 1981), anti-somatostatin (Pelletier et al., 1975; Pévet et al., 1980b), anti-UMO5R (Pévet et al., 1980b). Moreover, by means of serial section it has been demonstrated in the rat that different antibodies stained the same pineal elements near the perivascular space (Pévet et al., 1980b). It appears thus that the α-MSH-like compound, the LHRH-like compound, the somatostatin-like compound and the AVT-like compound detected immumocytochemically are all present in the same structure. More exactly, as it was also demonstrated that the staining potentialities of all antibodies were affected after adsorption to heterologous peptides, it seems that all antibodies react with a unique compound, possibly a protein or a protein complex. The nature of this compound of which, as demonstrated by means of *in vitro* techniques, the synthesis by the pineal is controlled by sympathetic innervation (Pévet, unpublished results) "could be a pineal endocrine principle, an enzyme, or any structural protein" (Pévet et al., 1980b). Recently Rix et al. (1981) using different antibodies against angiotensin, have obtained a staining pattern similar to that already described. Examining adjacent semithin and ultrathin sections by immunocytochemistry and electronmicroscopy respectively, they have moreover demonstrated that the staining reaction was located in the flocculent material present in the extracellular perivascular compartment. Although in our own study we have been able to observe some stained cell bodies (for example fig. 3 in Pévet et al., 1980b), it is very probable that the observations of Rix et al. (1981) are also applicable, at least partly, to our own results. The nature of the flocculent material, its origin and especially its responsibility in a possible false positive immunoreaction have thus to be determined now before pursuing this approach of the pineal peptide problem.

B. *Ultrastructural Aspect*

The observations reported in this part concern exclusively the major cell type of the pineal: the pinealocyte or its phylogenetic homologues when non-mammalian vertebrates are concerned. More than one way of production and release of proteinaceous substances is taking place in these cells. Indeed, two processes involved in the synthesis and release of pineal proteic/peptidic agents have been identified at the ultrastructural level and are characterized either by the formation of granular vesicles (GV) by the Golgi saccules, or by the formation of vacuoles containing flocculent material of moderate electron density by the cisterns of the rough endoplasmic reticulum (RER), the so-called ependymal-like secretory process. For an exact definition of these two processes and details on the demonstration of their proteic nature as well as on their possible relationships, we refer to previous review articles (Pévet, 1979; 1981a,b; 1982a,b).

The identification of these secretory processes permits now by a qualitative and quantitative electron microscopic analysis of the different cell organelles, to study, although their chemical nature has not yet been determined, the regulatory mechanisms of the pineal protein/peptide production and especially to identify the substances (hormones, neurotransmitters, etc.) involved in the control of their production.

(1) *The influence of noradrenaline on the processes of protein/peptide secretion*

The idea that the activity of the pineal gland is controlled by the sympathetic innervation via its neurotransmitter release is now well accepted. This opinion, however, is based essentially if not exclusively on biochemical and physiological results concerning indoleamines, especially melatonin. Information on a possible effect of the sympathetic innervation on the processes of protein/peptide secretion is very rare.

In *in vitro* experiments after application of noradrenaline (NA) to the medium, an increase in the formation of GV by the Golgi apparatus has been demonstrated in the pinealocytes of the hamster, rabbit and rat (Karasek, 1974; Romijn and Gelsema, 1976; Haldar-Misra and Pévet, 1982a). NA appears thus to stimulate the formation of GV at least in these three species. This observation is confirmed by different other results. In rat, for example, pineal tissue grafted under the kidney capsule, a situation in which the pineal is completely denervated, a drastic decrease in number of GV was observed (Aguado et al., 1977). Moreover, after pineal denervation by superior cervical ganglionectomy, Romijn (1975) and Lin et al. (1975) described a decrease in number of GV in the pinealocytes of the rabbit and the hamster respectively. This observation however is contested by Sheridan (1975) who observed in the hamster under apparently the same experimental conditions as well as after destruction of the noradrenergic innervation by 6-hydroxydopamine injections an increase in number of GV. To date there is no explanation for this discrepancy.

In *in vitro* experiments, after application of NA to the medium, a decrease in the number of GV is observed in mouse pinealocytes (Haldar-Misra and Pévet, 1982a). Thus, contrary to the observations in the hamster, rabbit and rat (see above), NA appears in the mouse to inhibit the process of protein secretion characterized by the formation of GV. Other findings in the mouse, demonstrating a direct correlation between a high NA content in the pineal and a low number of GV in the pinealocytes or, on the contrary, between a low pineal NA concentration and a large number of GV, support this interpretation. Benson and Krasovich (1977), for example, have observed a circadian rhythm in the number of GV in the ending process of the mouse pinealocytes characterized by a high number during the light phase and a low number during the dark phase. The low number of GV is thus directly related to the nocturnal discharge of NA from the sympathetic nerve terminals. Moreover, Krasovich and Benson (1979) have also observed an increase in number of GV after pharmacological blockage of NA synthesis.

All these results indicate clearly that NA has a clear influence on the process of protein/peptide secretion characterized by the formation of GV. This effect, however, appears to be species specific (for more details see Haldar-Misra and Pévet, 1982a) and this could possibly explain some contradictory observations reported in the literature when the relationship light/dark periodicity and pineal function is considered.

Concerning the ependymal-like secretory process the observations are very scarce. Haldar-Misra and Pévet (1982a) have observed in the cultured pineal of rat that the addition of NA to the medium led to the formation of an extended complex of cisterns of the RER and of vacuoles containing flocculent material originating from the cisterns. This result, however, needs a more detailed analysis.

From the above data it can be concluded that NA strongly influences the processes of protein/peptide secretion in the pinealocytes. This, therefore, emphasizes again the importance of the sympathetic innervation for the activity of the pineal (see review in Ariëns Kappers, 1981). However, as NA induces the formation of melatonin which, in turn, can also act on the process of protein secretion (see below), it should be realized that we are facing a very complex system of regulation.

(2) *The influence of hormones of the sexual axis on the process of protein/peptide secretion*

The proteic/peptidic agents synthesized via the processes studied at present are believed to act on the resproductive axis (see Introduction). Consequently, as the function of the pineal, like that of every endocrine organ, depends on feedback systems (see details in Ariëns Kappers, 1978), an effect of the hormones of the sexual axis is very probable. Many studies in the literature deal with effects of sexual and hypophysial hormones on the pineal (see review in Vollrath, 1981) but surprisingly few concern specifically an effect on the processes of protein/peptide secretion.

After castration Clementi et al. (1965) and Karasek et al. (1976a) observed an increase in number of GV in the rat pinealocytes. According to the last-mentioned authors, the increase was due to the plasmatic elevation of gonadotropins induced by gonadectomy. This interpretation was later demonstrated to be true. Administration of gonadotropins to rats provoked indeed similar changes either *in vivo* (Karasek and Marek, 1978) or *in vitro* (Karasek et al., 1978). A similar gonadotrophin-induced increase in number of GV in the pineal of snakes and lizards has also been demonstrated by Petit (1976) and Vivien (1965). Karasek et al. (1982b) have also observed a very large increase in number of GV (100%) in the pinealocytes of the female rat at days 21 and 22 of pregnancy. They related this increase to the hormonal changes which occur during the late stage of gestation. All these results led Karasek to suggest the existence of a feedback mechanism between the pineal gland and the hypophysis: "This mechanism may consist in enhancement of the synthetic activity of the pineal gland by gonadotropins causing a subsequent increased elaboration of pineal antigonadotropic factors" (Karasek, 1979). Yet, this concept is supported by new findings by these authors. Indeed, Karasek et al. (1982a) observed that when compared to phenotypically normal mice, the pinealocytes of the snell dwarf mouse (a mutant with hereditary hypopituitarism) contain a very low number of GV but that this number was increased after a pituitary graft.

Considering other hormones of the gonadal axis and especially those like LHRH, oestrogen, progesterone and testosterone which are known to be taken up by the pineal (see details and references in Vollrath, 1981), relevant studies are even more rare. It appears, however, that LHRH does not act directly on this process of protein secretion. Indeed, in *in vitro* experiments Karasek et al. (1978) were not able to observe any effect of this peptide. After administration of oestradiol, morphological features of increased pinealocyte activity have been also observed (Karasek and Marek, 1980).

The activity of the ependymal-like secretory process seems also to be especially influenced by the endocrine activity of the gonadal axis. In the mole a parallelism between this synthetic activity of the pinealocytes and the hormonal production of the gonadal axis has been established (Pévet and Smith, 1975). Formation of accumulations of proteinaceous material in the pinealocytes of the garden dormouse has been observed during the natural period of anoestrus (winter, during hibernation) and also during experimentally induced anoestrus (Roux et al., 1977, 1981). In the hedgehog, vacuoles containing flocculent material have been described in the cisterns of the RER, both at the beginning of the period of sexual quiescence (Pévet, 1976; Pévet and Saboureau, 1973) and at the beginning of the period of an experimentally provoked decrease of sexual activity (Pévet and Saboureau, 1974). After castration (Karasek et al., 1976a,b) or after administration of gonadotropic hormones (Karasek and Marek, 1978) an increase in the number of vacuoles containing flocculent material is observed in the rat pinealocytes. Very interesting, in the last days of pregnancy when, as noted before, a strong increase in the number of GV was observed, a decrease in the number of these vacuoles was described (Karasek et al., 1982b). According to Karasek these results could be explained, as demonstrated by a study of the effect of prolactin on rat pinealocytes (Karasek et al., personal communication), by the fact that the ependymal-like secretory process is implicated in the regulation of the prolactin secretion and that characterized by the formation of GV in the regulation of gonadotropins.

(3) *Relationships between indoleamines and the processes of protein/peptide secretion*

The pinealocyte in mammals or the rudimentary photoreceptor cell — its phylogenetic homologue — in birds and reptiles is engaged in both the metabolism of indoleamines and the production of proteinaceous substances (see details in Collin, 1979; Collin and Oksche, 1981; Pévet, 1981b,c, 1982b), a situation which permitted Ueck and Wake (1977, 1979) to assert that these cells belong to the paraneurons according to the definition of Fujita (1979). Consequently, as in other paraneurons, a physiological relationship between indoleamines and proteins has to be taken into consideration. The intracellular relationship of indoleamines with protein synthesis in pineal cells has been essentially studied in non-mammalian species, the results concerning mammals being yet fragmentary. However, these studies have clearly shown that in the hamster, lizard, mouse and parakeet, serotonin coexists with a proteinaceous compound in the GV, an observation which recalls the APUD concept of Pearse (1969) and which would indicate that serotonin could change membrane permeability and thus allow or control the release of peptidergic hormones All these results have been reviewed recently in detail by Collin (1979), Collin and Oksche (1981) and Pévet (1981b,c, 1982b), and we refer to these articles for more details.

In the pineal gland, however, the problem of the relationship of indoleamine with protein presents another aspect which, in our opinion, is also of great physiological importance. It is known indeed that the pineal gland synthesizes and releases indoleamines but also concentrates circulating indoleamines, at least serotonin and melatonin (Wurtman et al., 1964). Consequently, the indoleamines released from the pineal and probably also from the retina, Harderian gland, intestin, etc. (details and references in Pévet et al., 1980a) may reach (intact?

modified?) the pineal via the general circulation and then act on the pineal secretory activities. Unfortunately, only very few data related to this problem have been reported in the literature. Injections of serotonin during the period of sexual activity in the hedgehog provokes, in addition to an inhibition of the sexual activity, the formation in the pinealocytes of vacuoles containing flocculent material (Pévet and Saboureau, 1974) characteristic of the so-called ependymal-like secretory process (Pévet, 1976). Melatonin, chronically administered via implantation of melatonin-containing beeswax pellets causes in hamster pinealocytes an increase in the number of GV. A single injection of 25 μg of melatonin induces a decrease in number of GV in the ending of the mouse pinealocyte processes. In contrast, melatonin administered in daily intraperitoneal doses of 50 μg over a period of 5 days increases markedly the number of GV at least during the light period of a light/dark cycle (Benson and Krasovich, 1977). Pinealocytes of melatonin-treated rats are characterized by a prominent Golgi apparatus and especially by an increased number of microtubules (Freire and Cardinali, 1975), structures which are known to be implicated in secretory processes in a variety of cells.

In *in vitro* experiments melatonin as well as other 5-methoxyindoles such as 5-methoxytryptophan, 5-methoxytryptamine, 5-methoxytryptophol, induce in mouse pinealocytes, depending of the time of the treatment and of the presence or absence of NA, an increase or a decrease in the number of GV (Haldar-Misra en Pévet, 1982b,c). 5-Methoxyindole-3-acetic acid, however, provokes always independent of the experimental conditions, an increase in GV number (Haldar-Misra et al., 1982c).

These ultrastructural results demonstrate clearly that indoleamines, especially 5-methoxyindoles, participate in the regulation of the activity of the processes of protein/peptide secretion. These results are probably of great importance when the physiology of the pineal is concerned. However, this does not mean that all indoleamines are working at the same time. We know for example that in hamsters kept under laboratory conditions, 5-methoxytryptophol is synthesized by the pineal during the light phase (Pévet et al., 1980a) while melatonin in hamsters as well as in other species is produced during the dark period (Klein et al., 1981). Moreover, we also know that in hamsters kept under natural environmental conditions the synthesis of 5-methoxytryptophan by the pineal, retina and Harderian gland is high in winter and very low in summer, an opposite evolution being observed for melatonin and 5-methoxytryptamine, at least in the pineal (see details in Balemans et al., 1982).

IV. CONCLUSION

From the experiments described in this review it is now clear that a lot of proteic or peptidic substances isolated from various pineal extracts and known to be very active in some bioassays, cannot be considered as pineal factors. They are indeed not synthesized by the pineal itself (peptides and proteins of classes I and II).

The problem of structural identification of one or more true pineal proteic or peptidic factors (hormones?) remains as yet unsolved. However, the ultrastructural identification of processes of protein/peptide secretion in specific pineal cells and the observation that the activity of these processes is regulated by the sympathetic innervation as well as by the 5-methoxyindoles, demonstrate clearly the reality of the existence of such proteic/peptidic pineal factors, a reality which should not be doubted anymore.

REFERENCES

Aguado, L.I., Benelbaz, G.A., Gutierrez, L.S., and Rodriguez, E.M., 1977, Ultrastructure of the rat pineal gland grafted under the kidney capsule, Cell Tiss. Res., 176:131.

Altschule, M.D., 1957, Some effects of aqueous extracts of acetone-dried beef-pineal substance in chronic schizophrenia, N. Eng. J. Med., 257:919.

Amundson, B.C., and Wheaton, J.E., 1979, Effects of chronic LHRH treatment on brain LHRH contant, pituitary and plasma LH and ovarian follicular activity in the anestrous ewe, Biol. Reprod., 20:633.

Ariëns Kappers, J., 1978, Localization of indoleamine and protein synthesis in the mammalian pineal gland, J. Neural Transm., suppl. 13:13.

Ariëns Kappers, J., 1981, Evolution of pineal concepts, in: "The Pineal Organ - Photobiology, Biochronometry, Endocrinology", A,Oksche and P.Pévet, eds., Elsevier/North-Holland Biomed. Press, Amsterdam.

Balemans, M.G.M., Pévet, P., Van Benthem, J., Haldar-Misra, C., Smith, I., and Hendriks, H., 1982, Seasonal rhythmicity in the capacity of HIOMT to synthesize different 5-methoxyindoles in the pineal, the retina and the Harderian gland of the golden hamster (Mesocricetus auratus), submitted.

Barry, J., 1979, Immunofluorescence study of the preoptico-terminal LRH tract in the female squirrel monkey during the estrous cycle, Cell Tiss. Res., 198:1.

Benson, B., and Krasovich, M., 1977, Circadian rhythm in the number of granulated vesicles in the pinealocytes of mice. Effect of sympathectomy and melatonin treatment, Cell Tiss. Res., 184:499.

Benson, B., Matthews, M.J., and Rodin, A.E., 1972, Studies on a non-melatonin pineal antigonadotrophin, Acta Endocrinologica, 69:257.

Berelowitz, M., Kronheim, S., and Pimstone, B.L., 1978, Somatostatin in tissues, synaptosomes and biological fluids as measured by radioimmunoassays, in: "Hypothalamic Hormones - Chemistry, Physiology and Clinical Applications", D.Gupta and W.Voelter, eds., Verlag Chemie, Weinheim, N.Y.

Bianchi, P., and Osima, B., 1960, Bull. Soc. ital. Biol. sper., 36:1647, cited in Reiter and Vaughan, 1977.

Blask, D.E., Vaughan, M.K., Reiter, R.J., Johnson, L.D., and Vaughan, G.M. 1976, Prolactin-releasing and release-inhibiting factor. Activities in the bovine, rat and human pineal gland. In vitro and in vivo studies, Endocrinology, 99:152.

Bowie, E.P., 1980, Immunocytochemical evidence for the presence of an albumin-like substance in the rat pineal gland, Histochemistry, 66:83.

Bowie, E.P., and Herbert, D.C., 1976, Immunocytochemical evidence for the presence of arginine vasotocin in the rat pineal gland, Nature (Lond.), 261:66.

Buijs, R.M., and Pévet, P., 1980, Vasopressin and oxytocin containing fibres in the pineal gland and subcommissural organ of the rat, Cell Tiss. Res., 205:11.

Buijs, R.M., Pévet, P., and Swaab, D.F., 1982, "Chemical Transmission in the Brain", Progr. Brain Res., 55, in press.

Carson, R.S., Matthews, C.D., Findley, J.K., Symons, R. G., and Burger, H.G., 1977, Biological and immunological luteinizing hormone-releasing hormone (LHRH) activity of the ovine pineal, Neuroendocrinol., 24: 221.

Changaris, D.G., Keil, L.C., and Severs, W.B., 1979, Angiotensin II immunohistochemistry of the rat brain, Neuroendocrinol., 25:257.

Chazov et al., 1972, Dokl. Acad. Nauk. SSSR, 27:246, cited in Reiter and Vaughan, 1977.

Cheesman, D.W., 1970, Structure elucidation of a gonadotropin-inhibiting substance from the bovine pineal gland, Biochem. biophys. Acta, 207:247

Cheesman, D.W., and Fariss, B.L., 1970, Isolation and characterization of a gonadotropin-inhibiting substance from the bovine pineal gland, Proc. Soc. exp. Biol. Med., 133:1254.

Clementi, F., Muller, E., and Zanoboni, A., 1965, Pineal function and modification of its ultrastructural aspects, in: Proc. 2nd Intern. Congr. Endocrinology, London, 1964, Excerpta Medica, Int. Congress Series no. 83.

Collin, J.P., 1979, Recent advances in pineal cytochemistry. Evidence of the production of indoleamines and proteinaceous substances by rudimentary photoreceptor cells and pinealocytes of amniota, in: "The Pineal Gland of Vertebrates including Man", J.Ariëns Kappers and P. Pévet, eds., Prog. Brain Res., vol. 51, Elsevier, Amsterdam.

Coculescu, M., Oprescu, M., and Zagrean, L., 1977, Angiotensin I-like immunoreactive substance in pineal gland of normal and Brattleboro rats with hereditary diabetes insipidus, Rev. Roum. Morph. Embryol. Physiol., 14:47.

Collin, J.P., and Oksche, A., 1981, Structural and functional relationships in the mammalian pineal gland, in: The Pineal Gland, Vol. I, Anatomy and Biochemistry, R.J.Reiter, ed., CRC Press, Boca Raton, USA.

Dilman and Anisimov, 1975, Bull. Exp. Biol. Med., 80:1371, cited in Reiter and Vaughan, 1977.

Dogterom, J., Snijdewint, F.G.M., Pévet, P., and Buijs, R.M., 1978, On the presence of neuropeptides in the mammalian pineal gland, EPSG-Newsletter, Suppl. 1:15.

Dogterom, J., Snijdewint, F.G.M., Pévet, P., and Buijs, R.M., 1979, in: The Pineal Gland of Vertebrates including Man, J.Ariëns Kappers and P.Pévet, eds., Prog. Brain Res., 52, Elsevier, Amsterdam.

Dogterom, J., Snijdewint, F.G.M., Pévet, P., and Swaab, D.F., 1980, Studies on the presence of vasopressin, oxytocin and vasotocin in the pineal gland, subcommissural organ and foetal pituitary gland: Failure to demonstrate vasotocin in mammals, J. Endocr., 84:115.

Duraiswami, S., Franchimont, P., Boucher, D., and Thieblot, M., 1976, Immunoreactive luteinizing hormone releasing hormone (LH-RH) in the bovine pineal gland, Horm. Metab. Res., 8:232.

Ebels, I., Benson, B., and Matthews, M.J., 1973, Localization of a sheep pineal antigonadotropin, Analyt. Biochem., 56:546.

Ebels, I., Moskowska, A. et Scemama, A., 1965, Etude in vitro des extraits épiphysaires fractionnés. Résultats préliminaires, C.R. hebd. Séanc. Acad. Sci., 260:5120.

Ebels, I., Versteeg, D.H.G., and Vliegenthart, J.F.G., 1965, An attempt to isolate arginine vasotocin from sheep and bovine pineal body, Proc. Roy. Netherl. Acad. Sci., Ser. B, 68:1.
Engel, P., 1935, Ueber die antigonadotrope Wirkung des Epiphysans. Wiener Klin. Wochenschr., 38.
Fernstrom, J.D., Fischer, L.A., Cusack, B.M., and Gillis, M.A., 1980, Radioimmunologic detection and measurement of nonapeptides in the pineal gland, Endocrinology, 106:243.
Fisher, L.A., and Fernstrom, J.D., 1981, Measurement of nonapeptides in pineal and pituitary using reversed phase, ion-pair liquid chromatography with post column detection by radioimmunoassay, Life Sci., 28 (13):1471.
Freire, F., and Cardinali, D.P., 1975, Effects of melatonin treatment and environmental lighting on the ultrastructural appearance, melatonin synthesis, norepinephrine turnover and mictotubule protein content of the rat pineal gland, J. Neural Transm., 37:237.
Fujita, T., 1976, The gastro-enteric endocrine cell and its paraneuronic nature, in: "Chromaffin, Enterochromaffin and Related Cells".A Naito Foundation Symposium, R.E.Coupland and T.Fujita, eds., Elsevier, Amsterdam-Oxford-New York.
Geelen, G., Allevard-Burguburu, A.M., Gauquelin, G., Xiao, Y.Z., Frutoso, J., Gharib, Cl., Sempore, B., Meunier, C., and Augoyard, G., 1981, Radioimmunoassay of arginine vasopressin, oxytocin and arginine-vasotocin-like material in the human pineal gland, Peptide, 2:459.
Gradwell, P.B., Millar, R.B., and Symington, R.B., 1976, Failure to demonstrate high concentrations of luteinizing hormone-releasing hormone in the bovine pineal body, S.A. Med. J., 50:217.
Haldar-Misra, C., and Pévet, P., 1982a, The influence of noradrenaline on the process of protein/peptide secretion in the mammalian pineal organ. Comparative in vitro studies, Cell Tiss. Res., 224:33.
Haldar-Misra, C., and Pévet, P., 1982b, Effect of melatonin on pineal peptide/protein synthesis: An ultrastructural study in the mouse pineal in vitro, Gen. Comp. Endocr., 46:356.
Haldar-Misra, C., and Pévet, P., 1982c, The influence of different 5-methoxyindoles on the process of protein/peptide secretion in the mouse pineal gland. An in vitro study, submitted.
Hofstätter, R., 1914, Ueber organotherapeutische Versüche mit Epigandol und Pineal-tabletten, Med. Klin., 10: 1460.

Hofstätter, R., 1938, Pinealtherapie bei prämenstrueller Anfälligkeit, Zbl. Gyn., 62:1192.

Holder, J.C., and Guerné, J.M., 1976, Une préparation particulièrement sensible à l'action de la vasotocine: le bulbe aortique de l'anguille d'eau douce, C.R. Acad. Sci., 283:1767.

Holder, F.C., Pollatz, M., Schroeder, M.O., Guerné, J.M., Vivien-Roels, B., Pévet, P., Buijs, R.M., and Dogterom, J., 1982, A specific and sensitive bioassay for arginine-vasotocin: description, validation and some applications in lower and higher vertebrates, Gen. Comp. Endocr., in press.

Jackson, I.M.D., Saperstein, R., and Reichlin, S., 1977, Thyrotropin releasing hormone (TRH) in pineal and hypothalamus of the frog: Effect of season and illumination, Endocrinology, 100:97.

Joseph, S.A., 1976, Seasonal variation in luteinizing hormone releasing hormone (LHRH) content of rat pineal gland, Anat. Rec., 184:439 (abstract).

Karasek, M., 1974, Ultrastructure of rat pineal gland in organ culture: influence of norepinephrine, dibutyry cyclic adenosine 3-5-monophosphate and adenohypophysis, Endokrinologie, 64(1):106.

Karasek, M., 1979, Ultrastructural study of pineal-adenohypophysial relationships in rats, in: "The Pineal Gland of Vertebrates including Man", J.Ariëns Kappers and P.Pévet, eds., Prog. Brain Res., 52:195.

Karasek, M., Bartke, A., King, T.S., Hansen, J.T., and Reirer, R.J., 1982a, Effects of hereditary hypopituitarism and ectopic pituitary transplants on pinealocytes of the mouse: A quantitative ultrastructural study, Endocrinology, in press.

Karasek, M., and Marek, K., 1978, Influence of gonadotropic hormones on the ultrastructure of rat pinealocytes, Cell Tiss. Res., 188:133.

Karasek, M., and Marek, K., 1980, Influence of estradiol on the ultrastructure of rat pineal gland, Acta Med. Pol., 21:355.

Karasek, M., Lewinska, I., Lewinski, A., Hansen, J.T., and Reiter, R.J., 1982b, Ultrastructure of rat pinealocytes, Cytobios, in press.

Karasek, M., Marek, J., and Kunert-Radek, J., 1978, Ultrastructure of rat pinealocytes in vitro: influence of gonadotropic hormones and LHRH, Cell Tiss. Res., 195:547.

Karasek, M., Pawlikowski, M., Ariëns Kappers, J., and Stepien, H., 1976a, Influence of castration followed by administration of LH-RH on the ultrastructure of rat pinealocytes, Cell Tiss. Res., 167:325.

Karasek, M., Pawlikowski, M., Pévet, P., and Stepien, H., 1976b, Ultrastructure and fluorescence histochemical studies of the rat pineal gland after castration, Ann. Med. Sect. Pol. Acad. Sci., 21:57.

Kastin, A.J., Nissen, C., Nikolics, K., Medzihradszky, K., Coy, D.H., Teplan, I., and Schally, A.V., 1976, Distribution of ^{3}H-α-MSH in rat brain, Brain Res. Bull., 1:19.

Kastin, A.J., Lawrence, S.P., and Coy, D.H., 1981, Radioimmunoassay of MIF/Tyr-MIF-1-like material in rat pineal, Pharmac. Biochem. Behav., 13:901.

King, J.D., and Millar, R.P., 1981, Decapeptide luteinizing hormone releasing hormone in ovine pineal gland, J. Endocr., 91:405.

Klein, D.C., Auerbach, D.A., Namboodiri, M.A.A., and Wheler, G.H.T., 1981, Indole metabolism in the mammalian pineal gland, in: "The Pineal Gland , Vol. I., Anatomy and Biochemistry", R.J.Reiter, ed., CRC Press, Boca Raton, USA.

Krasovich, M., and Benson, B., 1979, Effects of reserpine and p-chlorophenylalanine on the circadian rhythm of granulated vesicles in the pinealocytes of mice, Cell Tiss. Res., 203:457.

Legros, J.J., Louis, F., Demoulin, A., and Franchimont, P., 1976, Immunoreactive neurophysins and vasotocin in human pineal glands, J. Endocrinol., 69:289.

Lin, H.S., Hwang, B.H., and Tseng, C.Y., 1975, Fine structural changes in the hamster pineal gland after blinding and superior cervical ganglionectomy, Cell Tiss. Res., 158:285.

Matthews, M.J., Benson, B., and Rodin, A.E., 1971, Antigonadotropic activity in a melatonin-free extract of human pineal glands, Life Sci., 10:1375.

Milcou, St., Milcou, I., and Nanu, L., 1963a, Le rôle de la glande pinéale dans le métabolisme des glucides, Ann. Endocrinol., 24:233.

Milcu, I., Nanu, L., Marcean-Petrescu, R., and Milcu, St.-M., 1976, Precedeu de obtinere a unui preparat opoterapic de glanda pineală, Brevet OSIM,nr 63:948.

Milcu, S.M., Pavel, S., and Neacşu, C., 1963b,Biological and chromatographic characterization of a polypeptide with pressor and oxytocic activities isolated from bovine pineal gland, Endocrinology, 72:563.

Millar, R.P., and Tobler, C., 1981, Structural and functional differences in pineal and hypothalamic luteinizing hormone-releasing hormone, in: "Neuropeptides: Biochemical and Physiological Studies", R.P.Millar, ed., Churchill Livingstone, New York.

Millar, R.P., Denniss, P., Tobler, C., and Symington, R. B., 1981, Immunological, biochemical and functional

differences in pineal and hypothalamic luteinizing hormone-releasing hormone, in: "Pineal Function", C.D.Matthews and R.F.Seamark, eds., Elsevier/North-Holland Biomed. Press, Amsterdam.
Møller, M., Fahrenkrug, J., and Ottesen, B., 1981, The presence of vasoactive intestinal peptide (VIP) in nerve fibres connecting the brain and the pineal gland of the cat, EPSG-Newsletter, suppl. 3:45.
Moszkowska, A., Citharel, A., L'Heritier, A., Ebels, I., and Laplante, E., 1974, Some new aspects of a sheep pineal gonadotropin inhibiting activity in in vitro experiments, Experientia, 30:964.
Moszkowska, A., Scemama, A., Lombard, M.N., and Héry, M., 1973, Experimental modulation of hypothalamic content of the gonadotropin releasing factor by pineal factors in the rat, J. Neural Transm., 34:11.
Neacşu, C., 1971, Structure-activity data on a pineal peptide with oxytocic and vasopressor activities, Excerpta Medica, Int. Congr. Series, 241:275, Excerpta Medica, Amsterdam.
Neacşu, C., 1972, The mechanisms of antigonadotropic action of polypeptide extracted from a bovine pineal gland, Rev. Roum. Physiol., 9:161.
Negro-Vilar, A., Sanchez-Franco, F., Kwiatkowski, M., and Samson, W.K., 1980, Failure to detect radioimmunoassayable arginine vasotocin in mammalian pineals. Brain Res. Bull., 4:789.
Noteborn, H.P.J.M., Ebels, I., Pévet, P., Reinharz, A.C., Neacşu, C., and Salemink, C.A., 1982, Comparison of some peptidergic and proteic ovine pineal fractions with a bovine E5 pineal fraction, J. Neural Transm., in press.
Nürnberger, F., and Korf, H.W., 1981, Oxytocin- and vasopressin-immunoreactive nerve fibres in the pineal gland of the hedgehog, Erinaceus europaeus, L., Cell Tiss. Res., 220:87.
O'Donohue, T.L., Miller, R.L., and Jacobowitz, D.M., 1979, Identification, characterization and stereotaxic mapping of intraneuronal α-melanocyte stimulating hormone-like immunoreactive peptides in discrete regions of the rat brain, Brain Res., 176:101.
O'Donohue, T.L., Miller, R.L., Pendleton, R.C., and Jacobowitz, D.M., 1980, Demonstration of an endogenous circadian rhythm of α-melanocyte stimulating hormone in the rat pineal gland, Brain Res., 186:145.
Oliver, C., and Porter, J.C., 1978, Distribution and characterization of α-melanocyte-stimulating hormone in the rat brain, Endocrinology, 102:3.
Ota, M., Horiuchi, S., and Obara, K., 1975, Inhibition of ovulation induced with PMS and HCG by a melatonin-

free extract of bovine pineal powder, Neuroendocrinology, 18:311.

Orts, R.G., Liao, T.H., Sartin, J.L., and Bruot, B.C., 1980, Isolation, purification and amino acid sequence of a tripeptide from bovine pineal tissue displaying antigonadotropic properties, Biochem. Biophys. Acta, 628:201.

Parhon, C.I., Milcu, S.M., and Tomorug, E., 1940, Augmentation pondérale sous l'influence d'un extrait épiphysaire (l'épiphysormone), Bull. Soc. roumain. Endocr., 6:122.

Pavel, S., 1978, Arginine vasotocin as a pineal hormone, J. Neural Transm., 13:135.

Pavel, S., 1979, The mechanism of action of vasotocin in the mammalian brain, in: "The Pineal Gland of Vertebrates, including Man", J.Ariëns Kappers and P.Pévet, eds., Prog. Brain Res., 52:445, Elsevier, Amsterdam.

Pavel, S., and Petrescu, S., 1966, Inhibition of gonadotrophin by a highly purified pineal peptide and by synthetic arginine-vasotocin, Nature, 212:1054.

Pavel, S., Goldstein, R., and Calb, M., 1975, Vasotocin content in the pineal gland of foetal, newborn and adult male rats, J. Endocrinol., 66:283.

Pearse, A.G.E., 1969, The cytochemistry and ultrastructure of polypeptide hormone-producing cells of the APUD series and the embryologic, physiologic and pathologic implication of the concept, J. Histochem. Cytochem.,17:303.

Pelletier, G., Leclerc, R., Dubé, D., Labrie, F., Buciani, R., Arimura, A., and Schally, A.V., 1975, Localization of growth-hormone release inhibiting hormone (somatostatin) in the rat brain, Am. J. Anat., 142: 397.

Petit, A., 1976, Contribution à l'étude de l'épiphyse des Reptiles: le complèxe épiphysaire des lacertiliens et l'épiphyse des ophidiens: Etude embryologique, structurale; analyse qualitative et quantitative de la sérotonin dans les conditions normales et expérimentales. Thèse, Strasbourg.

Pévet, P., 1974, The pineal gland of the mole (Talpa europaea, L.). I. Fine structure of the pinealocytes, Cell Tiss. Res., 153:277.

Pévet, P., 1976, Correlations between pineal gland and sexual cycle. An electron microscopical and histochemical investigation on the pineal gland of the hedgehog, mole, mole-rat and white rat. Thesis, Amsterdam.

Pévet, P., 1977, The pineal gland of the mole (Talpa europaea, L.). IV. Effect of pronase on material present in cisternae of the granular endoplasmic reticu-

lum of pinealocytes. Cell Tiss Res., 182:215.
Pévet, P., 1979, Secretory processes in the mammalian pinealocytes under natural and experimental conditions, in: "The Pineal Gland of Vertebrates including Man", J.Ariëns Kappers and P.Pévet, eds., Prog. Brain Res., 52; Elsevier, Amsterdam.
Pévet, P., 1981a, Ultrastructure of the mammalian pinealocyte, in: "The Pineal Gland, Vol. I, Anatomy and Biochemistry", R.J.Reiter, ed., CRC Press, Boca Raton, USA.
Pévet, P., 1981b, Peptides in the pineal gland of vertebrates: ultrastructural, histochemical, immunocytochemical and radioimmunological aspects, in: "The Pineal Organ: Photobiology - Biochronometry - Endocrinology", A.Oksche and P.Pévet, eds., Elsevier/North-Holland Biomed. Press, Amsterdam.
Pévet, P., 1981c, Cytological aspects of indoleamine secretion in the pineal gland, in: "Melatonin - Current Status and Perspectives", N.Birau and W.Schloot, eds., Pergamon Press, Oxford.
Pévet, P., 1982a, Pineal peptides in the fetus and in young and adult mammals, in: Melatonin Rhythm", D.C.Klein, ed., Karger, Basel, in press.
Pévet, P., 1982b, The anatomy of the pineal gland of mammals, in: "The Pineal Gland", R.Relkin, ed., Elsevier/North-Holland, New York, in press.
Pévet, P. and Haldar-Misra, C., 1982, Effect of 5-methoxytryptamine on testicular atrophy induced by experimental or natural short photoperiod in the golden hamster (Mesocricetus auratus), J. Neural Transm., in press.
Pévet, P., and Saboureau, M., 1973, L'épiphyse du Hérisson (Erinaceus europaeus, L.) mâle. I. Les pinéalocytes et leurs variations ultrastructurales considérées au cours du cycle sexuel, Z.Zellforsch., 143: 367.
Pévet, P., and Saboureau, M., 1974, Effect of serotonin administration on the ultrastructure of pinealocytes during the period of maximal sexual activity of the male hedgehog (Erinaceus europaeus, L.), Experientia (Basel), 30:1069.
Pévet, P., and Smith, A.R., 1975, The pineal gland of the mole (Talpa europaea, L.). Ultrastructural variations observed in the pinealocytes during different parts of the sexual cycle, J. Neural Transm., 36: 227.
Pévet, P., and Swaab, D.F., 1979, Immumocytochemical evidence for the presence of an α-MSH-like compound in the rat pineal, J. Physiol., Paris, 75:75.

Pévet, P., Balemans, M.G.M., Legerstee, W.C., and Vivien-Roels, B., 1980a, Circadian rhythmicity of the activity of HIOMT in the formation of melatonin and of 5-methoxytryptophol in the pineal, retina, and Harderian gland of the golden hamster, J. Neural Transm., 49:229.

Pévet, P., Buijs, R.M., Dogterom, J., Vivien-Roels, B., Holder, F.C., Guerné, J.M., Reinharz, A., Swaab, D. F., Ebels, I., and Neacşu, C., 1981a, Peptides in the mammalian pineal gland, in: Pineal Function", C.D. Matthews and R.F.Seamark, eds., Elsevier/North-Holland, Amsterdam.

Pévet, P., Dogterom, J., Buijs, R.M., and Reinharz, A., 1979, Is it the vasotocin or a vasotocin-like peptide which is present in the mammalian pineal and subcommissural organ?, J. Endocrinol., 80:49.

Pévet, P., Ebels, I., Swaab, D.F., Mud, M., and Arimura, A., 1980b, Presence of AVT-, α-MSH-, LHRH- and somatostatin-like compounds in the rat pineal gland and their relationship with the UMO5R pineal fraction: An immunocytochemical study, Cell Tiss. Res., 200:341.

Pévet, P., Haldar-Misra, C., and Öcal, T., 1981b, The independency of an intact pineal gland of the inhibition by 5-methoxytryptamine of the reproductive organs in the male hamster, J. Neural Transm., 52:95.

Pévet, P., Juillard, M.T., Smith, A.R., Kappers, J.Ariëns, 1976, The pineal gland of the mole (Talpa europaea, L.). III. A fluorescence histochemical study. Cell Tiss. Res., 165:297.

Pévet, P., Neacşu, C., Holder, F.C., Reinharz, A., Dogterom, J., Buijs, R.M., Guerné, J.M., and Vivien-Roels, B., 1981c, The vasotocin-like biological activity present in the bovine pineal is due to a compound different from vasotocin, J. Neural Transm., 51:295.

Pévet, P., Reinharz, A.C., and Dogterom, J., 1980c, Neurophysins, vasopressin and oxytocin in the bovine pineal gland, Neurosci. Lett., 16:301.

Pévet, P., Smith, A.R., Van de Kar, L., and Van Bronswijk, H., 1975, Effect of castration on the rat pineal gland: a fluorescence histochemical and biochemical study, Experientia, 31:1237.

Piekut, D.T., and Knigge, K.M., 1981, Immunocytochemical analysis of the rat pineal gland using antisera generated against luteinizing hormone-releasing hormone (LHRH), J. Histochem. Cytochem., 29:616.

Powel, O., Krabanek, S., and Cannon, D., 1977, Substance P: radioimmunoassay studies, in: "Substance P", U.S. Von Euler and B.Pernow, eds., Raven Press, New York.

Redding, T.W., and Schally, A.V., 1973, The distribution halflife and excretion of tritiated luteinizing hormone-releasing hormone (LHRH) in rats, Life Sci., 12 23.

Redding, T.W., Kastin, A.J., Nair, R.M.G., and Schally, A.V., 1973, Distribution, half-life and excretion of ^{14}C- and ^{3}H-labeled L-Prolyl-L-Leucyl-Glycinamide in the rat, Neuroendocrinology, 11:92.

Reinharz, A., Valloton, M.B., Pévet, P. and Dogterom, J., 1981, Neurophysins and neurohormones in the mammalian pineal gland, in: "Pineal Function", C.D.Matthews and R.F.Seamark, eds., Elsevier/North-Holland, Amsterdam

Reiter, R.J., and Vaughan, M.K., 1977, Pineal antigonadotrophic substances: polypeptides and indoles, Life Sci., 21:139.

Rix, E., Hackenthal, E., Hilgenfeld, U., and Taugner, R., 1981, Neuropeptides in the pineal gland? A critical immunocytochemical study, Histochemistry, 72:33.

Romijn, H.J., 1975, The ultrastructure of the rabbit pineal gland after sympathectomy, parasympathectomy, continuous illumination and continuous darkness, J. Neural Transm., 36:183.

Romijn, H.J., and Gelsema, A.J., 1976, Electron miscoscopy of the rabbit pineal organ in vitro. Evidence of norepinephrine-stimulated secretory activity of the Golgi apparatus, Cell Tiss. Res., 172:365.

Rosenbloom, A.A., and Fisher, D.A., 1975, Radioimmunoassayable AVT and AVP in adult mammalian brain tissue: comparison of normal and Brattleboro rats, Neuroendocrinol., 17:354.

Rossier, J., Vargo, T.M., Minick, S., Ling, N., Bloom, F.E., and Guillemin, R., 1977, Regional dissociation of β-endorphin and enkephalin contents in rat brain and pituitary. Proc. Natl. Acad. Sci. USA, 74:5162.

Roux, M., and Richoux, J.P., 1981, Effets de l'énucléation oculaire bilaterale sur l'ultrastructure de l'épiphyse chez la femelle du Lérot (Eliomys quercinus, L.). Correlations avec l'axe hypothalamo-hypophyse-ovarien, Reprod. Nutr. Develop.,21:47.

Roux, M., Richoux, J.P., and Cordonnier, J.L., 1977, Influence de la photopériode sur l'ultrastructure de l'épiphyse avant et pendant la phase génitale saisonnière chez la femelle du Lérot (Eliomys quercinus). J. Neural Transm., 41:209.

Rudman, D., Del Rio, A.E., Garcia, L.A., Barnett, J., Bixler, T., and Hollins, B., 1970, Lipolytic substances in bovine, thyroid, parotid and pineal glands, Endocrinology, 87:27.

Rudman, D., and Scott, J.W., 1975, Melanotropic-lipolytic peptides of the pineal gland and other CNS regions, in:"Frontiers of Pineal Physiology", M.D.Altschule, ed., MIT Press, Cambridge, Mass.

Sheridan, M.N., 1975, Pineal gland fine structure, in: "Brain-Endocrine Interactions", vol. 2 (The Ventricular System), 2nd Int. Symp., Shizuko, pp. 324.

Smith, A.R., 1972, Conditions influencing the serotonin and tryptophan metabolism in the epiphysis cerebri of the rabbit. A fluorescence histochemical, microscopic and electrophoretic study. Thesis, Amsterdam.

Smith, A.R., Jongkind, J.F., and Kappers, J.Ariëns, 1972a, Distribution and quantification of serotonin-containing and autofluorescent cells in the rabbit pineal organ, Gen. Comp. Endocr., 18(2):364.

Smith, A.R., Kappers, J.Ariëns, and Jongkind, J.F., 1972b, Alterations in the distribution of yellow fluorescing rabbit pinealocytes produced by p-chlorophenylalanine and different conditions of illumination, J. Neural Transm., 33:91.

Smith, A.R., Pévet, P., Van de Kar, L., and Van Oostrom, R., 1975, Effect of gonadotropic hormones on the rat pineal gland. A fluorescence histochemical and biochemical study, J. Neural Transm., 36:217.

Swaab, D.F., 1982, Neuropeptides: their distribution and function in the brain. in: "Chemical Transmission in the Brain", R.M.Buijs, P.Pévet and D.F.Swaab, eds., Prog. Brain Res., 55, in press.

Thiéblot,L., Grizard, G., Dastugue, B., Gachon, A.M., et Thiéblot, Ph., 1979, Purification du facteur antigonadotrope de la glande pinéale, Ann. Endocrinol., 40:519.

Trentini, G.P., De Gaetani, C.F., Di Gregorio, C., and Botticelli, C.S., 1980, LHRH incorporation in normal and denervated pineal gland and in pineal gland of rats with constant estrous-anovulatory syndrome: a preliminary study, Endokrinologie, 76:6.

Uddman, R., Alumets, J., Håkanson, R., Lorén, I., and Sundler, F., 1980, Vasoactive intestinal peptide (VIP) occurs in nerves of the pineal gland, Experientia, 36:1119.

Ueck, M., and Wake, K., The pinealocyte - a paraneuron? A review. Arch. histol. jap.,40:suppl. 261.

Ueck, N., and Wake, K., 1979, The pinealocyte. A paraneuron, in:"The Pineal Gland of Vertebrates including Man", J.Ariëns Kappers and P.Pévet, eds., Prog. Brain Res. 52, Elsevier, Amsterdam.

Vaudry, M., Tonin, M.C., DeLarue, L., Vaillant, R., and Kraicer, J., 1978, Biological and radioimmunological evidence for melanocyte stimulating hormones (MSH) of extrapituitary origin in the rat brain, Neuroendocrinology, 27:9.

Vaughan, M.K., 1981, Arginine vasotocin and vertebrate reproduction, in: "The Pineal Gland, vol. II, Reproductive Effects", R.J.Reiter, ed., CRC Press, Boca Raton, USA.

Vaughan, M.K., and Blask, D.E., 1978, Arginine vasotocin - A search for its function in mammals, in:"The Pineal and Reproduction", R.J.Reiter, ed., Progr. Reprod. Biol. 4:99, Karger, Basel.

Vaughan, M.K., Reiter, R.J., McKinney, T., and Vaughan, G.M., 1974, Inhibition of growth of gonadal dependent structures by arginine vasotocin and purified bovine pineal fractions in immature mice and hamsters, Int. J. Fertil., 19:103.

Vivien, J.H., 1965, Signe de stimulation des activités secrétoires des pinéalocytes chez la couleuvre, Tropidonotus natrix, L. traitée par des principes gonadotropes, C.R. Acad. Sci., 260:5370.

Vivien-Roels, B., Guerné, J.M., Holder, F.C., and Schroeder, M.D., 1979, Comparative immunohistochemical, radioimmunological and biological attempts to identify arginine-vasotocin (AVT) in the pineal gland of reptiles and fishes, in: "The Pineal Gland of Vertebrates including Man", J.Ariëns Kappers and P.Pévet, eds., Prog. Brain Res., Elsevier-North-Holland Biomed. Press, Amsterdam.

Vivien-Roels, B., Pévet, P., Guerné, J.M., Holder, F.C., Meiniel, A., Dogterom, J., and Buijs, R.M., 1981, On the presence of arginine-vasotocin (AVT) in the pineal organ of non-mammalian vertebrates, in: "Pineal Function", C.D.Matthews and R.F.Seamark, eds., Elsevier/North-Holland, Amsterdam.

Vizsolyi, E., and Perks, A.M., 1976, Neurohypophysial hormones in fetal life and pregnancy. I. Pharmacological studies in the sheep (ovis aries), Gen. Comp. Endocrinol., 29:28.

Vollrath, L., 1981, The pineal organ, in: "Handbuch der mikroskopischen Anatomie des Menschen", VI/7, Springer, Berlin.

Vuolteenaho, O., and Leppäluoto, J., 1981, Immunoreactive ACTH in rat pineal: stress and dexamethasone lower the concentration. Acta Physiologica, 112:491.

Vuolteenaho, O., Vakkuri, O., and Leppäluoto, J., 1980, Wide distribution of β-endorphin-like immunoreactivity in extrapituitary tissues of rat, Life Sci., 27:57.

Weindl, A., and Sofroniew, M.V., 1982, Peptide neurohormones and circumventricular organs in the pigeon, in: "Cerebrospinal Fluid (CSF) and Peptide Hormones", E.M.Rodriguez and Tj.B.van Wimersma Greidanus, eds., Karger, Basel.

Wheaton, J.E., 1980, Immunoreactive luteinizing hormone releasin hormone (LH-RH) in ovine pineal glands, Horm. Metab. Res., 12:314.

White, W.F., Hedlund, M.T., Weber, G.F., Rippel, R.M., Johnson, E.S., and Wilber, J., 1974, The pineal gland: A supplementary source of hypothalamic releasing hormones, Endocrinology, 94,5:1422.

Wilber, J.F., Montoya, E., Plotnikoff, M.P., White, W,F., Gendrich, R., Renaud, L.P. and Martin, J.P., 1976, Gonadotropin-releasing hormone and thyrotropin-releasing hormone: distribution and effects in the central nervous system, Recent Prog. Horm.Res., 32: 117.

Wurtman, R.J., Axelrod, J., and Potter, L.T., 1964, The uptake of H^3-melatonin in endocrine and nervous tissues and the effects of constant light exposure, J. Pharmacol. Exp. Ther., 143:314.

Youngblood, W.W., Humm, J., and Kizer, J.S., 1979, TRH-like immunoreactivity in rat pancreas and eyes, bovine and sheep pineals, and human placenta: Non-identity with synthetic pyroglu-His-Pro-NH_2 (TRH), Brain Res., 163:101.

Acknowledgement: The author wishes to thank Miss J.Sels for her skillful secretarial aid.

PTERIDINES IN THE PINEAL AND EFFECTS OF THESE SUBSTANCES ON THE INDOLE METABOLISM OF THIS ORGAN*

I. Ebels, M.G.M. Balemans**, J. van Benthem**, H.P.J.M. Noteborn, and A. de Morée

Department of Organic Chemistry and **Zoological Laboratory, State University of Utrecht
The Netherlands

Pteridines are very light sensitive substances. The first compounds of this class were isolated as a yellow pigment from the wings of the brimstone butterfly by Wieland and Schöpf (1925) and as the white pigment from the wings of the cabbage butterfly by Schöpf and Wieland (1926). These pigments were named xanthopterin and leucopterin indicating the colour and source of the compounds. The real chemical structure of these substances was not solved until 1940 by Purrmann. The structure of a third insect pigment, isoxanthopterin was also elucidated by Purrmann in 1941. The well-known structure folic acid, and its derivatives which play a key role in metabolism, contains also a pteridine ring. For a survey of the most important literature in the pteridine field till 1969, see Blakly's "Biochemistry of folic acid and related pteridines".

In mammals three aromatic amino acid monooxygenases: phenylalanine-, tyrosine-hydroxylase and tryptophan-hydroxylase require a reduced pterin as cofactor (Nagatsu et al., 1981). It is generally thought that tetrahydrobiopterin (BH4) is the natural factor for these mammalian monooxygenases. These three hydroxylases play an important role in the biosynthesis of biogenic monoamines from aromatic amino acids, i.e.:
(i) the catecholamines dopamine, norepinephrine, adrenaline, and
(ii) the indoleamine serotonin.
In this respect BH4 is also very important for the biosynthesis of melatonin and other methoxyindoles in the pineal.

*Our research is supported by grants of the Netherlands Organisation for the Advancement of Pure Research (ZWO) through the Foundation for Biological Research (BION) no. 14.70.03 and through the Foundation for Medical Research (FUNGO) no. 13-35-33.

In this paper we will give a survey of:

I. The isolation of pteridines from sheep pineals
II. The influence of synthetic pteridines and that of low molecular weight pineal substances, present in fractions which show a pteridine-like fluorescence, on indole-metabolism of pineals
III. Some general remarks

I. THE ISOLATION OF PTERIDINES FROM SHEEP PINEALS

I.a. The Isolation of 6-L-Erythrobiopterin from an Aqueous Extract of Sheep Pineals

It was in 1973, that we first observed, that when an aqueous extract of sheep pineals was separated on Sephadex G-25 and Sephadex G-10 columns, besides the well-known fluorescence like indoles have, another high fluorescent peak was present, which could not be detected in a comparable aqueous extract of sheep cerebral cortex. This high peak showed an excitation maximum (e) at 360 nm and a fluorescence maximum (f) at about 440 nm. After ultrafiltration of the low molecular weight Sephadex G-25 fraction F2 - F3 of a pineal extract through different Amicon ultrafiltration membranes, it was found that the substance with these special fluorescence characteristics was located in the UM05F-fraction, which contains substances with a molecular weight (MW) < 500. When this UM05F sheep pineal fraction was further separated on a Sephadex G-10 column (142 x 1 cm), the substance with the above mentioned fluorescence was eluted between about 150 and 250 ml from that column. For the extraction and separation see Scheme 1 and Ebels (1979).

When these Sephadex G-10 fractions were tested in a bioassay by Moszkowska et al. (1976) using anterior pituitaries of male rats *in vitro*, an inhibition of the gonadotropic activity of these pituitaries was observed. In paper electrophoresis and paper chromatography experiments, the substance with special fluorescence characteristics, like pteridines, and the inhibiting activity in the bioassay *in vitro*, was always found in the same fraction. After preparative paper chromatography, elution of the special fluorescent band and lyophilization of the eluate, trimethylsilylation was carried out. Gas liquid chromatography (GLC) gave a peak with the same retention time as a synthetic preparation of 6-biopterin. With GLC-mass-spectrometry (GC-MS) a spectrum was obtained, identical with that of 6-biopterin.

Thin layer chromatography and the activity in a *Chrithidia fasciculata*-test revealed that we had isolated 6-L-erythrobiopterin (6-L-B), Fig. 1. from sheep pineals. For details see the experiments of Van der Have-Kirchberg et al. (1977). Since the time we knew that 6-L-B was present in our extracts, we have carried out all our experiments under dim red light-conditions: $\lambda > 600$ nm, because pteridines are very light sensitive substances. Moreover, some of the oxidation products, such as pterin-6-aldehyde and pterin-6-carboxylic acid, can

SCHEME 1. OUTLINE OF THE METHODS USED IN PART I.a.

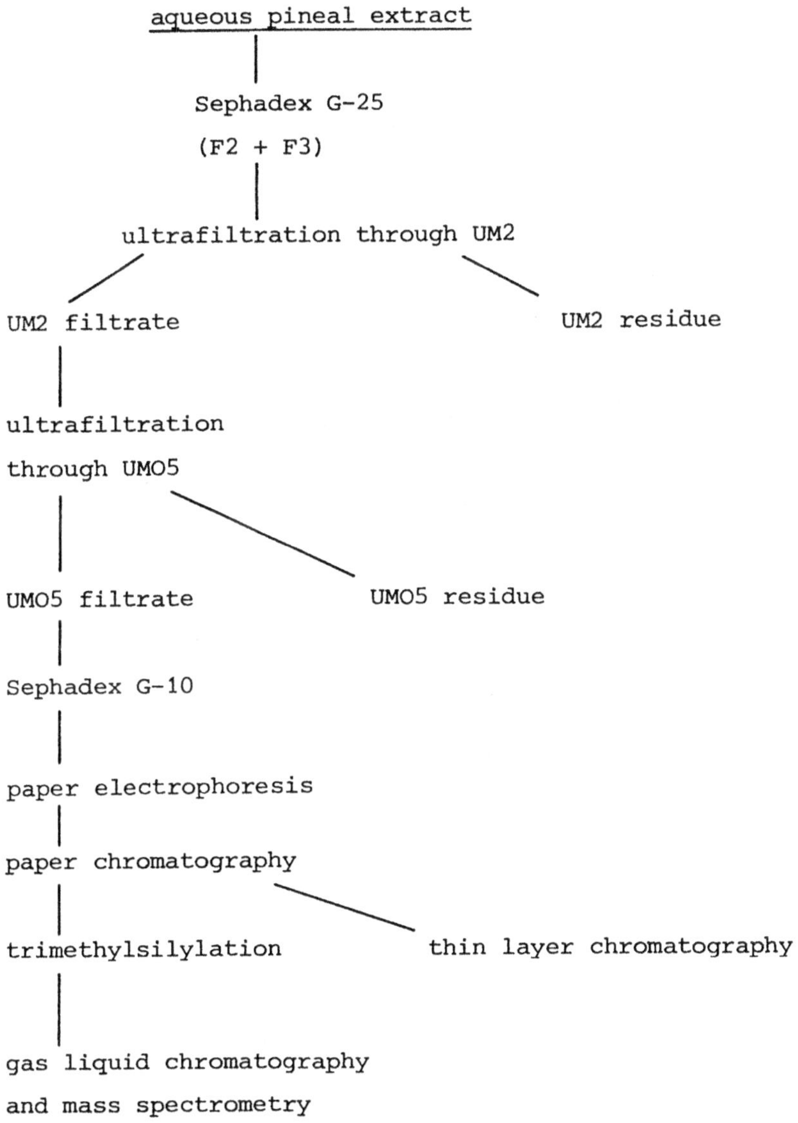

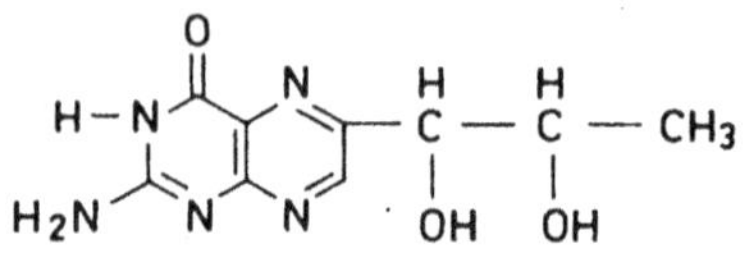

Figure 1. 6-L-erythrobiopterin

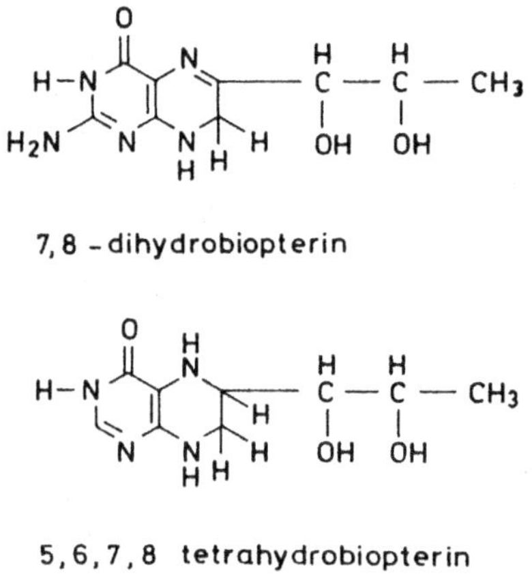

Figure 2. Reduced 6-L-erythrobiopterins

also influence metabolic processes, related to gonadal growth. For details see Ebels (1981) and Ebels, another chapter in this book.

The experiments of Fukushima and Nixon, (1979, 1980) have supported our results. These authors observed, using high performance liqui chromatography, that the pineal of the rat contains more biopterin than seven other organs tested and also much more than several other regions of the rat brain per gram wet tissue. By using an oxidation method, the authors concluded from their results, that biopterin is present in rat pineals in the reduced form, mostly as BH4, the cofactor for aromatic amino acid monooxygenases (Fig. 2).

Levine et al. (1979) studied the tryptophanhydroxylase-, tyrosinehydroxylase- and cofactor-activity, expressed as BH4, in different regions of the rat brain. These authors detected the highest amount of cofactor-activity in the pineal.

Kapatos et al. (1981) observed that rat pineal glands in organ culture synthesize and release biopterin and are able to maintain concentrations of biopterin, occurring *in vivo*, for up to 54 hours *in vitro*. These authors found also that the intracellular biopterin content is reduced 50 percent by treatment with L-norepinephrine. This is, according to these authors, an indication that biopterin levels are regulated by an adrenergic cyclic adenosine monophosphate-dependent mechanism. The observed decline in tissue biopterin content, produced mainly by inhibition of the biosynthesis, is not associated with either an increase in biopterin release or a shift in the reduction state of biopterin. Kapatos et al. (1981) postulate that the decline in biopterin might be related to the increase in the activity of N-acetyl transferase. These authors hope that a further study of the adrenergic cyclic AMP-induced inhibition of pineal biopterin biosynthesis may lead to the discovery of a new function of biopterin in the pineal gland and possibly to a new role for this compound in other tissues.

Ebels (1979, 1981, and another chapter in this book), has described, that different pterins show antigonadotropic activity in *in vitro* and *in vivo* experiments. Therefore, we have undertaken experiments with the aim to study whether pteridines are present in different pineal fractions, prepared by a Bensinger-extraction method, which show antigonadotropic activity.

I.b. Isolation of Neopterin, Pterin-6-Carboxylic Acid and Biopterin from Sheep Pineal Fractions, Prepared with a Bensinger-Extraction Method

Bensinger et al. (1973) published in abstract-form a method for aqueous acetone extraction of bovine pineals and the further removal of fats and indoleamines by chloroform/methanol extraction. A compensatory ovarian hypertrophy (COH)-inhibiting substance was subsequently extracted into isobutanol and further purified by gelfiltration on Sephadex G-15 and thin layer chromatography. Although the details of the method and all the results were not published in detail, the authors stated that the active principle was probably a small, non-tryptophan containing peptide.

Blask et al. (1976) observed, using the same method, that two partially purified bovine pineal fractions, apparently free of indoleamines, inhibited prolactin release from anterior pituitaries *in vitro*.

Ebels et al. (1978, 1979) compared an acetic acid extraction method and the Bensinger-extraction method. From these experiments was concluded that although both extracts were active in inhibiting COH, after unilateral ovariectomy, different Rf-values for the active paper chromatographic regions were obtained. Moreover, these authors could detect more COH-inhibiting activity when the Bensinger extraction method was used than with aqueous and acetic acid extraction methods, used earlier by the same group. Therefore, we started experiments to study the presence of pteridines which can have antigonadotropic activity, in pineal fractions prepared by the Bensinger extraction method.

The different fractions were first prepared according to Scheme 2. The isobutanol extract of sheep pineals was separated on a Sephadex G-15 column, and a sheep cerebral cortex isobutanol extract was separated in the same way. Absorbance-measurements of the Sephadex G-15 fractions at 280 nm are given in Fig. 3.

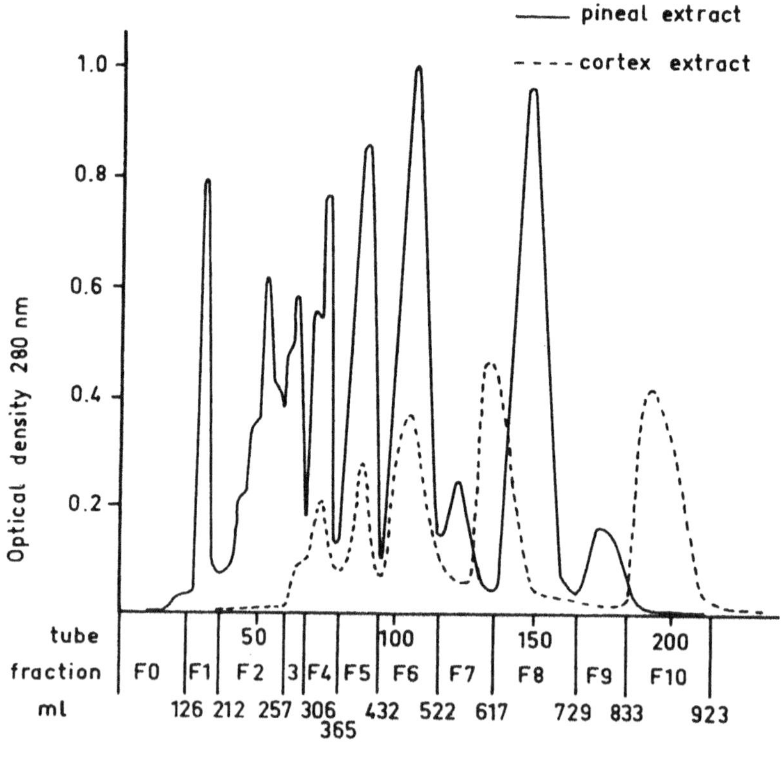

Figure 3. Comparison of the 280 nm absorbance-elution-profiles of the isobutanol extracts of sheep pineals and sheep cerebral cortex after gelfiltration on Sephadex G-15 columns (85 x 2.5 cm), equilibrated and eluted with 0.2 N acetic acid.

SCHEME 2. OUTLINE OF THE METHODS USED IN PART I.b.

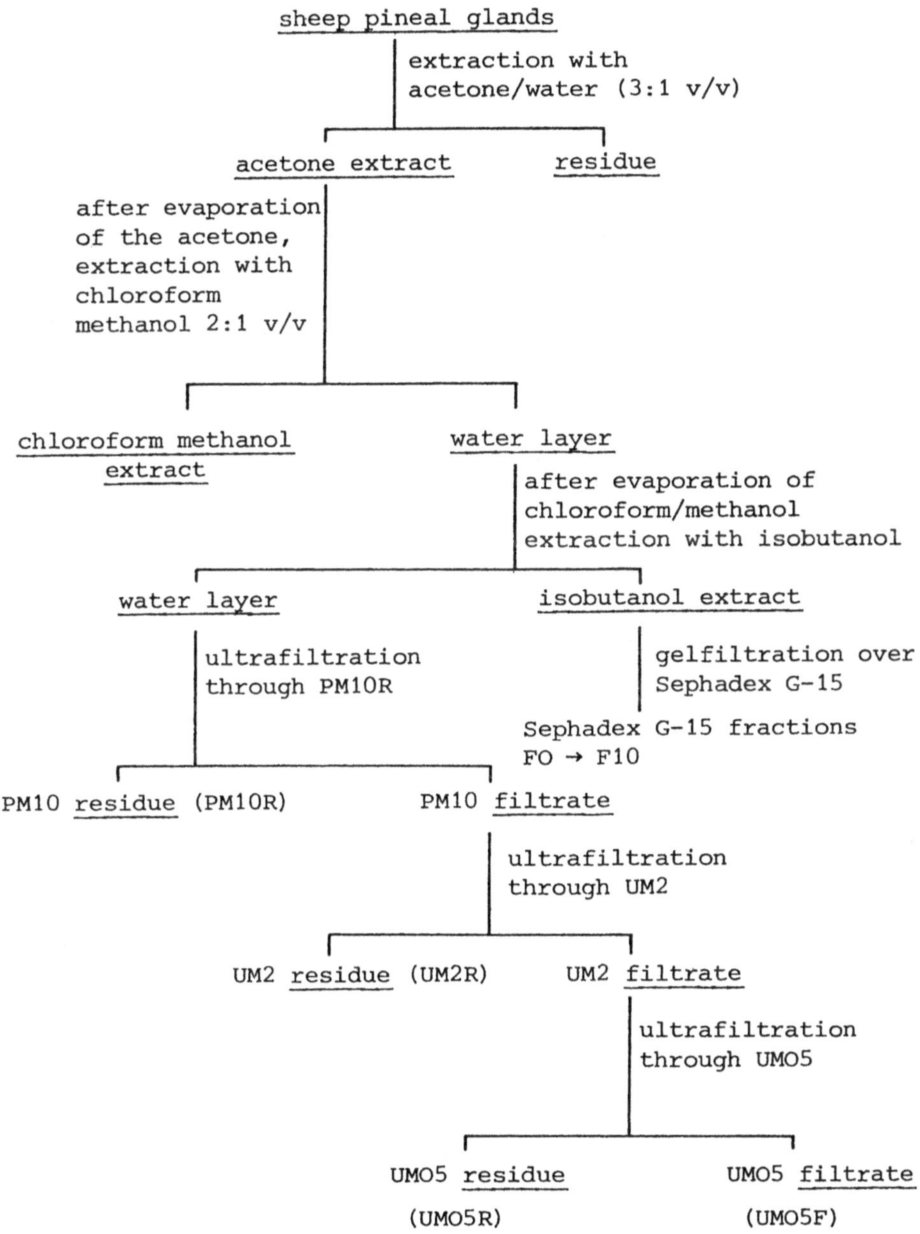

In fig. 4a and 4b the fluorescence measurements are presented. In both extracts several peaks with excitation and fluorescence maxima like indoles, pteridines and flavines could be detected. The UV-absorbance results are comparable to those obtained by Ebels et al. (1979) for bovine pineal extracts. It can be concluded that the curves of Fig. 4a and 4b give more detailed information about the different compounds, present in the extracts than the curves in Fig. 3. On this Sephadex G-15 column dextran blue 2000 was eluted from 115→225 ml and Cl^- from 325 → 420 ml.

In the pineal isobutanol extract Sephadex G-15 fraction F5, eluted between 384 and 465 ml a high peak is present with a pteridine-like fluorescence. This fraction is further studied with paper chromatography and after elution of the bright blue fluorescent band and lyophilization, this fraction was trimethylsilylated and studied with GLC and GC-MS as described before in part I.a for biopterin.

Table 1 shows some Rf-values of the isolated pineal compound and some synthetic pteridines.

Table 2 gives the retention time of the isolated compound and some pteridines after silylation.

From Table 1 and 2 can be concluded that the Rf-value in paper chromatography and the retention time in GLC-experiments is corresponding to those of synthetic 6-D-neopterin.

Fig. 5 gives the mass spectrum of the isolated pteridine-like compound of the Sephadex G-15, F5-fraction, which is identical to the spectrum of 6-neopterin (Fig. 6) and in agreement with the spectrum published by Lloyd et al. (1971).

The UMO5R-fraction of the water layer, which is obtained after the isobutanol extraction (see Scheme 2) is also studied, while Ebels et al. (1979) detected in a comparable bovine pineal fraction COH-inhibiting activity. This fraction was studied with the same techniques as described before. Table 3 gives the Rf-values of a pineal UMO5 residue (UMO5R) and some synthetic pteridines.

Table 4 shows the retention time in GLC-experiments after trimethylsilylation of the paper chromatography fraction Rf = 24 (Table 3, UMO5R).

From the GLC experiments can be concluded that in the UMO5R a substance is present with about the same retention time as 6-neopterin, while the Rf-value of neopterin is more than twice that of the bright yellow band (Rf = 24) in paper chromatography. The mass spectrum which was obtained with GC-MS was the same as given in Fig. 5.

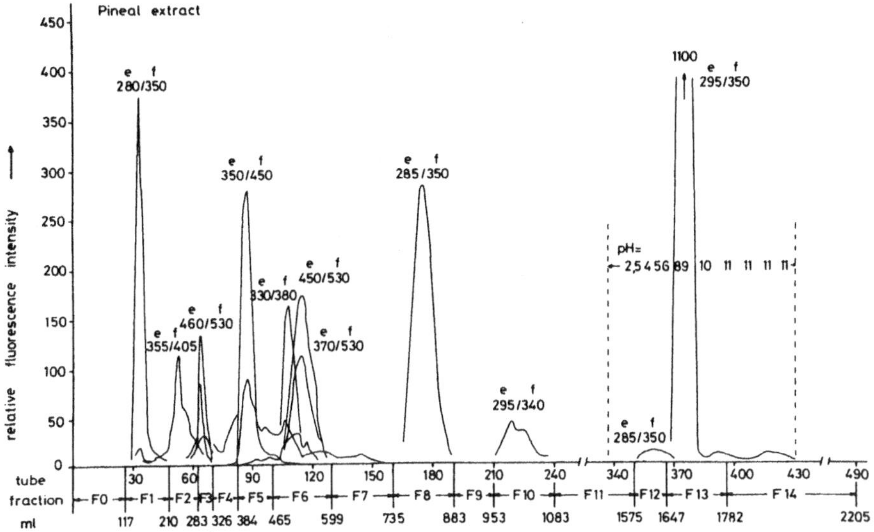

Figure 4.a. Separation of a sheep pineal isobutanol-extract by gelfiltration on Sephadex G-15 (85 x 2.5 cm), equilibated and eluted with 0.2 N acetic acid.
e: excitation; f: fluorescence

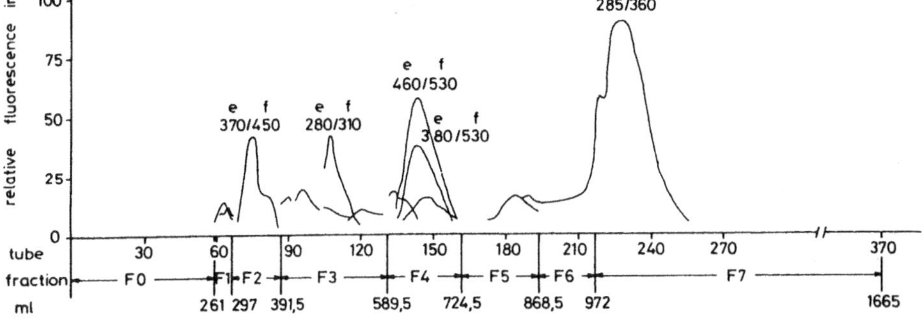

Figure 4.b. Separation of a sheep cerebral cortex isobutanol-extract by gelfiltration on Sephadex G-15 (85 x 2.5 cm), equilibrated and eluted with 0.2 N acetic acid.
e: excitation; f: fluorescence

Table 1.

Rf-values ($x10^2$) of the main spot with a pteridine-like fluorescence of the sheep pineal Sephadex G-15, F5 fraction (Fig 4a) and of some synthetic pteridines, in two solvent-systems.

Compound or fraction	Solvents B-A-W	E-W
pterin-6-carboxylic acid	12	72
6-D-neopterin	13	60
6-L-erythrobiopterin	28	65
Sephadex G-15, F5, (eluted between 384 and 465 ml)	15	64

B-A-W = n-butanol-acetic acid-water, 4:1:1

E-W = ethanol-water, 1:3

Table 2.

Retention time in minutes in gas liquid chromatography of the main fluorescent compound of a sheep pineal, Sephadex G-15, F5 and some synthetic pteridines on a Dexsil-column (2.10 m x 2 mm).

Compound	Retention time (min)
6-D-erythroneopterin	20.98
6-L-erythrobiopterin	16.44
7-L-erythrobiopterin	14.34
pterin-6-carboxylic acid	18.76
Sephadex G-15, F5-fraction eluted from 384 → 465 ml (Fig. 4a)	7.40, 16.72, 21.08
paper chromatography band (Rf, 64) see Table 1.	20.52

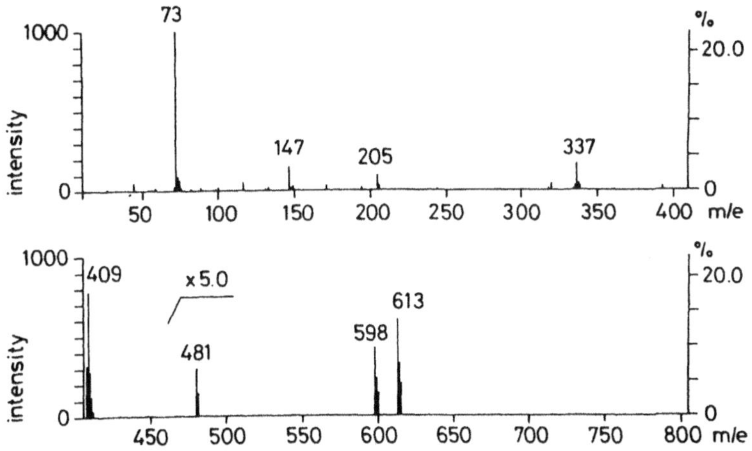

Figure 5. Mass spectrum of 6-neopterin, isolated from a sheep pineal Sephadex G-15, F5-fraction (see Fig. 4.a), after trimethylsilylation.

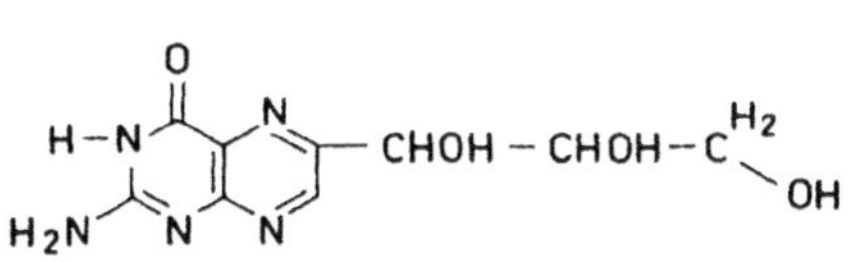

Figure 6. 6-neopterin

Table 3.

Rf-values ($x10^2$) of compounds, present in a sheep pineal UM05-residue (UM05R) and of some synthetic pteridines.

Compound or fraction	Rf-value ($x10^2$) in ethanol-water 1:3
isoxanthopterin	31
pterin-6-aldehyde	34
xanthopterin	52
6-D-erythroneopterin	59
6-L-erythrobiopterin	65
pterin-6-carboxylic acid	72
folic acid	78
riboflavin	22
UM05R	23, 45, 65, 73, 81, 86
UM05R*	44, 74, 82
UM05R**	24

* This UM05R fraction was irradiated with normal laboratory light during about 3 hours.

** The bright yellow fluorescent band (rf, 23) was eluted, lyophilized and re-chromatographed.

Table 4.

Retention time in minutes in gas liquid chromatography of an UMO5R, paper chromatographic band (Rf, 24, see Table 3) and of some synthetic pteridines, on a Dexsil-column (2.00 m x 2 mm).

Compound	Retention time (min)
leucopterin	11.32
pterin	5.60
6-D-erythroneopterin	31.44
pterin-6-carboxylic acid	26.62
UMO5R-paper chromatographic fraction (Rf, 24)	3.28, 31.68

Table 5.

The relative intensities (RI) of the excitation (e) and fluorescence (f) maxima of the isobutanol- and water-layers of different ultrafiltration-fractions corresponding to 5 grams of sheep pineals

Fraction	isobutanol-layer		water-layer	
	e/f in nm	R.I.	e/f in nm	R.I.
UM2R MW > 1000 and < 10.000	300/350	43	290/345	200
	350/415	22	350/430	64
	350/450	12		
	370/520	37	380/520	60
	450/520	58	450/520	92
UMO5R MW > 500 and < 1000	300/350	17	290/345	16
	340/390	12		
UMO5F MW < 500	300/350	470	300/355	400
	355/445	185	355/455	360
	450/525	88	450/525	67

From these experiments can be concluded, that neopterin can be isolated from UMO5R and from the Sephadex G-15, F5-fraction, two fractions prepared with a Bensinger-extraction method. In UMO5R, neopterin is probably bound. From the experiments we carried out till now, we have the impression that we can detect neopterin in the isobutanol-extract, when higher molecular weight material is extracted into the isobutanol (see Fig. 4a, F1). When that peak is absent, then we could isolate neopterin from the water layer. Therefore, we have the impression that neopterin can be bound to higher molecular weight material.

Another pteridine could be isolated from a Sephadex G-15 fraction, obtained when the column was eluted with NH_4OH after the acetic acid elution. This compound has an Rf-value of about 60 in the solvent ethanol:water 1:3. Fig. 7 gives the mass spectrum, which was obtained after trimethylsilylation. This spectrum is identical to that of synthetic pterin-6-carboxylic acid (Fig. 8) and in good agreement with the spectrum published by Lloyd et al. (1971). From that pineal fraction was obtained another mass spectrum after trimethylsilylation which can be explained by a structure like "biopterin-glucopyranose-phosphate". All the main peaks of a biopterin mass spectrum were found and those of α-D-glycopyranose-6-phosphate (see Zinbo and Sherman, 1970) were observed. This is again an indication of a pterin which can most probably be bound to other compounds.

As it was observed, using the extraction- and separation-methods presented in Scheme 2, that a fluorescence, like pteridines have, could be detected in the isobutanol- and the water layer, we have changed our last separation-steps in order to remove first the substances with MW > 500, present in the water layer, after the chloroform/methanol extraction, by ultrafiltration and to carry out subsequently the isobutanol-extraction on the UMO5F-fraction (MW < 500), with the aim to separate "free" and "bounded" pteridines.

The different pineal fractions were now prepared according to Scheme 3. The isobutanol extracts and water layers were compared and their excitation (e) and fluorescence (f) maxima were determined. The results are summarized in Table 5.

In the UM2R water layer the compounds with an indole-like fluorescence were located, while in UMO5F these fluorescence characteristics are present as well in the water layer as in the isobutanol fraction. In UMO5R only little fluorescence is found in both layers.

In the UMO5F-fraction compounds with an indole-, a pteridine- and a flavine-like fluorescence could be located in the water- and in the isobutanol-layer (see also Ebels, 1981).

The elution profiles after gelfiltration of the UMO5F-water layer and the UMO5F-isobutanol layer we published before (Ebels, 1981). In

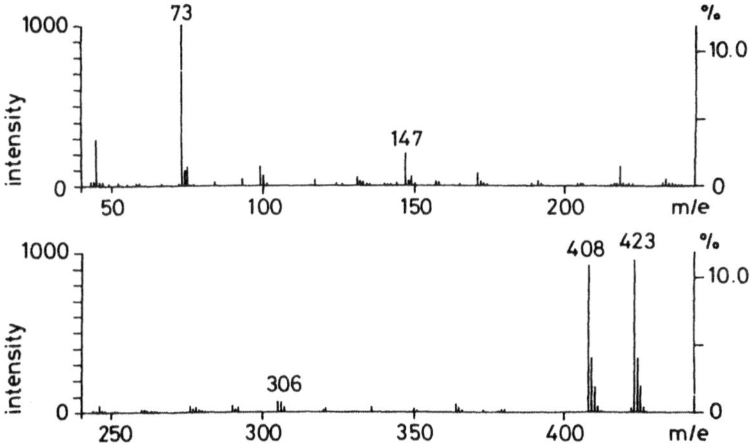

Figure 7. Mass spectrum of pterin-6-carboxylic acid, isolated from an UMO5 Sephadex G-15 fraction, (see Scheme 2), after trimethylsilylation.

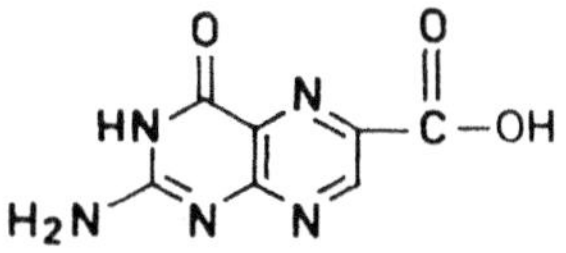

Figure 8. Pterin-6-carboxylic acid.

SCHEME 3. OUTLINE OF THE METHODS USED IN PART I.b.

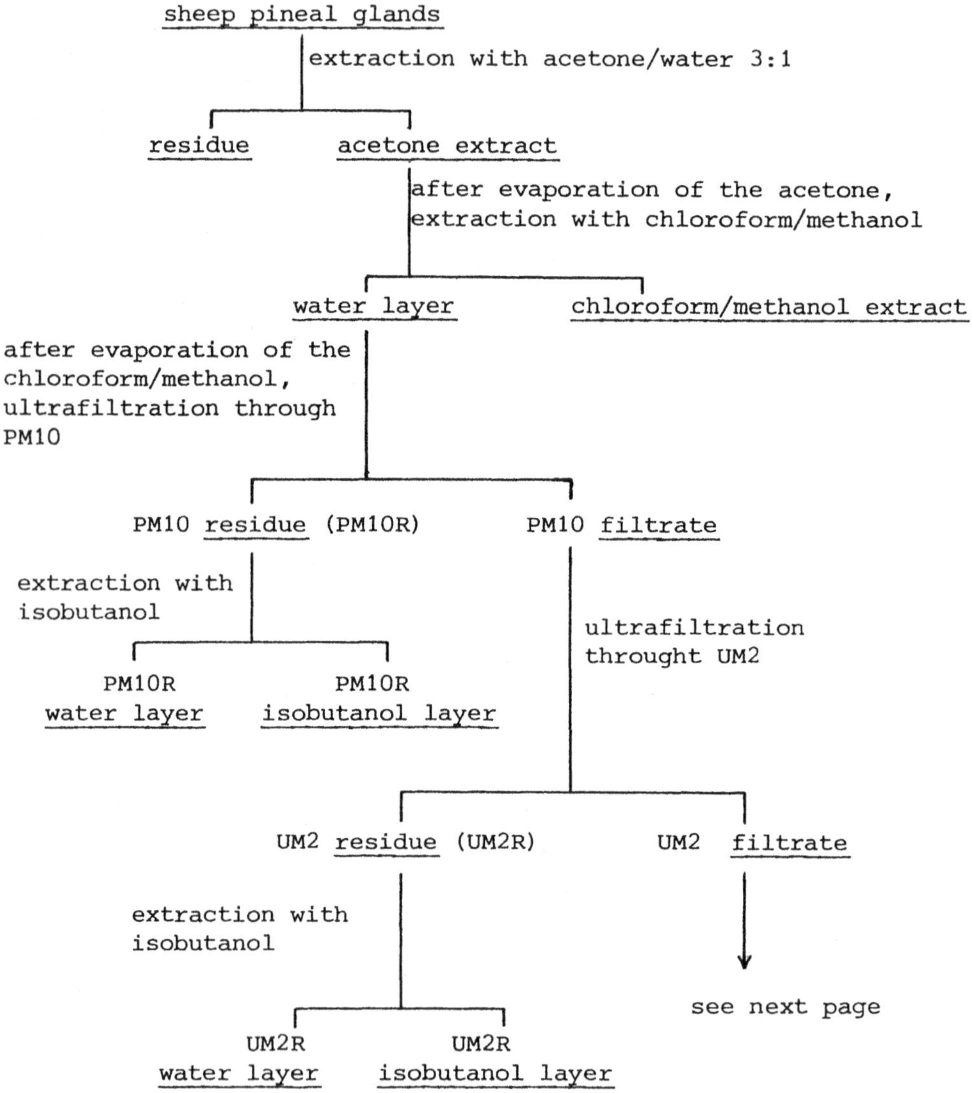

SCHEME 3. OUTLINE OF THE METHODS USED IN PART I.b. (continued)

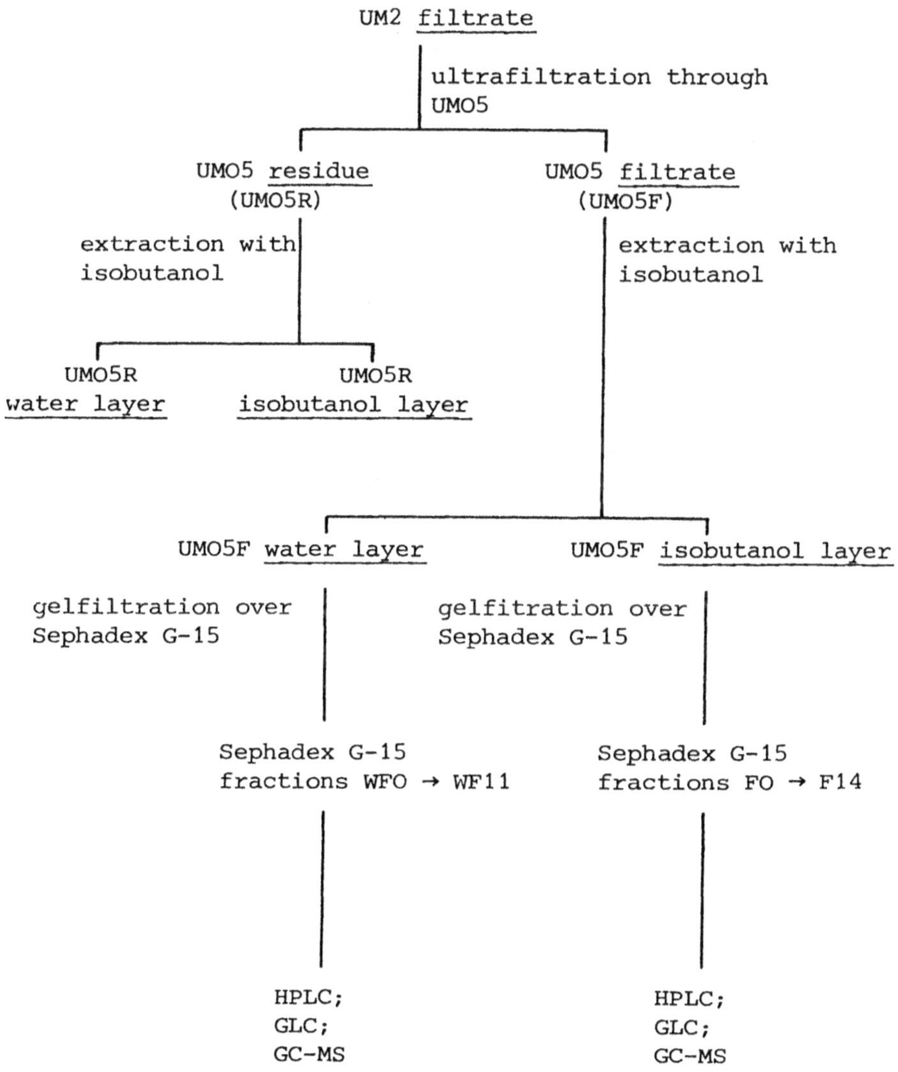

the water layer the different fluorescence peaks were about twice as high as in the isobutanol layer. In the isobutanol layer however, one very high peak with a pteridine-like fluorescence could be detected, eluted between 347 and 408 ml (F4), and for the water layer a comparable fluorescent peak was eluted between 303 and 358 ml (WF2). In the Sephadex G-15 fraction F2 (eluted from 202 → 230 ml) of the isobutanol layer a fluorescence could be observed with e/f = 350 nm/ 430 nm. In that Sephadex G-15 fraction applied to a high performance liquid chromatography column (HPLC) a peak was observed with a retention time identical to that of pterin-6-carboxylic acid; a Partisil PXS 10/25 SCX; col. no 1E (Whatman Inc.) was used.

The main peaks obtained in HPLC-experiments are summarized in Table 6. The isobutanol-Sephadex G-15 fraction F4 was further studied with GLC and GC-MS after trimethylsilylation, and a spectrum identical with that of 6-biopterin was obtained.

Although the retention time of the highest peaks in HPLC-experiments were comparable to those of respectively pterin-6-carboxylic acid and 6-biopterin, only from the isobutanol Sephadex G-15 fraction F4, a mass spectrum, identical to that of synthetic biopterin was obtained, besides the mass spectra of three other compounds with molecular weight peaks M/e of 309; 282; 541. From the water layer Sephadex G-15 WF2-fraction no mass spectrum of biopterin was obtained, but from that fraction the mass spectra of four other compounds were determined with M/e = 304; 367; 533; 628. All these mass spectra could not be identified till now.

With these results it is again clearly demonstrated, that HPLC is a very nice analytical method, but that the final step for structure elucidation of compounds must always be a GC-MS-experiment! From all these experiments can be concluded that in different ultrafiltration fractions of the water layer and in several Sephadex G-15 fractions of sheep pineals, prepared according to a Bensinger-extraction method, neopterin, pterin-6-carboxylic acid and biopterin could be isolated and identified. Moreover, in the fractions with the highest pteridine fluorescence several other compounds could be detected, but the structure of those is till now unknown. Thus several fractions are still mixtures of compounds! As comparable bovine pineal fractions showed COH-inhibitory activity and pteridines can have an antigonadotropic activity in <u>in vitro</u> and <u>in vivo</u> experiments (Ebels, 1979, 1981, and another chapter of this book) it will be worthwhile to study, whether the pteridines detected in the above described pineal fractions can be responsible for the observed antigonadotropic effects.

II. THE INFLUENCE OF SOME SYNTHETIC PTERIDINES AND LOW MOLECULAR WEIGHT PINEAL FRACTIONS ON INDOLE METABOLISM OF PINEALS

Cremer-Bartels and Hollwich (1978) were the first authors who

Table 6.

HPLC-retention times of the main peaks with a pteridine-like fluorescence, in some Sephadex G-15 sub-fractions of UM05F (isobutanol- and water-layer)

Isobutanol-layer			water-layer		
fraction	retention time in min.	co-chromatographically identical with	fraction	retention time in min.	co-chromatographically identical with
F2	1.3	pterin-6-carboxylic acid	WF4	1.5	pterin-6-carboxylic acid
F4	6.8	6-biopterin		6.4	6-biopterin

indicated an influence of pteridines (triaminophenylpteridine or triamterene) on hydroxyindole-O-methyltransferase (HIOMT), the enzyme responsible for all O-methylating processes in indole metabolism in the pineal and the retina (Balemans, 1979). A decrease of HIOMT-activity was observed when low concentrations of the substrate N-acetylserotonin were tested. These authors postulate that naturally occurin pteridines might be involved in the regulation of HIOMT in dependence of light.

Cremer-Bartels (1979) showed the effect of two synthetic pteridines: triamterene and xanthopterin on multiple forms of HIOMT in the retina and the pineal gland. These authors, using flat-bed-gel-electrofocusing technique, found several HIOMT enzymes, which differ in the pH of the isoelectric point. The pattern of the concentrations of the enzymes differed in dependence on light adaptation. The blocking effects of synthetic pteridines varied depending on light adaptation and species. Moreover, it was observed that retina extracts can inhibit the activity of purified pineal HIOMT.

Cremer-Bartels and Ebels (1980) have shown that pterin-6-aldehyde one of the photolytic products of folic acid, can decrease the activities of bovine pineal HIOMT, purified by flat-bed-gelelectrofocusing, with isoesectric points at pH 5.0, 5.2 and 5.7. These authors suggest that pteridines may be involved in non-retinal effects of light on pineal melatonin biosynthesis, besides the retinal pathway.

Ebels and Cremer-Bartels (1982) observed that low molecular weig substances in partially purified pineal fractions, which show a pteridine-like fluorescence, can also influence the activity of HIOMT, isolated from avian pineals and retina.

The results described demonstrated that this mechanism of inhibiting HIOMT activity seems to be independent of the species and the organ. A blocking effect was found in this in vitro system, whether retinal or pineal HIOMT-preparations were used.

In other experiments, with dialyzed and non-dialyzed bovine pineal extracts, was shown that low molecular weight substances, probably accompanying the high molecular-weight HIOMT-enzyme preparation can suppress the activity of the enzyme. It is however, not known at this moment if these "suppressor" molecules are pteridines. However, it must be considered that in the experiments, described above, only the activity of HIOMT at one point in a 24 h period was tested. In the experiments described in the following part, the influence of pteridines on the capacity of methylation in pineals is determined every 2 or every 4 hours during a 24 h period.

Balemans (1979, 1981); Balemans et al. (1980) studied the influence of three synthetic pterins (2-amino, 4-oxo-pteridines) on the diurnal fluctuations of HIOMT activity in the isolated pineal gland of 42-day old male Wistar rats, during the month of October. The method used to determine the HIOMT-activity of the pineal (Balemans et al., 1978) permits the separate determination of the synthesis of 5-methoxyindoles. The results of these experiments showed that reduced neopterin stimulated the methylation of the substance 5-hydroxytryptophan (5-HTP) 5-hydroxytryptamin (5-HT) and 5-hydroxyindole-3-acetic acid (5-HIAA), during the night, HIOMT-action on the combination N-acetylserotonin with 5-hydroxytryptophol (N-Ac-5-HT/5-HTL) was shifted to a later moment in the dark period.

Pterin-6-aldehyde stimulated HIOMT-action on 5-HT during the day-time, HIOMT action on the substrates 5-HTP, 5-HIAA and N-Ac-5-HT/5-HTL was shifted towards an earlier period.

Isoxanthopterin did not exert any influence on diurnal variation in the HIOMT-activities studied in this period of the year. From these experiments was concluded that reduced neopterin and pterin-6-aldehyde can influence the activity and the circadian rhythmicity of 5-methoxyindole synthesis. In later experiments has been shown that the above described, circadian rhythmicity of HIOMT activity could only be influenced by pteridines, when the maximal HIOMT-activity was measured at 24.00 o'clock (in June and in the period from half September till the beginning of October). There was no or only a small influence detectable, when the maximal HIOMT-activity was determined at 04.00 o'clock (in April and in the last half of October). These alterations in sensitivity of HIOMT-activity, also striking in the month of December and January (Ebels et al., in preparation, see Fig. 9a and 9b) may be of physiological importance. For details see also Ebels (1981).

Recently, Balemans et al. (in press) studied changes in the circadian rhythmicity of HIOMT activity in the synthesis of 5-methoxyindoles in the pineal gland of 28 day old male Wistar rats, exposed to white, red and green light. The results of these experiments showed that red light, like pterin-6-aldehyde shifts the peak of activity of HIOMT to the light period, whereas green light, like reduced neopterin, shifts that peak of activity to the dark period. In juvenile rats (Miline, 1949) and in birds (Benoit, 1964, 1972) red and green light show opposite effects on gonadal growth, red light showing a stimulatory and green light an inhibitory effect. From these experiments (Balemans et al., in press) it appears that red and green light show also opposite effects on HIOMT circadian activity. Therefore, Balemans et al. concluded that it seems likely that a relationship exists between light of different wavelenghts, pterins, indolemetabolism and gonadal development.

The above described experiments show two different approaches to study the influence of pteridines on the activity of HIOMT. In the first part only one point of the 24 h period and in the second

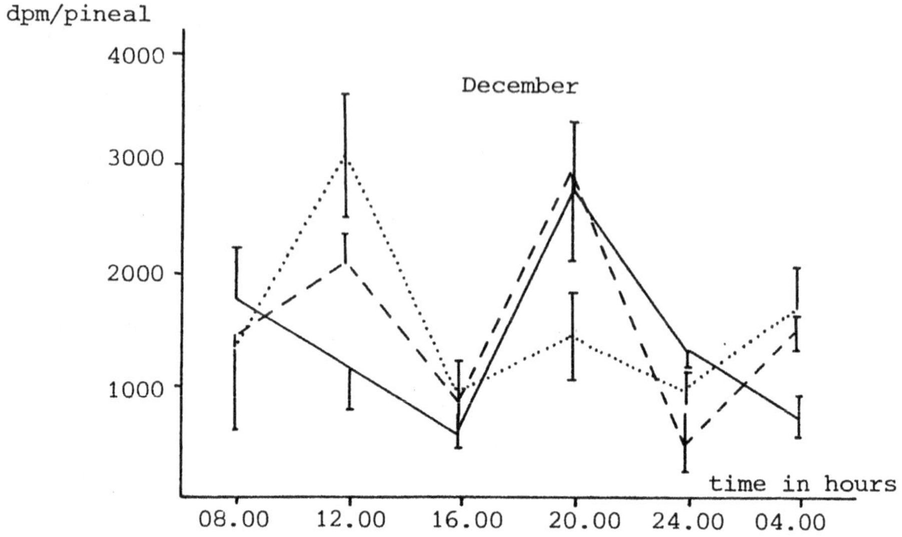

Figure 9a: The day/night rhythm of the HIOMT-activity in synthesizin melatonin/5-methoxytryptophol (---), and the influence of pterin-6-aldehyde (....) and reduced neopterin (——) on th enzyme activity, in pineals of 28-day old male Wistar rat *in vitro*, in December.

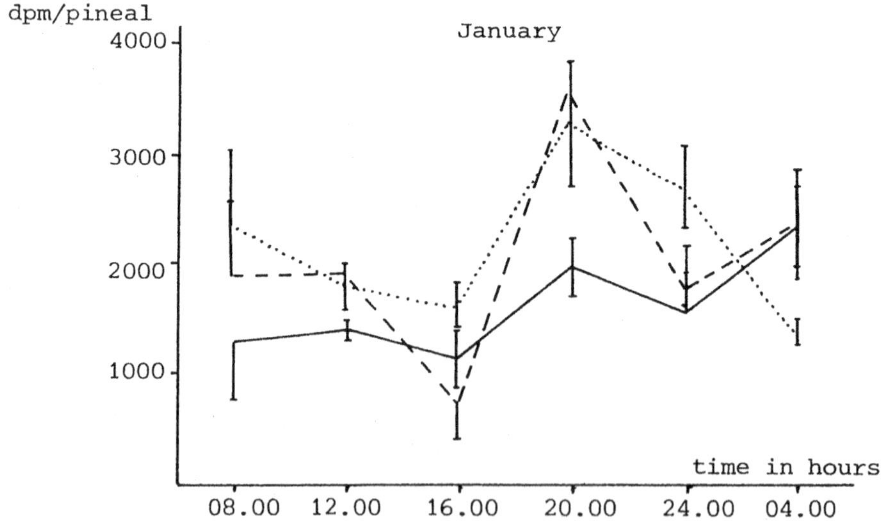

Figure 9b: gives the curves of comparable experiments carried out in January.

part the influence of pteridines on the circadian rhythmicity of the enzyme activity was studied. In the last part it is observed that pterin-6-aldelhye can shift the peaks of activity, this means that in that model it cannot be said if it is stimulatory or inhibitory. Therefore, by testing only one point the then observed inhibition can only be concluded for the time examined.

From the above described results has been concluded that pteridines can influence the HIOMT activity of the pineal. At the other side the biosynthesis of pteridines can be influenced by indoles.

Katoh et al. (1982) described that rat brain sepiapterin reductase, which is involved in the biosynthesis of tetrahydrobiopterin, was directly inhibited by a catecholamine, L-norepinephrine and by the indoleamines; serotonin, N-acetylserotonin and melatonin. These authors postulate that there might be a regulatory system in the formation of norepinephrine or serotonin in the brain by the amount of melatonin.

Much more work has to be carried out before we will really understand how the biosynthesis of pteridines and that of indoles are related.

SOME GENERAL REMARKS

The following diseases show a disorder of pteridine metabolism: Schaub et al. (1978) described a patient with atypical phenylketonuria, due to defective dihydrobiopterin biosynthesis. The same group have published several cases; see also Ebels (1981). Nagatsu et al. (1981) found that in the striatum of Parkinsonian human brain both biopterin concentrations and tyrosine hydroxylase activity was greatly reduced compared to age-matched control brains. Moreover, the same group observed that the dopamine beta-hydroxylase in the cerebrospinal fluid of Parkinsonian patients was greatly decreased as compared with that of control patients (Nagatsu et al., 1982). Ziegler and Kokolis (1979) found very high levels of BH4 in all tumour patients; see also Ebels (1981). Recently Ziegler et al. (1982a,b) published the presence of a glycoprotein in tumour-bearing patients, which binds pterins. The authors discuss the possibility that this glycoprotein may cause the disorder in pteridine-metabolism in these patients. Whether this effect can be influenced by pineal substances remains to be studied. From the work of Lapin (1974, 1976, 1978, 1979) we know that the pineal can influence neoplastic growth, see also Lapin and Ebels (1981); Lapin and Frowein (1981).

Kapatos and Kaufman (1981) observed that peripherally administered reduced pterins do enter the brain. These authors concluded from their experiments, that the content of tetrahydrobiopterin (BH4) in rat brain was doubled by peripherally administered BH4, the natural L-stereoisomer being more effective than the unnatural D-configuration. The model pteridine, 6-methyltetrahydropterin was ten times more efficient than BH4, crossing the bloodbrain barrier. These authors suggest that administration of tetrahydropterins might provide a pharmacological technique for increasing the biosynthesis of catecholamines and serotonin in the brain and periphery, and could perhaps be useful in the treatment of disorders resulting from deficits in biogenic amines biosynthesis.

Levine et al. (1981) concluded from their experiments, that the hydroxylase cofactor BH4, and its biosynthetic system are localized in dopaminergic nerve terminals in the striatum. This conclusion was based on the nearly equivalent loss of tyrosinehydroxylase and BH4 and its initial biosynthetic enzyme, guanosine triphosphate cyclohydrolase, after injection of 6-hydroxydopamine into the substantia nigra. These authors found moreover, that the phosphorylated form of tyrosine hydroxylase is saturated with cofactor and therefore optimally active. Conversely, the nonphosphorylated enzyme would be relatively inactive and would not contribute significantly to the endogenous synthesis of catecholamines. It will be very interesting to carry out comparable experiments with pineal tissue in the future, in order to locate the high amount of cofactor BH4, the hydroxylases and its biosynthetic system.

Different methods are used for the determination of pteridines, see Ebels (1981). The most sensitive and specific method seems to be the method reported by Nagatsu et al. (1981). Using a specific antiserum against 6-L-erythrobiopterin, a radioimmunoassay (RIA) has been developed to measure the biopterin concentrations in urine, serum, cerebrospinal fluid and tissues. The limit of sensitivity was 0.5 p mol/tube. For details see also Ebels (1981). This RIA may be of great importance for a better understanding and the significance of the high amount of pineal biopterin for different metabolic processes, especially those in which catecholamines and indoles play an important role.

SUMMARY

1. 6-L-erythrobiopterin was isolated and identified from sheep pineals by means of rather simple and mild extraction and separation methods, from a low molecular weight fraction MW < 500.
2. A second pterin neopterin was isolated and identified, from two different fractions, prepared with a Bensinger isobutanol extraction method. This pterin can probably be present in sheep pineal fractions as a complex which is to date still unidentified.
3. Other substances in pineal fractions, displaying a pteridine-like fluorescence were detected; they are isolated but not yet identified.
4. Pterin-6-carboxylic acid was isolated and identified from an UM05R fraction, MW > 500 and < 1000, indicating that a bound pterin may be present in that fraction.

5. Pteridines can influence the indole metabolism of rat pineals *in vitro*, and probably reverse.
6. Pterin-6-aldehyde and low molecular weight sheep pineal fractions, displaying a pteridine-like fluorescence, can decrease the activity of isolated HIOMT.
7. A disorder of pteridine metabolism is found in atypical phenylketonuria, Parkinson disease and in neoplastic growth.
8. The radioimmunoassay for 6-L-erythrobiopterin will be of great importance for pterin-metabolism in general and hopefully for pineal-biopterin metabolism particularly.

REFERENCES

Balemans, M.G.M., 1979, Indole metabolism in the pineal gland of the rat. Some regulatory aspects, *Progr. Brain Res.*, 52:221.

Balemans, M.G.M., 1981, Indole metabolism in the pineal gland, the harderian gland and the retina in mammals *in*: "The Pineal Organ," A. Oksche and P. Pévet, eds., Elsevier/North-Holland, Amsterdam-New York-Oxford.

Balemans, M.G.M., Noordergraaf, E.M., Bary, F.A.M., and Van Berlo, M.F., 1978, Estimation of the methylating capacity of the pineal gland. With special reference to indole metabolism, *Experientia*, 34:887.

Balemans, M.G.M., Van Benthem, J., Legerstee, W.C., de Morée, A., Noteborn, H.P.J.M., and Ebels, I., 1980, The influence of some pterins on the circadian rhythmicity of hydroxyindole-O-methyl transferase in the pineal gland of 42-day old male Wistar rats, *Reprod. Nutr. Develop.*, 20(4A):1051.

Balemans, M.G.M., Ebels, I., Hendriks, H.G., and Van Berlo, M.F., 1982, Changes in the circadian rhythmicity of hydroxyindole-O-methyl transferase (HIOMT) activity in the synthesis of 5-methoxyindoles in the pineal gland of 28-day old male Wistar rats, exposed to white, red and green light, *J. Neural Transm.*, in press.

Benoit, J., 1964, The role of the eye and of the hypothalamus in the photostimulation of gonads in the duck, *Ann. N.Y. Acad. Sci.*, 117:204.

Benoit, J., 1972, Etude quantitative de l'action de la lumière visible sur les fonctions génitales et endocriniennes et autres fonctions vegetatives des vertébrés. Mécanisme physiologique de cette action, *Lux*, 69:1.

Bensinger, R., Vaughan, M.K., and Klein, D.C., 1973, Isolation of a non-melatonin lipophilic antigonadotrophic factor from the bovine pineal gland, *Fed. Proc. Am. Socs. Exp. Biol.*, 32:225.

Blakly, R.L., 1969, The biochemistry of folic acid and related pteridines, North Holland Publishing Company, Amsterdam-London.

Blask, D.E., Vaughan, M.K., Reiter, R.J., Johnson, L.Y., and Vaughan, G.M., 1976, Prolactin-releasing and release-inhibiting factor activities in the bovine, rat and human pineal gland . *In vitro* and *in vivo* studies, *Endocrinology*, 99:152.

Cremer-Bartels, G., 1979, The effect of pteridines on multiple forms of hydroxyindole-O-methyl transferase (HIOMT) in the retina and the pineal gland. Progr. Brain Res., 52:231.

Cremer-Bartels, G., and Hollwich, F., 1978, Effect of triaminophenylpteridine on hydroxyindole-O-methyl transferase of rat pineal gland and bovine retina, J. Neural Transm., Suppl. 13:360.

Cremer-Bartels, G., and Ebels, I., 1980, Pteridines as nonretinal regulators of light-dependent melatonin biosynthesis, Proc. Natl. Acad. Sci. U.S.A. 77:2415.

Ebels, I., 1979, A chemical study of some biologically active pineal fractions, Progr. Brain. Res., 52:309.

Ebels, I., 1981, Pteridines in the pineal organ, in: "The Pineal Organ," A. Oksche and P. Pévet, eds., Elsevier/North-Holland, Amsterdam-New York-Oxford.

Ebels, I., Benson, B., Bria, C.F., McDonnell, D., Chang, S.Y., and Hruby, V.J., 1978, Location by paper chromatography of compensatory ovarian hypertrophy (COH) inhibiting activity in acetic acid extracts from bovine pineals, J. Neural Transm., 42:275.

Ebels, I., Benson, B., Bria, C.F., Richardson, D., Larsen, B.R., and Hruby, V.J., 1979, Location by paper chromatography of compensatory ovarian hypertrophy (COH) inhibiting activity in isobutanol extracts of bovine pineals, J. Neural Transm., 45:43.

Ebels, I., and Cremer-Bartels, G., 1982, Inhibition of avian mammalian hydroxyindole-O-methyl transferase (HIOMT) with low molecular weight fractions of mammalian pineal glands, Life Sci., 40:1369

Fukushima, T., and Nixon, J.C., 1979, Reverse-phase high performance liquid chromatographic separation of unconjugated pterins and pteridines, in: "Chemistry and Biology of Pteridines," R.L. Kisliuk and G.M. Brown, eds., North-Holland-Amsterdam.

Fukushima, T., and Nixon, J.C., 1980, Analysis of reduced forms of biopterin in biological tissues and fluids, Anal Biochem., 102:176.

Kapatos, G., and Kaufman, S., 1981, Peripherally administered reduced pterins do enter the brain, Science, 212:955.

Kapatos, G., Kaufman, S., Weller, J.L., and Klein, D.C., 1981, Biosynthesis of biopterin: Adrenergic cyclic adenosine monophosphate-dependent inhibition in the pineal gland, Science, 213: 1129.

Katoh, S., Sueko, T., and Yamada, S., 1982, Direct inhibition of brai sepiapterin reductase by a catecholamine and an indoleamine, Biochem. Biophys. Res. Comm., 105:75.

Lapin, V., 1974, Influence of simultaneous thymectomy and pinealectomy on the growth and formation of metastases of the Yoshida sarcoma in rats, Exp. Path., 9:108.

Lapin, V., 1976, Pineal gland and malignancy, Osterr. Z. Onkol., 3:51

Lapin, V., 1978, Effects of reserpine on the incidence of 9,10-dimethyl-1,2-benzanthracene-induced tumours in pinealectomized and thymectomized rats, Oncol., 35:132.

Lapin, V., 1979, Pineal influence on tumors, Progr. Brain Res., 52: 523.

Lapin, V., and Ebels, I., 1981, The role of the pineal gland in neuroendocrine control mechanisms of neoplastic growth, J. Neural Transm., 50:275.

Lapin, V., and Frowein, A., 1981, Effects of growing tumours on pineal melatonin levels in male rats, J. Neural Transm., 52:123.

Levine, R.A., Kuhn, D.M., and Lovenberg, W., 1979, The regional distribution of hydroxylase cofactor in rat brain, J. Neurochem., 32:1575.

Levine, R.A., Miller, L.P., and Lovenberg, W., 1981, Tetrahydrobiopterin in striatum: Localization in dopamine nerve terminals and role in catecholamine synthesis, Science, 214:919.

Lloyd, T., and Weiner, N., 1971, Isolation and characterization of a tyrosine hydroxylase cofactor from adrenal medulla, Mol. Pharmac., 7:569.

Miline, R., 1949, L'influence de la lumière sur la maturation sexuelle, Medicinski Pregled., (Med Progr.) 3:86.

Moszkowska, A., Hus-Citharel, A., L'Héritier, A., Zurburg, W., and Ebels, I., 1976, Separation of pineal extracts by gelfiltration. V. Location by paper chromatography, J. Neural Transm., 38:239.

Nagatsu, T., Yamaguchi, T., Kato, T., Sugimoto, T., Matsuura, S., Akino, M., Tsushima, S., Nakazawa, N., and Ogawa, H., 1981, Radioimmunoassay for biopterin in body fluids and tissues, Anal. Biochem., 110:182.

Nagatsu, T., Wakui, Y., Kato, T., Fujita, K., Kondo, T., Yokochi, F., and Narabayashi, H., 1982, Dopamine beta-hydroxylase activity in cerebrospinal fluid of Parkinsonian patients, Biomed Res., 3:95.

Purrmann, R., 1940a, Über die Flügelpgmente der Schmetterlinge VII. Synthese des Leukopterins und Natur des Guanopterins, Ann. Chem., 544:182.

Purrmann, R., 1940b, Die Synthese des Xanthopterins. Uber die Flügelpigmente der Schmetterlinge X, Ann. Chem., 546:98.

Purrmann, R., 1941, Konstitution und Synthese des sogenannten Anhydroleukopterins. Über die Flügelpigmente der Schmetterlinge XII, Ann. Chem., 548:284.

Schaub, J., Däumling, S., Curtius, H.-Ch., Niederwieser, A., Bartholomé, K., Viscontini, M., Schircks, B., and Bieri, J.H., 1978, Tetrahydrobiopterin therapy of atypical phenylketonuria due to defective dihydrobiopterin biosynthesis, Arch. Disease Childhood, 53:674..

Schöpf, C., and Wieland, H., 1926, Über das leukopterin das weisse Flügelpigmentder Kohlweiszlinge (Pieris brassicae und P. Napt.) Ber. Deut. Chem. Ges., 59:2067.

Van der Have-Kirchberg, M.M.L., de Morée, A., Van Laar, J.F., Gerwig, G.J., Versluis, C., Ebels, I., Hus-Citharel, A., L'Héritier, A., Roseau, S., Zurburg, W., and Moszkowska, A., 1977, Separation of pineal extracts by gelfiltration. VI. Isolation and identification from sheep pineals of biopterin: comparison of the isolated compound with some synthetic pteridines and the biological activity in in vitro and in vivo bioassays. J. Neural.

Transm., 40:205.

Wieland, H., and Schöpf, C., 1925, Über den gelben Flügelfarbstoff des Citronenfalters (*Gonepteryx rhamni*), *Ber. Deut. Chem. Ges.*, 58:2178.

Ziegler, I., and Kokolis, N., 1979, *In vivo* metabolism of deutero-L-phenylalanine and deutero-L-tyrosine and levels of tetrahydro-biopterin in the blood of tumour bearing organisms, *in*: "Chemistry and Biology of Pteridines," R.L. Kisliuk, and G.M. Brown, eds., Elsevier/North-Holland, New York-Oxford-Amsterdam.

Ziegler, I., Maier, K., and Fink, M., 1982, Pteridine-binding α1-acid glycoprotein from blood of patients with neoplastic diseases, *Cancer Res.*, 42:1567.

Ziegler, I., Maier, K., and Wilmans, W., 1982, Blood levels of a pteridine-binding α1-acid glycoprotein in cancer patients, *Cancer Res.*, 42:1574.

Zinbo, M., and Sherman, W.R., 1970, Gas chromatography and mass spectrometry of trimethylsilyl sugar phosphates, *J. Am. Chem. Soc.*, 92:2105.

PEPTIDIC AND PROTEIC SUBSTANCES ISOLATED FROM PINEALS AND THEIR RELATION TO THE HYPOPHYSIAL-HYPOTHALAMIC-GONADAL AXIS

I. Ebels*

Department of Organic Chemistry
State University of Utrecht
The Netherlands

At the ultrastructural level, processes implicated in the synthesis and release of peptidic/proteic compounds have been clearly identified in pineal specific cells (Collin, 1979, 1981; Collin and Oksche, 1981; Pévet, 1979, 1981a,b, 1982a,b). Thus the synthesis and release of peptidic/proteic substances cannot be discarded. The isolation, purification and characterization of this class of pineal compounds have however presented a formidable challenge, since extraction methods for peptide purification have failed to yield large quantities of pure compounds, sufficient for structure elucidation. Moreover, a simple and specific bioassay, indispensable for the isolation of biologically active compounds has not been available till now. Different observations however support the presence of pineal peptidic and proteic substances showing a definite biological activity.

The research group of Thiéblot has worked since about 1950 on the isolation and purification of (a) pineal peptidic/proteic substance(s). For a survey of this work see Thiéblot and Thiéblot (1981).

Milcu et al. (1963) isolated a proteic product from the pineal with hypoglycemic properties that they termed "pinealin".

Reiss et al. (1963a,b) found two dramatically opposed actions of different fractions of aqueous extracts of pineal glands on the

*Our research is supported by grants of The Netherlands Organization for the Advancement of Pure Research (ZWO) through the Foundation for Medical Research (FUNGO) No. 13-35-33 and for the Foundation for Biological Research (BION) No. 14.70.03.

gonadal system. The supernatant of an aqueous extract of pineals, deproteinized by trichloroacetic acid (TCA) to a final concentration of 3.3%, freed from TCA by repeated ether extraction, and lyophilized was used. This fraction contained a substance inhibiting gonadotropin activity, while the TCA precipitate after being dried by repeated washing with acetone, contained a stimulatory principle. While studying different pineal gland fractions, one fraction reduced whereas another increased the phosphorus turnover of the pituitary. The same authors observed that the inhibitory pineal extract stimulated the ^{32}P incorporation in the brain and decreases considerably that of the liver. One of the most interesting aspects of these investigations was the observation of the opposite roles played by the pineal and pituitary in their metabolic actions on the brain. When these authors extracted rat cerebrum and cerebellum, freeze dried cattle brain, rat muscles, spleen, liver and kidney, the above described effects could not be observed with a comparable fraction obtained from these tissues. Therefore it seems that the observed effects are rather specific for the pineal.

Neacșu (1972) isolated a peptidic fraction from bovine pineals, which he termed the E5-fraction. This fractions blocks spermatogenesis in gonadotropin-treated male frogs and acts on pituitary LH and prolactin levels in intact rats (Vaughan et al., 1982).

Ota et al. (1975) observed an inhibition of ovulation induced with pregnant mare serum (PMS) and human chorionic gonadotropin (HCG) by a melatonin-free extract of bovine pineal powder. The authors hold that perhaps two substances which possess anti-ovulatory properties are present in the pineal: one having a molecular weight (MW) of more than 100.000 and one with a MW less than 10.000.

Slama-Scemama et al. (1979) have shown that high- and lower molecular weight ovine pineal fractions, prepared under dim red light conditions stimulated the gonadotropic releasing activity of the medial basal hypothalamus. For details see Ebels and Benson (1978) and Benson and Ebels (1981).

Several studies indicate also the presence of some of the hypophysial and hypothalamic hormones in the pineal, detected in general with radioimmunological methods. These compounds, however, can according to Pévet (see this book) originate:

- from other organs than the pineal because these compounds are present in cells such as nerves, which connect the pineal with other brain structures.
- from other organs than the pineal, with a secondary uptake by the gland from the general circulation
- from the pineal, synthesized by specific cells in the gland.

Only compounds of the last group can be considered as true pineal "hormones".

The present paper will give a survey of the results, obtained in the last decades with different pineal fractions. Especially effects of different partially purified pineal fractions on the reproductive system will be described. The following subjects will be discussed.

I. Experiments *in vitro*, using a bioassay or/and radioimmunoassay
I.1. Experiments using anterior pituitaries, and a bioassay with immature female mice
I.2. Experiments using anterior pituitaries and a radioimmunoassay (RIA) for the detection of prolactin
I.3. Experiments with rat and mice hypothalami, using a bioassay with immature female mice
II. Experiments *in vivo*, using hemicastrated and castrated animals
II.1. Effects of bovine and ovine pineal extracts on compensatory ovarian hypertrophy (COH)
II.2. The isolation of the tripeptide threonylseryllysine from bovine pineals using the COH-test
III. Effects of a partially purified bovine pineal fraction E5
IV. Comparison of ovine peptidic/proteic pineal fractions with a E5-fraction, isolated from bovine pineals

I. EXPERIMENTS *IN VITRO*, USING A BIOASSAY OR/AND RADIOIMMUNOASSAY

I.1. Experiments Using Anterior Pituitaries, and a Bioassay with Immature Female Mice

It has been known for many years that either fresh pineal glands of rats and sheep, or acetone-dried sheep pineal powder, can diminish or inhibit the secretion of anterior hypophyses of adult male rats *in vitro* (Moszkowska, 1956, 1958, 1964 1965; Moszkowska and Heersche, 1962). When an extract of sheep pineal powder was gelfiltrated on a Sephadex G-25 column, two subfractions were obtained which demonstrated opposite effects on the follicle stimulating activity of anterior hypophysis of the male rat *in vitro*. The low molecular weight Sephadex G-25 fraction F2 increased the follicle stimulating activity. The subfraction Sephadex G-25 F3, could diminish the follicle stimulating activity of the anterior hypophysis (Ebels et al., 1965). In further purification steps only a slight antigonadotropic activity was found in one experiment, while in other experiments no activity was observed at all.

Different regions of sheep hypothalamic tissue, cerebral cortex and cerebellum were extracted and separated with the same methods as employed in the experiments with sheep pineal bodies. In these experiments no antigonadotropic activity was found in the Sephadex G-25 F3-fraction of an extract of the median eminence region of sheep hypothalami. Significant stimulation was observed with the Sephadex G-25 F3-fraction of sheep cerebral cortex extracts and neither stimulatory nor inhibitory activity was found in the low molecular weight Sephadex G-25 F3 of cerebellum.

In other experiments was shown that the antigonadotropic activity of sheep pineal fraction F3 differed from the activity of synthetic melatonin, 5-methoxytryptophol, serotonin and arginine vasotocin (Moszkowska and Ebels, 1968, 1971). The conclusion was drawn that the antigonadotropic activity of sheep pineal Sephadex G-25 F3 seemed to be fairly specific for the pineal body, but that it is unstable under the experimental conditions used at that time (Ebels et al., 1970; Moszkowska and Ebels, 1971).

Therefore, simple and milder extraction and separation methods were developed (Balemans et al., 1970; Ebels et al., 1972a,b; Zurburg and Ebels, 1975; Ebels and Horwitz-Bresser, 1976). With these methods a substance showing special fluorescence characteristics could be isolated from a low molecular weight Sephadex G-25 fraction. This substance showed an excitation maximum at 360 nm and a fluorescence maximum at 440 nm, and could not be detected in similar amounts of sheep cerebral cortex treated in the same way (Zurburg and Ebels, 1974).

It has been shown in later experiments that the substance(s) with that special fluorescence, was associated with an antigonadotropic activity in vitro, using anterior hypophysis of male rats. After ultrafiltration through Amicon membranes, it was observed that the antigonadotropic activity and the special fluorescence was located in the UM05F-fraction (MW < 500). After gelfiltration of UM05F on Sephadex G-10 this fluorescent compound and the antigonadotropic activity was found in the same fraction. From that Sephadex G-10 fraction 6-L-erythrobiopterin was isolated and identified using subsequently preparative paper-electrophoresis and paper-chromatography, gasliquid-chromatography (GC) after trimethylsilylation and GC-mass-spectrometry.

It has been shown in later experiments that isolated and synthetic 6-L-erythrobiopterin preparations can diminish the gonadotropic activity of anterior hypophyses of male Wistar rats in vitro and moreover decrease radioimmunologically detectable plasma LH-levels of immature pinealectomized and hemicastrated male rats in vivo, when measured 24 h after a single injection (for details see Ebels, 1981).

From the work of Fukushima and Nixon (1979, 1980) we know that the pineal gland of rats contain much more biopterin than in seven other organs and several different regions of the brain. Moreover, these authors stated that biopterin in the pineal is mostly present in the reduced form: tetrahydrobobiopterin. This compound is the natural cofactor of tryptophan hydroxylase, phenylalanine- and tyrosine hydroxylase. We have observed later on that reduced biopterin is a very potent inhibitor of the gonadotropic activity of anterior pituitaries of male rats in vitro. Therefore, we believe that 6-L-erythrobiopterin in the reduced form is (partly) responsible for the antigonadotropic activity, found in low molecular weight sheep pineal fraction in in vitro and in vivo experiments (Ebels, 1981). Pteridines are very unstable, light sensitive compounds which may explain that pineal fractions may lose their antigonadotropic activity during isolation procedures. For details see Ebels (1979, 1981) and another chapter on pteridines in this book.

I.2. Experiments In Vitro Using Anterior Pituitaries and a RIA for the Detection of Prolactin

Bartke et al. (1975) induced pineal mediated gonadal regression in male hamsters by exposure to "short days" and observed that early recrudescence was observed after treatment with exogenous prolactin.

Blask et al. (1976) have demonstrated that hemi-pituitaries, incubated with a crude acidic extract of bovine pineals after neutralizing to pH 7.4 with NaOH, released significantly more prolactin into the incubation medium, than the corresponding control hemi-pituitaries, incubated with rat cortical extracts. It was also shown that, when incubated with bovine pineal extracts, hemi-pituitaries released significantly more prolactin into the medium than their corresponding halves, treated with rat hypothalamic extract.

On the other hand, bovine pineal fractions, prepared according to Bensinger et al. (1973), which show the most potent inhibition of compensatory ovarian hypertrophy (COH) in mice, further separated by thin-layer chromatography, gave two subfractions, which could inhibit the release of prolactin from hemi-pituitaries. From these results is concluded that crude bovine pineal extracts contain prolactin-releasing activity (PRF) and prolactin-inhibiting activity (PIF). These PRF and PIF activities can both act directly on the anterior pituitary in vitro. These results were confirmed and extended by Chang et al. (1979), using other extraction and separation methods. Dilute acetic acid extracts of bovine pineals were gelfiltrated on Sephadex G-25 and the low molecular weight Sephadex G-25 fractions were separated further by ultrafiltration, using the Amicon membranes PM10, UM2 and UM05. These last fractions were incubated with rat hemi-pituitaries and the effects on prolactin secretion determined. The UM05F-fraction (MW < 500) shows inhibition of prolactin release in this system. The PM10R-fraction (MW > 10.000) could stimulate the release of prolactin in the same system.

However, the UM05R-fraction (MW > 500 and < 1000) did not show any activity in this in vitro-system, but could reduce plasma prolactin in vivo in urethane-anesthesized rats. Larsen and Benson (1979) using high performance liquid chromatography as a final step, obtained the same results with a more purified prolactin-inhibiting fraction from bovine pineals.

From these experiments can be concluded that pineal extracts contain PRF- and PIF-activities, acting directly on the pituitary, but that there is also a partially purified low molecular weight pineal fraction, which can decrease the plasma prolactin levels, by possibly stimulating the secretion of a hypothalamic prolactin inhibiting factor.

The results of Matthews et al. (1978), Bex et al. (1978) Reiter and Ferguson (1979), and Benson and Matthews (1980), in which the transplantation of pituitaries beneath the kidney capsules either partially or totally prevented gonadal atrophy in light-deprived hamsters, suggest a possible role for prolactin and pineal prolactin regulating factors in pineal mediated gonadal atrophy. Therefore, the pineal regulation of prolactin secretion may be more important than was thought in previous years. See also Leadem and Blask (1982).

Benson and Matthews (1980) concluded from the results of their and other's experiments, that the pineal may secrete a prolactin regulating substance. Such a substance could act by either regulating the synthesis and/or release of hypothalamic prolactin releasing factors or prolactin release inhibiting factors such as dopamine. Recently, Martens et al. (1981) showed in a microsuperfusion-system of neurointermediate lobes of Xenopus laevis, that dopamine can inhibit the release of all newly synthesized peptides except the precursor, leading to an enhanced level of these peptides remaining in the lobe. Cyclic AMP stimulated the release of all peptides, except the precursor, leading to reduced levels of lobe peptides. A comparable mechanism may exist in the hypothalamus for gonadotropic releasing substances regulated by pineal substances. Benson et al. (in press) have found more experimental data to support the last hypothesis. See also Benson and Ebels (1981).

I.3. Experiments In Vitro with Rat and Mice Hypothalami Using a Bioassay with Immature Female Mice

In 1973, Moszkowska et al. showed the modulation of gonadotropin-releasing factor content in the rat hypothalamus by pineal factors in vitro and in vivo. The same group demonstrated an effect on the hypothalamus in vitro by addition of sheep pineal fractions. Those fractions were prepared by gelfiltration on Sephadex G-25 of an aqueous pineal extract and subsequent ultrafiltration through the Amicon membranes UM2 and UM05 of the low molecular weight Sephadex G-25 fractions (see Citharel et al., 1973). Mice hypothalami incubated with an UM2R-fraction (MW > 1000) showed a highly significant decrease in the secretion of hypophysiotropic activity in vitro as compared to the controls. When the same fraction was tested with anterior hypophyses of male mice, no effect on the gonadotropin was observed. Therefore it was concluded that this UM2R-fraction has an influence on the hypothalamus. Hypothalami incubated with the sheep pineal fraction UM05R (MW > 500 and < 1000) showed increased hypophysiotropic activity. No significant influence on the anterior hypophyses alone were observed with UMO5R-fractions (Citharel et al., 1973; Ebels et al., 1975).

All these experiments were carried out under normal laboratory light conditions. However, in later experiments, using an isobutanol extraction method (Ebels et al., 1979) and ultrafiltrating the remaining water layer through Amicon membranes PM10, UM2 and UMO5, the PM10R-fraction (MW > 10.000) and UM2R-fraction (MW > 1000 and < 10.000) showed a stimulating effect on the hypophysiotropic activity on the median eminence of the hypothalamus. Further ultrafiltration of this PM10R-fraction on the filters XM100 and PM30, and testing of those residues have shown that XM100R (MW > 100.000) and PM30R (MW > 30.000 and < 100.000) contain the stimulating activity. Nearly no activity was left in the PM10R fraction (MW < 30.000 and > 10.000) of the PM30 filtrate. All these experiments were carried out in red light (λ > 600 mm) and as much as possible in darkness. This was for us the first time that we found biologically active high molecular weight pineal fractions. Later on, comparable results were obtained with ultrafiltrated fractions, obtained by direct ultrafiltration of aqueous ovine pineal extracts. It may be that the experimental lighting conditions, used in the last years to protect the light-sensitive pteridines in our pineal extracts during the isolation procedure, preserve high molecular weight complexes which increase the gonadotropic-releaseing activity of male rat medial basal hypothalami. For details, see Slama et al. (1979). It may be that the pineal fractions XM100R and PM30R contain active complexes, which can be destroyed by normal light. This supposition is supported by the results of Frowein and Lapin (1979) who have compared the effects of sham-pinealectomy in newborn and adult rats performed under white and red light. These authors observed that melatonin biosynthesis is altered by hitting the pineal with white light _in situ_ during sham-operation. The melatonin synthesis was reduced irreversibly for at least 2 months. It is left unaltered when surgery is performed under red light.

Deguchi (1981) observed that direct illumination of cultured chicken pineals suppressed the night-time increase of N-acetyltransferase activity. The sensitivity curve of the enzyme activity for light of different wavelengths resembled the absorption spectrum of rhodopsin. In the near future more attention will be paid to these high molecular weight complexes.

II. EXPERIMENTS _IN VIVO_, USING HEMI-CASTRATED AND CASTRATED ANIMALS

II.1. Effects of Pineal Extracts on Compensatory Ovarian Hypertrophy (COH)

Moszkowska (1951) first showed that aqueous pineal extracts can inhibit (COH) and the formation of _corpora lutea_ in guinea pigs after unilateral ovariectomy (UO). Later it was shown that daily administration of a pineal extract (Reiss et al., 1963a) or of melatonin (Sorrentino, 1968), blocked COH after UO in the rat. Especially the research group of Benson and co-workers have carried out numerous experiments in this field. The results of the earlier work of this group are summarized in a review by Benson and Orts (1972) in which the hypothesis was postulated that pineal antigonadotropic substances act by altering the sensitivity of the hypothalamo-hypophysial axis to the feedback effects of gonadal steroids. A support for this hypothesis was found in experiments, in which it was shown that fractions of aqueous extracts of bovine pineal glands, partially purified to exclude melatonin, inhibited the post-castration rise in LH in orchidectomized rats. In later studies it was shown by the same group that similar LH-inhibitory properties can be observed in aqueous extracts of rat pineals. The media of hemi-pituitaries incubated with cerebral cortex and cerebellar, did not demonstrate similar antigonadotropic activity. Inactivation of the anti-COH substance in rat pineal incubation media and bovine pineal extracts, was produced by proteolytic enzymes (Matthews and Benson, 1973). These authors postulated already at that time that the non-melatonin pineal antigonadotropin (PAG) is a small peptide in the molecular weight range of 500 - 1000, based on ultrafiltration experiments.

From studies carried out with ovine pineal glands, using simple and mild extraction methods, it could be concluded that the ovine UMO5-residue, (MW > 500 and < 1000) contains (a) substance(s) which inhibited COH and reduced ventral prostate weights in mice; the same fraction retarded vaginal opening in young female mice. These studies showed clearly that (an) antigonadotropic substance(s) could be separated from melatonin, since this indoleamine, as well as 5-methoxytryptophol and 5-hydroxytryptophol are eluted from Sephadex G-25 and Sephadex G-10 columns after the fraction in which COH-inhibiting activity is localized.

Similarly prepared melatonin-free fractions, obtained from extracts of bovine pineal glands delayed vaginal opening time, reduce the incidence of estrous vaginal smears in rats and reduce serum LH-levels in long term orchidectomized rats.

Benson et al. (1976a,b) studied also the effects of antigonadotropic fractions derived from bovine pineal extracts on fertility and ovulation. A gonadal site of action was ruled out by Benson et al. (1976a) and the effects on ovulation and fertility were postulated to be the result of inhibition of LH-secretion. In other studies the antigonadotropin PAG was localized to pineal parenchyma and vasotocin-like activity to the pineal stalk by the same group. In later years Benson et al. attempted final purification of bovine PAG. Throughout these studies it was learned that the sulfur-containing amino acid taurine was closely associated with the COH-inhibiting factor. Further studies showed that the COH-inhibiting substance could be separated from taurine, hypotaurine and reduced and oxidized glutathione, by paper-chromatography (see also Benson and Ebels, 1981).

In 1973 Bensinger et al. described in an abstract a method for aqueous-acetone extraction of bovine pineal and the further removal of fats and indoleamines by chloroform/methanol extraction. Although the details of the methods and all the results were not published in detail the authors stated that the active principle was probably a small, non-tryptophan-containing peptide. We have extracted also bovine pineals according to the method of Bensinger et al. (1973) and confirmed the presence of COH-inhibiting activity (Ebels et al., 1979). However, further separation by gelfiltration and paper-chromatography revealed different Rf-values for the inhibiting activity than the COH-inhibiting substances obtained from acetic and aqueous extracts of ovine and bovine pineals. In later studies the antigonadotropic substance(s) obtained by both acid and isobutanol extraction could be localized using high pressure liquid chromatography. Since the amount of residue obtained are extremely small, a large amount of pineals will be needed to obtain enough pure material for structural determination by mass-spectrometry. For more details of the many experiments carried out by Benson et al., see the reviews written in recent years by Ebels and Benson (1979) and by Benson and Ebels (1981).

II.2. The Isolation of the Tripeptide Threonylseryllysine from Bovine Pineal Glands

Orts et al. (1978) isolated from bovine pineals a tripeptide, using glacial acetic acid extracts, separated by Sephadex gelfiltration, ultrafiltration and paper-electrophoresis. The antigonadotropic activity was followed through the purification steps using the compensatory ovarian hypertrophy test with female mice. The sequence was determined to be threonylseryllysine (Thr-Ser-Lys). The tripeptide was synthesized by Larsen et al. (1979) and the biological effects were studied. Studies on mice and rats have shown that Thr-Ser-Lys causes a weak antigonadotropic activity (Vaughan et al., (1980).

Although the effects on prolactin observed with Thr-Ser-Lys are consistent with those of the pineal prolactin regulating substance, partially purified by Larsen and Benson (1979), the elution times, using high performance liquid chromatography, are different, which implicate different chemical structures.

This tripeptide is of a smaller molecular weight than most pineal antigonadotropic factors prepared by the group of Benson et al. in the last years, suggesting that the use of glacial acetic acid as Orts et al. did, may have effectively cleaved this tripeptide from a larger parent molecule.

III. EFFECTS OF A PARTIALLY PURIFIED BOVINE PINEAL FRACTION E5

Milcu et al. (1963) have published the biological and chromatographic characterization of a polypeptide with pressor and oxytocic activities, isolated from bovine pineal glands. These authors concluded from the results of their experiments that their pineal polypeptide was neither arginine- or lysine vasopressin, nor oxytocin, but that the biological and chromatographic characteristics are similar to those of arginine vasotocin or a compound with similar structure. Ebels et al. (1965) attempted to isolate a peptide with pressor and oxytocic activities from sheep and bovine acetone desiccated pineal bodies by gelfiltration and column chromatography. These authors could not confirm the occurrence of arginine vasotocin in the pineal organ. Moszkowska and Ebels (1968) studied the antigonadotropic action of synthetic arginine vasotocin. These authors concluded from their experiments that arginine vasotocin can act on the gonads or on the gonadotropic hormones and not on the secretion of the anterior hypophysis _in vitro_, as was observed in _in vitro_ experiments with acetone-dried sheep pineal powder and with a Sephadex G-25 F3 fraction (Moszkowska, 1965; Ebels et al., 1965).

Neacşu (1972) used a bovine peptidic/proteic fraction E5, prepared according to the method published by Milcu et al. (1963) and proposed a composition for the peptide E5, having a molecular weight of about 2000. Under the experimental conditions this author used, the "polypeptide" E5 can inhibit, when injected prior to human chorionic gonadotropin, the activity of this hormone on spermatogenesis in male frogs. It was concluded that E5 influenced rather LH than FSH.

In another publication Neacsu (1972) postulated that the E5-peptide has a larger ring than oxytocin and vasopressin and a larger side chain of five amino acids but also ending by Pro-Arg-Gly(NH_2) as AVT and AVP, although no experimental details of methods and results were presented for this statement.

Since the time Cheesman (1970), Cheesman and Fariss (1970) published the isolation and identification of AVT from bovine pineals, a large number of publications appeared in which the effects of synthetic AVT were described. For a survey of this literature see: Vaughan and Blask (1978); Pavel (1979); Vaughan (1981) and Reiter (1980, 1981).

Recently Vaughan et al. (1982) tested again an E5-fraction prepared by Dr. Neacşu, Bucharest, Romania. These authors concluded from their experiments that: pituitary levels of prolactin were significantly diminished after five subcuteneous injections of 5 μg of this partially purified bovine pineal peptide fraction E5, in both intact and castrated rats. E5 treatment did not significantly affect the castration-induced changes in plasma LH, FSH or pituitary LH, but did partially block the pituitary fall in FSH in castrated animals.

Pévet et al. (1980) using radioimmunological methods, concluded from their experiments, that the pineal does not contain AVT, but an AVT-like peptide, which may be responsible for the antigonadotropic activity, observed with the bovine E5-fraction. See also Pévet (1979, 1981b and this book).

It can be concluded that the presence of AVT or and AVT-like peptide in the mammalian pineal is still debated. Vaughan (1981) wrote in one of her reviews: "that after 2 decades of research can be concluded that the nonapeptide AVT directly or indirectly affects the reproductive system of all classes of vertebrates, but that there is considerable controversy whether an antigonadotropic peptide of the pineal is AVP, AVT, oxytocin, E5 or another unidentified peptide".

Although different biological effects are obtained with bovine "E5-fractions", it is not proven that this fraction is a pure peptide, because a description with details of the exact experimental conditions used for the preparation of the E5 "peptide" from bovine pineals is still lacking. Therefore, in the next part of this paper we will describe a comparative study, in which different peptidic/proteic ovine pineal fractions were compared with a bovine E5-fraction, prepared by Dr. Neacşu, Bucharest, Romania.

IV. COMPARISON OF SOME PEPTIDIC AND PROTEIC PINEAL FRACTIONS WITH A BOVINE E5 FRACTION

The bovine pineal E5-fraction was prepared by Dr. C. Neacşu, sent to Dr. P. Pévet, Netherlands Institute for Brain Research, Amsterdam and there stored at -80°C until use in our institute in Utrecht. The ovine pineals were collected by Ersco, San Mateo, California in the winter months, shipped on dry ice and stored at -20°C until use.

Using rather simple and mild extraction and separation methods, three ovine pineal fractions: XM300R, PP7.2' and PP7.2S were prepared according to Scheme 1. From the results of the experiments can be concluded that these sheep pineal fractions and the bovine pineal E5-fraction contain peptidic and proteic substances as measured with the "micro-Lowry" method. However, the proteic content of the fractions is between 60 and 70% for the ovine pineal fractions and below 40% for the bovine E5-fraction, using serum albumin as a standard, and the same method under the same experimental conditions. This may mean that the above mentioned fractions contain besides their proteic part, another structure. At this moment we have no other explanation for this fact. All the above mentioned fractions show excitation and emission maxima just like indoles show. These fluorescence characteristics are sensitive to normal laboratory light and most stable in red light ($\lambda > 600$nm).

SCHEME 1. OUTLINE OF THE PREPARATION OF DIFFERENT OVINE PINEAL FRACTIONS.

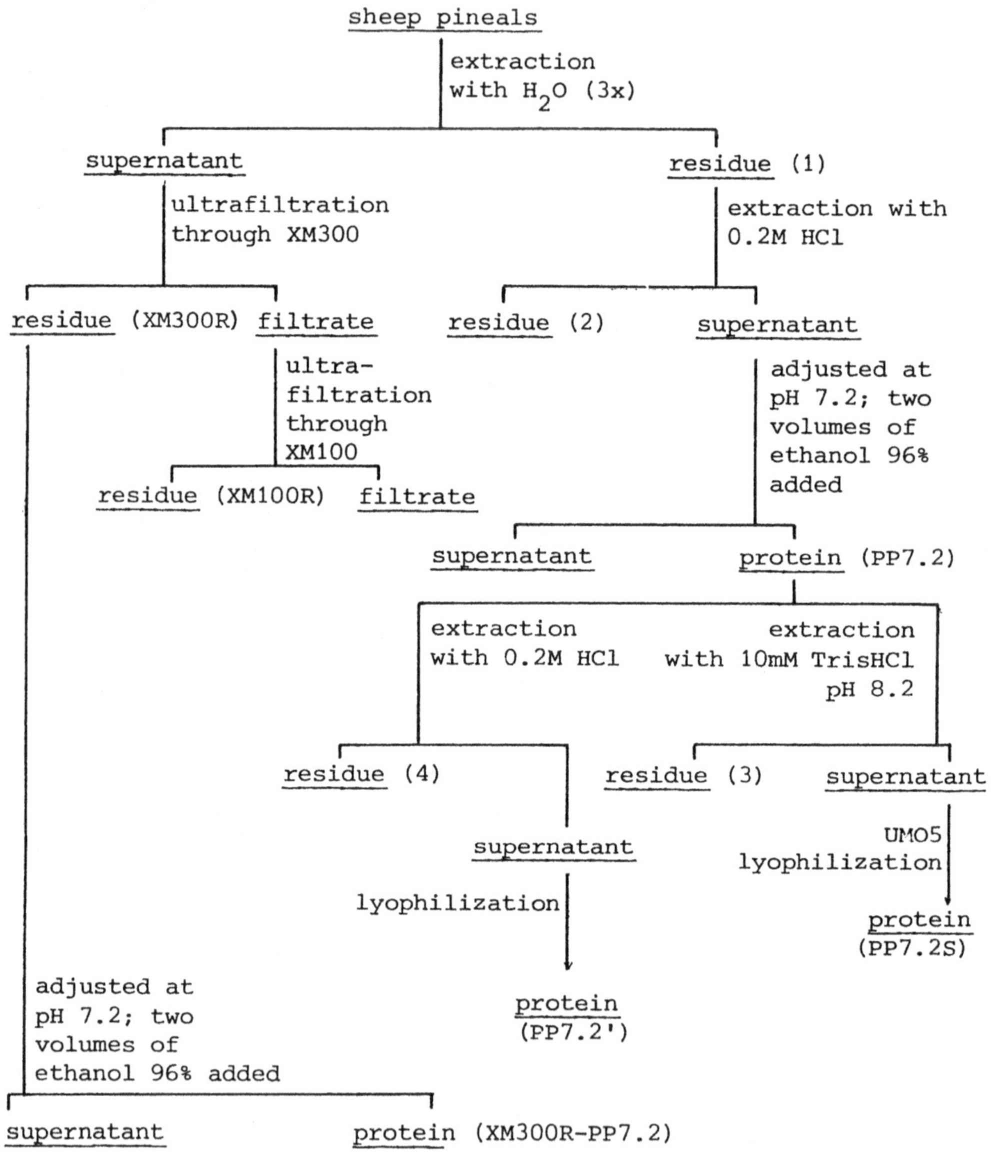

Several years ago, Ariëns Kappers et al. (1974) mentioned that in rabbit and rat pineals protein(s) are probably present which contain much tryptophan. In gel-electrophoresis experiments the tryptophan-rich protein band showed also a specific yellow autofluorescence (for details see also Smith and Ariëns Kappers, 1975).

In the near future special attention will be paid to the indole-like fluorescence in our fractions, since it has been shown that tryptophan is very important for the biological activity of somatostatin (Hirst et al., 1980) and of pancreozymin (Rajh et al, 1980).

From the staining pattern obtained with Coomassie blue of the bovine pineal fraction E5 and the ovine pineal fraction PP7.2 using disc gel-electrophoresis, it can be concluded that a similarity exists between the patterns of the bovine pineal fraction E5 and the ovine fraction PP7.2S. However, using sodium-dodecyl sulphate slab gel-electrophoresis, unexpected high molecular weight proteic bands were located in the gel. This justifies the conclusion that the bovine pineal fraction E5 contains proteinaceous material, which was not detected with the disc gel-electrophoresis method.

When the ovine fraction XM300R - PP7.2 and PP7.2S were treated in the same way a number of proteic/peptidic bands were also visible. The protein bands can be divided into 5 major groups:$>$ 94K; $>$ 68K; $>$ 43K; $>$ 30K and $>$ 21K.

Data from HPLC experiments in combination with those of gel-electrophoresis experiments suggest however that polypeptidic substances of a lower molecular weight are also present in the bovine E5- and the ovine PP7.2-fractions. In a RIA developped by Dr. A. Reinharz, Geneva, Switzerland, use was made of an antibody against the carboxy terminal tripeptide AVP or AVT. The ovine fractions XM300R-PP7.2 and PP7.2 are recognized by this antibody. The dilution curve is parallel to the E5 and to the standard (here AVP). This implies that (a) peptide(s) present in all these fractions are most probably terminating by Pro-Arg-Gly(NH_2). It thus appears that (a) peptide(s) ending by the same carboxy terminal end as AVP or AVT is present in the E5-fraction obtained from bovine pineals as well as in the ovine pineal fractions XM300R-PP7.2 and PP7.2 (for details of these experiments see Noteborn et al., (in press)).

The present findings do not allow to conclude that the blockage of spermatogenesis in gonadotropin-treated male frogs; the inhibition of uterine weights and ^{32}P incorporation in gonadotropin-treated female mice (Neacşu, 1972); the activity on pituitary levels of LH and prolactin in intact rats (Vaughan et al., 1980, 1982) and the high activity in compensatory ovarian hypertrophy test (unpublished results) present in the bovine pineal fraction E5, are due to pineal peptidic antigonadotropic "hormones" (Pévet et al., 1981a,b), because the bovine pineal fraction E5 contains also a large amount of proteic compounds MW > 14.300 daltons), which could be responsible for biological activities. It remains, however, possible that the same peptide is present in the bovine pineal fraction E5 and the ovine pineal fractions studied because it is difficult to detect a similarity in the bands < 14.3K due to diffusion in this front region in our slab gel-electrophoresis experiments. It has been demonstrated that the ovine pineal fractions share several chemical and immunological properties with the bovine pineal fraction E5. However, the structure of the compound(s) and their biological activity remain to be elucidated.

SUMMARY

1. The antigonadotropic activity *in vitro* of a low molecular weight (MW < 500) ovine pineal fraction, using male rat anterior pituitaries and a bioassay, is most probably caused by reduced 6-L-erythrobiopterin.
2. High and lower molecular weight ovine pineal fractions, prepared under red light conditions, can increase the gonadotropin releasing activity of male rat medial basal hypothalamus *in vitro*.
3. A low molecular weight pineal fraction (MW > 500 and < 1000) contains a substance, which inhibit compensatory ovarian hypertrophy, reduces ventral prostate weights in mice, retards vaginal opening in young female mice and can decrease plasma LH-levels.
4. A tripeptide Thr-Ser-Lys with antigonadotropic capability is isolated from bovine pineals and is synthesized later on.
5. Crude and partially purified bovine pineal fractions contain prolactin regulating substances *in vitro*.
6. A low molecular weight bovine pineal fraction (MW > 500 and < 1000) can decrease plasma prolactin levels *in vivo*.
   The pineal regulation of prolactin secretion may therefore be more important than was thought in previous years.
7. Comparison of some ovine peptidic/proteic pineal fractions with a bovine pineal fraction E5 has learned that several chemical and immunological properties are shared. The results obtained with an antibody against the carboxy terminal tripeptide of AVP or AVT indicates that a peptide(s) ending by Pro-Arg-Gly(NH_2) is probably present in the bovine E5 fraction, as well as in some ovine pineal fractions.
8. Although the presence of peptidic/proteic substances in biologically active pineal fractions cannot be discarded, the chemical nature of the biological active substances remain to be elucidated.

REFERENCES

Ariëns Kappers, J., Smith, A.R., and de Vries, R.A.C., 1974, The mammalian pineal gland and its control of hypothalamic activity, Progr. Brain Res., 41:149.

Balemans, M.G.M., Ebels, I., and Vonk-Visser, D.M.A., 1970, Separation of pineal extracts on Sephadex G-10. I. A spectrofluorimetric study of indoles in a cockerel pineal extract., J. Neurovisc. Relat., 32:65.

Bartke, A., Croft, B.T., and Dalterio, S., 1975, Prolactin restores plasma testosterone levels and stimulates testicular growth in hamsters exposed to short daylength, Endocrinology, 97:1601.

Bensinger, R., Vaughan, M.K., and Klein, D.C., 1973, Isolation of a non-melatonin lipophilic antigonadotrophic factor from the bovine pineal gland. Fed. Proc. Fed. Am. Socs. Exp. Biol., 32:225.

Benson, B., and Ebels, I., 1981, Other pineal peptides and related substances physiological implications for reproductive biology, in: "The Pineal Gland," Vol. II, Reproductive Effects, R.J. Reiter ed., CRC Press, Inc. Boca Raton, Florida.

Benson, B., Larsen, B.R., Findell, P,R., and Orstead, K.M., 1982, Participation of pineal peptides in reproduction, in press.

Benson, B., and Matthews, M.J., 1980, Possible role of prolactin and pineal prolactin-regulating substances in pineal-mediated gonadal atrophy in hamsters, Hormone Res., 12:137.

Benson, B., and Orts, R.J., 1972, Regulation of ovarian growth by the pineal, in: "Regulation of Organ and Tissue Growth," Academic Press, New York.

Bénson, B., Matthews, M.J., Hadley, M.E., Powers, S., and Hruby, V.J., 1976a, Differential localization of antigonadotropic and vasotocic activities in bovine and rat pineal, Life Sci., 19:747.

Benson, B., Matthews, M.J., and Hruby, V.J., 1976b, Characterization and effects of a bovine pineal antigonadotropic peptide, Am. Zool., 16:17.

Bex, F., Bartke, A., Goldman, D.B., and Dalterio, S., 1978, Prolactin, growth hormone, luteinizing hormone receptors, and seasonal changes in testicular activity in the golden hamster, Endocrinology, 103:2069.

Blask, D.E., Vaughan, M.K., Reiter, R.J., Johnson, L.Y., and Vaughan, G.M., 1976, Prolactin-releasing and release-inhibiting factor activities in the bovine, rat and human pineal gland. In vitro and in vivo studies, Endocrinology, 99:152.

Chang, N., Ebels, I., and Benson, B., 1979, Preliminary characterization of bovine pineal prolactin releasing (PPRF) and release-inhibiting factor (PPIF) activity, J. Neural Transm., 46:139.

Cheesman, D.W., 1970, Structural elucidation of a gonadotropin-inhibiting substance from the bovine pineal gland, Biochim. Biophys, Acta, 207:247.

Cheesman, D.W., and Fariss, B.L., 1970, Isolation and characteri-

zation of a gonadotropin-inhibiting substance from the bovine pineal gland, Proc. Soc. Exp. Biol. Med., 133:1254.
Citharel, A., Ebels, I., L'Héritier, A., and Moszkowska, A., 1973, Epiphyseal-hypothalamic interaction. An in vitro study with sheep pineal fractions, Experientia, 29:718.
Collin, J.-P., 1979, Recent advances in pineal cytochemistry. Evidence of the production of indoleamines and proteinaceous substances by photoreceptor cells and pinealocytes of Amniota, Brain Res., 52:271.
Collin, J.-P., 1981, New data and vistas on the mechanisms of secretion of proteins and indoles in the mammalian pinealocyte and its phylogenetic precursors; the pinealin hypothesis and preliminary comments on membrane traffic, in: "The Pineal Organ," A. Oksche and P. Pévet, eds., Elsevier/North-Holland, Amsterdam-New York-Oxford.
Collin, J.-P., and Oksche, A., 1981, in: "The Pineal Gland," Vol. I, Anatomy and Biochemistry, R.J. Reiter, ed., CRC Press Inc., Boca Raton, Florida.
Deguchi, T., 1981, Rhodopsin-like photosensitivity of isolated chicken pineal gland, Nature, 290:706.
Ebels, I., 1979, A chemical study of some biologically active pineal fractions, Brain Res., 52:309.
Ebels, I., 1981, Pteridines in the pineal organ, in: "The Pineal Organ," A. Oksche and P. Pévet, eds., Elsevier/North-Holland, Amsterdam-New York-Oxford.
Ebels, I., and Benson, B., 1978, A survey of the evidence that unidentified pineal substances affect the reproductive system in mammals, in: "The Pineal and Reproduction," Progress in Reprod. Biol. Vol. 4, R.J. Reiter, ed., S. Karger, Basel.
Ebels, I., and Horwitz-Bresser, A.E.M., 1976, Separation of pineal extracts by gelfiltration. IV. Isolation, location and identification from sheep pineals of three indoles identical with 5-hydroxytryptophol, 5-methoxytryptophol and melatonin, J. Neural Transm., 38:31.
Ebels, I., Balemans, M.G.M., and Tommel, D.K.J., 1972a, Separation of pineal extracts on Sephadex G-10. III. Isolation and comparison of extracted and synthetic melatonin, Anal. Biochem., 50: 234.
Ebels, I., Balemans, M.G.M. and Verkleij, A.J., 1972b, Separation of pineal extracts on Sephadex G-10. II. A spectrofluorimetric and thin-layer chromatographic study of indoles in a sheep pineal extract, J. Neurovisc. Relat., 32:270.
Ebels, I., Citharel, A., and Moszkowska, A., 1975, Separation of pineal extracts by gelfiltration. III. Sheep pineal factors acting either on the hypothalamus, or on the anterior hypophysis of mice and rats in in vitro experiments, J. Neural Transm., 36:281.
Ebels, I., Moszkowska, A., and Scemama, A., 1965, Etude in vitro des extraits epiphysaires fractionnés. Résultats préliminaires, C.R. Acad. Sci. Paris, 260:5126.

Ebels, I., Moszkowska, A., and Scemama, A., 1970, An attempt to separate a sheep pineal extract fraction showing antigonadotropic activity J. Neurovisc. Relat., 32:1.

Ebels, I., Benson, B., Bria, C.F., Richardson, D., Larsen, B.R., and Hruby, V.J., 1979, Location by paperchromatography of compensatory ovarian hypertrophy (COH) inhibiting activity in isobutanol extracts of bovine pineals, J. Neural Transm., 45:43.

Frowein, A., and Lapin, V., 1979, Effects of sham-pinealectomy, performed under white and red light, on the melatonin content of the rat pineal glands, Experientia, 35:1681.

Fukushima, T., and Nixon, J.C., 1979, Reverse-phase high-performance liquid chromatographic separation of unconjugated pterins and pteridines, in: "Chemistry and Biology of Pteridines," R.L. Kisliuk, and G.M. Brown, eds., North-Holland, Amsterdam.

Fukushima, T., and Nixon, J.C., 1980, Analysis of reduced forms of biopterin in biological tissues and fluids, Anal. Biochem., 102:176.

Hirst, B.H., Shaw, B., Meyers, C.A., and Coy, D.H., 1980, Structure-activity studies with somatostatin: The role of tryptophan in position 8, Reg. Peptides, 1:97.

Larsen, B.R., and Benson, B., 1979, Purification of bovine pineal prolactin inhibiting factor, Anal. Rec., 193:598.

Larsen, B.R., Benson, B., and Hruby, V.J., 1979, Synthesis and effects on prolactin of a bovine pineal tripeptide, in: "Peptides, Structure and Biological Function," Pierce chemical co., Rockford, Illinois.

Leadem, C.A., and Blask, D.E., 1982, A comparitive study of the effects of the pineal gland on prolactin synthesis, storage and release in male and female blind-anosmic rats, Biol. Reprod., 26:413.

Martens, G.J.M., Jenks, B.G., and Van Overbeeke, A.P., 1981, Microsuperfusion of neurointermediate lobes of Xenopus laevis: Concomitant and coordinately controlled release of newly synthesized peptides, Comp. Biochem. Physiol., 69c:75.

Matthews, M.J., and Benson, B., 1973, Inactivation of pineal antigonadotropin by proteolytic enzymes, J. Endocrinol., 56:339.

Matthews, M.J., Benson, B., and Richardson, D.L., 1978, Partial maintenance of testes and accessory organs in blinded hamsters by homoplastic anterior pituitary grafts or exogenous prolactin, Life Sci., 23:1131.

Milcu, S., Milcu, J., and Nanu, L., 1963, Le Rôle de la glande pinéale dans le métabolisme des glucides, Ann. d'Endocr., 24:233.

Moszkowska, A., 1951, Contribution à l'etude de l'antagonisme épiphyso-hypophysaire, J. Physiol., Paris, 43:827.

Moszkowska, A., 1956, L'antagonisme épiphyso-hypophysaire. Etude in vitro par la méthode de E. Wolff, C.R. Acad. Sci., Paris, 243:315.

Moszkowska, A., 1958, Etude in vitro du rôle de l'epiphyse dans l'excrétion d'hormones gonadotropes hypophysaires, C.R. Acad Sci., Paris, 247:1659.

Moszkowska, A., 1964, Quelques arguments en faveur de la spécificité zoologique de l'activité antigonadotrope de l'épiphyse, Ann. Endocr. 25(suppl.):79.
Moszkowska, A., 1965, Contribution à l'étude du mécanisme de l'antagonisme épiphyso-hypophysaire, Prog. Brain. Res., 10:564.
Moszkowska, A., and Ebels, I., 1968, A study of the antigonadotropic action of synthetic arginine vasotocin, Experientia, 24:610.
Moszkowksa, A., and Ebels, I., 1971, The influence of the pineal body on the gonadotropic function of the hypophysis, J. Neurovisc. Relat., Suppl 10:160.
Moszkowska, A., and Heersche, J.N.M., 1962, in: "Int. Congr. Physiol. Sci.," Leiden, No 518, Excerpta Med. Int. Congr. Series 48: XXII.
Moszkowska, A., Scemama, A., Lombard, M.M., and Héry, M., 1973) Experimental modulation of hypothalamic content of the gonadotropic releasing factors by pineal factors in the rat, J. Neural. Transm., 34:11.
Neacşu, C., 1972, Structure-activity data on a pineal peptide with oxytocic and vasopressor activities, Proc. Sci. Int. Symposium, Liège, Sept. 28 - Oct. 1, 1971; Excerpta Med. Int. Congr. Series, No 241:275.
Neacşu, C., 1972, The mechanism of antigonadotropic action of a polypeptide extracted from a bovine pineal gland, Rev. Roum. Physiol., 9:161.
Noteborn, H.P.J.M., Ebels, I., Pévet, P., Reinharz, A.C., Neacşu, C. and Salemink, C.A., 1982, Comparison of some peptidic and proteic ovine pineal fractions with a bovine pineal E5 fraction, J. Neural Transm., in press.
Orts, R.J., Laio, T.-H., Sartin, J.L., and Bruot, B., 1978, Purification of a tripeptide with anti-reproductive properties isolated from bovine pineal glands, Physiologist, 21:87 .
Ota, M., Horiuchi, S., and Obara, K., 1975, Inhibition of ovulation induced with PMS and HCG by a melatonin-free extract of bovine pineal powder, Neuroendocrinology, 18:311.
Pavel, S., 1979, The mechanism of action of vasotocin in the mammalian brain, Progr. Brain Res., 52:445.
Pévet, P., 1979, Secretory processes in the mammalian pinealocyte under natural and experimental conditions, Progr. Brain Res., 52:149.
Pévet, P., 1981a, Ultrastructure of the mammalian pinealocytes, in: "The Pineal Gland," Vol. I, Anatomy and Biochemistry, R.J. Reiter, ed., CRC Press, Boca Raton, Florida.
Pévet, P., 1981b, Peptides in the pineal gland of vertebrates. Ultrastructural, histochemical, immunocytochemical and radioimmunological aspects, in: "The Pineal Organ," A. Oksche, and P. Pévet, eds., Elsevier/North-Holland, Amsterdam-New York-Oxford.
Pévet, P., 1982a, Pineal peptides in the fetus and in young and adult mammals, in: "Melatonin Rhythm," D.C. Klein, ed., S. Karger, Basel, in press.

Pévet, P., 1982b, The anatomy of the pineal gland of mammals, in: "The Pineal Gland," R. Relkin, ed., Elsevier/North-Holland, New York, in press.
Pévet, P., Ebels, I., Swaab, D.F., Mud, M.T., and Arimura, A., 1980, Presence of AVT-, α-MSH-, LHRH- and somatostatin-like compounds in the rat pineal gland and their relationship with the UMO5R pineal fraction, Cell. Tiss. Res., 206:341.
Rajh, H.M., Smyth, M.J., Renckers, B.A.M., Jansen, J.W.C.M., De Pont, J.J.H.H.M., Bonting, S.L., Tesser, G.J., and Nivard, R.J.F., 1980, Role of the tryptophan residue in the interaction of pancreazymin with its receptor, Biochim. Biophys. Acta, 632: 386.
Reiss, M., Davis, R.H., Sideman, M.B., Mauer, I., and Plichta, E.S., 1963a, Action of pineal extracts on the gonads and their function, J. Endocr., 27:107.
Reiss, M., Mauer, I., Sideman, M.B., Davis, R.H., and Plichta, E.S., 1963b, Pituitary-pineal-brain interrelationships, J. Neurochem., 10:851.
Reiter, R.J., 1980, "The Pineal Gland," Vol. 5, Eden Press Inc., St. Albans, U.S.A.
Reiter, R.J., 1981, "The Pineal Gland," Vol. 6, Eden Press Inc., St. Albans, U.S.A.
Reiter, R.J., and Ferguson, B.N., 1979, Delayed reproductive regression in male hamsters bearing intrarenal homografts and kept under natural winter periods, J. Exp. Zool., 209:175.
Slama-Scemama, A., L'Héritier, A., Moszkowska, A., Van der Horst, C.J.G., Noteborn, H.P.J.M., De Morée, A., and Ebels, I., 1979, Effects of sheep pineal fractions on the activity of male rat hypothalami in vitro, J. Neural Transm., 46:47.
Smith, A.R., and Ariëns Kappers, J., 1975, Effect of pinealectomy, gonadectomy, p-CPA and pineal extracts on the rat parvocellular neurosecretory hypothalamic system; a fluorescence histochemical investigation, Brain Res., 86:353.
Sorrentino, S. jr., 1968, Antigonadotropic effects of melatonin in intact and unilaterally ovariectomized rats, Anat. Rec., 160:432.
Thiéblot, L., and Thiéblot, Ph., 1981, "La Gland Pinéale," Physiologie et Clinique, S.A. Maloine, ed., Paris.
Vaughan, M.K., 1981, Arginine vasotocin and vertebrate reproduction, in: "The Pineal Gland," Vol. II, Reproductive Effects, R.J. Reiter, ed., CRC Press, Boca Raton, Florida.
Vaughan, M.K., and Blask, D.E., 1978, Arginine vasotocin - A search for its function in mammals, in: "The Pineal and Reproduction," Progress in Reproductive Biology, Vol. 4., R.J. Reiter, ed., S. Karger, Basel.
Vaughan, M.K., Johnson, L.Y., Pévet, P., Neacşu, C., and Reiter, R.J., 1980, Effect of a polypeptide (E5) extracted from bovine pineal glands on plasma and pituitary levels of luteinizing hormone (LH) and prolactin in normal and castrated adult male rats. Tenth Ann. Mtg., Soc. Neurosci., 16:457.

Vaughan, M.K., Richardson, B.A., Johnson, L.Y., Reiter, R.J., Pévet, P., and Neacşu, C., 1982, Effects of a bovine pineal peptidic fraction (E5) on plasma and pituitary levels of LH, FSH and prolactin, Experientia, in press.

Zurburg, W., and Ebels, I., 1974, Separation of pineal extracts by gel filtration. I. Isolation from sheep pineals of a substance with special fluorescence characteristics, J. Neural Transm., 35:117.

Zurburg, W., and Ebels, I., 1975, Separation of pineal extracts by gel filtration. II. Identification and isolation of two indoles from sheep pineal glands, J. Neural Transm., 36:59.

MOLECULAR ASPECTS OF NEUROENDOCRINE INTEGRATIVE PROCESSES IN THE PINEAL GLAND*

D.P.Cardinali,** Mónica N.Ritta,*** María I.Vacas,** P.R.Lowenstein,*** P.V.Gejman,*** C.González Solveyra and Elba Pereyra

Centro de Estudios Farmacológicos y de Principios Naturales (CEFAPRIN), Serrano 665/669, 1414 Buenos Aires, Argentina

INTRODUCTION

The mammalian pineal gland fulfills the criteria of a "neuroendocrine" transducer.[1] It translates a neural language provided by norepinephrine (NE) released at the synaptic biophase to a hormone language, melatonin and perhaps endocrine active peptides. The pinealocytes are also "endocrine-endocrine" transducers inasmuch as they convert an endocrine language, e.g. estradiol attaining the gland via the general circulation, to a different endocrine signal like melatonin. Additionally "endocrine-neural" transducer events occur in the pineal gland, as revealed by the significant modifications of the activity of the innervating sympathetic pathway after several hormone treatments.[1,2]

NEUROENDOCRINE TRANSDUCTION

Historically the quality and quantity of ambient illumination have been pointed out as the major environmental signal affecting the pineal gland. Light information is relayed neurally by the retino-hypothalamic tract and it is integrat-

* Supported by grant nº 6638 from Consejo Nacional de Investigaciones Científicas y Técnicas (CONICET), Argentina.

** Established Investigator, CONICET.

*** Research Fellow, CONICET.

ed with circadian information in the suprachiasmatic nucleus of the anterior hypothalamus (Fig. 1). Caudal projections from the suprachiasmatic nucleus connect via a multisynaptic pathway with the superior cervical ganglia (SCG).[3] Axons from the SCG innervate the pineal gland where they release NE in the vicinity of pineal cells. It is now accepted generally that darkness stimulates the release of NE from these nerves.[4,5]

The biochemical mechanisms by which NE released from pineal nerve endings accelerates melatonin synthesis have been analyzed in detail using organ cultures (Fig. 1). The catecholamine acts on β-adrenoceptors of the pineal cells and stimulates the activity of adenyl cyclase which results in an increased synthesis of cAMP.[6-8] In the subsequent induction of enzymes participating in melatonin synthesis e.g. serotonin N-acetyltransferase (SNAT) and hydroxyindole-O-methyl transferase (HIOMT), both RNA and protein are synthesized within the pinealocytes.[9,10] Presynaptic β-adrenoceptors modulating NE release are presumably present in pineal nerve endings.[11]

There is also information on the existence of α-adrenergic effects of NE in the pineal gland. These include in vivo[12] and in vitro[13,14] modification of SNAT activity and stimulation of phosphatidylinositol turnover. Either by acting directly on pineal nerve endings or through the modification of a hypothetical trans-synaptic feedback signal, α_2-adenoceptor agonists and antagonists modify NE release from pineal nerves.[16,17] Postsynaptic α-adrenoceptors that increased significantly in number following sympathetic denervation of the pineal gland were described by ^{3}H-dihydroergocryptine binding in rat pineal membranes.[16]

Most neurotransmitters act via membrane receptors to induce a variety of second messengers. For one receptor subclass the transducing event depends upon the activation of adenyl cyclase to produce a single output signal: cAMP. The link between the β-adrenoceptor-adenyl cyclase system and melatonin secretion in the pineal gland has extensively been reviewed.[8,18] The second major receptor sub-class is more ubiquitous in that it gives rise to various output signals including calcium, cyclic nucleotides and several arachidonic acid metabolites like prostaglandins (PGs), thromboxanes, and leukotrienes. Possibly all these signals are intimately related to the hydrolysis of phosphatidylinositol.[19,20,21]

The arachidonic acid metabolites seem to perform a variety of functions depending on the cell considered. In some cases they subserve the role of second messengers within the cells; in other cases they diffuse away from the cell and thus act as autacoids to modulate the surrounding cells.[21] This dual

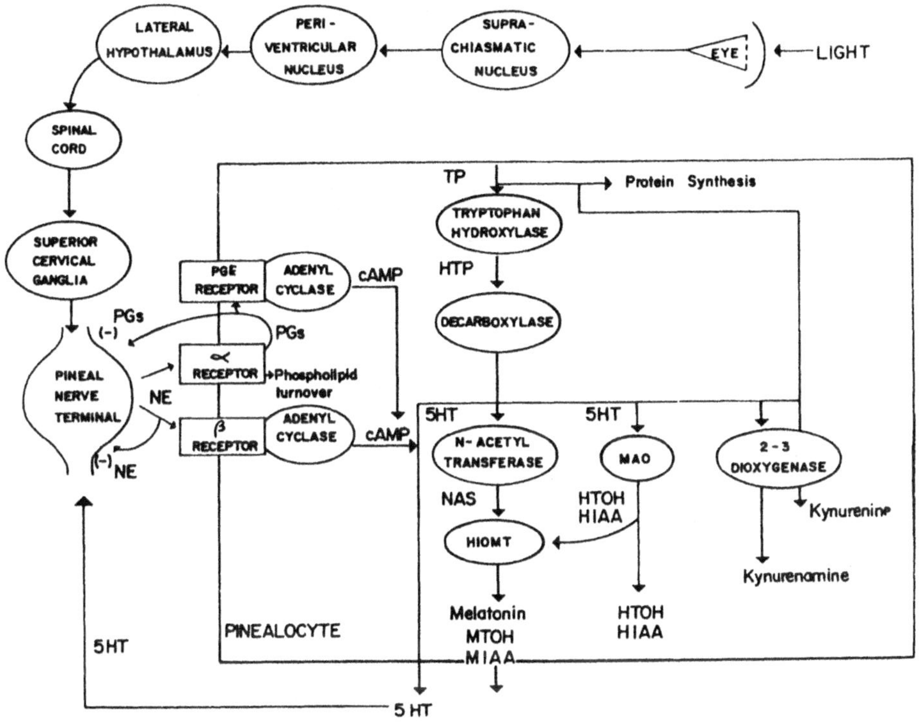

Fig. 1 Diagram representing the subcellular mechanisms underlying neuroendocrine-transducing events in the mammalian pineal. TP: tryptophan; HTP: 5-hydroxytryptophan; 5HT: serotonin; NAS: N-acetylserotonin; HTOH: 5-hydroxytryptophol; HIAA: 5-hydroxyindoleacetic acid; MTOH: 5-methoxytryptophol; MIAA: 5-methoxyindoleacetic acid.

role is of particular importance in the adrenergic neuroeffector junction, where PGs play both pre- and postsynaptic functions; for example, besides influencing the cellular responses to the neurotransmitter, the PGE_2 released from vascular smooth muscle functions as a negative feedback signal to depress transmitter release from the nerve endings.[22]

In 1976 Szabo and Friedhoff[23] reported that the injection of 10-50 mg/kg of the PG synthesis inhibitor indomethacin (Id) to adult male rats for 4-5 days decreased by 65-95% the activity of pineal SNAT when measured at the 4th h of dark during daily photoperiod. Essentially similar results were obtained by using smaller amounts of Id[24]. A single injection of 5 mg/kg to rats at the end of the light phase of daily photoperiod brought about a significant, 60%-depression of the dark-induced activation of pineal SNAT, and diminished significantly pineal melatonin levels by 27%. The treatment also decreased HIOMT activity by 60%. Although no attempt was made to characterize whether the depressed melatonin synthesis and levels after Id were due to a true impairment of pineal NE-driven phenomena or merely to a shift in the peak levels of melatonin at night, our results argued in favor of a modulatory role of PGs at any of the several anatomical stages of the retinal-pineal pathway (Fig. 1). Obviously the location(s) of the effect (pineal and/or extrapineal) could not be determined from those experiments.

In the pineal gland monoamine oxidase (MAO) type A activity, but not MAO type B activity, was found to depend on intact sympathetic nerves.[25,26] This observation led to the conclusion that the former is located in the nerve endings (or needs absolutely of their integrity) whereas the latter is located in the pinealocytes. Id treatment of rats increased by 55% pineal MAO type A and decreased by 37% MAO type B; pineal catechol-O-methyl transferase was not affected by Id.[24] These observations suggested that the drug enhanced by acting on pineal or extrapineal sites, the intraneuronal metabolism of NE.

Providing that all Id pineal effects can be attributed to inhibition of PG synthesis, the above discussed results suggested the occurrence of pre- and/or postsynaptic effects of PGs (or of some of the products of lipoxygenase e.g. leukotrienes, whose synthesis may increase after cyclooxygenase inhibition).[19-21] However besides cyclooxygenase, Id alters at submicromolar amounts several other relevant enzyme activities like phospholipase, phosphodiesterase or cAMP-dependent protein kinase.[27,28] Therefore as it is usual when only a single experimental approach is used to answer a posed question, the foregoing results were beset by the vagaries of unspecified site(s) and uncertain mechanisms of action of the drug used.

The following experiments were carried out in vitro in order to supersede some of these drawbacks. Specifically they sought to fulfill the following criteria to attribute a role for PGs in NE-driven melatonin biosynthesis: (a) NE should release physiologically significant amounts of PGs at the concentrations that increase melatonin production in vitro; (b) incubation of NE with various PG synthesis inhibitors should impair or block NE-induced melatonin release; (c) physiological concentrations of PGs should increase melatonin synthesis in vitro, as well as counteract the inhibition of melatonin release brought about by cyclooxygenase inhibitor; (d) specific recognition sites for PGs should be identified in pineal subcellular fractions.

The effects of NE on PG release by pineal glands in vitro

were examined in bovine and rat explants. By applying specific bioassays for PGE-and PGF- like materials, we observed that the basal release of PGs by bovine pineals in vitro was increased by 2-3-fold after adding 1-100 uM NE to the medium.[29] Obviously these experiments were not addressed to the question on the site of origin (pre- or postsynaptic) of the PGs released by NE; however in view of the very small population of sympathetic nerve endings in bovine pineal gland,[30] the possibility that NE could release PGs from the nerve varicosities rather than from the pinealocytes was considered remote.

More recently we undertook a series of experiments aiming to assess the adrenoceptor involved and site of origin of PGE_2 released by NE from rat pineal explants. Pineal glands incubated with 10 or 100 uM NE released 51%-and 415% more radioimmunoassayable PGE_2 to the medium than in the absence of NE. Phentolamine (10 uM) fully prevented NE effect at both concentration of the transmitter, whereas propranolol failed to affect it significantly. After pineal sympathetic denervation by superior cervical ganglionectomy (SCGx) the release of PGE_2 evoked by NE was significantly higher than in sham-operated controls, an effect also abolished by α-adrenoceptor (but not β-adrenoceptor) blocking agents (results submitted for publication). Thus the dose dependent, NE-stimulated PG release in rat pineal gland appears both to be mediated by a postsynaptic α-adrenoceptor and to originate at a postsynaptic site. The increased α-adrenergic response to NE in denervated rat pineals correlated well with the priorly reported increase in the number of α-adrenoceptor sites after SCGx.[17]

In mammals certain membrane phospholipids like phosphatidylinositol, phosphatidylcholine and phosphatidylethanolamine are the cellular repositories of arachidonic acid, the precursor molecule for PGs, thromboxanes and leukotrienes.[20,21] The number 2 position of phosphatidylinositol is higly enriched in arachidonic acid, and it would not be unreasonable to postulate that the acceleration of pineal phosphatidylinositol turnover brought about by NE via α-adrenoceptors leads ultimately to the liberation of this rate-limiting precursor for the synthesis of PGs and other products of cyclo-oxygenase and lipoxygenase. Moreover in those cases where the specificity of the receptor mechanism has been examined, it is clear that the PGs are released by the same receptor sub-class that gives rise to the phosphatidylinositol response and that gates calcium.[19]

In order to examine the possible occurrence of pineal PG receptors, radioligand binding studies were carried out in 900 g supernatants of bovine glands by using a charcoal suspension to separate free from bound ligand.[29]

^{3}H-PGF$_{2\alpha}$ and ^{3}H-PGE$_2$ high affinity binding to pineal protein components was augmented significantly by increasing calcium concentrating in medium and was decreased after preparation of homogenates in buffer not containing Id (a finding which suggests that the endogenous PGs formed during the homogenization and incubation procedure occupy the available pineal binding sites)[29]. Pineal PG binding, was maximal in the 27,000 g pellet, where mitochondria, plasma membrane fragments, lysosomes and large microsomal particles precipitate.

Analysis of the binding data by Scatchard plots revealed the presence of single populations of binding sites for PGE$_2$ and PGF$_{2\alpha}$, with Kd of about 1-2 nM. Comparison of the relative affinities of PGE$_1$, PGE$_2$ and PGF$_{2\alpha}$ suggested that both sites were separate entities, and that the PGE$_2$ binding required the 9-keto moiety while the PGF$_{2\alpha}$ binding required the 9 hydroxyl group.[29] Therefore, not only NE releases PGs from rat and bovine pineal glands via post-synaptic α -adrenoceptors but also the released PGs become bound to specific calcium-dependent recognition sites, probably in pineal membranes.

The next series of experiments sought the answer two questions: (a) do PGs affect pineal SNAT activity and melatonin accumulation and release in pineal organ cultures? (b) does blockade of PG synthesis alter the NE-evoked melatonin release from pineal explants ?

So that the effects of PGs on SNAT activity and melatonin accumulation and release could be examined pineal glands from male rats were incubated for 6 h in TC 199 medium.[31,32] Among the several PGs tested only PGE$_2$ increased the enzyme activity at all studied doses (1-1000 nM) the stimulation curve being bell-shaped with a maximum at 100 nM.[31] PGE$_1$ increased pineal SNAT only at the highest studied dose while PGF$_{2\alpha}$ was devoid of effect. Nanomolar concentrations of PGE$_2$ increased melatonin accumulation and release to the medium.[32] Differing from results on SNAT the maximal effect of PGE$_2$ was detectable on both pineal and medium concentrations of melatonin at 1 nM PGE$_2$, with gradual decreases at greater concentrations. Therefore actual secretion of melatonin may not accompany always the changes in SNAT activity.

To disclose a pre- or postsynaptic site of action of PGE$_2$, pineal glands from rats subjected to SCGx or sham operation 7 days earlier were incubated with PGE$_2$ and the activity of SNAT was determined.[31] PGE$_2$ caused a significantly greater stimulation of SNAT in pineal glands from SCGx than from control rats. These findings indicated that the stimulatory effects of PGE$_2$ on melatonin synthesis is predominantly postsynaptic, and that the pineal supersensitivity that develops

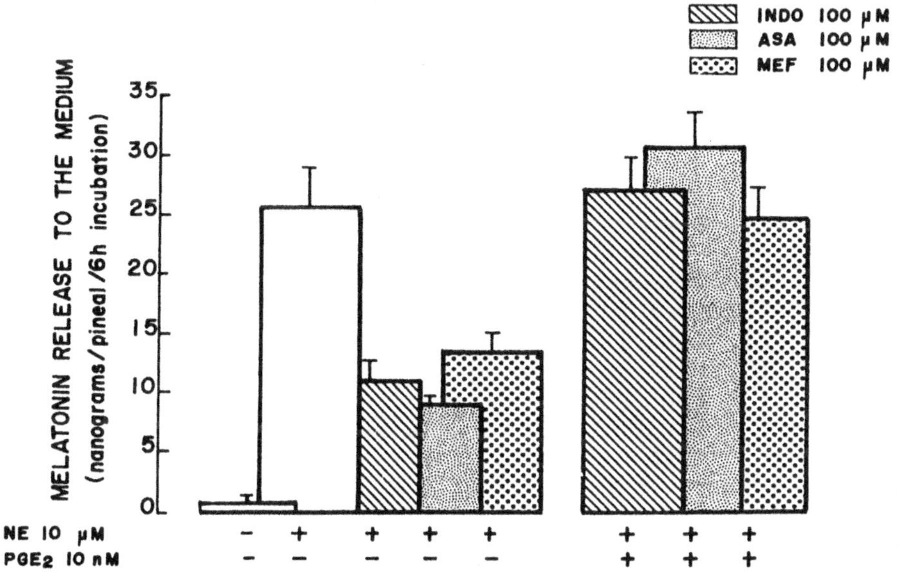

Fig. 2 Effect of indomethacin, acetylsalicylic acid and mefenamic acid on NE- and PGE-induced melatonin release by rat pineals in organ culture. Shown are the mean ± SEM (n= 6). Controls differed from all the remaining groups ($p < 0.01$, ANOVA). For experimental details see ref. 33.

after SCGx is not restricted to the adrenoceptor-mediated mechanisms. However, differing from the adrenergic supersensitivity phenomenon after SCGx in rat pineal,[12] supersensitivity to PGE_2 is accompanied by an increase in the maximal response to the agonist without any shift in the dose-response curve.[31]

The effect of 10 uM NE on melatonin release to the medium is shown in Fig. 2. Exposure to the adrenergic transmitter elicited a 13- to 20-fold increase in melatonin release. Addition of Id, acetylsalicylic acid or mefenamic acid (100 uM) brought about a significant impairment of melatonin release stimulation by NE.[32] However in spite of that at the inhibitor concentration used PG synthesis is totally blocked (as assessed by measuring radioimmunoassayable PGE_2 in medium, unpublished results) a residual 4-6- fold stimulation of melatonin release was observed. In every case the addition of 10 uM NE and 10 nM PGE_2 after the PGs synthesis inhibitor resulted in a stimulation of melatonin release that was indistinguishable from that found with NE alone (Fig. 2).

These results strongly suggest that PGE_2 is a physiological intermediary in NE-controlled melatonin biosynthesis in the pineal gland. Additionally they suggest that the PG link in melatonin biosynthetic mechanisms appears to be complementary rather than necessary inasmuch as supramaximal amounts of cyclo-oxygenase inhibitors failed to prevent fully melatonin release. A similar observation has been made by others in studies or NE-driven renin release in renal cortex.[33]

The molecular mechanisms through which PGs exert their effects in the brain remain undefined. PGs appear to be involved with the regulation of adenyl cyclase- cAMP system either directly through a receptor-mediated activation of adenyl cyclase or indirectly through PG-mediated effects on NE release.[22,34,35] In a series of in vitro experiments we assessed the effects of different PGs on pineal cAMP accumulation and its binding to intracellular receptors.[31] A dose-response experiment for the effects of PGE_2, PGE_1, $PGF_{2\alpha}$, 15-keto $PGF_{2\alpha}$ and PGI_2 indicated that only PGE_2 increased significantly the cAMP concentration in doses between 10 and 1000 nM. This effect of PGE_2 on cAMP was also observed in the pineal glands of rats subjected to SCGx (unpublished results), a finding which suggests the postsynaptic nature of the event. Incubation of pineal explants with 50 nM PGE_2 decreased the unoccupied cAMP-binding sites in pineal homogenates while the total number of binding sites remained unaltered.[31] Since a direct link between the adenyl cyclase-cAMP system and SNAT induction has been demonstrated [6,8,13,18], it seems feasible that the two observed effects of PGE_2, i.e. increased cAMP and melatonin synthesis, are causally related.

It has been well established that in addition to the negative feedback control exerted by NE itself a number of substances (e.g. PGs) released at the neuroeffector junction may play a role in modulating adrenergic neurotransmission.[22] PGE_2 locally mobilized by sympathetic nerve stimulation may counteract further release of NE by a negative feedback mechanism. Such an effect is exerted by an action on stimulus-secretion coupling and more specifically on the availability of calcium.[35]

In order to examine the presynaptic effect of PGE_2 in the rat pineal gland we studied the release of NE from pineal sympathetic nerves, after labelling the endogenous NE stores with ^3H-NE.[32] Only 100 nM PGE_2 was able to impair significantly NE release in this preparation. Neither $PGF_{2\alpha}$ nor PGE_1 exerted, at concentrations up to 100 nM, any effect on the potassium-elicited ^3H-NE release from the pineal glands in vitro.

Therefore the PGE_2 braking effect on NE release also appears to take place in the pineal neuroeffector junction. It should be stressed however that concentrations of PGE_2 100 times greater than those required to enhance melatonin secretion were needed to affect ^{3}H-NE release. Interestingly enough these are the PGE_2 concentrations attained in incubation medium after exposing bovine and rat pineal explants to NE.[29] Thus an attractive hypothesis concerning the role of PGE_2 at the pineal neuroeffector junction is that at low concentrations (i.e. 1-10 uM, close to the Kd of PG receptor sites[29]) PGE_2 is an intracellular intermediary (or messenger) involved in NE stimulation of melatonin secretion. Increasing amounts of PGE_2 mobilized towards the synaptic cleft would effectively inhibit NE release from nerve endings, therefore exerting a negative feedback on the whole system. Perhaps the simultaneous impairment of dark-induced melatonin synthesis and stimulation of MAO type A activity (an alleged pineal presynaptic parameter) reported by us in Id-treated rats,[24] were due to a dual effect of the drug on pre- and postsynaptic compartments by removing an inhibitory signal for the former and a stimulatory one for the latter (Fig. 1).

Another postsynaptic signal that may play a role in regulating NE release from pineal nerve endings is angiotensin II. Components of the renin-angiotensin system are present in the pineal gland. Renin[36] and angiotensin I[37] and II[38] are detectable in intact pineals and cultured pinealocytes, the pineal being the richest source of renin in the brain. Angiotensin II is known to increase adrenergic varicosities[39] and serotonin content [40] in the pineal gland. Proteolytic enzymes play an important role in the production of angiotensin II. Renin converts renin substrate to angiotensin I (a decapeptide), and a "converting enzyme" converts angiotensin I to angiotensin II (an octapeptide). Angiotensin converting enzyme also degrades peptides like bradykinin, insulin and enkephalin.[41]

Angiotensin converting enzyme was detectable in rat pineal gland and exhibited a circadian rhythm in activity with maximum at the end of the light phase of daily photoperiod.[42] SCGx or exposure to light for 6 days increased enzyme activity and obliterated morning-evening differences whereas injection of the β-agonist isoproterenol depressed the high levels observed in SCGx rats.[42] The significant increase in converting enzyme activity after SCGx strongly suggested that angiotensin II production may increase. Indeed in a preliminar set of experiments we observed an increase in immunohistochemically detectable angiotensin II after SCGx.[43] Conversely angiotensin I exhibited a negative correlation with converting enzyme activity during the diurnal cycle.[42]

The pineal renin-angiotensin system could be relevant for local control of organ's blood flow.[44] It may also facilitate the release, inhibit the uptake and accelerate the biosynthesis of catecholamines in pineal adrenergic nerves. Lastly it could have also an endocrine role, inasmuch as the pineal gland appears to participate in the regulation of hydration.[45]

ENDOCRINE-ENDOCRINE TRANSDUCTION

Although environmental lighting is the major input controlling pineal activity, hormones secreted by endocrine organs whose activity is modulated directly or indirectly by the pineal also affect the function of the gland.[46] Hormone effects on the pineal are exerted either directly or through functional changes in the neural input to the gland. Experiments carried out in animals injected with labeled steroids, or on subcellular fractions incubated with different radioactive hormones indicate that the pineal gland of various species (rat, sheep, cow, rhesus monkey) exhibits protein components which bind the hormone with high affinity and specifically. To date receptors for estradiol,[47] testosterone,[47,48] 5α-dihydrotestosterone (DHT),[48] progesterone,[49] prolactin[2] and melatonin[50] have been detected in pineal subcellular fractions. Autoradiographic studies indicated also nuclear binding sites for estradiol and androgens in rat pineals.[51] Enzymatic activities that convert testosterone to estrogens,[52] as well as progesterone [53,54] or testosterone[48] to 5α-reduced metabolites are present in rat pineal homogenates. Several in vivo and in vitro studies indicate that a number of pineal constituents and enzymatic activities are changed by hormone treatment in mammals. In both female humans and rats available data support the conclusion that melatonin synthesis and release are depressed at the time of mid-cycle LH peak (for references see ref. 46).

It is generally accepted that after transversing the plasma membrane of the cell steroid hormones bind with high affinity and specificity to receptor proteins and that due to this interaction an alteration of the receptor occurs, rendering the hormone-receptor complex significantly more nucleotropic.[55] The translocation of complexes to the nucleus is a direct consequence of this interaction. As first showed by Jensen and co-workers for the uterus[56], deplenishment of cytosol estrogen receptor (ER) is followed by an active mechanism of R replenishment, partly sensitive to protein synthesis inhibitors (for references see ref. 55). We have recently examined several aspects of this depletion-replenishment cycle of ER in the pineal gland. The effect of a single s.c. injection of 0.1-20 ug of estradiol 3 h earlier of cytosol ER and nuclear ER complex is shown in Fig. 3. A 2 ug-dose resulted in ER values close to those reported at proestrus during the estrous cycle.[58] Infant

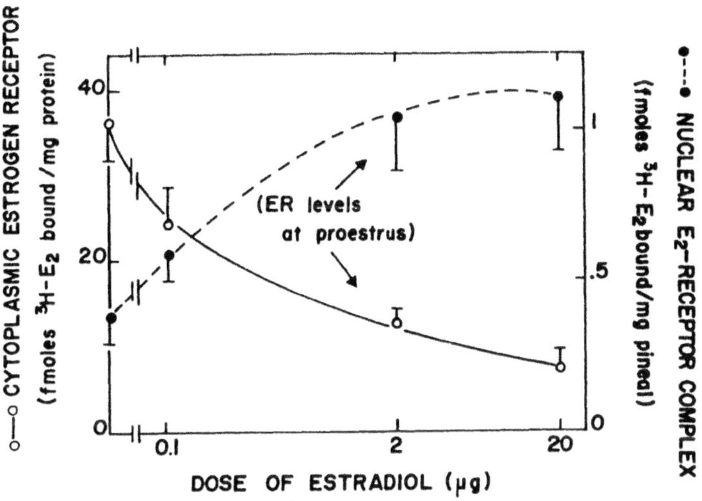

Fig. 3 Effect of estradiol (0.1-20 ug, 3 h earlier) on pineal ER in spayed rats. The concentration of cytosol ER [57] and nuclear ER complex [58] was estimated from single point binding assays of pooled pineal glands (17-28 mg). Shown are the means ± SEM (n= 4 in each group).

rats exhibited about half cytosol ER content as compared to adult spayed rats, while spayed 20-day-old rats showed intermediate levels (Fig. 4). In all 3 groups estradiol treatment depressed cytoplasmic ER for up to 18 h, and a significant increase was found 30 h after injection. These observations are in general agreement with the results obtained when the depletion-replenishment pattern of hypothalamic and adenohypophysial ER was examined in castrated rats.[59,60]

The observation that SCGx decreased both cytoplasmic and nuclear estrogen and androgen receptors[47,61] and inhibited hormone-induced increase of pineal protein synthesis[61] provided the first evidence that the activity of pineal sympathetic nerves controls gland's sensitivity to hormones. NE injection restored pineal cytosol estrogen and androgen receptor sites in SCGx rats; this effect was mediated by a β-adrenoceptor and changes in RNA synthesis.[47,61] We have further examined the interaction of neurotransmitters and ER by studying the effect of adrenergic drugs on the cycle of depletion-replenishment of cytosol ER. SCGx caused 1 week later a significant impairment of estradiol induced depletion of cytosol ER as well as of its nuclear accumulation and replenishment (Fig.5). Treatment with isoproterenol (1 mg/Kg, 19 and 3 h before, and simultaneously with estradiol) restored-cytosol ER levels, but either impaired or blocked the depletion and translocation to the nucleus induced by estradiol. Blockage of β-adrenoceptors by propranolol (20 mg/Kg) inhibited partially the estrogen-induced depletion of cytosol ER. Thus although the sympathetic input was needed to keep the cycle of receptor depletion-replenishment intact, repetitive injection of the β-agonist isoproterenol failed to restore this parameter in SCGx rats. Perhaps another nerve-originated input besides NE keeps pineal ER kinetics normal, or the schedule of injection of isoproterenol failed to reproduce the normal interaction of endogenous NE with its pineal β-adrenoceptors. An answer to this query was given by the examination of ER kinetics and estradiol biological effects in pineals of SCGx rats killed 8-18 h after surgery. During this time period, that elapses for about 10 h, adrenergic varicosities degenerate and the NE released exerts post-synaptic effects that are indistinguishable from those observed after stimulation of the sympathetic nerves.[62] In the rat pineal, degeneration of sympathetic nerves, as measured

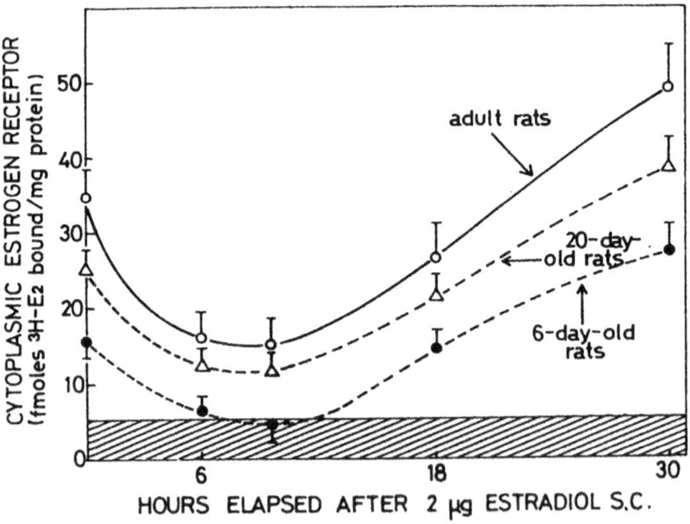

Fig. 4 Effect of estradiol on pineal cytosol ER in intact infant rats, or spayed 20- or 90-day -old rats. Other details as in Fig. 3.

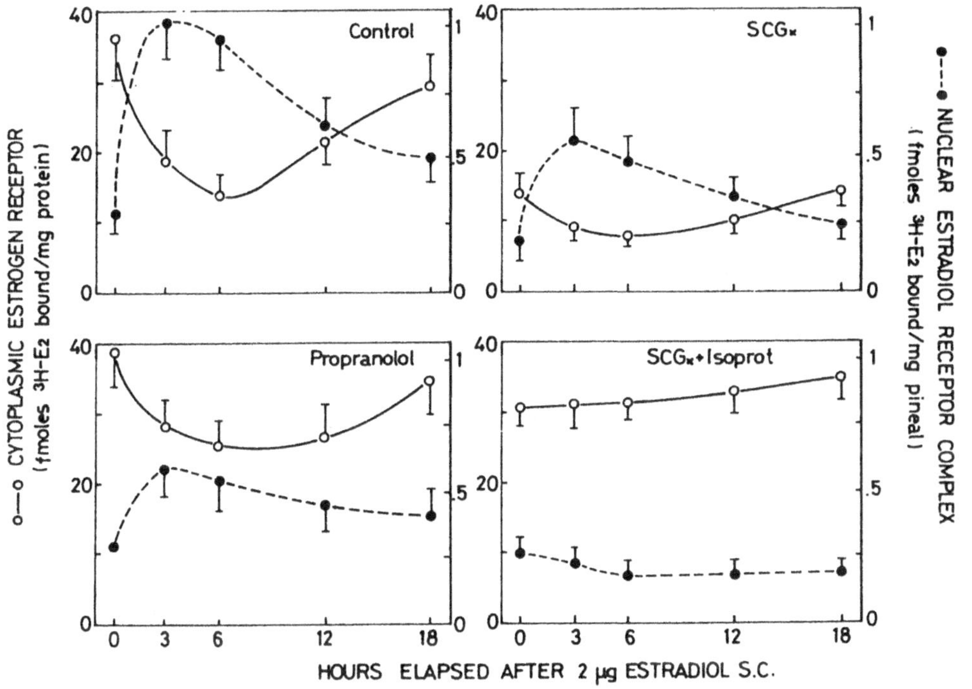

Fig. 5 Effect of estradiol on pineal ER in spayed rats subjected to SCGx 7 days earlier, and/or isoproterenol (1 mg/Kg) or propranolol (20 mg/Kg). Drugs were given as 3 s.c. injections 19 and 3 h before, and simultaneously with estradiol. Other details as in Fig. 3.

by the decay of organ's NE content or by the increase in pineal melatonin levels (unpublished results) starts at the 8-10 h after SCGx. Translocation of cytosol ER complexes to the nucleus was maximal in estradiol-treated rats 12 h after SCGx, while it was impaired prior or after the expected nerve ending degeneration period. Propranolol injection (but not phentolamine) prevented the effect of degeneration. Essentially similar results were obtained when estradiol effects on pineal melatonin synthesis was examined at the same time intervals after SCGx (results submitted for publication). Therefore these data strongly support the conclusion that the continuous, physiological interaction of NE with its receptors is needed to modulate pineal ER kinetics.

Another physiological example of neuronal modulation of pineal responsiveness to circulating hormone signals was given by studies at different times of the day in rats. In male rats pineal response to androgens (as well as to other hormones, e.g. ACTH.[63]) depends largely on the time of day when the hormone is injected. The extent of stimulation of pineal protein synthesis brought about by testosterone[64] parallels the daily rhythm in pineal NE content and turnover[4] and pineal testosterone receptors[65] in that all these reach maximal values at night. At this time pineal testosterone aromatization was high whereas testosterone 5α-reduction was low.[2] Recently time-dependent effects of testosterone on pineal electrical activity[66] and cAMP phosphodiesterases[67] have been reported.

In vitro addition of NE or cAMP mimicked the effect of night-time on pineal testosterone metabolism; they also depressed progesterone 5 α-reduction. During the degeneration of nerve terminals after SCGx an increase in testosterone arometization and a decrease in 5α-reduction was found (results submitted for publication). Collectively, the aforementioned results indicate that the overall effect of testosterone on the pinealocytes depends on the level of activity of the innervating sympathetic neurons through regulation of hormone binding sites and metabolism in target cells.

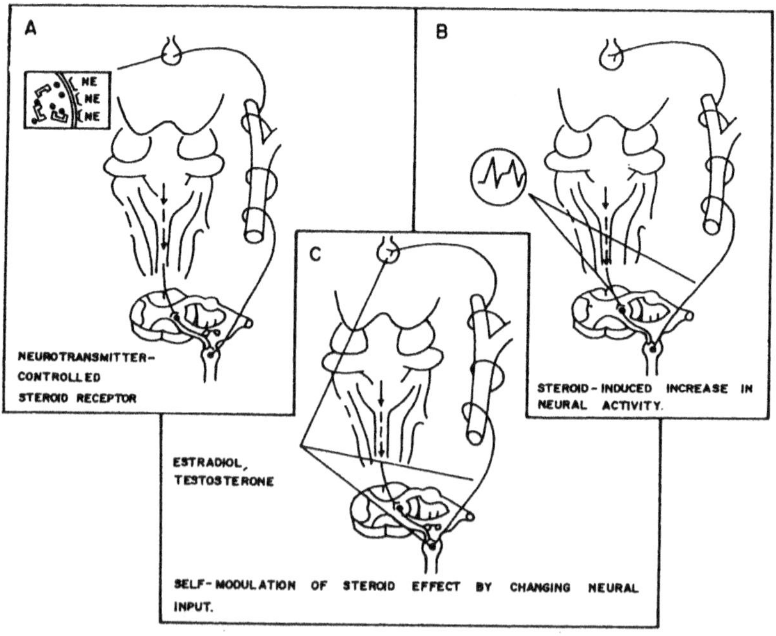

Fig. 6. A hypothetic self-modulatory mechanism for estradiol and testosterone effects in pineal gland.

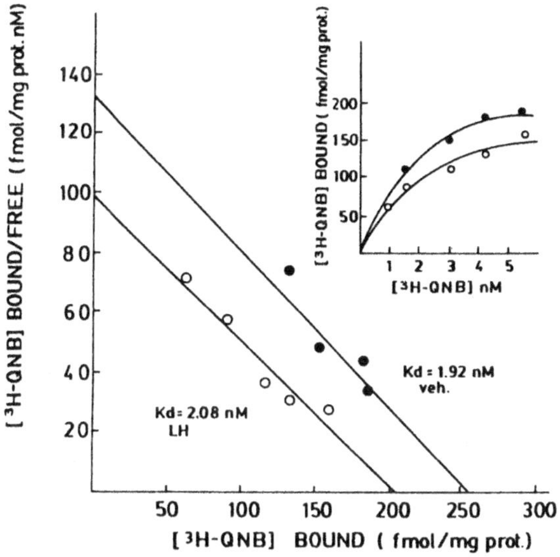

Fig. 7 Effect of LH (NIH-S-19) (100 ug, 3 and 18 h after orchidectomy) on quinuclydinil benzylate (QNB) binding in rat SCG; receptor studies were carried out by conventional techniques[74]. Differences in Bmax are significant ($p < 0.05$, analysis of covariance).

ENDOCRINE-NEURAL TRANSDUCTION

Several hormones affect NE turnover in pineal sympathetic nerves. Estradiol or testosterone treatment of castrated rats increased pineal NE turnover rate,[68] whereas LH or FSH injection depressed it.[69] Changes in intra- and extraneural metabolism of NE in the pineal gland may explain in part those effects.[26,69] Prolactin treatment increased pineal and SCG tyrosine hydroxylase activity in rats.[70] Since estradiol or testosterone treatment augments NE release from pineal nerve ending,[68] and inasmuch as NE is needed absolutely for the steroid hormone to affect pineal function (see above) a self-modulatory mechanism for hormone action can be hypothesized (Fig. 6).

Some hormones also affect the post-synapsis at the pinealocyte level; for example testosterone injection decreased significantly β-adrenoceptor density in rat pineal gland, as well as in MBH and cerebral cortex.[71] The inhibitory effect of testosterone on pineal β-adrenoceptors were not mimicked by estradiol, LH or FSH administration to castrated rats.[71,72] It occurred even in the absence of intact sympathetic nerve terminals indicating its intrinsic post-synaptic nature.

Sex steroid and gonadotropin injection affect NE turnover and several enzymatic activities related to NE synthesis and metabolism in rat SCG.[26,61,69,70,73]. For estradiol[73] and testosterone[70] specific receptor sites occur in SCG. Other relevant metabolic activities and constituents of SCG are also affected by sex steroids, gonadotropins, prolactin and corticosteroids (for references see ref. 70). For example LH treatment of castrated rats decreased muscarinic receptor binding in SCG (Fig. 7), an effect observed also after incubation of SCG slices with LH in vitro. These data strongly suggest that neuroendocrine integrative processes take place in the SCG. Such observations can be of physiological importance in view of the several neuroendocrine efferent pathways of SCG.[45,70,75-77]

REFERENCES

1. R.J.Wurtman, Neuroendocrine transducers and monoamines Fed.Proc. 32: 1769 (1973).
2. D.P.Cardinali, Molecular mechanisms of neuroendocrine integration in the central nervous system: An approach through the study of the pineal gland and its innervating sympathetic pathway, Psychoneuroendocrinology, in press.
3. R.T.Moore, The innervation of the mammalian pineal gland, Prog.Reprod.Biol., 4: 1 (1978).
4. M.Brownstein and J.Axelrod, Pineal gland: 24-hour rhythm in norepinephrine turnover, Science, 148: 163 (1974).
5. H.Nishino, K.Koizumi and C.Mc C.Brooks, The role of suprachiasmatic nuclei of the hypothalamus in the production of circadian rhythms, Brain Res, 112: 4 (1975).
6. D.C.Klein, G.R.Berg and J.L.Weller, Melatonin synthesis: Adenosine 3'5'-monophosphate and norepinephrine stimulate N-acetyltransferase, Science, 168: 979 (1970).
7. R.J.Wurtman, H.M.Shein and F.Larin, Mediation by β-adrenergic receptors of effect of norepinephrine on pineal synthesis of ^{14}C serotonin and ^{14}C melatonin, J.Neurochem, 18: 1683 (1971).
8. M.Zatz, Sensitivity and cyclic nucleotides in the rat pineal gland, J.Neural Transm, Suppl. 13: 97 (1978)
9. W.Lovenberg and J.J.Morrisey, Synthesis of RNA in pineal gland during serotonin-N-acetyltransferase induction, Biochem.Pharmacol., 27: 551 (1978).
10. W.Lovenberg and J.J.Morrisey, Protein synthesis in pineal gland during serotonin-N-acetyltransferase induction, Arch.Biochem.Biophys, 191: 1 (1978).
11. F.Pelayo, M.L.Dubocovich and S.Z.Langer, Possible role of cyclic nucleotides in regulation of noradrenaline re-

lease from rat pineal through presynaptic adrenoceptors, Nature, 274: 76 (1978).
12. H.J.Lynch, M.Ho and R.J.Wurtman, The adrenal medulla may mediate the increase in pineal melatonin synthesis induced by stress, but not that caused by exposure to darkness, J.Neural Transm. 40: 87 (1977).
13. D.C.Klein and J.Weller, Adrenergic-adenosine 3',5'-monophosphate regulation of serotonin-N-acetyltransferase activity to synthesis of ^{3}H-N-acetylserotonin and ^{3}H-melatonin in the cultured rat pineal gland, J.Pharmacol. Exp.Ther., 189: 516 (1973).
14. L.Alphs, A.Heller and W.Lovenberg, Adrenergic regulation of the reduction in acetyl coenzyme A: arylamine N-acetyltransferase activity in the rat pineal, J.Neurochem., 34: 83 (1980).
15. T.L.Smith, J.Eichberg and G.Hanser, Postsynaptic localization of the alpha receptor-mediated stimulation of phosphatidylinositol turnover in pineal gland, Life Sci., 24: 2179 (1979).
16. F.Pelayo, M.L.Dubocovich and S.Z.Langer, Regulation of noradrenaline release in the rat pineal gland through a negative feedback mechanism mediated by presynaptic α-adrenoceptors, Eur.J.Pharmacol., 45: 317 (1977).
17. M.I.Vacas, P.R.Lowenstein and D.P.Cardinali, Dihydroergocryptine binding sites in bovine and rat pineal glands, J.Auton.Nerv.System, 2: 305 (1980).
18. D.C.Klein, D.A.Auerbach, M.A.A.Namboodiri, G.H.T.Wheler, Indole metabolism in the mammalian pineal gland, in: "The Pineal Gland. Vol. I. Anatomy and Biochemistry", R.J.Reiter, ed., CRC Press, Boca Raton, Fla. (1981), p. 199.
19. M.J.Berridge, Phosphatidylinositol hydrolysis: A multifunctional transducing mechanism, Mol.Cell.Endocr., 24: (1981).
20. E.G.Lapetina, Regulation of arachidonic acid production: Role of phosphalipases C and A_2, Trends Pharmacol.Sci., 3: 115 (1982).
21. L.S.Wolfe, Eicosanoids: Prostaglandins, thromboxanes, leukotrienes and other derivatives of carbon-20 unsaturated fatty acids, J.Neurochem., 38: 1 (1982).
22. T.C.Westfall, Neuroeffector mechanisms, Annu.Rev.Physiol., 42: 383 (1980).
23. R.Szabo and A.J.Friedhoff, Decrease of serotonin-N-acetyltransferase activity in rat pineal organs after treatment with prostaglandin synthesis inhibitor indomethacin, Prostaglandins, 11: 503 (1976).
24. M.N.Ritta and D.P.Cardinali, Effect of indomethacin treatment on monoamine metabolism and melatonin synthesis of rat pineal gland, Hormone Res., 12: 305 (1980).

25. N.H.Neff and H.Y.Yang, Another look at the monoamine-oxidase inhibitor drugs, Life Sci., 14: 2061 (1974).
26. M.I.Vacas and D.P.Cardinali, Effects of castration and reproductive hormones on pineal serotonin metabolism in rats, Neuroendocrinology, 28: 187 (1979).
27. R.Flowers, Drugs which inhibit prostaglandin synthesis, Pharmacol.Rev., 26: 33 (1974).
28. H.S.Kantor and M.Hampton, Indomethacin in submicromolar concentrations inhibits cyclic AMP dependent protein kinase, Nature, 276: 841 (1978).
29. D.P.Cardinali, M.N.Ritta, N.S.Speziale and M.F.Gimeno, Release and specific binding of prostaglandins in bovine pineal gland, Prostaglandins, 18: 577 (1979).
30. M.Møller and Th. van Veen, Fluorescence histochemistry of the pineal gland, in:"The Pineal Gland. Vol.I Anatomy and Biochemistry", R.J.Reiter, ed., CRC Press, Boca Ratón Fla (1981) p. 69.
31. M.N.Ritta and D.P.Cardinali, Prostaglandin E_2 increases adenosine 3',5'-monophosphate concentration and binding site occupancy, and stimulates serotonin-N-acetyltransferase activity in rat pineal glands in vitro, Mol.Cell. Endocr., 23: 151 (1981).
32. D.P.Cardinali, M.N.Ritta, C.Gonzalez Solveyra and E. Pereyra, Role of prostaglandins in rat pineal neuroeffector junction. Changes in melatonin and norepinephrine release in vitro, Endocrinology, in press.
33. S.Suzuki, R.Franco-Saenz and P.J.Mulrow, The role of renal prostaglandins in the renin response to isoproterenol in the rat in vitro, Endocrinology, 108: 1654 (1981).
34. C.R.Partington, M.W.Edwards and J.W.Daly, Regulation of cyclic AMP formation in brain tissue by α-adrenergic receptors: Requisite intermediacy of prostaglandins of the E series, Proc.Nat.Acad.Sci.USA, 77: 3024 (1980).
35. P.Hedqvist, Basic mechanisms of prostaglandin action on autonomic neurotransmission, Annu.Rev.Pharmacol.Toxicol. 17: 249 (1977).
36. S.Hirose, H.Yokosawa, I.Inagami and J.Workman, Renin and prorenin in hog brain: ubiquitous distribution and high concentration in the pituitary and pineal, Brain Res., 191: 489 (1980).
37. D.G.Changaris, L.M.Demers, L.C.Keil and W.B.Severs, Immunopharmacology of angiotensin I in brain, in: "Central Actions of Angiotensin and Related Hormones" J.P. Buckley and C.Ferraro, eds., Pergamon, New York (1977) p 233.
38. D.G.Changaris, L.C.Keil and W.B.Severs, Angiotensin II immunohistochemistry of the rat brain, Neuroendocrinology, 25: 257 (1978).

39. N.M.Panagiotis and G.F.Hungerford, Response of pineal sympathetic nerve processes and endings to angiotensin, Nature, 211: 374 (1966).
40. I.Haulica, G.Petrescu, M.Ulnitu, V.Rosca and S.Slatineanus, Influence of angiotensin II on dog pineal serotonin content, Neurosci.Lett., 18: 329 (1980).
41. B.Chertow, The role of lysosomes and proteases in hormone secretion and degradation, Endocr.Rev. 2: 137 (1981).
42. V.E.Nahmod,M.S.Balda, C.J.Pirola, S.Finkielman, P.V.Gejman and D.P.Cardinali, Circadian rhythm and neural regulation of rat pineal angiotensin converting enzyme, Brain Res. 236: 216 (1982).
43. V.E.Nahmod, E.F.Lazcano, C.J.Pirola, M.S.Balda, A.Alvarez, P.V.Gejman and D.P.Cardinali, Efecto inhibitorio del simpático sobre la actividad del sistema renina-angiotensina en la pineal de la rata, Medicina (Buenos Aires) 40: 770 (1980) (abs).
44. M.J.Peach, Renin-angiotensin system: Biochemistry and mechanisms of action, Physiol.Rev., 57: 313 (1977).
45. P.V.Gejman, D.P.Cardinali, S.Finkielman and V.E.Nahmod, Changes in drinking behavior caused by superior cervical ganglionectomy and pinealectomy in rats, J.Auton.Nerv. System, 4: 249 (1981).
46. D.P.Cardinali, Hormone effects on the pineal gland, in: "The Pineal Gland. Vol.I. Anatomy and Biochemistry", R.J.Reiter, ed., CRC Press, Boca Raton Fla (1981) p 243.
47. D.P.Cardinali, C.A.Nagle and J.M.Rosner, Control of estrogen and androgen receptors in the rat pineal gland by catecholamine transmitter, Life Sci. 16: 93 (1975).
48. D.P.Cardinali, C.A.Nagle and J.M.Rosner, Metabolic fate of androgens in the pineal organ: Uptake, binding to cytoplasmic proteins and conversion of testosterone into 5α-reduced metabolites, Endocrinology, 95: 179 (1974).
49. M.I.Vacas, P.R.Lowenstein and D.P.Cardinali, Characterization of a cytosolprogesterone receptor in bovine pineal gland, Neuroendocrinology, 24: 84 (1979).
50. M.I.Vacas and D.P.Cardinali, Binding sites for melatonin in bovine pineal gland, Hormone Res., 13: 121 (1980).
51. W.E.Stumpf and M.Sar, Steroid hormone target cells in the periventricular brain: Relationship to peptide hormone producing cells, Fed.Proc., 36: 1973 (1977).
52. D.P.Cardinali, C.A.Nagle and J.M.Rosner, Aromatization of androgens to estrogens by the rat pineal gland, Experientia, 30: 1222 (1974).
53. D.P.Cardinali, C.A.Nagle and J.M.Rosner, Gonadal steroids as modulators of the function of the pineal gland, Gen. Comp.Endocr., 26: 50 (1975).

54. I.Hanukoglu, H.J.Karavolas and R.W.Goy, Progesterone metabolism in the pineal gland, brain stem, thalamus and corpuscallosum of the female rat, Brain Res., 125: 313 (1977).
55. G.Litwack, ed. "Biochemical Actions of Steroids", vol. 6, Academic Press, New York (1979).
56. E.V.Jensen, M.Numata, P.I.Brecher and E.R.De Sombre, Hormone-receptor interaction as a guide to biochemical mechanism, in: "The Biochemistry of Steroid Hormone Action", R.M.S. Smellie, ed., Academic Press, New York (1971) p. 133.
57. M.Ginsburg, B.D.Greenstein, N.J.MacLusky and P.J.Thomas, An improved method for the study of high affinity steroid binding: Oestradiol binding in the brain and pituitary, Steroids, 23: 773 (1974).
58. D.P.Cardinali, Nuclear receptor-estrogen complex in the pineal gland. Modulation by sympathetic nerves, Neuroendocrinology, 24: 333 (1977).
59. I.Lieberburg, N.MacLusky and B.S.McEwen, Cytoplasmic and nuclear estradiol-17β binding in male and female rat brain: Regional distribution, temporal aspects and metabolism, Brain Res., 193: 487 (1980).
60. T.G.Muldoon, Regulation of steroid hormone activity, Endocr.Rev., 1: 339 (1980).
61. D.P.Cardinali, E.Gomez and J.M.Rosner, Changes in ^{3}H-leucine incorporation into pineal proteins following estradiol or testosterone administration: Involvement of the sympathetic superior cervical ganglion, Endocrinology, 94: 849 (1976).
62. N.Emmelin and U.Trendelenburg, Degeneration activity after parasympathetic or sympathetic denervation, Rev.Physiol. Biochem.Exp.Pharmacol., 66: 148 (1972).
63. P.Schotman, J.Allart and W.H.Gispen, Pineal protein synthesis highly sensitive to ACTH-like neuropeptides, Brain Res., 219: 121 (1981).
64. C.A.Nagle, D.P.Cardinali and J.M.Rosner, Testosterone effects on protein synthesis in the rat pineal gland. Modulation by the sympathetic nervous system, Life Sci., 16: 81 (1975).
65. C.A.Nagle, D.P.Cardinali and J.M.Rosner, Diurnal rhythm in tissue radioactivity uptake after ^{3}H-estradiol and ^{3}H-testosterone administration to castrated rats, Steroids Lip.Res., 5: 107 (1974).
66. P.Seem, C.Demaine and L.Vollrath, The effects of sex hormones,prolactin and chorionic gonadotrophin on pineal electrical activity in guinea pigs. Cell.Mol.Neurobiol. 1: 259 (1981).

67. J.T.Epplen, H.Kaltenhauser, W.Engel and J.Schmidtke, Patterns of cyclic AMP phosphodiesterases in the rat pineal gland: Sex differences in diurnal rhythmicity, Neuroendocrinology, 34: 46 (1982).
68. D.P.Cardinali, C.A.Nagle, E.Gomez and J.M.Rosner, Norepinephrine turnover in the rat pineal gland. Acceleration by estradiol and testosterone, Life Sci., 16: 1717 (1975).
69. D.P.Cardinali and M.I.Vacas, Norepinephrine turnover in pineal gland and superior cervical ganglia. Changes after gonadotrophin administration to castrated rats, J.Neural Transm. 45: 273 (1979).
70. D.P.Cardinali, M.I.Vacas and P.V.Gejman, The sympathetic superior cervical ganglia as peripheral neuroendocrine centers, J.Neural.Transm., 52: 1 (1981).
71. M.I.Vacas, P.R.Lowenstein and D.P.Cardinali, Testosterone decreases β-adrenoceptor sites in rat pineal gland and brain, J.Neural Transm. 53: 49 (1982).
72. M.I.Vacas and D.P.Cardinali, Effect of estradiol on α- and β-adrenoceptor density in medial basal hypothalamus, cerebral cortex and pineal gland of ovariectomized rats, Neurosci.Lett., 17: 73 (1980).
73. D.P.Cardinali, M.I.Vacas, C.E.Valenti and C.Gonzalez Solveyra, Pineal gland and sympathetic cervical ganglia as sites for steroid regulation of photosensitive neuroendocrine pathways, J.Steroid Biochem., 11: 951 (1979).
74. L.T.Williams and R.J.Lefkowitz, "Receptor Binding Studies in Adrenergic Pharmacology", Raven Press, New York (1978).
75. D.P.Cardinali, M.I.Vacas, A.L.Fortis and F.J.Stefano, Superior cervical ganglionectomy depresses norepinephrine uptake, increases the density of α-adrenoceptor sites and induces supersensitivity to adrenergic drugs in rat medial basal hypothalamus, Neuroendocrinology, 33: 199 (1981).
76. M.Pisarev, D.P.Cardinali, G.Juvenal, M.I.Vacas, M.Barontini and R.Boado, The role of the sympathetic nervous system in the control of the goitrogenic response in the rat. Endocrinology, 109: 2202 (1981).
77. D.P.Cardinali, M.Pisarev, M.Barontini, G.Juvenal, R.Boado and M.I.Vacas, Efferent neuroendocrine pathways of sympathetic superior cervical ganglia. Early inhibition of pituitary-thyroid axis after ganglionectomy, Neuroendocrinology, in press.

THE RESPONSES OF MELATONIN RHYTHMS TO ENVIRONMENTAL LIGHTING

R.J. Wurtman, M.H. Deng and P. Ronsheim

Laboratory of Neuroendocrine Regulation
Department of Nutrition and Food Science
Massachusetts Institute of Technology
Cambridge, MA 02139

Two very powerful generalizations characterize present conceptions of normal pineal function (1-3): In almost all species examined thus far in laboratory situations, the pineal secretes more melatonin at nighttime than during the day, and melatonin secretion is suppressed when the animals are placed in a lighted environment. Since nighttime for humans coincides with the daily dark period, investigators have often tacitly assumed that the second generalization explains the first, - that is, that the circadian increases in melatonin secretion occurs nocturnally because the absence of light terminates the retina-mediated suppression of pineal sympathetic activity, thereby increasing the organ's stimulation by norepinephrine.

This formulation - that melatonin secretion occurs nocturnally because nighttime coincides with the absence of sunlight - overlooks the fact that many nocturnally-active species probably spend most or all of their lives in light intensities below those reportedly needed to suppress the pineal. Daytime is passed in dark burrows within the earth, while nighttime is passed under the light of the moon and the stars. Thus, full moonlight provides, on the earth's surface, about 0.05 μwatts/cm^2 (4,5), while the light intensity that, in a laboratory situation, causes a 50% inhibition of serotonin N-acetyltransferase (6) in rats [or a reduction in pineal melatonin in hamsters (7)] is about 0.5 μW/cm^2, or tenfold greater. In neither situation would light intensity be expected to be sufficient, by itself, to inhibit pineal sympathetic activity and melatonin synthesis. Moreover, when rats are exposed to "dim light" (0.1-0.3 μwatts/cm^2) for twelve hours daily and then to "darkness" (less than 0.01 μwatts/cm^2) or room light (greater than 45 μwatts/cm^2) for the other twelve hours, the dim light is neither inhibitory nor stimulatory to melatonin synthesis; rather, it is associated with either high or low rates of melatonin secretion, depending on the intensity present during the rest of the day (5). If albino rats live under light intensities less than or equal to 0.05 μwatts/cm^2, their propensity to secrete melatonin nocturnally clearly cannot be explained by postulating that their lighting environment cyclically inhibits pineal sympathetic outflow. How then are we to understand the nocturnal entrainment of melatonin secretion among nocturnal burrowing animals, and how can we relate their melatonin rhythms to natural ambient lighting?

Any analysis of these relationships requires consideration of those parameters of light which influence its biological consequences. They include, among other things, its spectrum, its intensity, and its timing.

The spectrum of sunlight at the earth's surface resembles that emitted by a theoretical "black body" heated to 5500 °K, as filtered through the atmosphere's ozone layer. It is continuous between 290 and 700 nanometers, and thus includes some mid-range ultraviolet, more long-wave ultraviolet, and major, nearly-equal proportions (in terms of percents of total energy) of visible irradiations. The effects of various visible and ultraviolet wave lengths on the rat's pineal have been examined by assessing their relative potencies in suppressing HIOMT activity. These potencies parallel their abilities to activate rhodopsin: yellow-green is most effective, blue and yellow less so, and red and ultraviolet not at all (8). This pattern isn't surprising, given the abundant evidence that more than 95% of the photoreceptive units in the rat's retina are rod-type cells. One might anticipate that the action spectrum for the photic suppression of melatonin secretion from the human's pineal will differ considerably from the rat's, since human retinas contain several types of photoreceptive cells and several photopigments. (Parenthetically, it might be noted that exposing people and laboratory animals to light spectra that differ considerably from the sun's constitutes an experiment which investigators may or may not realize that they are performing: It would seem wiser to standardize the light spectrum utilized for photobiologic studies, and to use one that resembles as closely as possible the spectrum present on the earth's surface.)

The intensities of light under which humans and other animals normally live can vary over 8 or more log units, depending on how members of each species choose to pass their time: One pressing task for the student of pineal function involves relating each species' circadian and circannual patterns of melatonin secretion to the light intensities that it normally receives. Humans spend portions of each day within four intensity ranges (Figure 1):

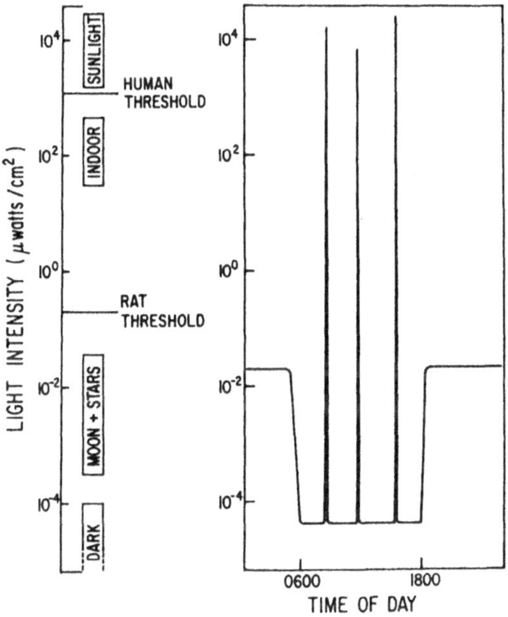

Figure 1. Ambient light intensities under various conditions, and intensities to which a nocturnal, burrowing rodent might be exposed on a typical day. (Daylight occurs for 12 hrs, from 0600 to 1800.) The threshold light intensities required in the laboratory to produce significant suppression of melatonin synthesis (albino rats, ref. 6) or secretion (human, ref. 8) are taken from the Literature.

Working out of doors in daytime they are exposed to 2,000 to 20,000 μwatts/cm^2, or even more; indoor lighting usually provides 50-500 μwatts/cm^2; the night light of the moon and stars varies between 0.001 μwatts/cm^2 and, at full moonlight, 0.05 μwatts/cm^2; and the light of a "dark" environment is probably below 0.001 μwatts/cm^2 or below. The light intensity needed to suppress nocturnal melatonin secretion from the normal human's pineal is about 1,000 μwatts/cm^2, or greater (9); this is well below that present out of doors in the daytime, but greater than that usually provided indoors. As discussed above, two of these ranges are above the intensities needed in the laboratory to suppress the rat's pineal (daytime outdoor light and interior light), and two are below (nighttime outdoor light and "darkness").

If albino rats are placed in a cage that allows them to choose between being exposed to ambient laboratory light or to the darkness of an artificial burrow (less than 0.01 μwatts/cm^2), and if the ambient light intensity is in the usual indoor range (66 μwatts/cm^2) for twelve hours and in the nocturnal moonlight range (0.03 μwatts/cm^2) for the other twelve hours, we find that the animals spend almost all of the "daylight" period in the dark burrow, and all of the "nighttime" in the outer cage (4,10). Since the light intensity at nighttime is <u>greater</u> than that in the burrow (0.03 vs. 0.01 μwatts/cm^2), nighttime actually coincides with the animals's daily <u>light</u> period. This schedule probably simulates the animal's schedule in the wild state, where rats are nocturnally active under the light of the moon and stars. It contrasts, of course, with the situation for humans - who are active under sunlight or indoor light, and inactive during the dark period, sleeping in darkness indoors. Melatonin secretion from the rat's pineal is now also maximal during the animal's <u>light</u> period. The generalization that peak melatonin secretion coincides with nighttime is thus preserved, but at the apparent expense of the generalization that it coincides with the daily dark period.

If an additional constraint is now imposed on the animals such that, once they enter the burrow each day, they are required to remain there until just after the onset of "moonlight," the amplitude of the pineal melatonin rhythm diminishes markedly, suggesting that either it has become free-running - and its peaks and valleys have been missed - or that it has in fact disappeared. Now animals spend their 24-hour days exposed alternately to two light intensities which are both below that shown experimentally to suppress melatonin's synthesis and secretion. If the rats are again allowed to leave the burrow at intervals, then the melatonin rhythm recurs and its peak is re-entrained to nighttime. Such sampling of the outside world need only be very brief, - a minute or two - a finding compatible with the evidence (11) that similarly-brief light pulses are adequate to suppress melatonin synthesis for hours, if delivered at times of day appropriate for the species. (Thus the importance of the timing of light exposure.)

Perhaps one stimulus that keeps the albino rat's melatonin rhythm entrained to the daily light cycle - and causes its secretion of melatonin to be maximal during nighttime - is these brief, daytime exposures to bursts of intense (for the albino rat) ambient light. This formulation allows the two seemingly-contrary generalizations to coexist (Figure 1). Melatonin secretion is again both nocturnal and suppressed by light; its entrainment to nighttime - the albino rat's daily light period - occurs because the animal is sporadically exposed to intense light during the day. (Of course the formulation ignores the fact that other sensory inputs, besides those generated by light, affect behavior and pineal function.) It will be interesting to see if similar explanations, relative to how each species actually chooses to live, can be adduced for the patterns of melatonin secretion found in other mammals besides albino rats and people. [That these patterns might exhibit great variety is suggested by the finding that melatonin is secreted nocturnally when sheep are exposed to long days (and short nights), but barely at all when the animals are in long nights (12).] In any case, an appreciation of light's photobiologically-important parameters, and of how species behave in relation to normal variations in their lighting environment, will probably be very important in unravelling the ways that mammals use their pineals.

ACKNOWLEDGEMENTS

These studies were supported by a grant from the National Institutes of Health (HD-11722). The experiments on animals with access to burrows were conceptualized and conducted by Dr. Harry Lynch.

REFERENCES

1. Wurtman, R.J., J. Axelrod, and D.E. Kelly, "The Pineal," Academic Press, New York, 1968.
2. Wurtman, R.J. and M. Moskowitz, Medical Progress: The Pineal Organ. N. Engl. J. Med. 296:1329-1333 (1977); 296:1383-1386 (1977).
3. Waldhauser, F. and R.J. Wurtman, The secretion and actions of melatonin, in: "Biochemical Actions of Hormones," G. Litwack, ed., vol. 10, Academic Press, New York (in press).

4. Lynch, H.J., R.J. Wurtman, and P. Ronsheim, Activity and melatonin rhythms among rats with recourse to dark burrows, in: "The Pineal and Its Hormones," R.J. Reiter, ed., Alan R. Liss, New York (1982).
5. Lynch, H.J., R.W. Rivest, P.M. Ronsheim, and R.J. Wurtman, Light intensity and the control of melatonin secretion in rats, Neuroendocrinology 33:181-185 (1981).
6. Minneman, K.P., H. Lynch, and R.J. Wurtman, Relationship between environmental light intensity and retina-mediated suppression of rat pineal serotonin-N-acetyltransferase, Life Sci. 15:1791-1796 (1974).
7. Brainard, G.C., B.A. Richardson, L.J. Pettibone, and R.J. Russel, The effect of different light intensities on pineal melatonin content, Brain Res. 233:75-81 (1982).
8. Cardinali, D.P., F. Larin, and R.J. Wurtman, Action spectra for effects of light on hydroxyindole-O-methyl transferase in rat pineal, retina and harderian gland, Endocrinology 91:877-886 (1972).
9. Lewy, A.J., T.A. Wehr, F.K. Goodwin, D.A. Newsome, and S.P. Markey, Light suppresses melatonin secretion in humans, Science 210:1267-1269 (1980).
10. Lynch, H.J., M.H. Deng, P. Ronsheim, and R.J. Wurtman, Daily rhythms in activity and melatonin secretion among rats in a naturalistic environment, Abstract, Soc. Neurosci. Minneapolis, Minn., November, 1982.
11. Illnerova, H., J. Vanecek, L. Wetterberg, and J. Saaf, Effect of one minute exposure to light at night on rat pineal serotonin-N-acetyltransferase and melatonin, J. Neurochem. 32:673 (1978).
12. Lincoln, G.A., O.F.X. Lameida, H. Klandorf, and R.A. Cunningham, Hourly fluctuations in the blood levels of melatonin, prolactin, luteinizing hormone, follicle-stimulating hormone, testosterone, tri-iodothyronine, thyroxine, and cortisol, in rats under artificial photoperiods, and the effects of cranial sympathectomy, J. Endocrinol. 92:237-250 (1982).

THE ROLE OF LIGHT AND AGE IN DETERMINING MELATONIN PRODUCTION IN THE PINEAL GLAND

Russel J. Reiter

Department of Anatomy
The University of Texas
Health Science Center at San Antonio
San Antonio, TX 78284

INTRODUCTION

Roughly two decades ago it was discovered that the photoperiodic environment to which animals are exposed determines the structure (Quay, 1956) and metabolism (Quay, 1963; Wurtman et al., 1963) of the mammalian pineal gland. Shortly thereafter, it was uncovered that the ability of the pineal to influence the reproductive system also depended on the photoperiod (Czyba et al., 1964; Hoffman and Reiter, 1965). This series of experiments provided not only an impetus for research on the pineal gland but also a renewed interest in the impact of light and darkness on the organism. In the intervening years the enthusiasm was translated into investigations that have definitively shown that the daily light:dark cycle provides the primary regulatory influence on the pineal gland. When the role of the photoperiod is discussed relative to the pineal, typically, indoleamine synthesis is considered (Cardinali, 1981) along with the hormonal effects of the pineal gland on the neuroendocrine-reproductive axis (Reiter, 1980). It is the purpose of the present survey to summarize the influence of light and darkness on indole metabolism within the pineal gland. Additionally, the resume will consider age of the organism as a factor in determining pineal melatonin production.

IMPACT OF LIGHT AND DARKNESS ON MELATONIN

Neural connections between eyes and pineal

In many non-mammalian vertebrates the pineal responds directly to light rays which reach the pineal gland after penetrating the skull. In mammals, however, it is light and darkness perceived by the lateral eyes which is critical in determining serotonin metabolism within the pineal gland. Because of this indirect action of photoperiod on the pineal, mammals have had to evolve a system of cummunicating information about the photoperiod, which is detected by the eyes, to the pineal gland. There were at least two means available whereby the retinas could signal the pineal gland. The retinas could either send a hormonal signal to the gland via the blood vascular system or it could transfer the information to the pineal over a neural route; during evolution mammals opted for the latter means. It is now known that photic information perceived by the

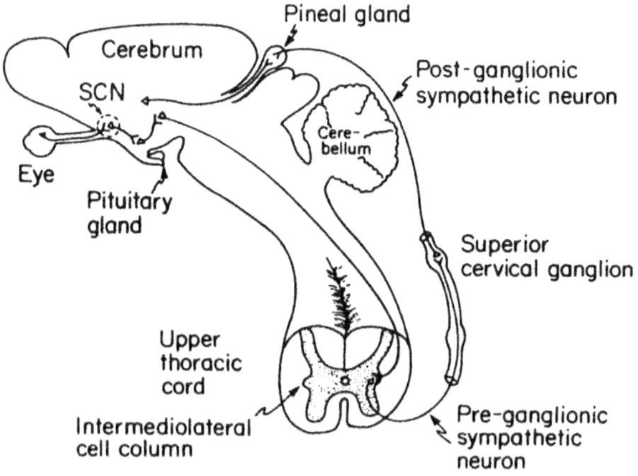

Fig. 1. Schematic representation of the neural connections between the eyes and the pineal gland as they are currently believed to exist in mammals. Information about the light:dark cycle reaches the pineal primarily via the peripheral sympathetic nervous system. Other fibers from the brain probably enter the pineal stalk and eventually terminate within the gland. SCN = suprachiasmatic nuclei.

retinas is transduced into a neural message which is subsequently transferred over a complex series of neurons to reach the pineal gland (Fig. 1). The neural signal reaches the hypothalamus by means of axons of ganglion cells which terminate in the suprachiasmatic nuclei (SCN); in the rat, fibers from a given retina end primarily, although not exclusively, in the contralateral SCN. Axons of SCN neurons theoretically project to the medial basal hypothalamus where another synapse is made and then the neural message is transferred to the lateral hypothalamic area (Moore, 1978). From this location long descending axons presumably carry the transduced photoperiodic message to the intermediolateral cell column of the upper thoracic cord; this cell column is composed of preganglionic sympathetic neurons whose axons emerge in the ventral roots of the thoracic spinal nerves. They pass up the sympathetic trunk to eventually make a terminal synapse in the superior cervical ganglia; these ganglia lie in the carotid sheath at the division of the common carotid into the internal and external carotid arteries. Postganglionic axons follow blood vessels into the skull and eventually, in the company of the arterial supply to the gland, the nerves arise in the vicinity of the pineal. Before they penetrate the gland they form one or two rather distinct nerve bundles, the nervi conarii (Kappers, 1960). After penetrating the capsule of the gland the nerve bundles break up into smaller fascicles which ramify among the parenchymal elements; they terminate primarily in the pericapillary spaces but occasionally endings are also found among parenchymal cells (Matsushima et al., 1981).

The sympathetic nerve endings which terminate within the pineal gland reportedly release norepinephrine (NE) especially during the dark phase of the light:dark cycle. The catecholamine subsequently acts on β-adrenergic receptors on the pinealocyte membrane (Zatz, 1981). NE as a neurotransmitter within the pineal has been proven in only a small number of animals, e.g., whereas it stimulates the metabolism of serotonin to melatonin in the rat pineal (Zatz, 1981) this has not been shown to be the case for the Syrian hamster (Lipton et al., 1982). The released neurotransmitter, be it NE or some other compound, interacts with β-receptors which are linked to an adenylate cyclase-cyclic AMP system (Klein et al., 1981). Via this mechanism the conversion of serotonin to melatonin is promoted (Cardinali, 1981). The two pineal indices which are most frequently measured are the increased acetylation of serotonin by the enzyme N-acetyltransferase (NAT) and the tissue levels of melatonin. Both these parameters increase during the daily dark period; the increases are often parallel. Once produced melatonin seems not to be stored within the gland but rather it is rapidly released. Hence, blood levels of melatonin are closely related to the production of the indole in the pineal gland (Wilkinson et al., 1981).

Under either artificial or natural photoperiodic conditions the daily period of light is associated with low levels of pineal NAT activity (Klein et al., 1981) and melatonin content (Lynch, 1971; Rollag et al., 1980a) and depressed titers of circulating melatonin (Wilkinson et al., 1977).

Light:dark cycle and patterns of melatonin production

Sometime during the daily dark period, pineal melatonin production rises and, as a consequence, blood levels of the constituent also increase (Panke et al., 1978; Brown et al., 1981). However, the pattern of the nocturnal pineal production of melatonin varies among species (Fig. 2). For example, when Syrian hamsters are kept under 14 hours of light and 10 hours of darkness (LD 14:10) daily, the first 4-6 hours of dark exposure is associated with essentially no rise in pineal melatonin content; thereafter, pineal melatonin levels increase rapidly to

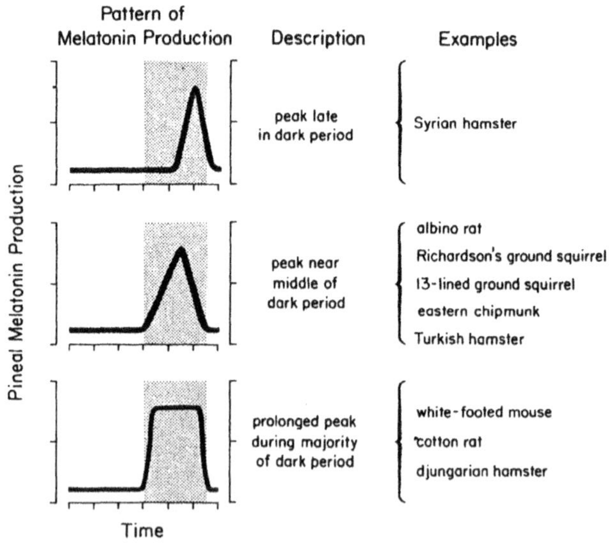

Fig. 2. Patterns of nocturnal melatonin production in rodents that have been studied to date. Whereas some species tend to exhibit a brief rise in pineal melatonin levels, in others elevated melatonin levels may be measured throughout the dark period.

reach a peak roughly 8 hours after darkness onset (Panke et al., 1979). Conversely, in the albino rat kept under the same photoperiodic conditions peak pineal melatonin levels are usually reached as soon as 4-5 hours after darkness onset with a gradual sustained drop thereafter (Johnson et al., 1982). In the Richardson's ground squirrel (*Spermophilus richardsonii*), a diurnally active rodent, pineal melatonin levels rise quickly after darkness onset and begin to drop soon thereafter (Reiter et al., 1981b). Finally, in a number of species melatonin levels rise rapidly during darkness and remain elevated during the majority of the dark period; examples include the Djungarian hamster (*Phodopus sungorus*) (Goldman et al., 1981), the cotton rat (*Sigmodon hispidus*) (Matthews et al., 1982), and the white-footed mouse (*Peromyscus leucopus*) (Petterborg et al., 1981). If the daily period of darkness to which these species are exposed is increased by 4 hours, i.e., to LD 10:14, melatonin levels remain elevated virtually throughout the daily dark period. In these species it appears as if the pineal gland may actually measure the length of the daily dark period, at least within the limits in which the system has been tested. In species such as the Syrian hamster where a single short term peak of melatonin occurs, prolonging the daily dark period seems not to greatly alter the duration or the timing of the nocturnal melatonin peak (Tamarkin et al., 1979; Reiter, 1981).

How these different patterns of melatonin relate to the ability of the pineal to inhibit reproduction remains unknown. In cotton rats, white-footed mice, and Djungarian hamsters it may be the increased and prolonged rise in melatonin which initiates gonadal involution when these species are placed under short day conditions (Petterborg et al., 1981). In the Syrian hamster, the mechanisms may be different inasmuch as the quantity of melatonin produced by the pineal gland of hamsters kept under artificially regulated long or short days varies rather minimally, yet under reduced photoperiods (less than 12.5 hours of light daily) gonadal regression ensues (Elliott, 1976). Perhaps in this species there are two rhythms, i.e., the melatonin rhythm and a rhythm in the sensitivity of the neuroendocrine-reproductive system to melatonin (Fig. 3). Under long day conditions these rhythms are out of phase and melatonin, although synthesized and secreted, does not inhibit reproduction. On the other hand, under short day conditions peak melatonin levels coincide with an increased sensitivity of the animal to the indole; as a result, reproductive collapse follows. This is an example of an internal coincidence model. Assuming there are, in fact, two rhythms (the melatonin cycle and the sensitivity fluctuation), what mechanisms are brought into play to bring them in and out of phase with each other remains unknown. Finally, under natural photoperiodic and temperature conditions the actual amount of melatonin produced within the pineal may vary seasonally (Brainard et al., 1983).

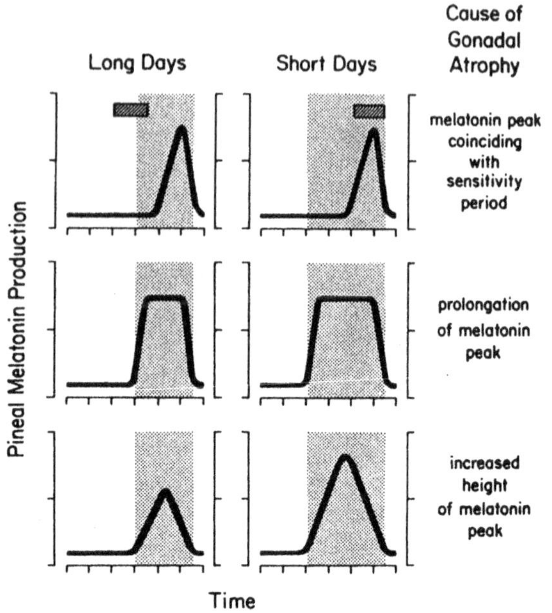

Fig. 3. Theoretical means whereby melatonin may induce its endocrine influences. As seen in the top panel, in some species there may be two rhythms, i.e., the melatonin rhythm and the cycle of sensitivity to melatonin (cross-hatched bar) Only when the melatonin peak coincides with the increased sensitivity is the indoleamine capable of exerting its effects. Alternatively, perhaps the amount of melatonin produced [due to a prolongation of the melatonin peak (middle panels) or to an increased height of the melatonin peak (bottom panels)] determine its action.

Influence of intensity on melatonin production

Animals seem to vary greatly in reference to the sensitivity of their pineal gland to light. In nocturnally active rats, Syrian hamsters and Djungarian hamsters born in captivity and raised under laboratory photoperiods, prolongation of normal artificial room light (intensity of 1000 lux or about 172 μW/cm² and much lower intensities as well) into the night prevents the normal nocturnal rise in pineal melatonin levels (Panke et al., 1979; Tamarkin et al., 1979). Likewise, the abrupt exposure of these animals to light at night when pineal melatonin levels are high leads to a precipitous decline in the melatonin content of the pineal gland (Rollag et al., 1980b; Brainard et al., 1982b).

In the case of the Syrian hamster, the intensity of light required to cause such a decline is very low, on the order of 1 lux (about 0.186 μW/cm^2.

There are some other species, however, which seem to be less sensitive to light in terms of the inhibition of nocturnal melatonin levels. For example, in the human it has been shown that the intensity of light at night required to inhibit serum melatonin titers (Vaughan et al., 1976; Lewy et al., 1980) is greater than 500 lux (about 86 μW/cm^2); this is much higher than the intensity which suppresses pineal melatonin production in the Syrian hamster pineal gland (Brainard et al., 1982b). When these two species are compared it is important to keep in mind that the human is usually diurnally active whereas the hamster is a nocturnal species. This could account for differential response to light. With this is mind, we examined pineal melatonin levels in two other diurnally active species, the Richardson's ground squirrel (Hurlbut et al.,1982; Reiter et al., 1982a) and the Eastern chipmunk (*Tamias striatus*) (Reiter et al., 1982b; 1982c), after their exposure to light during the normal dark period. As with the human, in neither species did normal room light suppress the rise in pineal melatonin normally associated with darkness (Fig. 4). Thus, these diurnal species responded in a manner similar to the human, i.e., they were considerably less sensitive to light than the nocturnal species previously investigated (Panke et al., 1979; Tamarkin et al., 1979; Rollag et al., 1980b; Brainard et al., 1982b). Do these differences really relate to the fact that the human, the Richardson's ground squirrel, and the Eastern chipmunk are diurnal while the other species examined are nocturnal? There is another difference between these two groups of animals. The diurnally active species were exposed to sun light during at least a portion of the life history while the rat and hamster were born in captivity and were raised under artificial lighting of the laboratory. Sun light has a much higher intensity (up to 100,000 lux) than do most artificial lights currently in use in animal rooms (approximately 1000-1500 lux). If the intensity of light to which animals are exposed during the day in fact determines what they interpret as darkness (Rivest et al., 1981; Lynch et al., 1981) then the results summarized above may not relate to whether the animals are diurnal or nocturnal but rather to their previous lighting history. Perhaps higher intensity light during the day, i.e., sun light, makes them insensitive to relatively lower intensity room light at night. Conversely, animals that are exposed to artificial room light during the day are much more sensitive to light at night. If this latter interpretation has validity, then perhaps raising rats or Syrian hamsters under natural sun light would also render them insensitive to room light at night. We have evidence that the diurnally active 13-lined ground squirrel (*Spermophilus tridecemlineatus*), born in captivity and exposed

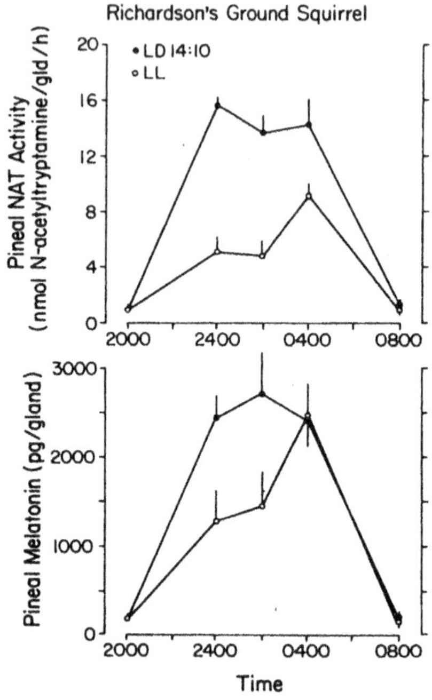

Fig. 4. Pineal N-acetyltransferase activity (top panel) and melatonin content (bottom panel) in the Richardson's ground squirrel. Animals were either kept in 10 hours of darkness (LD 14:10) during the night they were killed or they were kept in continual room light (LL) (approximately 4,000 lux or 925 μW/cm^2). Although light exposure at night slightly depressed the rhythms in NAT and melatonin they were not prevented by this treatment procedure.

throughout its life to artificial room light, responds to low intensity light at night in the same manner as does the rat and hamster (Reiter et al., unpublished observations).

Influence of wavelength on melatonin production

Besides light intensity, it is of interest to determine what wavelengths of light are most important in influencing pineal melatonin levels. These interactions have been only sparingly investigated. Recently, we used the Syrian hamster to examine which wavelengths of visible light are most efficient in depressing nocturnal pineal melatonin (Brainard and Reiter, 1982). During peak nocturnal melatonin production groups of hamsters were exposed to light sources with one half peak band widths of 339-371 nm (near ultraviolet), 435-500 nm (blue), 515-550 nm (green), 558-636 nm (yellow), and 635-770 nm (red). Animals were exposed to each color at an intensty of either 0.928 (about 5.4 lux) or 0.200 $\mu W/cm^2$ (about 1.16 lux). At the 0.928 $\mu W/cm^2$ intensity, the blue, green and yellow light sources led to a 70-80% reduction of pineal melatonin within 20 minutes; the near ultraviolet light caused a 40% drop while the red light source was ineffective in reducing pineal melatonin in the Syrian hamster. At 0.200 $\mu W/cm^2$, the blue and green light caused a 70-90% depression of pineal melatonin while near ultraviolet, yellow and red lights did not depress pineal melatonin levels during the 20 minute exposure period. Thus, at least in the Syrian hamster visible light of the blue and green wavelengths of the spectrum seem most efficient in suppressing pineal melatonin when the animals are abruptly exposed to these light sources at night.

To attempt to distinguish which wavelength of visible light, that is, blue or green, was most effective another experiment was performed in which animals were exposed to these wavelengths at the following intensities: 0.186, 0.074 or 0.019 $\mu W/cm^2$. Again, the animals were killed after 20 minutes exposure and pineal melatonin levels were estimated (Brainard et al., 1982a). At each of the intensities tested, blue light caused a 20-25% greater depression of pineal melatonin than did the green light, thus indicating that blue wavelengths of visible light are the most efficient wavelengths of those tested to date in depressing pineal melatonin levels in laboratory born Syrian hamsters.

IMPACT OF AGE ON MELATONIN

Although it has been assumed for years, with little substantive evidence that pineal function diminishes with age, only recently has specific information on pineal melatonin production in aging animals been reported. The circadian rhythm in pineal melatonin levels in 2-month-old and 18-month-old male and female hamsters were compared (Reiter et al., 1980b). During the daily light period the pineal gland of both young and old hamsters contained equivalent amounts of melatonin (roughly 100 pg/gland). During the 10 hour period of darkness pineal melatonin levels in the 2-month-old male and female hamsters rose to 835 and 887 pg/gland, respectively. In both sexes, peak melatonin levels were achieved at 0400 hours, 8 hours after darkness onset. By comparison, in the old animals peak melatonin levels reached only about 200 pg/gland. This marked reduction in pineal melatonin levels in aging Syrian hamsters could be related to a number of factors, the most likely of which may be the reduced sensitivity of the β-adrenergic receptors on the pinealocyte membrane to the sympathetic neurotransmitter within the gland (Greenberg and Weiss; Reiter et al., 1982d). Besides pineal levels of the indole, plasma titers of immunoreactive melatonin are also diminished in aging Syrian hamsters (Reiter et al., 1982e) (Fig. 5).

In a detailed study using albino rats, Reiter and colleagues (1981a) also found a marked reduction in pineal melatonin production during the aging process. In this study both pineal NAT activity levels and immunoreactive melatonin concentrations were compared in 2, 12 and 29-month-old female rats killed either during the day or during darkness. Whereas pineal NAT values varied little among the three age groups, melatonin values differed greatly. Although daytime values of pineal melatonin

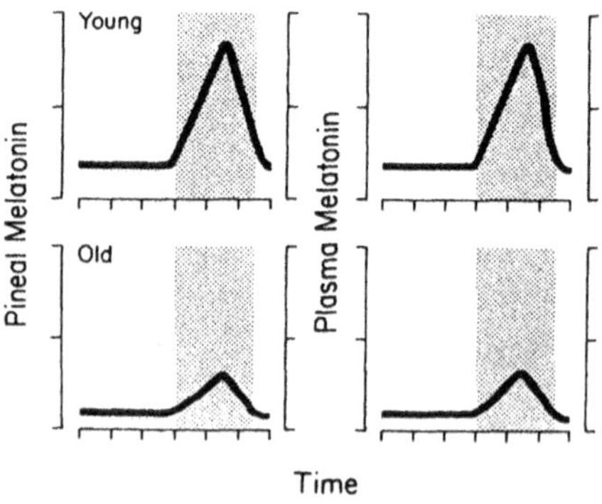

Fig. 5. Schematic diagram of pineal and plasma levels of immunoreactive melatonin in young and old rodents. Perhaps both daytime and nighttime levels of melatonin are reduced in old animals. It may be important that, although reduced, weak rhythms may still persist.

did not vary among the three groups, at night the 29-month-old rats exhibited a weak rise in pineal melatonin levels compared to that of the 2-month-old animals; the 12-month-old rats had nighttime melatonin levels between those of the other two groups. These findings suggest that there may be a gradual (as opposed to an abrupt) reduction in the production of pineal melatonin as these animals age.

Finally, in Mongolian gerbils the pineal production of melatonin also appears to be altered as animals become older (Reiter et al., 1980a; King et al., 1982). Of special interest in this species is that the pineal gland becomes heavily calcified and the nocturnal levels of melatonin, even in young animals, are much lower than in other rodents studied to date. Perhaps the weak melatonin rhythm in the pineal of this species is related to the high degree of pineal calcification.

CONCLUDING REMARKS

The patterns of nocturnal melatonin production vary considerably even among closely related rodent species. Whether these different patterns or alterations of these patterns during different light:dark cycles determine the action of melatonin in a given species remains to be determined. Both the intensity and the wavelength of light seem important in influencing pineal melatonin levels. The importance of light intensity in altering nocturnal pineal melatonin concentrations may be related to the activity patterns of the animals, i.e., whether they are nocturnal or diurnal, or to their previous lighting history. This remains an area of interest for future pineal research. Also, the role of different wavelengths of light in determining pineal melatonin synthesis will undoubtedly receive continued investigation in the future. Finally, it appears at least in rodents, that the age of the animal may be important in determining the ability of the pineal to produce one of its hormonal constituents, melatonin.

REFERENCES

Brainard, G. C., and Reiter R. J., 1982, The influence of different light spectra on pineal melatonin content in the Syrian hamster, Abstr. Endocrine Mtg., in press.

Brainard, G. C., Richardson, B. A., King, T. S., and Reiter, R. J., 1982a, The influence of low irradiances of blue and green light on pineal melatonin content in the Syrian hamster, Abstr. Neurosci. Mtg., in press.

Brainard, G. C., Richardson, B. A., Petterborg, L. J., and Reiter, R. J., 1982b, The effect of different light intensities on pineal melatonin content, Brain Res., 233: 75.
Brainard, G. C., Petterborg, L. J., Richardson, B. A., and Reiter, R. J., 1983, Pineal melatonin in Syrian hamsters: Circadian and seasonal rhythms in animals maintained under laboratory and natural conditions, Neuroendocrinology, in press.
Brown, G., Grota, L., and Niles, L., 1981, Melatonin: Origin, control of circadian rhythm and site of action, in: "Melatonin - Current Status and Perspectives", N. Birau and W. Schloot, eds., Pergamon, New York.
Cardinali, D. P., 1981, Melatonin. A mammalian pineal hormone, Endocr. Rev., 2: 327.
Czyba, J. C., Girod, C., and Durand, N., 1964, Sur l´antagonisme epiphyso-hypophysaíre et les variations saisonnieres de la spermatogenese chez le Hamster dore (*Mesocricetus auratus*). C. R. Soc. Biol., 158: 742.
Elliott, J., 1976, Circadian rhythms and photoperiodic time measurements in mammals, Fed. Proc., 35: 2339.
Goldman, B., Hall, V., Hollister, C., Reppert, S., Roychoudhury, P., Yellon, S., and Tamarkin, L., 1981, Diurnal changes in pineal melatonin content in four rodent species: Relationship to photoperiodism. Biol. Reprod., 24: 778.
Greenberg, L. H., and Weiss, B., 1978, β-adrenergic receptors in aged rat brain. Reduced number and capacity of pineal gland to develop supersensitivity, Science, 201: 61.
Hoffman, R. A., and Reiter, R. J., 1965, Pineal gland: Influence on gonads of male hamsters, Science, 148: 1609.
Hurlbut, E. C., King, T. S., Richardson, B. A., and Reiter, R. J., 1982, The effects of the light:dark cycle and sympathetically-active drugs on pineal N-acetyltransferase activity and melatonin content in the Richardson's ground squirrel, *Spermophilus richardsonii*, in: "The Pineal and Its Hormones", R. J. Reiter, ed., Alan R. Liss, New York.
Johnson, L. Y., Vaughan, M. K., Richardson, B. A., Petterborg, L. J., and Reiter, R. J., 1982, Variation in pineal melatonin content during the estrous cycle of the rat, Proc. Soc. Exp. Biol. Med., 169: 416.
Kappers, J. A., 1960, The development, topographical relations and innervation of the epiphysis cerebri in the albino rat, Z. Zellforsch., 52: 163.
King, T. S., Richardson, B. A., and Reiter, R. J., 1981, Age-associated changes in pineal serotonin N-acetyltransferase activity and melatonin content in the male gerbil, Endocr. Res. Commun., 8: 253.
Klein, D. C., Auerbach, D. A., Namboodiri, M. A. A., and Wheler, G. H. T., 1981, Indole metabolism in the mammalian pineal

gland, in: "The Pineal Gland, Vol. I., Anatomy and Biochemistry", R. J. Reiter, ed., CRC Press, Boca Raton.

Lewy, A. J., Wehr, T. A., Goodwin, F. K., Newsome, D. A., and Markey, S. P., 1980. Light suppresses melatonin secretion in humans, Science, 210: 1267.

Lipton, S. J., Petterborg, L. J., Steinlechner, S., and Reiter, R. J., 1982, *In vivo* responses of the Syrian hamster to isoproterenol or norepinephrine, in: "The Pineal and Its Hormones", R. J. Reiter, ed., Alan R. Liss, New York.

Lynch, H. J., 1971, Diurnal oscillations in pineal melatonin content, Life Sci., 10: 791.

Lynch, H. J., Rivest, R. W., Ronsheim, P. M., and Wurtman, R. J., 1981, Light intensity and the control of melatonin secretion in the rat, Neuroendocrinology, 33: 181.

Matsushima, S., Morisawa, Y., and Mukai, S., 1981, Functional morphology of sympathetic nerve fibers in the pineal gland of mammals, in: "The Pineal Gland, Vol. I., Anatomy and Biochemistry", R. J. Reiter, ed., CRC Press, Boca Raton.

Matthews, S. A., Evans K. L., Morgan W. W., Petterborg, L. J., and Reiter, R. J., 1982, Pineal indoleamine metabolism in the cotton rat, *Sigmodon hipidus*: Studies on norepinephrine, serotonin, N-acetyltransferase activity and melatonin, in: "The Pineal and Its Hormones", R. J. Reiter, ed., Alan R. Liss, New York.

Moore, R. Y., 1978, The innervation of the mammalian pineal gland, in: "The Pineal and Reproduction", R. J. Reiter, ed., Karger, Basel.

Panke, E. S., Reiter, R. J., Rollag, M. D., and Panke, T. W., 1978, Pineal serotonin N-acetyltransferase activty and melatonin concentrations in prepubertal and adult Syrian hamsters exposed to short daily photoperiods, Endocr. Res., Commun., 5: 311.

Panke, E. S., Rollag, M. D., and Reiter, R. J., 1979, Pineal melatonin concentrations in the Syrian hamster, Endocrinology, 104: 194.

Petterborg, L. J., Richardson, B. A., and Reiter, R. J., 1981, Effect of long or short photoperiod on pineal melatonin content in the white-footed mouse, Peromyscus leucopus, Life Sci., 29: 1623.

Quay, W. B., 1956, Volumetric and cytologic variation in the pineal body of *Peromyscus leucopus (Rodentia)* with respect to sex, captivity and day length, J. Morph., 98: 471.

Quay, W. B., 1963, Circadian rhythm in rat pineal serotonin and its modifications by estrous cycle and photoperiod, Gen. Comp. Endocr., 3: 473.

Reiter, R. J., 1980, The pineal and its hormones in the control of reproduction in mammals, Endocr. Rev., 1: 109.

Reiter, R. J., 1981, Chronobiological aspects of the mammalian pineal gland, in: "Biological Rhythms in Structure and

Function", H. V. Mayersbach, L. E. Scheving, and J. E. Pauly, eds., Alan R. Liss, New York.
Reiter, R. J., Johnson, L. Y., Steger, R. W., Richardson, B. A., and Petterborg, L. J., 1980a, Pineal biosynthetic activity and neuroendocrine physiology in the aging hamster and gerbil, Peptides 1, Suppl. 1: 69.
Reiter, R. J., Richardson, B. A., Johnson, L. Y., Ferguson, B. N., and Dinh, D. T., 1980b, Pineal melatonin rhythm: Reduction in aging Syrian hamsters, Science, 210: 1372.
Reiter, R. J., Craft, C. M., Johnson, J. E., Jr., King, T. S., Richardson, B. A., Vaughan, G. M., and Vaughan, M. K., 1981a, Age-associated reduction in nocturnal pineal melatonin levels in female rats, Endocrinology, 109: 1295.
Reiter, R. J., Richardson, B. A., and Hurlbut, E. C., 1981b, Pineal, retinal and Harderian gland melatonin in a diurnal species, the Richardson's ground squirrel (*Spermophilus richardsonii*). Neurosci. Letters, 22, 285.
Reiter, R. J., Hurlbut, E. C., Richardson, B. A., King, T. S., and Wang, L. C. H., 1982a, Studies on the regulation of pineal melatonin production in the Richardson's ground squirrel, in: "The Pineal and Its Hormones", R. J. Reiter, ed., Alan R. Liss, New York.
Reiter, R. J., King, T. S., Richardson, B. A., and Hurlbut, E. C., 1982b, Studies on pineal melatonin levels in a diurnal species, the Eastern chipmunk (*Tamias striatus*): Effects of light at night, propranolol administration or superior cervical ganglionectomy. J. Neural Transmis., in press.
Reiter, R. J., King, T. S., Richardson, B. A., Hurlbut, E. C., Karasek, M. A., and Hansen, J. T., 1982c, Failure of room light to inhibit pineal N-acetyltransferase activity and melatonin content in a diurnal species, the Eastern chipmunk (*Tamias striatus*), Neuroendocr. Letters, 4: 1.
Reiter, R. J., Trakulrungsi, W. K., Trakulrungsi, C., Vriend, J., Morgan, W. W., Vaughan, M. K., Johnson, L. Y., and Richardson, B. A., 1982d, Pineal melatonin production: Endocrine and age effects, in: "Neurobiology of the Melatonin Rhythm Generating System", D. C. Klein, ed., Karger, Basel.
Reiter, R. J., Vriend, J., Brainard, G. C., Matthews, S. A., and Craft, C. M., 1982e, Reduced pineal and plasma melatonin levels and gonadal atrophy in old hamsters kept under winter photoperiods, Exp. Aging Res., 8: 27.
Rivest, R. W., Lynch, H. J., Ronsheim, P. M., and Wurtman, R. J., 1981, Effect of light intensity on regulation of melatonin secretion and drinking behavior in the albino rat, in: "Melatonin - Current Status and Perspectives", N. Birau and W. Schloot, eds., Pergamon, New York.
Rollag, M. D., Panke, E. S., and Reiter, R. J., 1980a, Pineal melatonin content in male hamsters throughout the seasonal reproductive cycle, Proc. Soc. Exp. Biol. Med., 165: 330.

Rollag, M. D., Panke, E. S., Trakulrungsi, W., Trakulrungsi, C., and Reiter, R. J., 1980b, Quantification of daily melatonin synthesis in hamster pineal gland, Endocrinology, 106: 231.

Tamarkin, L., Reppert, S. M., and Klein, D. C., 1979. Regulation of pineal melatonin in the Syrian hamster, Endocrinology, 104: 385.

Vaughan, G. M., Pelham, R. W., Pang, S. F., Laughlin, L. L., Wilson, K. M., Sandock, K. J., Vaughan, M. K., Kaslow, S. H., and Reiter, R. J., 1976, Nocturnal elevation of plasma melatonin and urinary 5-hydroxyindole acetic acid in young men: Attempts at modification by brief changes in environmental lighting and sleep and by autonomic drugs, J. Clin. Endocr. Metab., 42: 752.

Wilkinson, M., Arendt, J., Bradtke, J., de Ziegler, D., 1977, Determination of a dark-induced increase of pineal N-acetyltransferase activity and simultaneous radioimmunoassay of melatonin in pineal, serum and pituitary tissue of the male rat, J. Endocr., 72: 243.

Wurtman, R. J., Axelrod, J., and Phillips, L. S., 1963, Melatonin synthesis in the pineal gland: Control by light, Science, 142: 1071.

Zatz, M., 1981, Pharmacology of the pineal gland, in: "The Pineal Gland, Vol. I, Anatomy and Biochemistry", R. J. Reiter, ed., CRC Press, Boca Raton.

DIFFERENTIAL REGULATION OF THE 24 HOUR PATTERN OF SERUM MELATONIN AND N-ACETYLSEROTONIN

G.M. Brown, L.J. Grota*, L. Harvey,
H.W. Tsui, and S.F. Pang**

Department of Neurosciences
McMaster University, Hamilton, Ontario
*Department of Psychiatry, University of Rochester
**Department of Physiology, University of Hong Kong

INTRODUCTION

While conducting a series of investigations with the objective of developing a highly specific radioimmunoassay for melatonin (MEL), we developed two assay systems one capable of measuring both MEL and N-acetylserotonin (NAS) activities, the other capable of measuring melatonin specifically. Results of investigations using these assays provided strong evidence that the 2 substances have different 24 hour patterns in serum. We have therefore undertaken the development of a highly specific radioimmunoassay for NAS and have used it to conduct a more detailed examination of this issue. Current findings indicate that dissociation of the 24 hour pattern of serum MEL and NAS occurs under conditions of short photoperiod. These findings suggest that different mechanisms may be involved in regulation of circulating NAS as compared to MEL.

DEVELOPMENT OF MANNICH ANTISERA TO MEL AND NAS

In 1974, we reported on production of antiserum capable of binding NAS and MEL equally and with high affinity, but discriminating against a large number of other indoleamines (Grota and Brown, 1974). The antigen used in these investigations was N-acetylserotonin coupled to bovine serum albumin using formaldehyde condensation (Mannich reaction). Subsequently, using MEL coupled to bovine serum albumin in the same manner, we succeeded in obtaining specific antisera which discriminated MEL from NAS and a wide variety of other indoles (Pang et al., 1977).

The two types of antisera appeared to differ primarily in their ability to discriminate between substitutions on the indole molecule in or around the five position. We reasoned that the site of conjugation for the NAS conjugate must be adjacent to the 5 position of the indole ring so that antibodies stimulated by the conjugate would not discriminate between groups in this region although they would discern changes at the amino end of the molecule. Our findings on cross-reactivity of the NAS conjugate were subsequently confirmed by Kennaway and co-workers (1977). Since the MEL conjugate stimulated antisera which showed discrimination at both the five and the amino positions, we inferred that this conjugate was coupled at a site remote from both of those positions. We have recently investigated this antigen further on the grounds that it might serve as a useful antisera for other indoleamines.

STUDIES ON MEL MANNICH ANTIGEN

Two approaches were used to determine the site of conjugation of MEL-M-BSA. In one approach we reacted model amines with formaldehyde and MEL, isolated the intermediates and chemically identified the reaction products. We found that carbon 2 was the most likely site of methylene bridge formation for the hapten MEL, based on nmr analysis of reaction products with glycine-ethyl-ester or piperdine (Grota et al., 1981). In the second approach we utilized cross-reactivity of antisera stimulated by the conjugate. The analysis showed that the presence of 5-methoxy, 6-unsubstituted, unsubstituted indole nitrogen, and N-acetyl(side chain) were all necessary for the binding of indolealkylamine to the antisera. See (Pang et al., 1977) for details. This approach was consistent with the chemical analysis which indicated that the methylene bridge coupling melatonin to BSA occurred at carbon 2 of the indole nucleus. Consistent with these observations are the data of Besselievre et al., (1980) and Wolinsky and Sundeen (1979) which show respectively that analogues of MEL with alkyl substitutions at the indole nitrogen do not form conjugates with protein in the formaldehyde reaction and that formaldehyde coupling of the indole to prolactin is unstable if it occurs at the indole nitrogen. Taking these data together, it appears that melatonin reacts with formaldehyde at the indole nitrogen initially but this binding forms a substitution which is somewhat unstable and the eventual stable bond is formed at Carbon 2 (Fig. 1).

One implication of our data is that the methylene bridge between hapten and the free amino groups on the protein occurs at different sites depending on the substitutions on the indole haptens. If a 5-hydroxy substitution is present, formaldehyde conjugation will occur at position 4. If there is a 5-methoxy substitution then coupling will occur at Carbon 2 provided that the

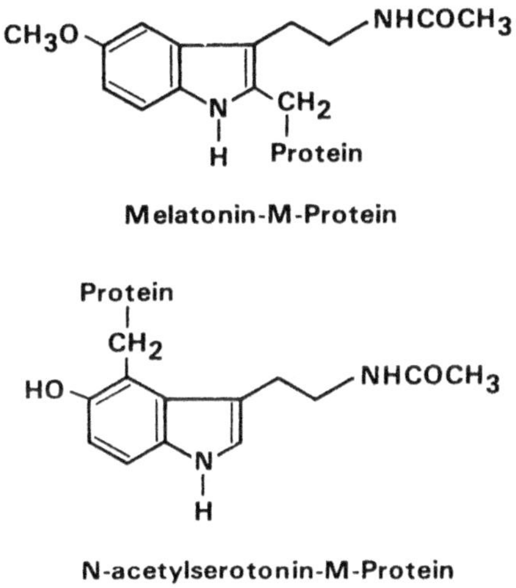

Figure 1

indole nitrogen is unsubstituted. The above holds true for N-acetylated compounds. It should be pointed out that the formaldehyde reaction occurs readily with free amino groups. Therefore, a free amino on the side chain of an indole will also react with formaldehyde to produce either polymerization of the indoles and/or coupling to the phenolic hydroxy groups of tyrosine in the protein. Furthermore, under certain conditions B-carboline may be formed if the amino group is free.

Based on the foregoing it is possible to devise a generalized strategy using the formaldehyde condensation reaction to produce antisera that will bind individual indolealkylamines: reversably block the terminal amino group and the 5-hydroxylated substitution before coupling with formaldehyde. This approach ensures that coupling occurs at Carbon 2. The blocks are removed after conjugation to provide a hapten - protein conjugate with a hapten of the desired configuration. The major advantage of this approach is that coupling of indolealkylamine to protein using this strategy will result in haptens where the site of coupling of hapten to protein occurs at relatively great distances from those groups which differ among the various indoleamines and which are the active metabolic sites *in vivo*. One should therefore be able to produce antisera that specifically bind individual biologically significant, indolealkylamines.

RADIOIMMUNOASSAY OF N-ACETYLSEROTONIN AND MELATONIN USING A SUBTRACTION METHOD

1. Validation

Using the two antisera described above, radioimmunoassays for MEL and for N-acetylserotonin plus melatonin were developed. For melatonin detailed cross-reactivity studies were completed, satisfactory parallelism was demonstrated, and satisfactory data on assay variation were presented (Pang et al., 1977). In addition, melatonin estimates in pineal were crossvalidated by GCMS (Pang et al., 1977). Crossvalidation of these radioimmunoassays in serum however, has been difficult because of lower serum levels of MEL. For human serum, crossvalidation with a number of other MEL radioimmunoassays has been accomplished (Wetterberg 1977, Wetterberg and Eriksson 1981); however, the comparison of results obtained using such similar techniques may not permit the identification of problems unique to radioimmunoassay. The recent development of a more sensitive negative chemical ionization gas chromatographic-mass spectrometry assay for MEL (Lewy and Markey 1978) permits the crossvalidation of the radioimmunoassay using a technique based on totally different physical-chemical principles. We have assayed 6 rat serum samples for MEL both by radioimmunoassay and gas chromatography-mass spectrometry and found a correlation of 0.983 (Grota et al 1981).

N-acetylserotonin in serum and pineal tissue has been estimated by a subtraction method in which assayed melatonin values are subtracted from the estimates of N-acetylserotonin plus melatonin (Pang et al., 1977). Validation of NAS assays are discussed below.

In an investigation by Yu et al., (1981), circulating melatonin following pinealectomy was found to be reduced but not abolished. This study, however, did not settle the issue of the source of circulating melatonin. Lewy et al., (1980) using GCMS assay for MEL, reported that pinealectomy eliminated circulating melatonin. In a separate study we have assessed the effect of pinealectomy on circulating melatonin using the melatonin assay which had been crossvalidated with GCMS (Grota et al., 1981). In those studies melatonin levels were reduced to below the limit of sensitivity in pinealectomized rats (Harvey et al., 1981) indicating that the pineal is the major (or only) source of circulating melatonin.

Studies on Regulation of Serum Melatonin

Our initial studies showed that MEL levels in albino rat pineals displayed a diurnal rhythm with high levels during dark and low levels during light (Pang et al., 1977). Rats were housed under a 12:12 light: dark cycle and sacrificed at 4 hour intervals. No rise was seen 2 h after dark onset and the peak occurred 6 h after dark onset. In another study the 24 h pattern of pineal and serum melatonin levels were compared (Brown and Grota, 1980). A similar night time rise was seen, but peak serum melatonin occurred later than peak pineal melatonin (10 h after dark onset vs 6 h).

We next examined the influence of feeding time on regulation of serum MEL and corticosterone (Holloway et al., 1979). Restriction of food availability to a two hour period during either light or dark had previously been shown by Krieger (1974) to have a significant influence on the adrenal rhythm with maximal elevation occurring just prior to food presentation. We have confirmed this finding. However, in contrast to corticosterone, serum MEL rhythms were not modified by the timing of food availability. Our findings are consistent with those of Moore and Traynor (1976) who reported no alteration in the pineal N-acetyltransferase rhythm in rats given 1 h access to food in the morning. These findings indicate that the MEL rhythm is more closely linked with the light /dark cycle than is the adrenal rhythm.

In several studies we examined the influence of shortening the photoperiod on the serum melatonin rhythm. In rats maintained in light:dark 12:12 and 2:22 or 1.5:22.5 cycles, a 24 hour rhythm was observed with the peak consistently occuring 18 h after light onset. (Brown et al., 1980a, b, Grota et al., 1982). If light onset were considered to be the synchronizer, the overall 24 hour patterns did not differ. Furthermore, in the hamster maintained under identical conditions, 24-hour serum melatonin patterns also did not differ (Brown et al., 1980 b, Brown et al., 1982), although in the short photoperiod overall melatonin levels were reduced. Thus, in both rat and hamster there is evidence that the 24 hour serum melatonin rhythm is cued by light onset.

These findings are of additional interest because of the pronounced endocrine effects produced in the hamster and the rat by exposure to short photoperiod (Brown et al., 1982). These changes have been attributed to an altered 24 hour pattern of melatonin and notably to an early rise in melatonin (Reiter 1980). We have found no evidence of such a change in the 24 hr. pattern of melatonin. Since our studies have been conducted at only certain intervals after exposure to short photoperiod, (3, 4 and 7 weeks in the rat, and 7 and 12 weeks in the hamster) it remains possible that 24 hour serum melatonin patterns may show alterations at other intervals.

Nonetheless, in the hamster when there is testicular regression and a drop in circulating testosterone induced by either 7 or 12 weeks of short photoperiod there is no evidence of an early rise in serum melatonin and, in fact, melatonin levels are reduced.

In the rat, olfactory bulbectomy increases the inherently low sensitivity of the reproductive axis to a short photoperiod. In bulbectomized rats, reduction of seminal vesicle and prostate but not testis weights relative to body weight can be detected after 3 and 7 weeks on a L:D 2:22 light cycle (Harvey et al., 1981). Absolute testis weights were reduced at 7 weeks but still showed histological evidence of sperm production. These effects were partially or totally reversed by pinealectomy. However, while pinealectomy reduced serum melatonin to undetectable levels, there were no significant differences in the height or duration of the melatonin peak between anosmic animals showing gonadal regression and intact controls.

Studies On Regulation of N-acetylserotonin

Diurnal rhythms of immunoreactive NAS and MEL have been examined in the serum of rats on a 12:12 light:dark cycle (Pang et al., 1980). In two separate experiments in which lights were on from 0600 to 1800 h, maximum levels of NAS were found at 800 h; with a minimum at 1200 and 1600 h. The patterning of NAS and of MEL differed somewhat as the MEL peak preceeded the NAS peak by 4 hours. In order to determine whether the circulating MEL and/or NAS originated at least in part from extra pineal sources, the 24 hour pattern was examined in pinealectomized animals on the same light/dark cycle. Concentrations of circulating NAS were significantly reduced after pinealectomy (Yu et al., 1981), suggesting that the pineal contributes a significant amount of this substance to the blood. However, NAS was not abolished, suggesting the existence of extra pineal sources capable of secreting this substance.

DEVELOPMENT OF SPECIFIC NAS ANTISERA

Because of the finding that N-acetylserotonin is present both in the pineal and circulation and because of limitations inherent in estimating NAS by a subtraction method we next undertook to develop a specific NAS radioimmunoassay.

Using MEL as a model substance, a variety of approaches for the development of specific antisera were explored. In addition to the approaches using formaldehyde condensation as described above, we have produced antisera using melatonin conjugated to albumin by means of diazotized p-aminobenzoic acid, Na-p-carboxybenzyl, and Na-propionic acid bridges (Wurzberger et al., 1976,

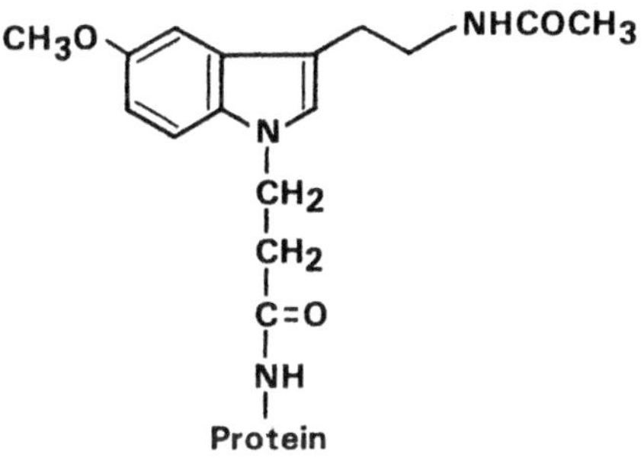

Melatonin-PA-Protein

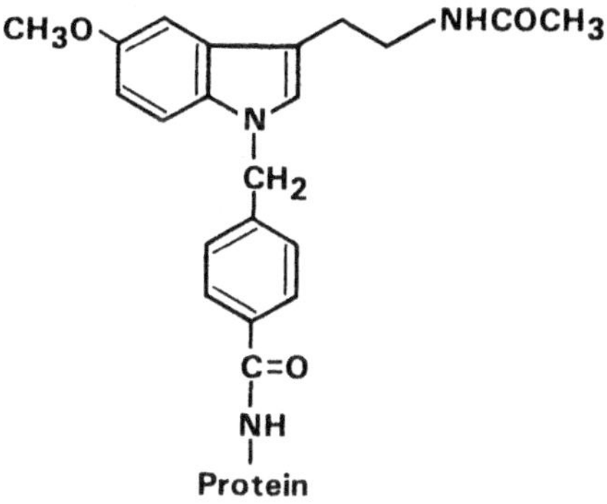

Melatonin-PCB-Protein

Figure 2

DeSilva and Snieckus 1978, Blair and Seaborn 1979, Brown and Grota 1980, Grota et al., 1981) (Fig. 2). Crossreactivity of compounds with structural similarity to melatonin was assessed using tritiated melatonin as the ligand. For these three classes of antisera, N-acetylserotonin and 6-hydroxymelatonin had the greatest crossreactivity which was 3% or less for each of the antisera. Thus, including the Mannich approach, we have assessed four different approaches, each of which is capable of producing antibodies to indole haptens which have sufficient specificity to be useful in radioimmunoassay. However, there are differences between these techniques. The Mannich reaction is carried out readily in aqueous solutions but the rate of reaction is low and it is difficult to control. Both the Na-p-carboxybenzyl and the Na-propionic acid derivatives can be prepared in crystalline form and conjugated to antigenic protein using water soluble carbodiimide. One disadvantage of this approach is the relatively poor solubility of these derivatives. Coupling via diazodized p-aminobenzoic acid is quick and done readily. However, we have found that this reaction yields numerous side products in model reactions and hence the antigen may be a heterogeneous one.

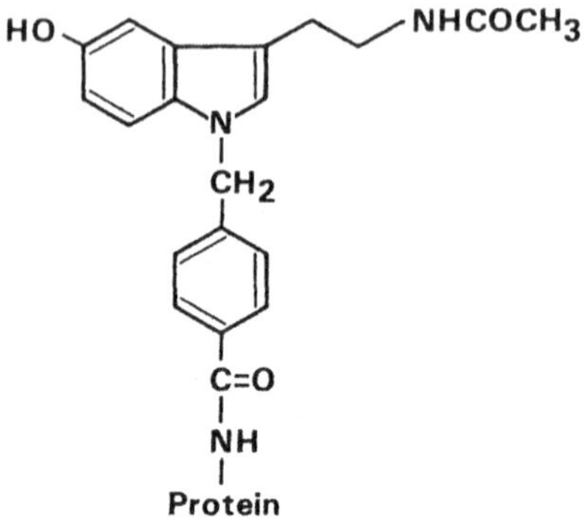

N-acetylserotonin-PCB-Protein

Figure 3

We have chosen to use the Na-p-carboxybenzyl (Pcb) approach for the preparation of specific antiserum to N-acetylserotonin (DeSilva and Snieckus 1978, Pang et al., 1982). Serum harvested from animals immunized with N-acetylserotonin - pcb - BSA (Fig. 3) was incubated with tritiated N-acetylserotonin (35Ci/mmol) obtained from Amersham. Indoleamines related to N-acetylserotonin were evaluated for their ability to inhibit binding of tritiated N-acetylserotonin. Of the naturally occuring substances tested 6-hydroxymelatonin had .5% crossreactivity while other indoleamines had less. Parallelism with the standard N-acetylserotonin was shown for rat, hamster, and rabbit serum, for rat and hamster serum extract and for rat brain, retina and pineal extract. Sensitivity of the assay was 10 - 25 pg/tube based on the dose of NAS producing 12% inhibition of binding (twice the standard deviation of the binding at zero concentration). Within and between assay co-efficient of variation for a rat serum control containing 142 picograms per tube were 8.4% and 7.7% respectively in 15 assays. An extracted sample of rat serum was chromatographed on high pressure liquid chromatography (Anderson et al., 1981). Radioimmunoassay of the aliquot gave only one peak which corresponded to authentic N-acetylserotonin and the radioimmunoassay value before and following chromatography agreed within 5%. To date, it has not been possible to cross validate this assay using GCMS as the appropriate GCMS methodology has yet to be developed.

STUDIES ON REGULATION OF SERUM N-ACETYLSEROTONIN

Because data from the scheduled feeding study indicated that the melatonin 24-hour rhythm was tightly coupled to the 24-hour light-dark rhythm we have used different light-dark rhythms to determine whether NAS and MEL in serum could be dissociated. In one study rats were maintained for 8 weeks under either a 12:12 light:dark cycle (LD 12:12) or under LD 2:22. At various times during a 24-hour period animals were killed and the serum examined for NAS and MEL. As observed earlier using the subtraction method (Pang et al., 1977), when animals were kept under a LD 12:12 cycle, both NAS and MEL have a crest late in the dark period. However, parenthetically, it should be noted that the crest for NAS is at a level of 2 to 2.5 ng/ml while for MEL a maximum level is reached at approximately 40 to 50 pg/ml. In contrast to these data we now found dissociation of the 24 hour patterns of NAS and MEL in serum among animals maintained under LD 2:22. For MEL there was no change in the rhythm relative to the LD 12:12 animals. However, the serum rhythm of NAS showed a shift to an earlier time period such that in both LD 12:12 and LD 2:22 the crest occurs approximately 6 to 8 hours after the onset of darkness. In addition, NAS levels were elevated with a maximum of approximately 3.6 ng/ml. These data are shown graphically in Fig. 4.

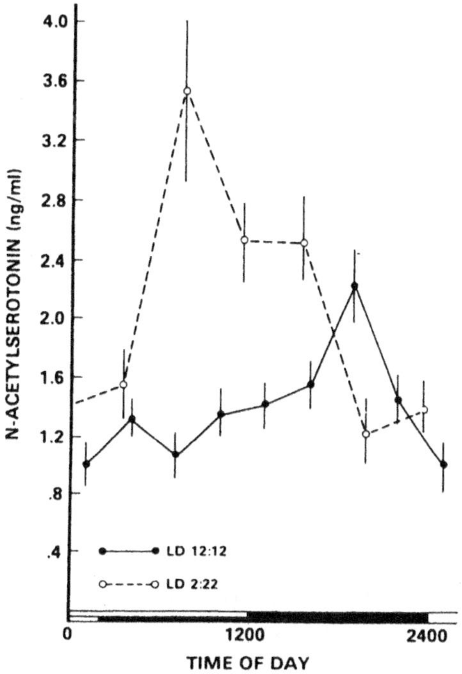

Figure 4 The 24 hour serum N-acetylserotonin rhythm in the rat differs in short and long photoperiods.

A second experiment examined the rate at which MEL and NAS in serum synchronized to 12 hour light dark cycles after a 30 day period in constant light. Constant light was used to eliminate the lighting cue and to reduce the production of NAS and MEL. As can be seen in Fig. 5, at the end of 30 days of constant light MEL levels were approximately 10 pg/ml. Within two hours of the first exposure to darkness serum MEL levels were elevated and there is evidence of synchronization of MEL levels to the dark periods, in that during darkness MEL levels are elevated above levels in the light. One data point differs for no known reason. In contrast to MEL, NAS shows no synchronization to periods of light and dark even after 6 days in the 12 hour light: dark cycle (Fig. 5). These two studies clearly show that the daily pattern of serum MEL can be dissociated from the pattern of serum NAS. Such data imply independent regulation of these two substances. The dissociation of these two substances in serum, even though they are related biosynthetically, may occur because the precursor NAS is 50 times more concentrated in serum than MEL. This fact means that the total serum variation for day and night MEL remains within the range of biological variability for NAS even at its lowest levels. Because these studies were done in only one species, the albino rat, we are pursuing essentially the same questions in the hamster to increase the generalizeability of these data. Only pilot data is available at this time but as for the rat, the patterning of MEL and NAS in short photoperiods appears different for the two hormones. Thus, in the hamster as well as the rat, serum levels of NAS and MEL can be dissociated.

At the present time the significance of the NAS rhythm is unknown. In fact, at this point we do not know with any certainty whether the NAS rhythm originates from the pineal or from another site of production as the enzyme N-acetyltransferase is found in a variety of tissues.

Nonetheless, the pronounced 24 hr rhythm of NAS in serum together with its dissociation from the MEL rhythm under short photoperiodic conditions in the rat suggests that synchrony and desynchrony of these rhythms may have a functional significance.

CONCLUSION

Using antigens in which melatonin and N-acetylserotonin are coupled to protein at or adjacent to the ring nitrogen we have produced antisera to both substances which are sufficiently specific to be useful in radioimmunoassays.

Radioimmunoassays for serum MEL and NAS have been developed and validated. Studies of serum melatonin have shown that this

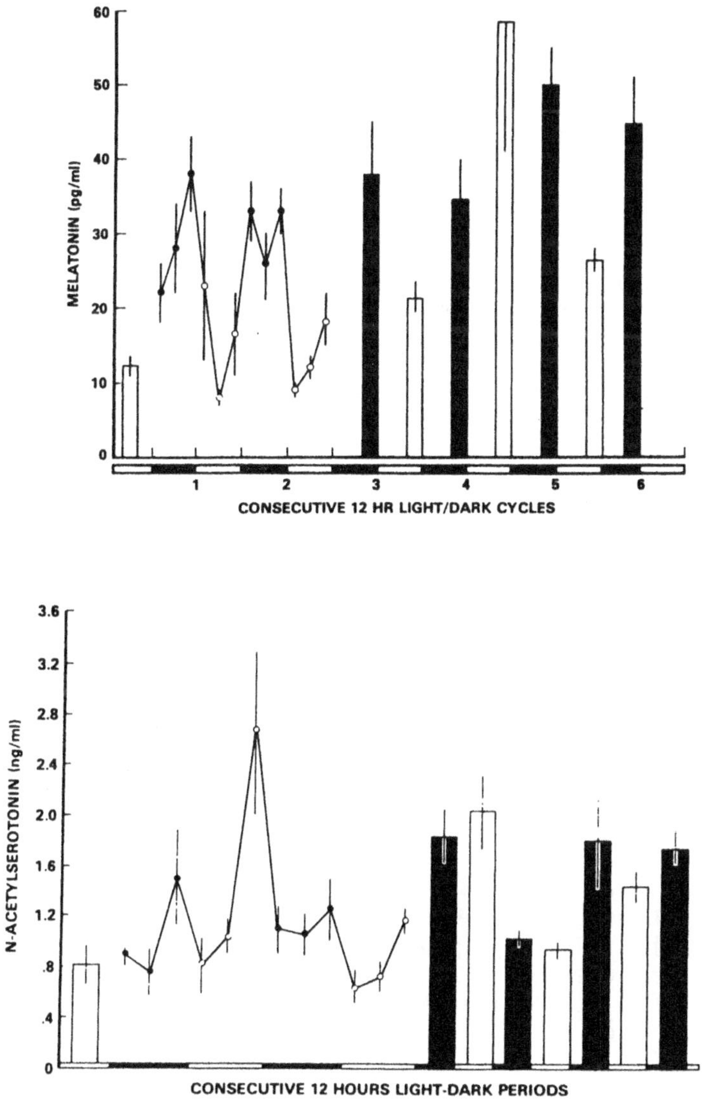

Figure 5 The 24 hour rhythm of serum melatonin but not N-acetyl-serotonin resynchronizes to the light cycle promptly at the end of 30 days of constant light.

substance is tightly regulated by the light/dark cycle. The peak in serum MEL was consistently found 18 h after light onset in both rat and hamster in both 12:12 light:dark cycles and in 2:22 cycles. These findings suggest that light onset serves as the synchronizer for the serum MEL rhythm. In contrast to MEL, the serum NAS rhythm in the rat shifts when the photoperiod is shortened, so that it peaks 6 to 8 hours after the onset of darkness. It is speculated that association and dissociation of serum NAS and MEL rhythms may have a functional significance.

ACKNOWLEDGEMENTS

Investigators wish to acknowledge technical assistance of L. Campbell, K. Caskey and M. Joshi together with the secretarial assistance of L. Koutalos.

REFERENCES

Anderson, G.M., Young, J.G., Ritter, D.K., Young, S.N., Cohen, D.J., and Shoywitz, B.A., 1981. Determination of indoles and catechols in rat brain and pineal using liquid chromatography with fluorometric and ampermetric detection. J. Chromatog, 223: 315.

Besselievre, R., Lemaitre, B.J., Husson, H.P., and Hartmann, L., 1980, Structural immunochemistry of melatonin-BSA binding, model of amino and indole groups cross-linking. Biomedicine, 33: 226.

Blair, I.A., and Seaborn, C.J., 1979, The synthesis of melatonin antigens. Aust. J. Chem., 32: 399.

Brown, G.M., and Grota, L.J., 1980, Use of Immunologic techniques in the examination of neurotransmitters and neuromodulators, In: "Physico-Chemical Methodologies in Psychiatric Research". I. Hanin and S. Koslow (Eds.), Raven Press, New York.

Brown, G.M., Grota, L.J., Bubenik, G., Niles, L., and Tsui, H., 1980a, Physiologic regulation of melatonin, in: "Advances in the Biosciences, Vol. 29, Melatonin: Current Status and Perspectives". N. Birau and W. Schloot, (Eds.), Pergamon Press Oxford.

Brown, G.M., Grota, L.J., and Niles, L., 1980b, Melatonin: origin, control of circadian rhythm and site of action, in: "Advances in the Biosciences, Vol. 29, Melatonin: Current Status and Perspectives". N. Birau and W. Schloot (Eds.), Pergamon Press Oxford.

Brown, G.M., Tsui, H.W., Niles, L.P. and Grota, L.J., 1982, Gonadal effects of the pineal gland. In: The Pineal Gland. C.D. Kennaway, and R.F. Seamark, (Eds.), Elsevier/North Holland Biomedical Press, Amsterdam.

De Silva, S.O. and Snieckus, V., 1978, Indole-N-alkylation of tryptamine via dianion and phthalimido intermediates. New potential indoleakylamine haptens. Canad. J. Chem., 56: 1621.

Grota, L.J., and Brown, G.M., 1974, Antibodies to indolealkylamines: serotonin and melatonin. Canad. J. Biochem., 52: 196.

Grota, L.J., Snieckus, V., DeSilva, S.O., Tsui, H.W., Holloway, W.R., Lewy, A.J., and Brown, G.M., 1981, Radioimmunoassay of melatonin in rat serum. Prog. Neuropsychopharmacol., 1981, 5: 523.

Grota, L.J., Holloway, Wm.R., and Brown, G.M., 1982, 24-hour rhythm of hypothalamic melatonin immunofluorescence correlates with serum and retinal melatonin rhythms. Neuroendocrinology, 34: 363.

Harvey, L., Brown, G.M. and Grota, L.J., 1981, Serum melatonin and photoperiod induced reproductive changes in the rat. Neuroendocrinology Letters, 3: 103 (abstract).

Holloway, W.R., Tsui, H.W., Grota, L.J., and Brown, G.M., 1979, Melatonin and corticosterone regulation: feeding time or the light:dark cycle. Life Science, 25: 1837.

Kennaway, D.J., Fritto, R.G., Phillipou, G., Matthews, C.D., and Seamark, R.F., 1977, A specific radioimmunoassay for melatonin in biological tissue and fluids and its validation by gas chromatography-mass spectroscopy. Endocrinology, 101: 119.

Krieger, D.J., 1974, Food and water restriction shifts corticosterone, temperature, activity and brain amine periodicity. Endocrinology, 95: 1195.

Lewy, A.J. and Markey, S.P., 1978, Analysis of melatonin in human plasma by gas chromatography negative chemical ionization mass spectrometry. Science, 201: 741.

Lewy, A.J., Tetsue, M., Markey, S.P., Goodwin, F.K. and Kopin, I.J., 1980, Pinealectomy abolishes plasma melatonin in the rat. J. Clin. Endocrinol. Metab., 50: 204.

Moore, R.Y. and Traynor, M.E., 1976, Diurnal rhythm in pineal N-acetyltransferase and hippocampal norepinephrine: effects of water deprivation, blinding and hypothalamic lesions Neuroendocrinology, 20: 250.

Pang, S.F., Brown, G.M., Grota, L.J., Chambers, J.W., and Rodman, R.L., 1977, Determination of N-acetylserotonin and melatonin activities in the pineal gland, retina, harderian gland, brain and serum of rats and chickens. Neuroendocrinology, 23: 1.

Pang, S.F., Yip, M.K., Liu, H.W., Brown, G.M., and Tsui, H.W., 1980, Diurnal rhythms of N-acetylserotonin and melatonin in the serum of male rats. Acta Endocrinol., 95: 571.

Pang, S.F., Brown, G.M., Campbell, S.L., Snieckus, V., DeSilva, S.O., Young, S.N. and Grota, L.J., 1981, A radioimmunoassay for N-acetylserotonin in biological tissues. J. Immunoassay, 2: 263.

Reiter, R.J., 1980, The pineal and its hormones in the control of reproduction in animals. Endocrine Reviews, 1: 109.
Wetterberg, L., 1977, Melatonin in serum. Nature, 269: 646.
Wetterberg, L. and Eriksson, O., 1981, Melatonin in human serum- a collaborative study of current radioimmunoassays in: "Advances in the Biosciences, Vol. 29, Melatonin-Current Status and Perspectives". N. Birau and W. Schloot, (Eds.), Pergamon, Oxford.
Wolinsky, J. and Sundeen, J.E., The reaction of 3-propylindole with aldehydes. Tetrahedron, 1979, 26: 5427.
Wurzberger, r.J., Kawashima, K., Miller, R.L. and Spector, S., 1976, Determination of rat pineal gland melatonin content by radioimmunoassay. Life Sci., 18: 867.
Yu, H.S., Pang, S.F., Tang, P.L. and Brown, G.M., 1981, Persistence of circadian rhythms of melatonin and N-acetylserotonin in the serum of rats after pinealectomy. Neuroend., 32: 262.

DIFFERENTIAL LOCALIZATION OF MELATONIN AND N-ACETYLSEROTONIN IN BRAIN

G.M. Brown, O. Pulido, L.P. Niles, S. Psarakis,
A. Porietis, G.A. Bubenik*, and L.J. Grota**

Department of Neurosciences
McMaster University, Hamilton, Ont.
*Department of Zoology, University of Guelph, Guelph, Ont.
**Department of Psychiatry, University of Rochester
Rochester, N.Y. 14624

INTRODUCTION

Until 1974 quantitative and qualitative localization of melatonin (M) had been performed mainly by bioassay (Ralph and Lynch, 1970). As this technique is relatively insensitive, the detection of M outside of the pineal gland was infrequent (Pang et al. 1974). This led to the conclusion that the epiphysis cerebri was the unique source of M and its derivatives (Reiter, 1973). This firm opinion persisted for more than 15 years after the discovery of M in 1958 (Lerner et al. 1958) despite numerous reports of extrapineal detection of hydroxyindole-O-methyl transferase (HIOMT) the enzyme essential for M synthesis (Quay, 1965; Cardinali and Rosner, 1971a; Cardinali and Wurtman, 1972) as well as the demonstration of synthesis of melatonin in the retina (Gern et al. 1978, Pevet et al. 1980).

IMMUNOHISTOLOGIC STUDY USING BISPECIFIC ANTISERUM

With the development of the first bispecific antiserum (binding M and its precursor N-acetylserotonin equally well) by Grota and Brown (1974), the way was opened for development of not only radioimmunoassay (RIA), but also for immunohistology of N-acetylated indolealkylamines (INAI). In 1974 we tested this antiserum on frozen sections of rat brain and pineal tissues. Using the modified double antibody technique developed by Coons and co-workers (1955) positive staining was first detected in the parenchymal cells of the pineal gland. In addition to this expected localization of immunoreactive N-acetylated indolealkylamines (INAI), however, positive staining was also detected in the granular layer of cerebellar cortex which was present in the same cross section of the rat brain as was the pineal gland. The staining was consistent using both fluorescein isothiocianate and UV light as well as peroxidase and diaminobenzidine reaction systems. Saturation of antiserum by an excess of M as well as replacement of first antiserum by a nonspecific serum abolished the staining (Bubenik et al. 1974). Consistent staining was also observed in the habenula (an area adjacent to the pineal gland) as well as in the fimbria hippocampi. However, in these tissues the saturation of antiserum by M did not completely abolish the staining and these areas were not further investigated.

The retina had been found to contain HIOMT (Cardinali and Rosner, 1971a) as well as to have the ability to synthesize M from serotonin in vitro (Cardinali and Rosner, 1971b). In addition, the retina has a close developmental relationship to the epiphysis cerebri (Kappers, 1976). Our successful localization of INAI in the outer nuclear layer of the retina (Bubenik et al. 1974) is therefore not surprising.

STUDIES USING SPECIFIC MELATONIN ANTISERUM

Two years later, specific anti-M antiserum became available, All tissues identified as INAI positive were re-tested. Using both the bispecific and the anti-M antisera, the proportion of immunoreactive melatonin (IM) and its precursor, N-acetylserotonin (NAS), in tissues were estimated. Thus the intensity of staining obtained with bispecific antiserum, minus the staining with monospecific anti-M antiserum indicated the amount of immunoreactive NAS (INAS). Predominantly M staining was detected in the optic nerves, chiasm and tract and the suprachiasmatic nucleus. Almost equal amounts of IM and INAS was estimated in the retina and the pinealocytes. In addition, to these CNS tissues, IM was also identified in the Harderian gland (Bubenik et al. 1976b) which is another tissue where HIOMT had previously been reported (Cardinali and Wurtman, 1972). The localization of melatonin in the retina was later confirmed by other colleagues using immunohistology (Vivien-Roels et al. 1981) TLC (Pevet et al. 1980) or RIA (Pang et al. 1977; Pang et al. 1980) employing a variety of different antisera. A similar confirmation of our findings came later for the localization in the pineal (Pang et al. 1977; Freund et al. 1977; Pevet et al. 1980; Vivien-Roels et al. 1981). In contrast, only INAS was detected in the granule and Golgi II cells of cerebellum, the axons of pontal reticular formation and the axons of the spinal tract of the trigeminal roots in the caudal pons and the rostral medulla oblongata (Bubenik et al. 1976a).

Because other detection techniques were not sufficiently sensitive at that time, immunohistology was also used in order to estimate possible circadian variation in these tissues as well as to study the ontogeny and the sites of possible M synthesis. Our conclusion, that extrapineal synthesis of melatonin may exist and that this might exhibit diurnal variation independent of the pineal gland was subsequently confirmed both by the presence of melatonin in tissues following pinealectomy and by documentation of synthesis in the retina in several studies (Gern et al. 1978; Gern and Ralph, 1979; Pang et al. 1980; Pevet et al. 1981; Holloway et al. 1980; Yu et al. 1981; Grota, Holloway and Brown, 1982).

Finally, the results of immunohistological investigations of retinas and Harderian glands of rats injected or implanted with M and killed during the daytime revealed a several fold increase in intensity of immunofluorescence as compared to control animals. The increase in intensity of staining vastly exceeded the highest levels observed during the night time. It has been hypothesized that the retina and the Harderian gland possess specific receptors which have spare capacity to pick up amounts of M which are higher than the normal physiologic levels produced locally (Bubenik et al. 1978). Receptors for M have been reported by several groups (Cohen et al. 1978; Niles et al. 1979; Cardinali et al. 1979; Holloway et al. 1979; Niles et al. 1980), however, so far none of these investigators have reported M receptors in the retina or the Harderian gland.

Why is there production of M in multiple sites? It has been speculated that M was originally produced in a number of organs for local neuromodulator (Bubenik and Purtill, 1980). However, with the increase in size of the pineal during phylogenetic development, the production of M there augmented to the point where it began to escape into the peripheral circulation and to assume hormonal properties (Bubenik et al. 1980; Quay, 1980). Thus the pineal gland became the main source of NAI in most mammalian species. However, local production may still continue in the retina, brain tissue and the intestines (Bubenik et al. 1977; Gern et al. 1978; Gern and Ralph, 1979; Pevet et al. 1981). These other organs are potential sources of circulating M if the primary source is lost (Kennaway et al. 1977; Yu et al. 1981). However, no circulating melatonin is detectable by GCMS following pinealectomy in the rat (Lewy et al, 1980).

STUDIES USING NAS SPECIFIC ANTISERA

With the recent availability of specific antisera binding N-acetylserotonin (NAS) (Brown and Grota, 1980; Pang et al., 1981), specific immunohistological localization of NAS has become possible. (Pulido et al. 1981). NAS-immunoreactivity has been demonstrated only in unfixed frozen sections. At the present time, no fluorescence has been observed in cryostat sections of tissues fixed by perfusion with: a) 4% Paraformaldehyde in PBS buffer pH 7.3 (Steinbusch, 1981): b) periodate-lysine paraformaldehyde (Mclean and Nakane, 1974) or c) 0.35% glutaraldehyde in Tris-HCL buffer (Willingham and Yamadi, 1979). We are aware that fixation would be advantageous for increased morphological detail and tissue preservation. Further experimental testing of several fixation combinations and concentrations are in progress. Despite this limitation, the application of NAS antisera to immunohistochemistry using unfixed frozen sections has proven to be a reliable and specific method for the localization of INAS in brain tissue.

Specificity of INAS Immunohistology

Specificity of the NAS antisera has been evaluated using the following parameters: 1) radioimmunoassay, 2) immunohistochemical controls using PBS and non-immune serum; 3) absorption tests using different concentrations of NAS; 4) inhibition experiments using different concentrations of different substances that might crossreact; 5) pharmacological treatments that might alter NAS dynamics in the brain; 6) comparison of NAS distribution with that of 5-HT and melatonin; 7) immunohistological evaluation of the same brain regions using melatonin antisera.

Each anatomically distinct area where positive INAS was identified, was evaluated for non-specific binding, cross-reactivity and melatonin immunoreactivity. Non-specific binding after exposure of slides to NRS was observed in the wall of the blood vessels, the ependymal lining, and the Pia mater. Areas with positive INAS showed strong inhibition of the fluorescence intensity after pre-incubation of the antiserum with concentrations of NAS over 1μg/100μl whereas no inhibition was seen with concentrations of 400μg/100μl of melatonin, serotonin, 5-hydroxytryptophol, 3-methoxytyramine, N-acetyltryptophan, dopamine, norepinephrine. These data indicate that this immunohistologic method is highly specific for immunoreactive N-acetylserotonin.

Mapping of INAS

Identification of INAS was confirmed in the granular layer of the cerebellum, tractus spinalis nervi trigemini and pontine reticular formation. INAS was also observed in the Purkinje cells and the neuronal cell bodies of the following area: colliculus superior and inferior, nucl. originis nervi trigemini, nucl. originis nervi facialis, nucl. cochlearis, nucl. raphae pontis, nucl. corporis trapezoidei, nucl. interpeduncularis, locus coeruleous, nucl. tractus mesencephali, superior olivary complex, nuclei pontis, nucl. ruber, nucl. reticularis gigantocellularis, nucl. reticularis pontis caudatis and oralis.

With the technique employed, INAS has been mostly visualized in neuronal cell bodies in the cerebellar cortex and brain stem (Fig 1, 2). Dendritic projections are distinguishable within some of the Purkinje cells. (Fig. 1) Stem processes are identified in some neurons (Fig. 2). Poorly defined filamentous structures are **seen** intermingled with neuronal cell bodies in areas such as nucl. pontis and nucl. interpenduncularis. Whether those fibers represent nerve processes, glial filaments or deposits of INAS within the extra cellular matrix remains to be determined.

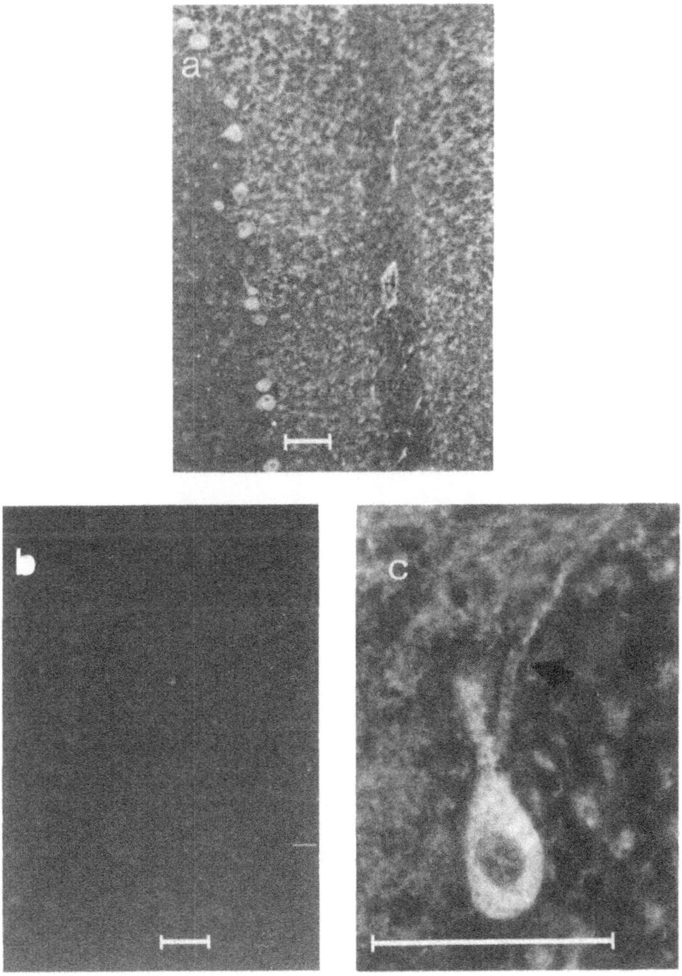

Figure 1: Photographs of frozen sections from rat cerebellar **cortex** after incubation with: a) NAS-antiserum b) NAS-antiserum with NAS, **c)** High magnification of a Purkinje cell showing specific NAS-immunofluorescence. A dendritic projection is indicated by an arrow. Bar = 50 um.

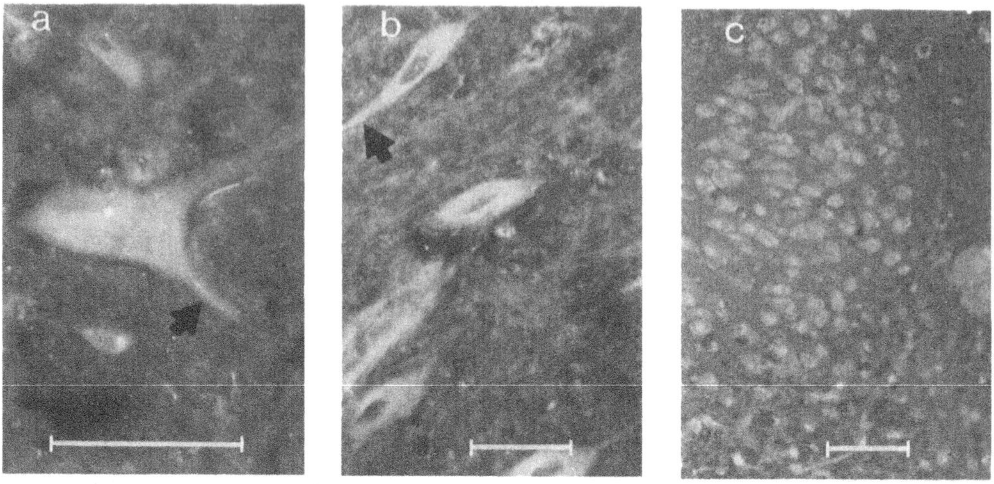

Figure 2: Photographs of frozen sections from rat brain stem showing specific NAS-immunofluorescence. a) Pontine Reticular Formation, b) Nucl. vestibularis, c) Nucl. Locus Coeruleus. Arrows indicate stem processes from neuronal cell bodies.
Bar - 75 um.

Relationship of INAS to Serotonin and Melatonin

Although serotonin appears to be present in many areas of the central nervous system (Saavedra, 1977, Steinbusch, 1981), major differences have been found between the distribution of INAS containing structures as compared to 5-HT.

INAS is present in significant amounts in the cerebellum where it is localized in the granular layer of the cortex, Purkinje cells and cerebellar nuclei. No melatonin containing structures have been visualized. Serotonin in cerebellum has been reported as present in very low amounts, being seen only in a few nerve terminals in the molecular layer of the cortex and in the cerebellar nuclei.

Although, there are regions in the brain stem where both INAS and 5-HT are visualized, in many of these regions, 5-HT is localized in nerve terminals whereas INAS is mostly identified in cell bodies i.e. nucl. originis nervi occulomotor, nucl. originis nervi trigemini, nucl. tractus mesencephali, nucl. originis nervi facialis, nucl. olivaris superior, nuclei pontis, nucl. reticularis giganto-cellularis, nucl. reticularis parvocellularis, nucl. reticularis pontis caudalis and oralis, nucl. vestibularis lateralis. Other areas such as the nucl. cochlearis dorsalis and ventralis do not have any or have very poor 5-HT innervations, whereas numerous INAS structures are visualized. Both INAS and 5-HT are visualized in cell bodies of the following regions: nucl. interpeduncularis, nucl. ruber, nucl. locus coeruleous, nucl. vestobularis medialis and nucl. raphae pontis. Whether in these regions both substances are present within the same cell or each substance has a different cell population, has yet to be investigated. No melatonin containing structures have been visualized in these regions.

These data, together with a lack of melatonin immunoreactivity in slides from the hindbrain indicate that major differences exist in the localization and distribution of INAS as compared to 5-HT and melatonin.

Localization of INAS in Hippocampus

Following preliminary observations of NAI localization in the hippocampus (G.A. Bubenik, unpubl.) the two-antisera technique was applied to this area (Porietis et al., 1978). Our data confirmed the presence of NAI's in hippocampus and identified the acetylindole there to be largely, if not exclusively, INAS. Presence of hippocampal NAS was subsequently confirmed by gas chromatography-mass-spectrometry (GCMS) (Brown et al., 1981). Serotonin localizations have also been widely reported in the hippocampus (Lidov et al., 1980; Steinbusch, 1981).

The recent advent of NAS-specific antisera (Brown and Grota, 1980; Pang et al., 1981) resulted in a continuation of these localizations in detail. In hippocampus, using both fluorescent and enzyme (Protein-A peroxidase) markers (Nakane and Kawaoi, 1974), INAS, but not IM, was localized within the pyramidal cell layers of CA1 and CA3, as well as in the polymorph and granular cell layers of the CA4/dentate region. In CA1 and CA3 the localization was in small units, submicron sized, external to but in close apposition to the large pyramidal cells. There is no evidence at this time to support localization within these cells. Occasional fibrous structures were observed coursing between the pyramidal cells in CA1/CA3. In the CA4/dentate region, NAS was localized in two distinct populations of structures, cell bodies and apparent processes/terminations. Cell bodies were identified in the polymorph cell layer, especially adjacent to the granule cell layer. The frequency of cell body staining was low. The occurence of small deposits of markers occurred more frequently throughout the polymorph cell layer, extending into the granule cell layer. Fibrous structures were seen in both the granule and polymorph cell layers, in the latter sometimes associated with stained cell bodies.

Earlier observations with the anti-melatonin sera failed to demonstrate any immunoreactive melatonin in the hippocampus. Likewise, the localization of serotonin as reported by others (Lidov et al., 1980; Steinbusch, 1981) describe a diffuse termination of 5HT neurons of brainstem origin in the basal dendritic fields of stratum oriens of CA1 and adjacent to the dentate granule cell layer. These authors do not describe 5HT-containing cell bodies within the hippocampus.

The serotonergic input to the hippocampus is largely from brainstem via the fimbria-fornix pathway. Lesions of these nuclei or of the fimbria-fornix typically result in degeneration of presumed 5HT neuron terminals in the hippocampus (Pasquier and Reinoso-Suarez, 1978). The pattern of neuronal degeneration within hippocampus reported by these authors does not resemble the distribution of INAS-containing terminals reported here. Furthermore, lesions of the fimbria-fornix, either unilateral or bilateral, do not affect the content or distribution of INAS-immunoreactivity within the hippocampus (Porietis et al., in preparation).

Development of hippocampal INAS

In pilot studies we identified INAS in the granular cell layer of the dentate gyrus as early as 20 days postconception. In an effort to examine the appearance of INAS in the CNS, the hippocampus was chosen as the logical place for its study as a function of cortical development.

The hippocampus is that horn-shaped part of the cerebral cortex that lies along the curvature of the lateral ventricle. This formation, which is continuous with the entorhinal cortex, is further divided into the hippocampus proper (Ammon's horn), the fascia dentata (dentate gyrus) and the subicular complex (Storm Mathisen, 1977).

The dentate gyrus was chosen for study because it is an easily distinguishable component of the hippocampus under the light microscope and it is feasible to follow its development as its formation is completed postnatally. Altman and his colleagues (O'Keefe and Nadel, 1978) have shown that its development lags well behind the rest of the hippocampus and the rest of the brain in general.

A specific NAS-antiserum and fluorescein-labelled second antibody were employed to investigate the presence of INAS in the dentate gyrus as a function of age. The hippocampus of rats at 18, 20, 22, 30 and 38 days postconception as well as that from adult animals were studied. Specificity of the NAS-antiserum had previously been evaluated, both by RIA and immunohistology, the results of which indicated that the NAS-antiserum was saturable using the NAS ligand, but not with melatonin, 6-hydroxymelatonin, N-acetyltryptamine, serotonin or other ligands (Pulido et al. 1981). The specificity of the NAS-antiserum was re-confirmed in the hippocampus by immunohistologic inhibition experiments using sections from adult, 22 and 38 day postconception rats treated with an NAS-antiserum that had previously been incubated with NAS, melatonin or serotonin. NAS-antiserum preincubated with NAS demonstrated immunofluorescence levels characteristic of non-specific binding as observed in tissues treated with non-immune serum. Both melatonin and serotonin in comparable concentrations failed to decrease the observed levels of fluorescence.

Regardless of age, INAS was observed primarily in the cell bodies of the granular cell layer. This distribution was different from that of serotonin which is present in the nerve terminals of the molecular layer and the hilar zone beneath the granular layer (Moore and Halaris, 1975). Furthermore, INAS appeared to be age-dependent reaching adult levels by 38 days postconception (Table 1). Our data showed characteristic changes during development in the rat which tend to parallel among other things concurrent changes in granular cell number. The low levels of INAS observed in the neonatal dentate are consistent with the small number of functional neurons present in the granular cell layer at birth. Bayer and Altman (1975) have shown using (^{3}H)-thymidine autoradiography and low-level X-irradiation that 85 percent of dentate granular cells originate after birth as compared to the majority of cells in the molecular and hilar layers of the dentate area that originate during embryonic life.

Crain et al., (1973) have likewise shown that synapse formation between the perforant path axons from the entorhinal cortex and the dendrites of the granular cells takes place concurrently with cell formation with the most active synaptogenesis occurring between 25 and 32 days postconception. Our findings also correlate with the appearance and subsequent increases of N-acetyltransferase in the hippocampus (Hsu and Mandell, 1981) suggesting, perhaps, a role for NAS in the normal functioning of these cells.

A more rigorous examination of the dentate gyrus is required in ordered to further elucidate this correlation. It may be more meaningful, perhaps, to examine NAS content per neuron rather than per tissue as the content per neuron may not be influenced by developmental changes in neuron numbers. Such information therefore may be more likely to reflect change or lack of change in NAS biosynthesis as a function of age. In order to accomplish this end it may be necessary to obtain a more accurate estimate of the total number of granular cells as a function of age. Bayer et al. (1982) have recently shown, using volumetric estimates, that the total number of granular cells increases linearly in rats between 1 month and 1 year of age thereby adding substantially to the existing population of neurons.

TABLE I

NAS-IMMUNOFLUORESCENCE AS A FUNCTION OF AGE IN THE DENTATE GYRUS OF THE HIPPOCAMPUS OF THE RAT

AGE (POSTCONCEPTION)	INTENSITY OF FLUORESCENCE (PHOTOMETER READING)
18 days	----
20 days	11.24 ± 2.0*
22 days	23.78 ± 2.0
30 days	23.32 ± 2.6
38 days	54.0 ± 7.8
Adult	56.37 ± 5.8

*standard Error of the Mean

To date, the function of the hippocampus is still unknown. Selective ablation of dentate granular cells in the hippocampus of infant rats produced by surgical destruction of the area results in increased locomotor activity (Bayer et al., 1973). Furthermore some functions of the hippocampus are reported to develop in synchrony with the maturation of the dentate gyrus (Douglas, 1975). Perhaps there is some function for NAS in hippocampal activity.

Pharmacologic studies on cerebellar INAS

The formation and metabolism of NAS in the CNS has not been determined. Synthesis of NAS in the brain may follow a pathway similar to that described for pineal tissue (Brownstein, 1975; Wurtman and Moskowitz, 1977), i.e., tryptophan→5-hydroxytryptophan→serotonin→NAS. To test this possibility, rats were treated with either parachlorophenylalanine (PCPA) or 6-fluorotryptophan, both of which are potent inhibitors of tryptophan hydroxylase, the enzyme necessary for the synthesis of serotonin from tryptophan (Deguchi and Barchas, 1973; Miller et al. 1970, McGeer, 1973).

Tryptophan hydroxylase has wide distribution in the CNS and the activity of this enzyme is highest in nuclei containing large amounts of serotonin (Brownstein et al., 1973; Saavedra, 1977). Tryptophan hydroxylase has also been found, to some extent, in all areas where NAS has been visualized. To inhibit the activity of this enzyme we used PCPA (320 mg/kg b.w.) and 6-fluorotryptophan (115 mg/kg b.w.). Semiquantitative evaluation of the NAS-fluorescence intensity in cerebellar cortex, using a photomultiplier-photometer (Photovolt 520M, Wild-Leitz, Canada) showed a significant ($p < 0.001$) inhibition of the NAS-fluorescence intensity in animals treated with PCPA (INAS-fluorescence: 52.4 ± 11) or 6-fluorotryptophan (INAS-fluorescence: 36.6 ± 5.2) as compared to the saline controls (INAS-fluorescence: 108.3 ± 11.5). Results are expressed as mean ± standard error of the mean. These data show that both tryptophan hydroxylase inhibitors significantly reduce cerebellar NAS-immuno-fluorescence intensity relative to saline treated controls.

In the pineal gland the rate limiting and final step in N-acetylserotonin (NAS) synthesis is the acetylation of serotonin by N-acetyltransferase (NATase, E. C. 2.3.1.5) (Brownstein, 1975). NATase has been demonstrated in a wide variety of brain areas, with the highest concentration in the cerebellum (Hsu and Mandel, 1981; Paul et al., 1975; Yu and Boulton, 1979). Brain NATase activity significantly increases following the administration of β-adrenergic agonists suggesting that extra-pineal brain NAS synthesis may be under noradrenergic control, as in the pineal gland (Friedhoff and Miller). This hypothesis is supported by our observations that NAS-containing structures in areas such as cerebellar cortex and cochlear nuclei are adjacent to a rich noradrenergic (NE) innervations (Pulido, 1981). There is also evidence that NE exerts a significant action on Purkinje cells (Freedman et al., 1976; Hoffer et al., 1973) and we have shown that Purkinje cells have high NAS-immunofluorescence. To test the possibility of a functional relationship between NAS and NE in the rat CNS, animals were treated with isoproterenol (10 mg/kg b.w.), a β-adrenergic agonist, and/or with propranolol (20 mg/kg b.w.), a β-adrenergic blocking agent, and NAS-immunofluorescence in the cerebellum assessed semiquantitatively.

We found significant ($p < 0.002$) increases in cerebellar NAS-immunofluorescence (155.8 ± 10.2) after l-isoproterenol treatment as compared to control fluorescence of 85.0 ± 9.1. The effect was completely blocked by the β-adrenergic antagonist, propranolol which had fluorescence of 108.5 ± 14.7. These data suggest that the synthesis of NAS in the cerebellum is mediated by NATase activity and probably regulated by a β-adrenergic mechanism which may be similar to that in the pineal gland. This interpretation is supported by the wide distribution of both typrophanhydroxylase and N-acetyltransferase in brain tissue and by the observation that the activity of N-acetyltransferase in the rat brain is increased by pretreatment with isoproterenol. (Friedhoff and Miller 1977).

^{3}H N-ACETYLSEROTONIN BINDING IN MAMMALIAN BRAIN

The specific binding characteristics of tritiated N-acetylserotonin (29.5 Ci/mmol, Amersham) were examined in synaptosomal fractions prepared from human, calf and rat brain tissues as previously described (Niles et al., 1982). Fresh or frozen (-20°C) tissues were homogenized in 10 volumes of ice cold 0.32M sucrose and following centrifugation (1000 x g for 10 minutes at 4°C) the nuclear pellet was discarded. The supernatant was spun at 20,000 x g for 20 min and the crude synaptosomal fraction (P_2) was suspended in 20 volumes of distilled water to allow lysis to occur and subsequently washed three times in distilled water with centrifugation at 48,000 g x 10 min. The final pellet was suspended in 80 - 100 volumes of 0.05M Tris-HCL buffer (pH 7.4) and used in receptor assays.

Binding assays were performed by incubating 1 ml aliquots of tissue in triplicate with ^{3}H NAS in the presence or absence of a large excess of unlabelled NAS (10^{-5}M). For inhibition experiments, various drugs in concentrations ranging from 10^{-10} - 10^{-5}M were also added to incubation tubes. After incubation at 37°C for 30 min., samples were rapidly filtered on GF/B filters by vacuum and then rinsed with 4 x 3ml aliquots of ice cold Tris buffer. Filters were equilibrated overnight in PCS cocktail and then counted by liquid scintillation overnight in PCS scintillation cocktail and then counted by liquid scintillation spectrometry. Tissue protein concentrations were determined as described by Lowry et al. (1951) using bovine serum albumin as standard. Less than 0.2% of ^{3}H NAS was non-specifically bound to filters in the absence of tissue.

Specific ^{3}H NAS binding, defined as that displaced by excess unlabelled NAS, was about 70% of total binding and was found in synaptosomal fractions from rat, calf and human brain as shown in Table 1. Scatchard analysis of binding data for fresh membranes from male or female rats indicated the presence of one population of high affinity binding sites with a dissociation constant (Kd) of 3-7 nM and a maximal binding site concentrations of 250-400 fmol/mg protein. In contrast to fresh tissues, frozen membranes produced higher levels of specific binding which yielded biphasic Scatchard plots. Graphic analysis (Rosenthal, 1967) indicated the presence of two receptor classes.

The high affinity (Kd_1) and low affinity (Kd_2) sites with their respective Bmax values are indicated in Table II for frozen membranes from rat, calf and human cerebellum. The similarity in the high affinity Kd and Bmax values for fresh and frozen membranes from rat brain suggests that the increased binding observed in frozen membranes is due to exposure of a low affinity, high capacity binding component.

TABLE II

REGIONAL DISTRUBUTION OF SPECIFIC [^{3}H] NAS AND [^{3}H] 5-HT BINDING IN THE RAT CNS

CNS REGION	SPECIFIC BINDING [^{3}H] NAS	[^{3}H] 5-HT
	Fmoles/mg protein	
STRIATUM	34.12 ± 0.86	197.91 ± 2.34
CORTEX	32.83 ± 4.94	185.13 ± 8.10
HYPOTHALAMUS	22.69 ± 2.30	167.92 ± 8.89
CEREBELLUM	22.43 ± 1.73	91.39 ± 27.07
PONS-MEDULLA	15.69 ± 2.00	48.54 ± 5.78
HIPPOCAMPUS	13.93 ± 0.34	141.49 ± 12.05
MIDBRAIN	12.55 ± 4.36	50.01 ± 4.38

Synaptosomal membranes were incubated in triplicate with 5nM of [^{3}H] NAS or [^{3}H] 5-HT at 37°C for 30 minutes. Nonspecific binding was measured in the presence of 10 µM nonradioactive NAS or 5-HT. The correlation coefficient (r) [^{3}H] 5-HT binding is 0.63. Means ± SEM are presented.

Regional Distribution of Binding

Regional binding studies carried out with fresh P_2 fractions from rat brain indicated that ^{3}H 5-HT binding was about 3-10 fold greater than ^{3}H-NAS binding in all the CNS areas examined However, similar relative levels of binding (94 - 100%) were found for both radioligands in the striatum and frontal cortex. With striatal binding taken as representative of 100% binding for both radioligands, there were marked differences in the relative binding observed in other brain areas. The relative binding levels of ^{3}H-NAS in the cerebellum pons-medulla and midbrain were 66%, 46% and 37% while corresponding values for ^{3}H-5-HT were 46%, 25% and 25%. Conversely, in the hypothalamus and hippocampus, ^{3}H-5-HT binding was higher at 85% and 71% respectively, as compared with 67% and 41% for ^{3}H-NAS. Thus, relatively more ^{3}H-5-HT binding was present in the limbic areas examined while more ^{3}H-NAS binding occurred in the cerebellum and brainstem.

Structural Characteristics of Binding

Inhibition experiments were carried out with fresh rat cerebellar synaptic membranes using a variety of 5-HT related compounds and other drugs (see Table III. Among the drugs tested, NAS and 5-HT had the highest affinity and were equipotent inhibitors of binding. The Hill number for NAS did not differ from unity but lower nH value for 5-HT suggests interaction at multiple binding sites. This is also suggested by the low Hill numbers for the 5-HT agonist, quipazine, which showed low affinity and the 5-HT antagonist, methysergide, which showed moderate binding affinity.

TABLE III

INHIBITION OF [^{3}H]NAS BINDING IN RAT CEREBELLUM

DRUG	IC_{50}	nH
NAS	10.19 ± 2.51	0.95 ± 0.17
5-HT	16.61 ± 7.82	0.78 ± 0.11
QUIPAZINE	2015	0.67
5-METHOXYTRYPTAMINE	268	0.28
MELATONIN	>10000	0.29
N-ACETYLTRYPTAMINE	ND	
METHYSERGIDE	173	0.57

Fresh cerebellar membranes were incubated with 2.5 nM [^{3}H]NAS and at least five concentrations (10^{-10} - 10^{-5}M) of each drug. The concentration (IC_{50}) of each drug which displaced 50% of bound radioligand was determined from logit-log inhibition plots. Hill coefficients (nH) represent the slopes of these plots. Means of 2-5 experiments are presented. ND = no displacement.

Replacement of the 5-hydroxyl group possessed by both NAS and 5-HT with 5-methody resulted in a 15-25 fold decrease in affinity as found for 5 methoxytryptamine. Melatonin, which has an N-acetyl group in addition to the 5-methoxy group showed even less binding affinity while N-acetyltryptamine, with only the N-acetyl group, failed to inhibit binding.

In summary, saturable, high affinity binding of ^{3}H-NAS is present in rat, calf and human brain. The regional distribution of this binding in the rat brain indicates that relatively more ^{3}H-NAS than ^{3}H-5-HT is bound in the cerebellum and brain stem, areas which have also been reported to contain INAS (Pulido et al., 1981). The pharmacology of ^{3}H-NAS binding indicates that the 5-hydroxyl group is necessary for high affinity binding which appears to occur primarily at central serotonergic binding sites (Niles et al., 1982). However, the complexity of binding suggests that ^{3}H-NAS also labels other non-serotonergic binding sites in the cerebellum.

CONCLUSION

A. Melatonin is distributed in pineal, harderian gland, hypothalamus, retina, and gut tisses. The function of melatonin in these tissues is unknown. Apart from the hypothalamus, melatonin is not found in the CNS. The distribution of IM follows previous localization of HIOMT (Axelrod et al. 1961), the enzyme converting NAS to melatonin.

B. A unique distribution of INAS has been defined in the central nervous system by two different immunohistological approaches. This distribution differs significantly from that of both melatonin and serotonin. The presence of NAS in brain tissue has been confirmed by GC-MS (Brown et al., 1981) and by radioimmunoassay (Pang et al. 1981). Studies on INAS in brain have demonstrated that tryptophan hydroxylase may be essential for its synthesis and that β-adrenergic stimulation increases INAS. Within the hippocampus the quantity of INAS correlates with development of the dentate gyrus Saturable high affinity binding of tritiated NAS in brain has been demonstrated; a binding which has a serotonergic component but which is relatively greater than [^{3}H] 5-HT binding in brain areas that are richest in INAS. Taken together, these findings suggest that N-acetylserotonin may have a function in brain separate from that of either melatonin or serotonin.

ACKNOWLEDGEMENTS

The expert technical assistance of S.L. Campbell and secretarial help of L. Koutalos is acknowledged. This work has been supported in part by the Medical Research Council of Canada and The Ontario Mental Health Foundation.

REFERENCES

Axelrod, J., MacLean, P. D., Albers, R. W. and Weissbach, H., 1961, Regional distribution of methyltransferase enzymes in the nervous system and glandular tissues, in: "Regional Neurochemistry" S. S. Kety and J. Elkes, eds. Pergamon Press, Oxford pp. 307-311.

Bayer, S. A., Brunner, R. L., Hine, R., and Altman, J., 1973, Behavioural effects of interference with the postnatal acquisition of hippocampal granule cells, Nature New Biology, 242: 222.

Bayer, S. A., and Altman, J., 1975, Radiation-induced interference with postnatal hippocampal cytogenesis in rats and its long-term effects on the acquisition of neurons and glia, J. Comp. Neurol., 163:1.

Bayer, S. A., Yackel, J.W., and Puri, R.S., 1982, Neurons in the rat dentate gyrus granular layer substantially increase during juvenile and adult life, Science, 216:890.

Brown, G. M. and Grota, L. J., 1980, Use of immunologic techniques in the examination of neurotransmitters and neuromodulators, in: "Physicochemical Methodologies in Psychiatric Research", I. Hanin and S. Koslow, eds., Raven Press, New York pp. 65-81.

Brown, G. M., Porietis, A. V., and Narasimhachari, N., 1981, Identification and quantification of N-acetylserotonin in rat brain regions by GCMS, Society for Neuroscience Abstracts, 7:580.

Brownstein, M. J., 1975, Minireview. The Pineal Gland. Life Sciences 16: 1363-1374.

Brownstein, M., Saavedra, J. A. and Axelrod, J., 1973, Control of pineal N-acetylserotonin by a beta adrenergic receptor. Mol. Pharmacology, 9: 605-611.

Bubenik, G. A., Brown, G. M., Uhlir, I. and Grota, L. J., 1974, Immunohistological localization of N-acetylindolealkylamines in pineal gland, retina and cerebellum. Brain Res. 81: 233-242.

Bubenik, G. A., Brown, G. M., and Grota, L. J., 1976, Differential localization of N-acetylated indolealkylamines in CNS and the Harderian gland using immunohistology. Brain Res, 118: 417-427.

Bubenik, G. A., Brown, G. M. and Grota, L. J., 1976b, Immunohistochemical localization of melatonin in the rat Harderian gland, J. Histochem. & Cytochem., 2411: 1173-1177.

Bubenik, G. A., Brown, G. M. and Grota, L. J., 1977, Immunohistological localization of melatonin in the digestive system of the rat, Experienta, 33:662-663.

Bubenik, G. A., Purtill, R. A., Brown, G. M. and Grota, L. J., 1978, Melatonin in the rat and the Harderian gland. Ontogeny, diurnal variations and melatonin treatment, Exp. Eye Res., 27: 323-334.

Bubenik, G. A., 1980, Localization of melatonin in the digestive tract of the rat. Effect of maturation, diurnal variation,

melatonin treatment and pinealectomy, Hor. Res., 12:313-323.
Bubenik, G. A. and Purtill, R. A., 1980, The role of melatonin and dopamine in retinal physiology, Can. J. Physiol. Pharmacol. 58:1457-1462.
Cardinali, D. P. and Rosner, J. M., 1971, Retinal localization of the hydroxyindole-O-methyl transferase (HIOMT) in the rat. Endocrinology, 89:301-303.
Cardinali, D. P. and Rosner, J. M., 1971 Metabolism of serotonin by the rat retina in vitro, J. Neurochem., 18:1769-1770.
Cardinali, D. P. and Wurtman, R. J., 1972, Hydroxyindole-O-methyl transferase in rat pineal, retina and the Harderian gland. Endocrinology, 91:247-252.
Cardinali, D. P., Vacas, M. I., and Boyer, E. E., 1979, Specific binding of melatonin in bovine brain, Endocrinology, 105:437-441.
Cohen, M., Roselle, D., Chabner, B., Schmidt, T. J. and Lipman, J., 1978, Evidence for a cytoplasmic melatonin receptor, Nature, 247 (5674):8894.
Coons, A. H., Leduc, E. H. and Connely, J. M., 1955, Studies on antibody production. I. A method for histochemical demonstration of specific antibody and its application to a study of the hyperimmune rabbit, J. Exp. Med., 102:49-59.
Crain, B., Cotman, C., Taylor, D., and Lynch, G., 1973, A quantitive electron microscopic study of synaptogenesis in the dentate gyrus of the rat, Brain Research, 63:195.
Deguchi, T. and Barchas, J., 1973, Comparative studies on the effect of parachlorophenylalanine on hydroxylation of tryptophan in pineal and brain of rat. In "Serotonin and Behaviour," J. Barchas and E. Usdin, Ed., Academic Press, New York, pp. 33-47.
Douglas, R. J., 1975, in: "The Hippocampus", 2: Plenum, New York.
Freedman, R., Hoffer, B. J., Puro, D. and Woodward, D. J. 1976: Noradrenaline modulation of the responses of the cerebellar Purkinje cells to afferent synaptic activity. Brit. J. Pharmac. 57:603-605.
Freund, D., Arendt, J. and Vollrath, L., 1977, Tentative immunohistochemical demonstration of melatonin in the rat pineal gland, Cell Tiss. Res., 181:239-244.
Friedhoff, A. J. and Miller, J. C. 1977: In vitro and in vivo studies of extra-pineal N-acetyltransferase of rat brain. Res. Comm. in Chem. Pathology and Pharmacol., 16:225-244.
Gern, W. A., Owens, D. W. and Ralph, C. L., 1978, The synthesis of melatonin by the trout retina, J. Exp. Zool., 206(2):263.
Gern, W. A. and Ralph, C. L., 1979, Melatonin synthesis by retina, Science, 204:183-184.
Grota, L. J. and Brown, G. M., 1974, Antibodies to indolealkylamines: serotonin and melatonin, Can. J. Biochem., 52:196-202.
Grota, L. J., W. R. Holloway and G. M. Brown, 1982, 24-hour rhythm of hypothalamic melatonin immunofluorescence correlates with serum and retinal melatonin rhythms. Neuroendocrinology, 34: 363-368.
Hoffer, B. J., Siggins, G. R., Oliver, A. P. and Bloom, F. E. 1973,

Activation of the pathway from locus coerulus to rat cerebellar Purkinje neurons: Pharmacological evidence of noradrenergic central inhibition. J. Pharmacol. Exp. Ther., 181: 553-569.

Holloway, W. R., L. J., Grota and G. M. Brown, 1978, Diurnal changes in melatonin in retinal, pineal gland, suprachic, colon, and duodenum of the rat. Neuroscience Abstracts, 4:347.

Holloway, W. R., Grota, L. J. and Brown, G. M., 1979, Immunocytochemical identification of melatonin binding material in the pineal gland of the rat, Neurosci. Abst., 5:448.

Holloway, W., L. J., Grota, and G. M. Brown, 1980, Quantitative determination of immunoreactive melatonin in the colon of the rat. J. Histochem. Cytochem., 28:255-262.

Hsu, L. L., and Mandell, A. J., 1981. Aromatic alkylamine N-acetyltransferase in regional tissues of developing rat brain, Brain Research, 206:187.

Kappers, A. J., 1976, The mammalian pineal gland, a survey, Acta Neurochir. (Wien) 34 (1-4):109-149.

Kennaway, D. J., Frith, R. G., Phillipou, G., Matthews, C. D. and Seamark, R. F., 1977, A specific radioimmunoassay for melatonin in biological tissue and fluids and its validation by gas chromatography-mass spectrometry, Endocrinology 101(1): 119-127.

Lerner, A. B., Case, J. D., Takahashi, Y., Lee, T. H. and Mori, W. 1958, Isolation of melatonin, the pineal gland factor that lightens melanocytes, J. Am. Chem. Soc., 80:2587.

Lidov, H. G. W., Grzanna, R., and Molliver, M. E., 1980, The serotonin innervation of the cerebral cortex in the rat - an immunohistochemical analysis, Neuroscience, 5:207.

Lowry, O. H., Rosebrough, N. J., Farr, A. L., and Randall, R. J., 1951, Protein measurement with the Folin phenol reagent, 193: 265.

McGreer, E., 1973, Tryptophan hydroxylase inhibitors other than PCPA, In: "Serotonin and Behaviour", J. Barchas and E. Usdin, Ed., Academic Press, New York, pp. 55-59.

McLean, I. W., and Nakane, P. K., 1974, Periodate lysine-paraformaldehyde fixative. A new fixative for immunoelectron microscopy. J. Histochem. Cytochem., 22:1077-1083.

Miller, F. P., Cox, R. H., Snodgrass, W. R. and Marckel, R. P., 1970, Comparative effects of p-chlorophenylalanine, p-chloroamphetamine and p-chloro-N-methylamphetamine on rat brain norepinephrine, serotonin and 5-hydroxylindole-3-acetic acid, Biochem. Pharmacol., 19:435-442.

Moore, R. Y., and Halaris, A. E., 1975, Hippocampal innervation by serotonin neurons of the midbrain raphe in the rat, J. Comp. Neurol., 165:171.

Nakane, P. K. and Kawaoi, A., 1974, Peroxidase-labelled antibody - a new method of conjugation, J. Histochem. Cytochem., 22:1084-1091.

Niles, L. P., Wong, Y. W., Mishra, R. K. and Brown, G. M., 1979, Melatonin receptors in brain, Europ. J. Pharmacol. 55(2):219-220.

Niles, L. P., Brown, G.M. and Mishra, R. K., 1980, Localization and characterization of melatonin receptors in CNS cytosol, Fed. Proc.,39:1008.

Niles, L. P., Brown, G. M., and Mishra, R. K., 1982, A serotonergic component of ^{3}H N-acetylserotonin binding in mammalian Brain, Prog. Neuropsychopharmacol., (In Press).

O'Keefe, J., and Nadel, L., 1978, in: "The Hippocampus as a cognitive map", Clarendon Press, Oxford.

Pang, S. F., Ralph, C. L. and Reilly, D. P., 1976, Melatonin in the chicken brain: its origin, diurnal variations and regional distributions, Gen. Comp. Endocrinol., 22:499-506.

Pang, S. F., Brown, G. M., Grota, L. J., Chambers, J. W. and Rodman, R. L., 1977, Determination of N-acetylserotonin and melatonin activities in the pineal gland, retina, Harderian gland, brain and serum of rats and chickens, Neuroendocrinology, 23:1-12.

Pang, S. F., Yu, H. S., Suen, H. C. and Brown, G. M., 1980, Melatonin in the retina of rats - a diurnal rhythm, J. Endocrinology, 87(1):89-95.

Pang, S. F., Brown, G. M., Campbell, S. L., Snieckus, V., deSilva, S. O., Young, S. N., and Grota, L. J. (1981), A radioimmunoassay for N-acetylserotonin in biological tissues, J. Immunoassay, 2:263-276.

Pasquier, D. A., and Reinoso-Suarez, F., 1978, The topographic organization of hypothalamic and brain stem projections to the hippocampus, Brain Res. Bull., 3:373.

Paul, S. M., Hsu, L. L. and Mandel, A. J., 1974, Extrapineal N-acetyltransferase activity in rat brain. Life Sci., 15: 2135-2143.

Pevet, P., Balemans, M. G. M., Legerstee, W. C. and Vivien-Roesl, B., 1980. Circadian rhythmicity of the activity of hydroxyindole-O-methyl transferase (HIOMT) in the formation of melatonin and 5-methoxytryptophol in the pineal, retina, and harderian gland of the golden hamster, J. Neural Transmission, 49:229-245.

Pevet, P., Balemans, M. G. M., and de Reuver, G. F., 1981. The pineal gland of the nole (Talga Europea L.) VII. Activity of hydroxy-indole-O-methyltransferase (HIOMT) in the formation of 5-methoxytryptophan, 5-methoxytrptamine, 5-methoxyindole-3-acetic acid, 5-methoxytryptophal and melatonin in the eyes and the pineal gland. J. Neurol. Trans., 51:271.

Porietis, A. V., Brown, G. M., and Grota, L. J., 1978, Immunohistochemical localization of N-acetylserotonin in rat hippocampus, Society For Neuroscience Abstracts, 4:226.

Pulido, O., Brown, G. M., and Grota, L. J., 1981, Localization of N-acetylserotonin (NAS) in the rat hindbrain by immunohistology, Prog. Neuropsychopharmacology, 5:573-576.

Quay, W. B., 1965, Retinal and pineal hydroxyindole-O-methyl transferase activity in vertebrates, Life Sci., 4:983-991.

Quay, W. B., 1980, Greater pineal volume at higher latitudes in rodentia: Exponetial relationship and its biological interpretation, Gen. Comp. Endocrinol., 41(3):340-349.

Ralph, C. L. and Lynch, H. J., 1980, A quantitative melatonin bioassay, Gen. Comp. Endocrinol., 15:334-338.

Reiter, R. J., 1973, Involvement of pineal indoles and polypeptides with the neuroendocrine axis, Prog. Brain Res., 39:281-286.

Rosenthal, H. E., 1967, A graphic method for the determination and presentation of binding parameters in a complex system, Analyt. Biochem., 20:525.

Steinbusch, H. W. M., 1981, Distribution of serotonin-immunoreactivity in the central nervous system of the rat-cell bodies and terminals, Neuroscience, 6:557.

Storm-Mathisen, J., 1977, Localization of transmitter candidates in the brain: the hippocampal formation as a model, Progress In Neurobiology, 8:119.

Vivien-Roels, B., Pevet, P., Dubois, M. P., Arendt, J. and Brown, G. M., 1981, Immunohistochemical evidence for the presence of melatonin the pineal gland, the retina and the Harderian gland, Cell Tiss. Res., 217:105-115.

Willingham, M. C. and Yamada, S. S., 1979, Development of a new primary fixative for electron microscopic immunocyto chemical localization of intracellular antigens in cultured cells. J. Histochem. Cytochem., 27:947-960.

Wurtman, R. J. and Moskowitz, M. A., 1977, The pineal organ. New Eng. J. Med., 296:1329-1333.

Yu, P. H. and Boulton, A. A., 1979, N-acetylation of tyramines: Purification and characterization of an arylamine N-acetyltransferase from rat brain and liver. Can. J. Biochem., 57: 1204-1209.

MELATONIN ACTION: SITES AND

POSSIBLE MECHANISMS IN BRAIN*

D.P. Cardinali,** María I. Vacas,** María I. Keller Sarmiento and E. Morguenstern***

Centro de Estudios Farmacológicos y de Principios Naturales (CEFAPRIN), Serrano 665/669, 1414 Buenos Aires, Argentina

INTRODUCTION

Knowing what the pineal gland does is a major step towards understanding its physiological function in the body or what it is for. What a hormone-secreting organ like the pineal does can be characterized by defining the nature and quantitative aspects of its input-output relationships, that is what the gland secretes and in response to what. Definition of specific target tissues and intracellular mechanisms of action for the hormone involved should be also included in the description.[1]

The pineal contains a number of biologically active compounds all or several of which could be the factor(s) responsible for the neuroendocrine effects of the gland. Indoles (often represented by melatonin[2]) and polypeptides (most of which are still unidentified) are the compounds usually considered within this context, although the requirements set by Bayliss and Starling's definition of a hormone, i.e. "a substance produced in one organ which after being transported by the blood (or CSF) acted upon a distant organ to alter its function" remain to be satisfied for most of them. In the case of melatonin current experimental evidence indicate that it

* Supported by grants from Consejo Nacional de Investigaciones Científicas y Técnicas (CONICET) and Fundación "Alberto J.Roemmers", Buenos Aires, Argentina.

** Established Investigator, CONICET.

*** Research Fellow, CONICET

is, in fact, a pineal hormone. Melatonin is synthesized and secreted by the pineal gland of all vertebrates tested so far, it circulates in blood with a characteristic circadian pattern and it influences the function of a variety of endocrine organs, mimicking several neuroendocrine effects attributed to the pineal gland.[2,3]

Since pineal's effects tend to be modulatory rather than primary (that is, to change the timing or amplitude of responses rather than to generate such responses) considerably more difficulties are often faced by the investigator to design successful experiments exhibiting melatonin activity, than, for example, experiments showing roles for pituitary hormones. In hamsters, massive daily doses of melatonin can be devoid of activity,[4] while very small doses, presumably physiological, given daily at a critical time have marked effects on neuroendocrine function.[3,5,6] Since the catabolism of melatonin remains fairly constant under a variety of circumstances, the logical locus for such dramatic changes in exogenous melatonin activity is at the target organ level, perhaps on receptor and/or post-receptor mechanisms.

The present article discusses the current knowledge on the sites and possible subcellular mechanism subserving the neuroendocrine activity of melatonin. The interested reader is referred to recent reviews[2,7,8] as well as to an accompanying one[9] for additional discussion of the neural and hormonal factors controlling melatonin synthesis and release in mammals.

METABOLIC FATE OF MELATONIN

Although the matter remains unsettled it is generally accepted that the blood rather than the CSF is the primary site of secretion of pineal melatonin. In rats the concentration of melatonin in the blood collected from the confluens sinuum is approximately 8 times greater than that in the trunk blood; moreover darkness induces increases in plasma melatonin levels without concomitant increases in CSF.[10] These observations together with the previous reports on CSF melatonin concentrations lower than the plasmatic ones in humans,[11] rhesus monkey[12] and sheep[13] have strengthened the opinion that the normal route of melatonin comprises the pineal capillaries draining into the surrounding venous sinuses. Still this remains to be demonstrated in a direct way.

In accordance with recent estimates in rat[14] and sheep,[15] melatonin has in blood a half-life of about 20 min. In neonatal rats melatonin half-life is considerably longer probably due to a low catabolic activity of the liver.[16] Rapid hepatic hydroxylation at position 6 accounts for most of melatonin inactivation in rat,[17] sheep,[18] monkey[19] and man.[17,18] Most (70%) of the 6-hydroxymelatonin is then conjugated to sulfate while a much smaller amount is conjugated with glucuronic acid; both conjugates are released from the body in urine and feces.[17] Other pathways for melatonin degradation include limited deacetylation by a hepatic aryl acylamidase to yield 5-methoxytryptamine[20] and cleavage of the pyrrole ring by indoleamine-2,3-dioxygenase to kynurenamine derivatives (presumably only in the brain).[21]

About 60-70% of circulating melatonin is bound to plasma proteins (mainly albumin) and none is bound to protein in CSF.[22] Plasma protein binding does not impede melatonin transport into the brain;[23] the mechanism accounting for the inability of CSF melatonin to attain equilibrium with plasma melatonin remains unknown. The active accumulation of circulating melatonin by rat and cat choroid plexus has been regarded as the first step for the secretion of the hormone into the CSF.[24] However, in the frog, the unidirectional flux across the IV ventricular plexus from the CSF to blood is considerably higher than that in the opposite direction.[25] The relevance of these observations in mammals awaits further exploration.

After systemic injections radioactive melatonin enters all tissues including the brain.[17] One hour after an i.v. injection to cats melatonin is concentrated in the pineal gland, iris-chroid and ovary.[26] Other endocrine organs (like testis, adrenal and thyroid) and the sympathetic nervous chain also take up ^{3}H-melatonin.[26] Labeled melatonin injected into the CSF is taken up unevenly by the brain and becomes concentrated within the hypothalamus and midbrain.[27,28] Melatonin was identified in the rat hypothalamus by either GCMS[29] or immunohistochemistry.[30] By the latter technique melatonin is also detectable in the retina, optic nerve and tract, suprachiasmatic nucleus and pineal gland.[30]

After a bolus intracisternal injection melatonin disappeared very rapidly from the brain, its decay being multiphasic.[28] Only 6% of the administered dose was present in the brain as unchanged melatonin 5 min after injection and only 0.8% remained after 20 min indicating the rapid exchange between CSF and vascular compartments. The concentration of melatonin in the hypothalamus was 4-5 times that in the rest of the brain.[28] In the CNS melatonin is readily oxidized to N-acetyl-5-methoxykynurenamine, which accounts for about 15% of total urinary melatonin metabolites.[21] The link between brain melatonin metabolism and hormone's mechanism of action remains unknown.

MELATONIN NEUROENDOCRINE ACTIVITY: CENTRAL OR PERIPHERAL ?

For many species at temperate latitudes light is the principal factor which initiates and synchronizes the timing of the reproductive seasonal rhythm.[3] Hamsters, voles, mice and gerbils, as well as rats under particular circumstances (e.g. anosmia), exhibit gonadal regression when exposed to short-day photoperiods, an effect needing the pineal gland and mimicked by melatonin administration in appropriate amounts and time schedules,[2,3] A number of aspects of the hypothalamic-hypophysial-gonadal axis are also depressed by melatonin treatment in sheep, steer, dog, rabbit, hare, monkey and weasel (for references see ref. 2). Additionally, other effects of short days on endocrine and behavioral function are also influenced by the pineal gland and melatonin.

The importance of target organ sensitivity in determining the effects of melatonin must be stressed. While appropriately timed (late evening) injections of small, physiological amounts of melatonin can induce gonadal atrophy in male hamsters maintained on long days,[3,5,6] similar injections can inhibit the testicular regression induced by short days.[31] Melatonin beeswax pellets or capsules that deliver a constant amount of melatonin prevent the short day-induced gonadal regression as well as that following evening melatonin injections; they can also stimulate gonadal growth in hamsters with regressed testes that are maintained on short days after capsule implantation.[3,32] Massive daily doses of melatonin, even at evening hours were without effect on the hamster reproductive system.[4] Thus a wide variety of circumstances (photoperiod length, dosage, time and way of administration during daily photoperiod) govern the presence or absence of melatonin antigonadal effects. The common factor for all of them can be the extent and length of exposure of the target cells to melatonin. As discussed below endogenous melatonin via down regulation of receptors or at post-receptor sites could modulate organ's sensitivity to subsequently injected exogenous melatonin in hamsters. This phenomenon can be a critical event in the sensitivity of the hamster CNS to interpret photoperiodic information in terms of neuroendocrine and behavioral responses.

Although there is little doubt that melatonin exerts antigonadal effects when appropriately injected, two groups of investigators published conflicting results with this view after active immunization of rats or hamsters against melatonin-protein conjugates.[30,33] Melatonin immunization did not prevent the dark-induced gonadal atrophy; moreover immunization induced testicular regression in hamsters kept in long photoperiods, a finding which was interpreted as indicating a progonadotropic rather than antigonadotropic effect of melatonin. Further experiments are needed to rule out a possible deleterious effect of the immunization procedure per se on the sensitivity of melatonin target organs, as well as to demonstrate that all the endogenous production of melatonin is effectively neutralized by that procedure.

The anatomical sites implicated in melatonin antigonadal activity may include one or a variety of tissues such as the CNS, the adenohypophysis or the gonads.[2] Since a significant reduction in gonadotropin output is observed in most experimental paradigms explored, it is reasonable to assume that melatonin affects neuroendocrine activity by acting centrally on the hypothalamic-hypophysial axis. Fraschini and co-workers [34] were the first to report that the stereotaxic implant of melatonin in rat medial basal hypothalamus (MBH) depressed pituitary and plasma LH levels. Subsequently several investigators observed that the cerebroventricular injections of melatonin decrease LH and FSH levels. With few exception (e.g.dogs [35]), melatonin administered intrapituitary as portal vessel infusions or systemically preceding an LHRH injection, failed to modify gonadotropin release.[2] These observations strongly favor a central rather than a pituitary site of action for the hormone.

In white-footed mice anterior hypothalamic melatonin-beeswax implants that released very small amounts (i.e. less than 100 ng/day) of melatonin induced significant reproductive organ collapse, an effect that was not reproduced by s.c. implantation of the pellets.[36] In hamsters exogenous melatonin failed to inhibit gonadal activity if the animals were priorly subjected to anterior hypothalamic deafferentation[37] or to lesion of the suprachiasmatic nucleus.[38] Thus in mice and hamsters melatonin appears also to act on central sites to depress gonadal growth.

The lack of direct effects of melatonin on the adenohypophysial function in rats is an age-related process. In neonatal but not in adult pituitaries melatonin (10^{-8}M) inhibited the stimulation of LH and FSH release brought about by LHRH in vitro. This effect of melatonin in infant rats was quite specific inasmuch as the pineal hormone did not alter either the basal and TRH-stimulated release of prolactin and TSH, or the basal and somatostatin-inhibited GH secretion.[39] In rats pinealectomy or constant light advances the time of puberty,[3] probably by advancing the maturation of gonadotropin control by the steroids,[40] while short photoperiods delay it. Melatonin injection of prepuberal rats disrupts partially steroid-induced gonadotropin release and inhibits reproductive organ growth. Perhaps the impending age-related adenohypophysial insensitivity to melatonin in rats is one of the signals to induce puberty onset.

Quay[41] first suggested that the pineal gland itself can be a target organ for melatonin, with resultant effects elsewhere in the body representing modification of the pineal release of other hormonally-relevant compounds. Indeed the presence of an intact pineal gland was needed in order for single daily afternoon injections of melatonin to cause gonadal atrophy in hamsters[3,6,42] or to affect prolactin release in underfed rats[43]. However repetitive injections of melatonin resembling the diurnal rhythm in hormone secretion to pinealectomized, ganglionectomized or suprachiasmatic-lesioned hamsters, reproduce the productive collapse seen in intact hamsters injected at the evening.[3,6,42] These observations constitute another evidence on the importance of target organ sensitivity as regulated by the pineal gland and melatonin itself. Data on brain melatonin receptors discussed below offer a basis for explaining these results.

Although the foregoing discussion imputes the hypothalamus (and depending upon age in rats, the adenohypophysis) as the most likely primary target organs for melatonin endocrine activity, it should be noted that the pineal hormone could also act peripherally. A few studies have shown that the injection of melatonin into hypophysectomized rats or mice inhibited HCG-induced gonadal and accessory organ growth, perhaps by inhibiting gonadotropin uptake.[44-46] Similarly melatonin diminished the growth of genital accessory organs in response to testosterone in rats.[47] Melatonin (10^{-6}M) depressed the in vitro biosynthesis of androgens in rat[48] and duck[49] testes as well adrenal steroid metabolism.[50] In lower concentrations (10^{-8}M or higher) melatonin also decreased the motility of the spayed rat uterus.[51] Inasmuch as alleged hormone receptors are detectable in the gonads and uterus[52,53] perhaps the melatonin-mediated effects of the pineal gland on the reproductive system in rodents may be the consequence of an overall influence on CNS, pituitary and peripheral structures.

Biosynthetic melatonin capacity as well as immunoassayable or immunohistochemically detectable melatonin have been shown in the mammalian retina,Harderian glands and certain parts of the gut.[30,54,55] In the latter the distribution of melatonin corresponds closely to the localization of serotonin-producing argentaffin cells; this observation led to the hypothesis that melatonin-synthesizing cells are part of the APUD system.[55a] Thus melatonin may have not only a classical endocrine role but also local neuromodulatory functions in certain tissues.

MELATONIN RECEPTORS

Early work supported the existence of a saturable melatonin uptake mechanism in brain. The intracisternal injection of excess amounts of non-labeled melatonin to rats resulted in a significant depression of the subsequent brain uptake of radioactive melatonin.[28] The logical corolary of those observations in vivo was to search for melatonin receptors in brain subcellular fractions. As soon as a radioactive melatonin of very high specific activity was available this goal could be achieved.

In 1978 two Laboratories published the first description of the occurrence of specific binding sites for melatonin in its target tissues. Cohen and co-workers[52] observed cytosol binding sites for this hormone in rat and hamster peripheral tissues like the ovary, uterus, testes, liver and eyes with dissociation constant (Kd)= 6 nM for high affinity sites and 550 nM for low affinities sites. No melatonin binding was detectable by the dextran-charcoal procedure in brain. Structural specificities of the high affinity sites suggested a requirement for the indole group and decreased affinity for indole derivatives lacking the N-acetyl group. Cardinali et al[56] found melatonin binding activity in crude membrane fractions of bovine MBH with Kd= 12 nM and binding site concentration of about 15 fmoles/mg prot. Binding affinities of melatonin analogues indicated that the 5-methoxy and N-acetyl groups were needed for full expression of binding to MBH membranes and that binding was abolished by the presence of a 6-hydroxy group. In a subsequent and more detailed description of these binding sites ionic requirements, subcellular localization and protein nature were studied.[57] Binding was increased by adding calcium, was inhibited by either sodium or potassium, and was located mainly in the crude mitochondrial pellet (P_2). Other areas of the bovine brain also exhibited melatonin binding activity, being 73% and 34% that of MBH in the occipital and cerebellar cortexes, whereas it was undetectable in amygdala, striatum or pons.

Differing from Cohen et al's report, Brown and co-workers[30] did find, by using the dextran-charcoal technique, cytoplasmic melatonin binding in rat, human and bovine brain. The Kd observed varied from 9 nM (hypothalamus) to 302 nM (midbrain). As in bovine striatum[57] no saturable binding of melatonin to rat striatal membranes was observed[30].

The possibility that cytosol and membrane melatonin binding sites co-exist in the same tissue is supported by observations on the rat liver.[53] In this tissue membrane and cytosol binding sites were found, the former being calcium dependent. Kd values were 8 nM for membrane binding sites and 20 nM for cytosol binding sites. Concentration of sites in liver membranes was about 10% that in the cytosol. Among the various tissues examined the highest membrane melatonin receptor concentration was found in the rat hypothalamus and pituitary gland.[53] Obviously additional studies are needed to establish whether one or both melatonin binding sites (cytoplasmic or membrane) are the physiologic receptors. In this sense it is interesting to note that the correlation of structure-activity relationships for melatonin analogues in amphibian skin[58] with that for binding to bovine MBH membranes[56,57] or to rat and hamster peripheral tissues[52] is fairly good, suggesting that a common receptor may mediate melatonin effects on brain and periphery. Unfortunately no data have yet been published on structure-activity relationships for melatonin binding in brain cytosol.

A number of reports indicated that both pineal biochemistry and morphology are affected by the exogenous injections of melatonin (for references see ref. 2). If the effect of melatonin is exerted directly on pineal cells rather than indirectly via the pineal sympathetic nerves or through a second hormone acting on the pineal gland, melatonin receptors should be present in the gland. Such specific melatonin binding sites were detected by conventional radiochemical techniques in bovine pineal membranes[59] and by a semi-quantitative immunohistological procedure in rat pineal glands.[30] Membrane pineal binding sites differed in a number of properties from those of MBH, including their high Kd (700 nM), ionic dependence and structural requirements; whether or not they represent pinealocyte autoreceptors remains to be defined.

Another potentially relevant site of action for melatonin is the retina. The melatonin forming enzyme hydroxyindole-O-methyl transferase is present in the mammalian retina, and the rat retina in vitro is capable of converting serotonin into melatonin.[54,55] Since these initial observations several publications have strengthened the concept that the mammalian retina is a physiological, significant site of synthesis and action for melatonin (for references see ref. 60). Melatonin has been shown to affect the aggregation of melanosomes in trout and guinea pig pigmented-epithelium; recently presumptive melatonin receptors exhibiting a Kd= 1-2 nM and a structure requirement resembling those of other brain areas have been identified in retinal tissue preparations.[61]

For all the aforementioned melatonin binding sites the rigorous demonstration of their receptor nature is still lacking. However circumstantial evidence has been accumulated in favor of this view for melatonin binding sites in brain membranes. A first example is given by starvation experiments in rats. Protein deprivation in these species is accompanied by significant increases of the number of melatonin receptor sites in brain membranes without changing their Kd for the radiolabeled compound.[62] Since several investigators have shown an increase in the sensitivity of the reproductive system to exogenous and presumably endogenous pineal melatonin in underfed rats,[3,43] binding and biological data correlate under these circumstances.

Another example is given by the daily rhythm of melatonin sensitivity in hamster and rat brains. As Tamarkin et al.[5] first showed, injection of melatonin is highly effective to mimic short photoperiods in hamsters if repeated daily at evening hours, but not during forenoon. Melatonin effect was blocked after pinealectomy or pineal denervation but in this case it was found that three injections per day restored the gonadal response; in contrast to the intact animals, the pinealectomized hamsters did not display a diurnal sensitivity to melatonin,[3,6,42] Pinealectomy also blocked the effect of daily evening injections on prolactin release in underfed rats.[43] These observations suggested that the diurnal rhythm in sensitivity to melatonin in hamsters and rats is influenced somewhat by the pineal itself. The possibility that the affinity or number of presumptive melatonin membrane receptors in hamster and rat brains could change accordingly at times when melatonin is biologically active or inactive was explored by us.[63] Interestingly enough receptors are maximally concentrated at a time during the lighting cycle when exogenous melatonin is effective in inducing gonadal atrophy; conversely the receptor number is low at a time when melatonin injection is uneffective in inhibiting reproductive physiology. In rat pineal the binding capacity of melatonin also increased markedly during the light phase of the day.[30] Suppression of melatonin rhythmicity by pinealectomy, pineal denervation or continuous exposure to light eliminated both the antigonadal effect of daily evening melatonin injection [3,42,43] and the morning-evening differences in brain membrane receptor number.[62] The hypothesis explaining these results implies a rhythm in melatonin receptors maintained in part by circulating melatonin (Fig. 1). Exposure to high melatonin levels during the dark phase of daily photoperiod causes desensitization of the neuroendocrine system by down-regulation of receptor binding sites whereas during daytime the number of receptors increases because melatonin levels are low, and restoration of sensitivity occurs at the end of the light phase.[3,63]

It is important to note that when the effects of daily night-time injections of melatonin are assessed in hamsters, the animals injected at the initiation or 2 h before the end of the dark phase showed gonadal involution, whereas after 2 or 4 h of dark the hormone failed to affect gonadal function.[64] Although the hypothesis was put forth that injections of

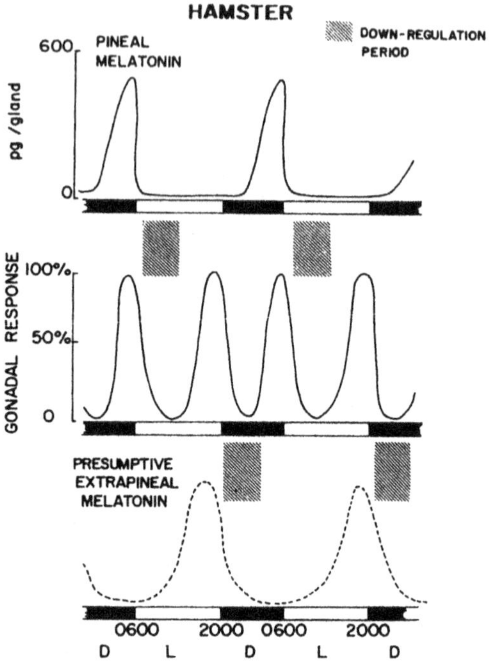

Fig. 1 A hypothesis to explain the diurnal rhythm in sensitivity of the hamster neuroendocrine system to single daily injections of melatonin. Gonadal response (inhibition) is found only after late evening or late night injections (central panel).[51,64] The lack of effect of early light or middle night melatonin injections is attributed to down-regulation of receptors elicited by high pineal or extrapineal melatonin secretion, respectively. For other details see text.

melatonin near the middle of the dark phase were ineffective to inhibit gonadal growth because circulating melatonin levels were already elevated at this time, subsequent observations did not support such a view. Differing from other mammals like sheep or calves, pineal and plasma melatonin in hamsters remain low for at least 4-6 h after the outset of darkness.[3,30]

Why then a subsensitivity to the hormone occurs at earlier times in the scotophase ? A hypothesis is advanced in Fig. 1. Down regulation of receptors (and therefore subsensitivity to melatonin) is considered to be the consequence of increased melatonin secretion by pineal or extrapineal sources. Thus a hypothetical extrapineal site of origin of melatonin is postulated as responsible for the subsensitivity of hamster reproductive system to exogenous melatonin when pineal melatonin secretion is low. Plasma levels of melatonin may or may not reflect changes in melatonin production by those presumptive extrapineal sites. In rats an increase in plasma melatonin (but not in pineal melatonin) preceding the initiation of the dark period was reported[10]; additionally the retinal capacity to synthesize melatonin shows peak values at late evening and early night.[2,65] Obviously many aspects of the interaction of melatonin with its brain receptors remains to be examined. Alternative explanation to those given in Fig. 1 are also possible, like for example to consider the subsensitivity to melatonin as depending partly on intrinsic properties of the hamster brain rather than exclusively on circulating melatonin levels. In addition it is not known whether true decreases of melatonin binding occur in hamster brain at the initiation of the dark phase. It should be also stressed that hamster and rat data may not be easily extrapolated to other species, in which a down-regulation of melatonin receptors is not readily apparent. For example constant release s.c. implants of melatonin or MBH implants in white-footed mice produce inhibition of gonadal function.[36]

POSSIBLE MECHANISM OF ACTION OF MELATONIN IN BRAIN

A great deal of attention have concentrated on melatonin-sensitive events at the hypothalamic level; this is not unreasonable considering the role that the hypothalamus plays in the control of pituitary function.

The following mechanisms have been advanced to explain the action of melatonin on the hypothalamus: action upon serotonin receptors[27]; binding to specific receptors in MBH;[56,57] interaction with benzodiazepine receptors;[66] inhibition of hypothalamic tubulin synthesis (with concomitant impairment of fast axonal transport and induction of ultrastructural changes in nerve endings of MBH);[67] reduced number of estradiol receptors;[68] MAO inhibition;[69] depressed synaptosomal uptake and increased release of serotonin;[70] depressed release of dopamine;[71] increased GABA content;[72] increased release of

Table 1. Similarities between melatonin and indomethacin.

1. Structural

Antibodies to indomethacin-HSA used in melatonin radioimmunoassay.[87]

2. In vivo effects

Inhibition of post-castration and steroid-induced LH release by systemic injection in female rats.[2,3,83]

Depression of serum LH levels by cerebro-ventricular injection or MBH implant.[2,3,34,83]

No effect on LH response to LHRH injection in adult rats.[2,3,83]

Depression of PGE_2 release by CNS.[78]

3. In vitro effects

Depression of PGE_2 release by MBH explants[85] and of basal and oxytocin-induced uterine motility in rats.[51,81]

Inhibition of thromboxane synthesis and ADP-induced aggregation in human platelets.[51,82]

neuropeptides;[73-75] increased number of immunoreactive LHRH cells; inhibition of PGE_2,[77-78] and cAMP synthesis;[79] stimulation of cGMP synthesis;[79] increased rate of neuronal discharge in rostral hypothalamus and reduced catecholamine inhibition.[80] Of course many of these mechanisms are probably not mutually exclusive.

Very few biological effects of melatonin on brain tissue have been obtained in vitro at the concentrations of the Kd of the presumptive receptor sites above described. One of them is the inhibition of PG (particularly PGE_2) synthesis in MBH of rats. Based on the structural similarities between melatonin and the inhibitor of PG synthesis indomethacin (N-p [chlorobenzoyl] -2-methyl-5-methoxyindole-3-acetic acid), Gimeno et al.[81] and Leach and Thornburn[82] put forth independently the hypothesis that melatonin impairs PG biosynthesis in target tissues. Melatonin (10^{-8} M or higher) inhibited the spontaneous motility of the spayed rat uterus as well as its reactivity towards added oxytocin (both PG-mediated effects); it also inhibited PGE and PGF release to the medium.[81] Thromboxane release from platelets[82] and platelet aggregation (a thromboxane-dependent phenomenon)[51,82] are also inhibited by melatonin. Melatonin (10^{-8} M or higher) depressed the spontaneous release of PGE_2 from rat MBH, as well as that elicited by adding NE to the medium.[77] Melatonin treatment of rabbits prevented the release of E prostaglandins into the cerebrospinal fluid after cervical stimulation.[78]

There is considerable evidence from pharmacological studies employing cerebroventricular and systemic injections of PGs and PG synthesis inhibitors to suggest that PGs are involved in the regulation of adenohypophysial hormone secretion. PGs (particularly of the E series) were found to stimulate pituitary hormone release; PGs appear to affect LH, FSH and prolactin release by acting on neural sites, TSH by acting at the pituitary level, and GH and ACTH by affecting both pituitary and hypothalamic sites.[83] The best characterized effect of PGE_2 on hypothalamic neurohormone release is that on LHRH. PGE_2 induces LHRH release to the rat portal blood in vivo,[84] and into the culture medium of rat MBH in vitro.[85] PG synthesis blockage by indomethacin prevents NE-induced LHRH release in vitro.[85] An increase in PGE_2 synthesis of rat MBH occurs simultaneously with LH release in estradiol-primed rats.[86]

Since the administration of monoaminergic blockers did not affect PGE_2-stimulated LH release in rats, the hypothesis was put forth that PGE_2 acts directly on the LHRH secreting neurons rather than presynaptically.[83] Of course one can visualize other presumptive sites of action of PGE_2 on the LHRH secreting mechanism. For example the effect of PGE_2 on LHRH release may be exerted presynaptically on neurons impinging on neurosecretory cells that use a non- monoaminergic transmiter like opioids,GABA or glutamate; alternately PGE_2 may affect primarily glial rather than LHRH containing cells in MBH.

Since melatonin inhibits basal and NE- stimulated PGE_2 release by MBH in vitro[77] and by brain tissue in vivo[78] the possibility should be considered that melatonin effect on pituitary hormone release is mediated by PG synthesis inhibition. Indeed there are striking similitudes between the neuroendocrine effects of melatonin and indomethacin (Table 1). However as far as LHRH secretion is concerned the data are conflicting with the view that melatonin depresses neurohormone secretion, inasmuch as increases, rather than decreases of neuropeptide release by MBH have been observed after exposure to melatonin in vitro.[73-75] On the other hand melatonin treatment resulted in increase of LHRH content of neurosecretory neurons, as assessed by immunohistochemistry,[76] a finding

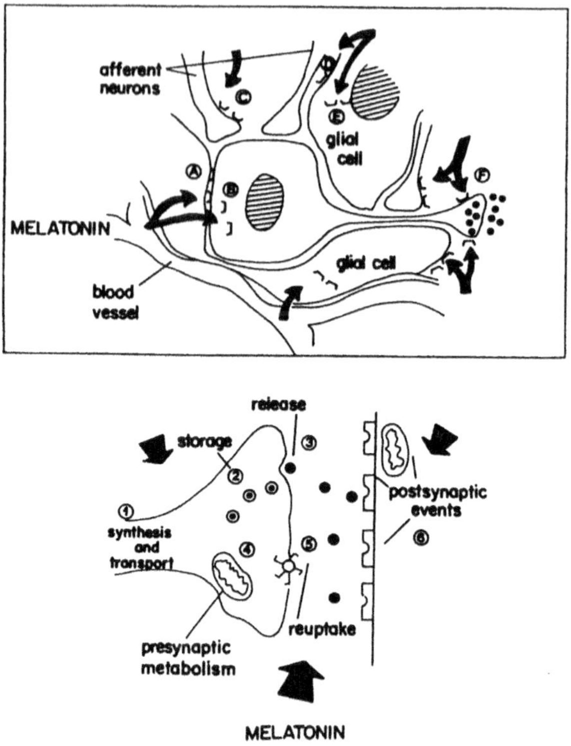

Fig. 2 Possible cellular and subcellular melatonin sites of action in MBH. Melatonin may affect the neurosecretory cell by interacting with membrane (A,F) or cytoplasmic (B) receptors; it may also change the activity of neurons impinging on neurosecretory cells (C), or act on the glial cells (D,E). In a hypothetical target neuron melatonin could affect synthesis and transport of neurotransmitter (1), its storage (2), release (3), presynaptic metabolism (4) or reuptake (5), or any of the postsynaptic events (6).

compatible with inhibition rather than stimulation of LHRH release. Therefore, studies on MBH fragments are beset by the vagaries of the cellular complexity of the preparation used and are difficult to interpret in terms of the cellular component or subcellular compartment affected (Fig. 2).

Another effect of melatonin exhibiting kinetics properties and structural-activity relationships compatible with the interaction with melatonin receptors is the modifications in cyclic nucleotide synthesis. Injections of 1 ug of melatonin in the cisterna magna of rabbits elicited a significant rise in cGMP

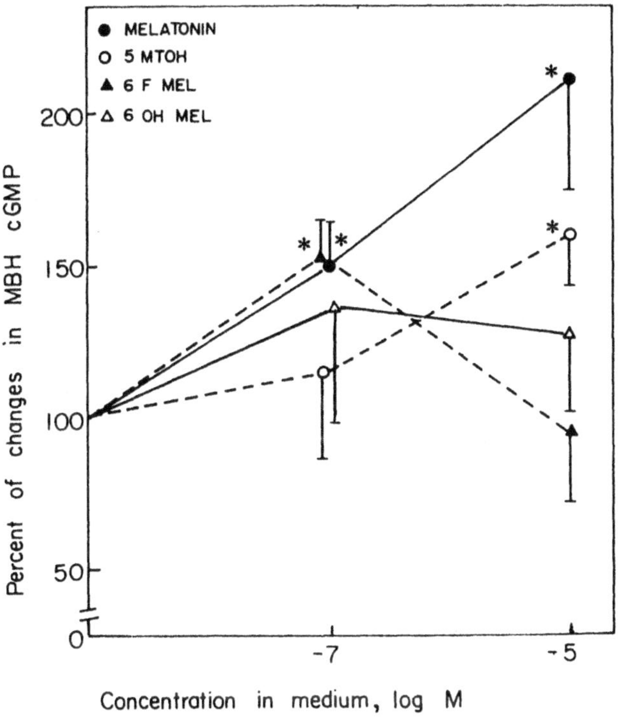

Fig. 3 In vitro effect of melatonin and 5-methoxytryptophol on cGMP accumulation in rat MBH.[79] Explants were incubated for 10 min at 37 C in KRB buffer containing 1 mg/ml glucose and 1 mM theophylline.cGMP concentration was determined by RIA (basal amounts: 0.23 ± 0.02 pmol/mg protein) (*) $p < 0.05$, ANOVA.

levels of the CSF.[88] High concentrations (more than 10^{-4} M) of melatonin also increased cGMP levels in the rat testis[89] and human monocytes.[90] In addition the inhibition by melatonin of tyrosinase activity in hair follicle cultures in vitro is mimicked by cGMP.[91]

In 1981 two reports were published indicating effects of physiological concentrations of melatonin on cGMP synthesis in vitro. Vessely[92] found that melatonin (10^{-8} M) increased by 2 to 3-fold guanylate cyclase activity in rat adenohypophysis, testis, ovary, thyroid and liver. Vacas et al.[79] observed that melatonin (10^{-8} M or higher) increased significantly cGMP in

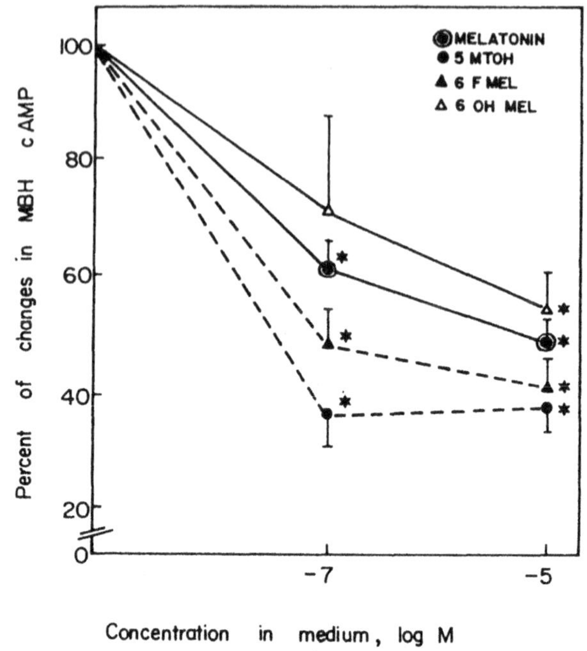

Fig. 4 Effect of melatonin, 5-methoxytryptophol (MTOH), 6-fluoromelatonin (6F-MEL) and 6-hydroxymelatonin (6-HO-MEL)on cAMP levels in rat MBH. Experimental details as in Fig. 3. Basal cAMP concentration: 128 ± 13 pmol/mg prot. (*) $p < 0.05$, ANOVA.

rat MBH (Fig. 3). This effect of melatonin was shared to a less extent by the melatonin analogues 6-fluoromelatonin and 5-methoxytryptophol (Fig. 3) while 6-hydroxymelatonin, the biologically inactive metabolite devoid of activity in receptor studies,[57] did not affect cGMP levels.[79]

Melatonin also decreased MBH cAMP accumulation;[79] in this case melatonin activity was less potent than that of methoxytryptophol or 6-fluoromelatonin, but greater than that of 6-hydroxymelatonin (Fig. 4). As shown in Fig. 5 this effect of melatonin on MBH cAMP synthesis was also observed in the cerebral cortex.

The relationships of cGMP and cAMP changes induced by melatonin in MBH to the mechanism of action of the hormone is not known. Strong evidence exists on the function of cAMP as a second messenger in transducing specific neurotransmitter actions onto target cells.[93] Other studies suggest that cGMP is involved in physiological processes at synapses in both peripheral and central nervous tissue. cGMP has been shown to accumulate in the superior cervical ganglion and different brain areas in response to electrical stimulation, applied acetylcholine or other depolarizing agents;[93] of much interest for the physiological significance of the melatonin-induced reduction in MBH cAMP synthesis is the demonstration that PG-synthesis inhibitors greatly reduced the cAMP accumulation elicited by NE in incubated rat hypothalamic and cerebral cortex slices,[94] The cAMP levels could be restored by addition of very low concentrations of PGE_2; the effect was found for α-adrenoceptors, and suggested a calcium-dependent mediation of PGE_2 in the stimulation of cAMP by α-adrenergic agonists. It is tempting to speculate that many of the central actions of melatonin could be explained in terms of PGE_2 synthesis inhibition, having a physiological role in the CNS as co-regulator of monoaminergic pathways.

Because of its nature as a rather atypical microtubule-active drug[95] melatonin could affect contractile protein-dependent mechanisms in brain. Indeed treatment with the hor-

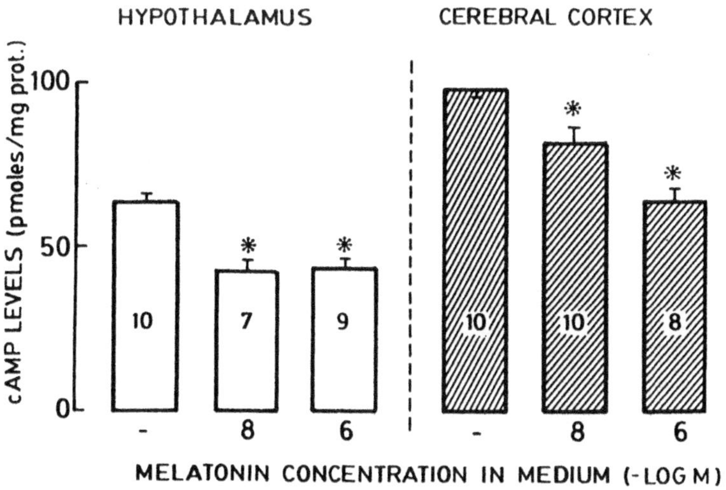

Fig. 5 In vitro effect of melatonin on cAMP accumulation in rat MBH and cerebral cortex. Details of the experiment as in Fig. 3. (*) $p < 0.05$, ANOVA.

mone[67] or induction of endogenous hormone release by injection of β -adrenergic agonist[96] depressed tubulin concentration in MBH. Additionally melatonin locally applied to retinal cells[67] or to sciatic nerve fibers[97] impaired fast axonal transport, a microtubule-dependent process. Peripherally, some effects of melatonin, e.g. inhibition of vasopressin-stimulated water transport[98], interference with antimitotic drug activity,[95] have been also explained in terms of melatonin interaction with tubulin. Further data are needed to establish the general validity of this interaction.

REFERENCES

1. R.J. Wurtman and D.P.Cardinali, The pineal organ, in: "Textbook of Endocrinology", 5th.ed., R.Williams, ed. W.B.Saunders, Philadelphia (1974) p. 832.
2. D.P.Cardinali, Melatonin. A mammalian pineal hormone, Endocrine Rev. 2: 327 (1981).
3. R.J.Reiter, The pineal and its hormones in the control of reproduction in mammals, Endocrine Rev. 1: 109 (1980).
4. H.J.Chen, Melatonin: failure of pharmacological doses to induce testicular atrophy in the male golden hamster, Life Sci. 28: 767 (1981).
5. L.Tamarkin, W.Westrom, A.Hamill and B.Goldman, Effect of melatonin on the reproductive systems of male and female Syrian hamsters: A diurnal rhythm in sensitivity to melatonin, Endocrinology 99: 1534 (1976).
6. E.L.Bittman, Hamster refractoriness: The role of insensitivity of pineal target tissues, Science 202: 648 (1978).
7. D.P.Cardinali, Hormone effects on the pineal gland, in: "The Pineal Gland. Vol. I. Anatomy and Biochemistry", R.J.Reiter, ed., CRC Press, Boca Raton, Fla. (1981) p. 243.
8. S.A.Binkley, Pineal biochemistry: Comparative aspects and circadian rhythms, in: "The Pineal Gland, Vol. I, Anatomy and Biochemistry", R.J.Reiter, ed., CRC Press, Boca Raton, Fla. (1981) p. 155.
9. D.P.Cardinali, M.N.Ritta, M.I.Vacas, P.R.Lowenstein, P.V.Gejman, C.Gonzalez Solveyra and E.Pereyra, Molecular aspects of neuroendocrine integrative processes in the pineal gland, this volume.

10. B.Withachumnarnkul and K.M.Knigge, Melatonin concentration in cerebrospinal fluid, peripheral plasma and plasma of the confluens sinuum of the rat, Neuroendocrinology, 30: 382 (1980).
11. J.Arendt, L.Wetterberg, T.Heyden, P.C.Sizonenko and L. Paunier, Radioimmunoassay of melatonin: human serum and cerebrospinal fluid, Horm.Res., 8: 67 (1977).
12. S.M.Reppert, M.J.Perlow, L.Tamarkin and D.C.Klein, A diurnal melatonin rhythm in primate cerebrospinal fluid, Endocrinology, 104: 295 (1979).
13. M.D.Rollag, R.J.Morgan and G.D.Niswender, Route of melatonin secretion in sheep, Endocrinology 102: 1 (1978).
14. F.P.Gibbs and J.Vriend, The half-life of melatonin elimination from rat plasma, Endocrinology, 109: 1796 (1981).
15. D.J.Kennaway, C.D.Matthews and R.F.Seamark, Pineal function in pregnancy: Studies in sheep and man, in: "Pineal Function", C.D.Matthews and R.F.Seamark, eds. Elsevier, Amsterdam, (1981), p. 123.
16. U.Weinberg, F.J.Gasparini and E.D.Weitzman, The developmental pattern of in vitro rat liver melatonin degrading activity, Endocrinology, 108: 1081 (1981).
17. I.J.Kopin, C.M.B.Pare, J.Axelrood and H.Weissbach, The fate of melatonin in mammals, J.Biol.Chem., 236: 3072 (1961).
18. A.J.Fellenberg, G.Phillipou and R.F.Seamark, Urinary 6-sulphatoxy melatonin excretion and melatonin production rate: Studies in sheep and man, in: "Pineal Function", C.D.Matthews and R.F.Seamark, eds.; Elsevier, Amsterdam (1981), p. 143.
19. M.Tetsuo, M.J.Perlow, M.Mishkin and P.Markey, Light exposure reduces and pinealectomy virtually stops urinary excretion of 6-hydroxymelatonin by Rhesus monkey, Endocrinology, 110: 997 (1982).
20. O.Beck and G.Jonsson, In vivo formation of 5-methoxytryptamine from melatonin in rat, J.Neurochem., 36: 2013 (1981).
21. F.Hirata, O.Hayaishi, T.Fokuyama and S.Senoh, In vitro and in vivo formation of two new metabolites of melatonin, J.Biol.Chem., 249: 1311 (1974).
22. D.P.Cardinali, H.J.Lynch and R.J.Wurtman, Binding of melatonin to human and rat plasma proteins, Endocrinology, 91: 1213 (1972).
23. W.M.Pardridge and L.J.Mietus, Transport of albumin-bound melatonin through the blood-brain barrier, J.Neurochem., 34: 1761 (1980).

24. G.P.Trentini, B.Mess, C.F.de Gaetani and C.Ruzsás, Pineal-brain relationship, Prog.Brain Res., 52: 341 (1979).
25. A.P.Smulders and E.M.Wright, Role of choroid plexus in transport of melatonin between blood and brain, Brain Res. 191: 155 (1980).
26. R.J.Wurtman, J.Axelrod and L.T.Potter, The uptake of H^3-melatonin in endocrine and nervous tissues and the effects of constant light exposure, J.Pharmacol.Exp. Ther., 143: 314 (1964).
27. F.Anton-Tay and R.J.Wurtman, Regional uptake of ^{3}H-melatonin from blood or cerebrospinal fluid by rat brain, Nature,221: 474 (1969).
28. D.P.Cardinali, M.T.Hyyppä and R.J.Wurtman, Fate of intracisternally injected melatonin in the rat brain, Neuroendocrinology, 12: 30 (1973).
29. S.H.Koslow and A.R.Green, Analysis of pineal and brain indolealkylamines by gas chromatography-mass spectrometry, Adv.Biochem.Psychopharmacol., 7: 33 (1973).
30. G.Brown, L.Grota, G.Bubenik, L.Niles and H.Tsui, Physiologic regulation of melatonin, Adv.Biosci., 29: 95 (1981).
31. F.W.Turek and P.Pappas, Inhibition of short-day induced testicular regression in the hamster by daily melatonin injections, Experientia, 36: 1426 (1980).
32. S.H.Loose and F.W.Turek, Melatonin treatment prevents the termination of the gonadal refractory condition normally observed in hamsters exposed to long days, in: "Pineal Function", C.D.Matthews and R.F.Seamark, eds., Elsevier, Amsterdam (1981) p. 67.
33. K.M.Knigge and M.N.Sheridan, Pineal function in hamsters bearing melatonin antibodies, Life Sci., 19: 1235 (1976).
34. F.Fraschini, R.Collu and L.Martini, Mechanisms of inhibitory action of pineal principles on gonadotropin secretion, in: "The Pineal Gland", G.E.W.Wolstenholme and J.Knight, eds., Churchill Livingstone, London, (1971) p, 259.
35. K.Yamashita, M.Mieno, T.Shimizu and E. Yamashita, Inhibition by melatonin of the pituitary response to luteinizing hormone releasing hormone in vivo, J. Endocrinol., 76: 487 (1978).
36. J.D.Glass and G.R.Lynch, Evidence for a brain site of melatonin action in the white-footed mouse, Peromyscus leucopus, Neuroendocrinology, 34: 1 (1982).
37. R.J.Reiter, "The Pineal. Vol. 6", Eden Press, Quebec (1981).

38. B.Rusak, Suprachiasmatic lesions prevent an antigonadal effect of melatonin, Biol.Reprod., 22: 148 (1980).
39. J.E.Martin and C.Sattler, Selectivity of melatonin pituitary inhibition for luteinizing hormone-releasing hormone, Neuroendocrinology, 34: 112 (1982).
40. M.R.Faigón, D.P.Cardinali and J.Moguilevsky, Pinealectomy advances the time of development of steroid positive feedback on luteinizing hormone release in immature female rats, Brain Res, in press.
41. W.B.Quay, "Pineal Chemistry", C.Thomas, Springfield, Ill, (1974).
42. B.Goldman, V.Hall, C.Hollister, P.Roychoudhury, L. Tamarkin and W.Westrom, Effects of melatonin on the reproductive system in intact and pinealectomized male hamsters maintained under various photoperiods, Endocrinology, 104: 82 (1979).
43. D.E.Blask, C.A.Leadem and B.A.Richardson, Nutritional status, time of day and pinealectomy: Factors influencing the sensitivity of the neuroendocrine-reproductive axis of the rat to melatonin, Horm.Res., 14: 104 (1981).
44. L.J.Hipkin, Effect of 5-methoxytryptophol and melatonin on uterine weight responses to human chorionic gonadotropin, J.Endocrinol., 48: 287 (1970).
45. R.Alonso, L.Prieto, C.Hernández and M.Mas, Antiandrogenic effects of pineal gland and melatonin in castrated an intact prepubertal male rats, J.Endocrinol., 79: 77 (1978).
46. G.P.Trentini, A.R.Botticelli, B.C.Sannicola and C.Barbanti Silva, Decreased ovarian LH incorporation after melatonin treatment, Horm.Metab.Res., 8: 234 (1976).
47. L.Debeljuk, V.M.Feder and O.A.Paulucci, Effects of melatonin on changes induced by castration and testosterone in sexual accessory structures of male rats, Endocrinology, 87: 1358 (1970).
48. L.C.Ellis, Inhibition of rat testicular androgen synthesis in vitro by melatonin and serotonin, Endocrinology, 90: 17 (1972).
49. D.P.Cardinali and J.M.Rosner, Effects of melatonin, serotonin and N-acetylserotonin on the production of steroids by duck testicular homogenates, Steroids, 18: 25 (1971).
50. T.F.Ogle and J.I.Kitay, In vitro effects of melatonin and serotonin on adrenal steroidogenesis, Proc.Soc. Exp.Biol.Med., 57: 103 (1978).
51. M.F.Gimeno, M.N.Ritta, A.Bonacossa, M.Lazzari, A.L. Gimeno and D.P.Cardinali, Inhibition by melatonin of prostaglandin synthesis in hypothalamus, uterus and platelets, Adv.Biosci., 29: 147 (1981).

52. M.Cohen, D.Roselle, B.Chabner, T.J.Schmidt and M.Lippman, Evidence for a cytoplasmic melatonin receptor, Nature, 274: 894 (1978).
53. U.Lang and P.C.Sizonenko, Tissue and cellular location of melatonin receptors "Internat.Symposium on Melatonin", Bremen, F.R.G., Sept. 28-30 (1980). abst.143.
54. D.P.Cardinali and J.M.Rosner, Metabolism of serotonin by the rat retina in vitro, J.Neurochem. 18: 1769 (1971).
55. D.P.Cardinali and R.J.Wurtman, Hydroxyindole-O-methyl transferases in rat pineal, retina and Harderian gland, Endocrinology, 91: 247 (1972).
55a. C.D.Matthews and A.Y.S.Leong, A possible role for the pineal gland and melatonin in the diffuse neuroendocrine system (DNES) Adv.Biosci., 29: 77 (1981).
56. D.P.Cardinali, M.I.Vacas and E.E.Boyer, High affinity binding of melatonin in bovine medial basal hypothalamus, IRCS Med.Sci., 6: 357 (1978).
57. D.P.Cardinali, M.I.Vacas and E.E.Boyer, Specific binding of melatonin in bovine brain, Endocrinology, 105: 437 (1979).
58. E.A.Messenger and A.E.Warner, The action of melatonin on single amphibian pigment cells in tissue culture, Br.J.Pharmacol., 61: 607 (1977).
59. M.I.Vacas and D.P.Cardinali, Binding sites for melatonin in bovine pineal gland, Horm.Res., 13: 121 (1980).
60. C.L.Ralph, Melatonin production by extra-pineal tissues, Adv.Biosci., 29: 35 (1981).
61. W.A.Gern, T.A.Gorell and D.W.Owens, Melatonin and pigment cell rhythmicity, Adv.Biosci., 29: 223 (1981).
62. D.P.Cardinali and M.I.Vacas, Molecular endocrinology of melatonin: Receptor sites in brain and peripheral organs, Adv.Biosci., 29: 237 (1981).
63. M.I.Vacas and D.P.Cardinali, Diurnal changes in melatonin binding sites of hamster and rat brains. Correlation with neuroendocrine responsiveness to melatonin, Neurosci.Lett., 15: 259 (1979).
64. L.Tamarkin, N.G.Lefebvre, C.W.Hollister and B.D.Goldman, Effect of melatonin administered during the night on reproductive function in the Syrian hamster, Endocrinology, 101: 631 (1977).
65. L.J.Grota, W.R.Holloway and G.M.Brown, 24-Hour rhythm of hypothalamic immunofluorescence correlates with serum and retinal melatonin rhythms. Neuroendocrinology, 34: 363 (1982).
66. P.J.Marangos, J.Patel, F.Hirata, D.Sondhein, S.M.Paul, P.Skolnick and F.K.Goodwin, Inhibition of diazepam binding by tryptophan derivatives including melatonin and its brain metabolite N-acetyl-5-methoxy Kynurenamine, Life Sci, 29: 259 (1981).

67. D.P.Cardinali and F.Freire, Melatonin effects on brain. Interaction with microtubule protein, inhibition of fast axoplasmic flow and induction of crystaloid and tubular formations in the hypothalamus, Mol.Cell. Endocr., 2: 317 (1975).
68. E.J.Roy and M.A.Wilson, Diurnal rhythm of cytoplasmic estrogen receptors in the rat brain in the absence of circulating estrogens, Science, 213: 1523 (1981).
69. R.L.Urry and L.C.Ellis, Monoamine oxidase activity of the hypothalamus and pituitary alterations after pinealectomy, changes in photoperiod or addition of melatonin in vitro, Experientia, 31: 891 (1975).
70. D.P.Cardinali, C.A.Nagle, F.Freire and J.M.Rosner, Effects of melatonin on neurotransmitter uptake and release by synaptosome-rich homogenates of the rat hypothalamus, Neuroendocrinology, 18: 72 (1975).
71. N.Zisapel and M.Laudon, Dopamine release induced by electrical field stimulation of rat hypothalamus in vitro: inhibition by melatonin, Biochem.Biophys,Res. Comm., 104: 1610 (1982).
72. F.Anton-Tay, Melatonin: Effects on brain function, Adv. Biochem.Psychopharmacol. 11: 315 (1974).
73. L.W.L.Kao and J.Weisz, Release of gonadotrophin-releasing hormone (Gn-RH) from isolated, perfused medial basal hypothalamus by melatonin, Endocrinology, 100: 1723 (1977).
74. S.B.Richardson, C.S.Hollander, J.A.Prasad and T.Hirooka, Somatostatin release from rat hypothalamus in vitro: effects of melatonin and serotonin, Endocrinology, 109: 602 (1981).
75. C.S.Hollander, J.A.Prasad, S.Richardson, Y.Hirooka and S.Suzuki, Melatonin modulates hormonal release from organ cultures of rat hypothalamus, J.Neural Transm, suppl.13: 369 (1978).
76. J.Leonardelli, G.Tramu and E.Hermand, Mélatonine et cellules a gonadolibérine (LH-RH) de l'hypothalamus du rat, C.R.Soc.Biol., 172: 481 (1978).
77. D.P.Cardinali, M.N.Ritta, A.M.Fuentes, M.G.Gimeno and A.L.Gimeno, Prostaglandin E release by rat medial basal hypothalamus in vitro. Inhibition by melatonin at submicromolar concentrations, Eur.J.Pharmacol., 67: 151 (1980).
78. C.M.Leach, J.A.Reynoldson and G.D.Thornburn, Release of E prostaglandins into the cerebrospinal fluid and its inhibition by melatonin after cervical stimulation in the rabbit, Endocrinology, 110: 1320 (1982).
79. M.I.Vacas, M.I.Keller Sarmiento and D.P.Cardinali, Melatonin increases cGMP and decreases cAMP levels in rat medial basal hypothalamus in vitro, Brain Res. 225: 207 (1981).

80. C.Demaine and H.C.Kann, Modification of the electrical activity of hypothalamic neurones by pineal indoles, Prog.Brain Res., 52: 373 (1979).
81. M.F.Gimeno, A.Landa, N.S.Speziale, D.P.Cardinali and A.L.Gimeno, Melatonin blocks in vitro generation of prostaglandin by the uterus and hypothalamus, Eur.J. Pharmacol., 62: 309 (1980).
82. C.M.Leach and G.D.Thornburn, A comparison of the inhibitory effects of melatonin and indomethacin on platelet aggregation and thromboxane release, Prostaglandins, 20: 51 (1980).
83. S.R.Ojeda, A.Negro-Vilar and S.M.McCann, Role of prostaglandins in the control of pituitary hormone secretion in: "Physiopathology of Endocrine Diseases and Mechanisms of Hormone Action", R.J.Soto, A.de Nicola and J.Blaquier, eds., Alan R.Liss Inc., New York (1981).
84. R.L.Eskay, J.Warberg, R.S.Mical and J.C.Porter, Prostaglandin E_2-induced release of LHRH into hypophysial portal blood, Endocrinology, 97: 816 (1975).
85. S.R.Ojeda, A.Negro-Vilar and S.M.McCann, Release of prostaglandin E_S (PGE) by hypothalamic tissue. Evidence for their involvement in catecholamine-induced LHRH release, Endocrinology, 104: 617 (1979).
86. D.P.Cardinali, J.F.Pardal, M.F.Gimeno and A.L.Gimeno, Effects of estradiol on norepinephrine and prostaglandin efflux in medial basal hypothalamus of ovariectomized rats, J.Neural Transm., 53: 39 (1982).
87. L.Levine and L.J.Riceberg, Radioimmunoassay for melatonin, Res.Comm.Chem.Pathol.Pharmacol. 10: 693 (1975)
88. D.Rudman, Injection of melatonin into cisterna magna increases concentration of 3',5'-cyclic guanosine monophosphate in cerebrospinal fluid, Neuroendocrinology, 20: 235 (1976).
89. T.Kano and T.Miyachi, Directaction of melatonin on testosterone and cyclic GMP production using rat testis tissue in vitro, Biochem.Biophys.Res.Comm. 72: 969 (1976).
90. J.A.Sandler, R.I.Clyman, V.C.Manganiello and M.Vaughan, The effect of serotonin (5-hydroxytryptamine) and derivatives on guanosine 3',5'-monophosphate in human monocytes, J.Clin.Invest. 55: 431 (1975).
91. B.Weatherhead and A.Logan, Cyclic nucleotides and the mediation of melatonin-induced inhibition of melanogenesis in mammals, Adv.Biosci., 29: 263 (1981).

92. D.L.Vessely, Melatonin enhances guanylate cyclase activity in a variety of tissues, Mol.Cell.Biochem. 35: 55 (1981).
93. P.Greengard, "Cyclic Nucleotides, Phosphorylated Proteins and Neuronal Function. Distinguished Lecture Series of the Society of General Physiologists. Vol. 1", Raven Press, New York, (1978).
94. C.R.Partington, M.W.Edwards and J.W.Daly, Regulation of cyclic AMP formation in brain tissue by α-adrenergic receptors: Requisite intermediancy of prostaglandins of the E series. Proc.Nat.Acad.Sci.USA, 77: 3024 (1980).
95. D.P.Cardinali, Molecular biology of melatonin: Assessment of the "microtubule hypothesis of melatonin action", Adv.Biosci. 29: 247 (1981).
96. D.P.Cardinali, M.I.Vacas, A.L.Fortis and F.J.Stefano, Superior cervical ganglionectomy depresses norepinephrine uptake, increases the density of α-adrenoceptor sites, and induces supersensitivity to adrenergic drugs in rat medial basal hypothalamus, Neuroendocrinology, 33: 199 (1981).
97. M.R.Prevedello, M.N.Ritta and D.P.Cardinali, Fast axonal transport in rat sciatic nerve, Inhibition by pineal indoles, Neurosci.Lett. 13: 29 (1979).
98. M.S.Haswell, W.A.Gern and C.L.Ralph, Melatonin inhibition of vasopressin-stimulated water transport in toad urinary bladder, J.Exp.Zool., 211: 407 (1980).

SEASONAL REPRODUCTIVE EVENTS RELATED TO THE PINEAL GLAND

Russel J. Reiter

Department of Anatomy
The University of Texas
Health Science Center at San Antonio
San Antonio, TX 78284

INTRODUCTION

Under natural environmental conditions reproduction, and especially birth of the young in mammals, is carefully regulated so that these events coincide with the seasons of the year which will most likely ensure the propagation of the species. The most important physiological event in the annual reproductive cycle is obviously the actual delivery of the young. To optimize the survival of the newborn, animals in their natural habitat characteristically deliver their young in the spring or early summer. It is at these times that environmental temperatures and food availability are such that the young will have the maximal chance of survival. In order for these species to ensure spring delivery of the newborns they must synchronize their breeding habits accordingly. Obviously, if the animals were sexually competent the entire year they would breed indiscriminately with regard to season. Thus, evolution has provided them with a system to restrict their reproductive habits to specific times of the year. In many mammals it is the environmental photoperiod which determines their reproductive capability; when this is the case, the pineal gland serves as the intermediary between the changing daylengths and the neuroendocrine-reproductive system. Dependence on the photoperiodic fluctuations allows animals to not only be in the proper sexual condition at a given time but to actually anticipate the upcoming season and to make the necessary physiological adjustments. In the absence of the pineal gland photosensitive species simply cannot make the necessary alterations and, consequently, they become continual as opposed to seasonal breeders (Reiter, 1973/74). This affords them no advantage under natural conditions and, if it persisted, the species as a whole could become extinct. The data illustrating photoperiodic-pineal-reproductive interactions in rodents are summarized below.

DAYLENGTH MEASUREMENT BY PHOTOSENSITIVE SPECIES

Mammals are generally divided into two broad categories in terms of their breeding habits; thus, there are long day breeders and short day breeders. In the present discussion the long day breeders with short gestation periods primarily will be considered. Many of the rodents fall into this category. Much of what has been learned about these interactions has been derived from studies on the Syrian hamster (*Mesocricetus auratus*). As a result, the following resume will consider this species in some depth.

The photoperiodic response of the Syrian hamster seems to depend on a circadian rhythm of sensitivity to light. When light is present during the insensitive phase of the rhythm the hamster interprets the day as "short". It has been established that when the daylength is less than 12.5 hours a day, the animal measures the day as "short" and the pineal gland suppresses the reproductive system (Elliott, 1976). Conversely, if light is present during the sensitive phase of the circadian rhythm the daylength is interpreted as long and high reproductive activity is maintained. Slight experimental perturbations of the circadian rhythm of sensitivity to light may prevent gonadal regression in animals kept under what would normally be interpreted as a long day (Eskes and Zucker, 1978).

A resonance paradigm has been used to examine the photosensitivity rhythm of the Syrian hamster. The paradigm uses a fixed duration of light with different lengths of the dark period to generate a cycle of 24 hours or more. In the experiments in question, the authors (Elliott et al., 1972; Stetson et al., 1975) used 6 hours of light followed by one of several different lengths of darkness. Hamsters exposed to light:dark (LD) cycles of either 6:18 or 6:42 experienced gonadal involution. On the other hand, when the animals were maintained under either LD 6:30 or LD 6:54 they interpreted these as long photoperiods and active gametogenic activity was maintained. Besides being capable of determining the reproductive state of the animal, light also entrains the circadian locomotor system of this species. When light is present during the latter half of the hamster's 24 hour circadian locomotor rhythm, the day is interpreted as long and the gonads do not regress.

This type of mechanism for maintaining high reproductive function in the hamster may have adaptive value for the animal in its natural habitat. Even a brief exposure to light during the photosensitive phase of the animal is most likely to occur during the long solar days (spring and summer); this likely assists in adjusting the animals reproductive state to the appropriate seasons. Judging from the studies of Elliott et al (1972) and Stetson and co-workers (1975), light exposure during the photosensitive phase need not be a daily occurrence.

In general, what the data summarized in the above paragraphs have been interpreted to mean is that hamsters must be exposed to 12.5 or more hours of light daily in order to maintain their reproductive systems in an active state. This is not actually the case; what they need is either at least one 12.5 hour light period or two (or more) shorter intervals of light during a 12.5 hour period. Interestingly, 15 minute periods of light at 6 hour intervals throughout a 24 hour period not only prevent gonadal atrophy in hamsters but also curtail the daily rise in N-acetyltransferase activity (NAT), the alleged rate limiting enzyme in melatonin production, in the hamster pineal gland (Rudeen and Reiter, 1979; 1980). Similar observations in reference to the state of the reproductive organs were made by Hoffman and Melvin (1974). The implication is that melatonin is the pineal hormone responsible for gonadal regression in short day exposed hamsters. Although there is general acceptance of this conclusion (Reiter, 1980a), there are some who doubt it (Brown et al., 1981).

REPRODUCTIVE CHANGES INDUCED BY PHOTOPERIOD AND PINEAL

In the laboratory setting with controlled photoperiods and temperature, the maintenance of hamsters in less than 12.5 hours of light daily causes reproductive collapse in both males and females of the species. These changes are clearly a manifestation of pineal hormonal activity since pinealectomy prevents darkness-induced gonadal involution in both sexes (Hoffman and Reiter, 1965; 1966). In males kept under short day conditions spermatogenesis is arrested (Gaston and Meanker, 1967; Reiter, 1967) while females under similar conditions do not ovulate; additionally, their ovaries exhibit a lack of antral follicles and a marked proliferation of the interstitial tissue (Reiter, 1968).

In male hamsters, the gonadally-related hormones seem to change in a predictable manner during darkness-induced testicular atrophy (Fig. 1). During regression, there is usually a gradual reduction in circulating levels of luteinizing hormone (LH), follicle stimulating hormone (FSH), and prolactin (Reiter, 1973; Berndtson and Desjardins, 1974; Turek et al., 1975; 1976; Tamarkin et al., 1976). However, these changes are not as uniform as might be predicted on the basis of the markedly

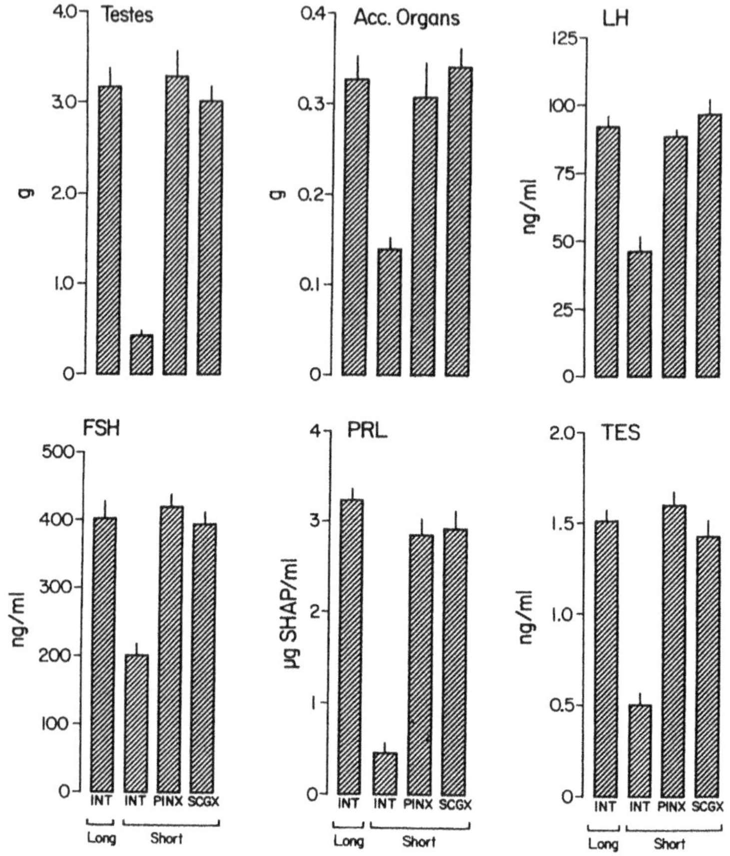

Fig. 1. Testicular and accessory sex organ weights (combined weight of seminal vesicles and coagulating glands) and plasma levels of luteinizing hormone (LH), follicle stimulating hormone (FSH), prolactin (PRL) and testosterone (TES) in male Syrian hamsters exposed to either long or short days. INT = intact; PINX = pinealectomy; SCGX = superior cervical ganglionectomy.

atrophic reproductive system (Reiter, 1980b). More consistently, the concentrations of gonadotrophins within the pituitary decrease in male hamsters maintained under reduced photoperiodic conditions (Reiter and Johnson, 1974a). The most consistent and dramatic changes in male hamsters may be those of prolactin. Both serum and pituitary levels of this constituent decrease dramatically as a result of pineal-induced testicular involution

(Reiter, 1980b). Treatments which restore circulating prolactin levels also markedly stimulate spermatogenic and endocrine activity of the testes in hamsters maintained under short day conditions (Bex et al., 1978; Matthews et al., 1978; Chen and Reiter, 1980). These observations have led to the tentative conclusion that prolactin is a key hormone in maintaining the normal spermatogenic and endocrine activity of the testes.

The hormonal changes which occur in females experiencing pineal-induced gonadal atorphy are markedly different from those that occur in males under the same conditions. Rather than being depressed, pituitary levels of LH and FSH actually increase (Reiter and Johnson, 1974b) while there are daily afternoon surges of each of these constituents in the blood (Goldman and Brown, 1979). Normally, the gonadotrophin surges occur on the afternoon of every fourth day. The net effect of the change in hormonal patterns in females in an acutal increase in circulating titers of LH and FSH. Despite this, the ovaries and uteri of these animals functionally involute although the size of the ovaries may actually increase due to the proliferation of the interstitial elements of the organ (Reiter, 1968). The significance of the daily afternoon rises in LH and FSH in the presence of the depressed sexual function remains unknown.

In both males and females of this species, surgical removal of the pineal gland prevents darkness from inducing both the morphological involution and the physiological inactivation of the neuroendocrine-gonadal axis. Indeed, all of the inhibitory effects of darkness on the reproductive system of photosensitive species seem to be overcome by either the surgical ablation of the gland or by its sympathetic denervation (Reiter and Sorrentino, 1970; Reiter 1980b).

PHOTOPERIOD, PINEAL AND SEASONAL REPRODUCTION

Animals maintained in their natural habitat are exposed to seasonal fluctuations in daylength. The degree to which the actual daily photoperiod changes throughout the year depends on the latitude at which the measurements are made. When Syrian hamsters are maintained under natural environmental conditions at about 30°N latitude the longest days in the summer approach 14 hours while the shortest days in the winter are of roughly 10 hours duration. Since, as noted above, hamsters require about 12.5 hours of light daily to maintain an active reproductive system, their exposure to the natural seasonal functions described above also are associated with appropriate adjustments in the reproductive system (Reiter, 1973) (Fig. 2). For descriptive purposes, the annual cycle of reproductive capability has been divided into the following four phases: the inhibition

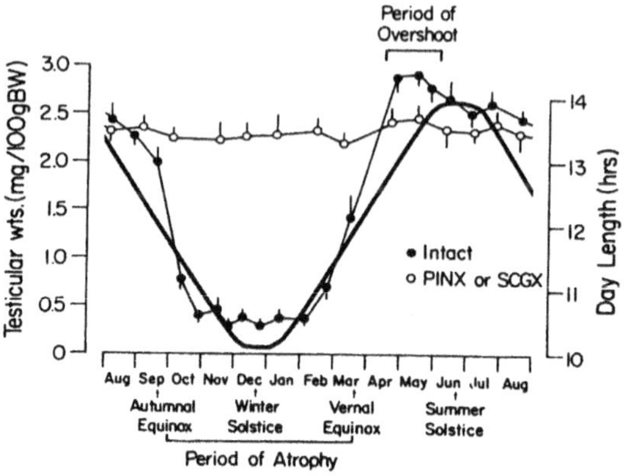

Fig. 2. Seasonal changes in daylength and testicular weights in male hamsters kept under natural photoperiodic conditions at about 30° N latitude. The seasonal changes in reproduction are negated if the animals are either pinealectomized (PINX) or superior cervical ganglionectomized (SCGX).

phase, the sexually quiescent phase, the restoration phase, and the sexually active phase (Reiter, 1975a) (Fig. 3).

Inhibition phase

In the fall of the year daylengths eventually fall below the 12.5 hours of light daily required to maintain the gonads in a highly active state. Additionally, inasmuch as Syrian hamsters are believed to be hibernators they presumably spend progressively more time in lightless underground burrows in preparation for hibernation. At any rate, the increased dark exposure stimulates the pineal gland and reproductive involution ensues. The end organ responses as well as the hormonal changes which occur under naturally decreasing daylengths are the same as those described above which follow the exposure of hamsters to artifically short

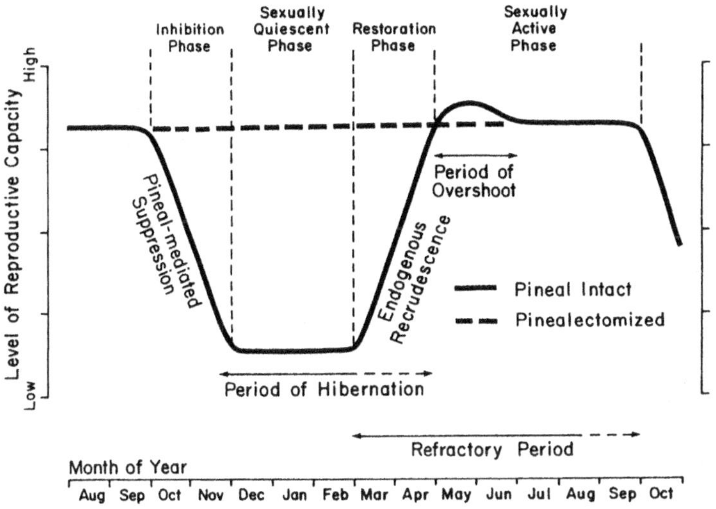

Fig. 3. Various stages of the annual cycle of reproduction. Gonadal atrophy coincides with the period of hibernation.

days of the laboratory (Reiter, 1974; Stetson and Tate-Ostroff, 1981). The generalized reduction in sexual physiology during the inhibition phase of the cycle does not occur in the absence of the pineal gland even though the days may be short and the temperatures may be cold (Reiter, 1975b; 1980b). The specific mechanisms involved in pineal-induced atrophy of the reproductive system under these conditions have not been elucidated; it has, however, been proposed that the antigonadotrophic hormone of the pineal gland, possibly melatonin, increases the sensitivity of the hypothalamo-pituitary unit to the inhibitory feedback effects of gonadal steroids on LH and FSH (Turek and Campbell, 1979). The results on which this conclusion was based, however, were obtained from castrated animals kept under short days in the laboratory. Hence, whether a change in feedback sensitivity helps to explain reproductive regression in intact hamsters in the field remains to be demonstrated. Also, the changing feedback sensitivity theory does not take into account the role of prolactin reduction in causing testicular involution in Syrian hamsters.

The period of time required for the gonads of hamsters to undergo maximal involution during the fall of the year is believed to be on the order of 8 weeks. Hence, the inhibition phase presumably monopolizes about a 2 month period in the case of the hamster; the duration of this phase undoubtedly varies in other species. By mid to late autumn the gonads are atrophic and the animals presumably are physiologically prepared for hibernation (Reiter, 1981).

Sexually quiescent phase

Once the gonads are totally involuted the animals enter a prolonged period during which the testes of the males and the ovaries of the females are minimally active. This is the sexually quiescent phase and it coincides with the short, cold days of the winter. During this phase the animals are obviously incapable of sexual reproduction; this prevents the birth of young at a time when they would be exposed to inclement weather conditions and reduced food availability. It probably allows for successful hibernation since high levels of gonadal steroids seem to impede hibernatory processes (Hall and Goldman, 1980).

During the sexually quiescent phase the pineal gland actively suppresses the growth and function of the neuroendocrine-reproductive axis. Thus, if hamsters with infantile reproductive organs are either pinealectomized or moved to the laboratory and exposed to long days the gonads initiate regeneration (Reiter, 1969); indeed, they become sexually competent within a relatively short time. During the period of sexual quiesence infantile reproductive organs are ensured by virtue of the fact that the animals remain in virtually the lightless environment of their subterranean burrows. Even if the animals were to emerge above ground during this period, the daylengths would be less than the minimum 12.5 hours daily required to maintain the sexual organs in a highly active state.

At some point after the winter solstice the gonads begin regenerating; by definition, this terminates the sexually quiescent phase. Apparently the neuroendocrine-gonadal axis, by mechanisms which are yet to be elucidated, becomes refractory to the inhibitory influence of the pineal gland and gonadal restoration ensues (Reiter, 1975b).

Restoration phase

In the case of Syrian hamsters, the restoration phase, i.e., the regrowth of the reproductive organs, is accomplished while the animals are still in the total darkness of their burrows (Reiter, 1975c). Throughout this phase of the cycle the pineal gland continues to synthesize and presumably secrete melatonin (Rollag et al., 1980; Brainard et al., 1983). Assuming melatonin is the pineal antigonadotrophic agent it is certainly not a cessation of melatonin production which results in the initiation of the restoration phase. Rather the recrudescence is most likely due to the fact that the reactive site no longer responds to the inhibitory signal produced by the pineal, i.e., the system becomes refractory. It was initially proposed that it was the increasing daylengths in the late winter and spring which caused full recrudescence of the reproductive organs (Turek et al., 1975). However, increasing daylengths is insufficient to cause the regrowth (Reiter, 1980c); furthermore, the reconstitution of a functional reproductive system is not impeded even if hamsters are totally deprived to light by surgical removal of the pineal gland (Reiter, 1975c). Thus, Reiter (1975c) feels that the recrudescence is a consequence of an endogenous phenomenon that is independent of the photoperiod. Since, as noted above, gonadal steroids impede hibernation, the restoration of functional reproductive organs, in addition to ensuring the production of mature gametes, also serves to terminate hibernation. Once the gonads reach maturity the restoration phase of the cycle is complete and the animals emerge above ground, ready to mate (Reiter, 1980b).

Sexually active phase

The restoration of reproductive maturity initiates the sexually active phase of the annual cycle. Shortly after the gonads recrudesce the animals mate and, because of the short gestation period of many of the animals in question, the young are born soon thereafter. Hence, the young are born in the spring , the time that their chances of survival are the greatest.

The refractoriness which is initiated late in the sexually quiescent phase and which ensures endogenous gonadal recrudescence extends into the sexually active phase (Fig. 3). Hence, if sexually mature male or female hamsters are deprived of light shortly after their gonads have regenerated, the sexual organs do no involute (Reiter, 1975b; 1975c). Because of the continuance of the refractory condition into the sexually active phase, long days are really not required at this time of the year to maintain the high functional state of the reproductive organs. Long days of the summer, however, are important in that they interrupt the refractory condition and, thereby, render the hypothalamo-pituitary-gonadal axis sensitive to the inhibitory of short days during the subsequent fall (Reiter, 1975b; Stetson et al., 1976). Indeed, the animals must be exposed, during the summer months, to 12 or more weeks of long days before the refractory condition is interrupted (Fig. 4).

Throughout the sexually active phase the females probably experience several pregnancies. As fall approaches, however, the reproductive system is shut down and the annual cycle is repeated. Obviously, both light and darkness play major roles in determining the annual fluctuation in reproduction in photoperiodic species. The changes are mediated by the pineal gland

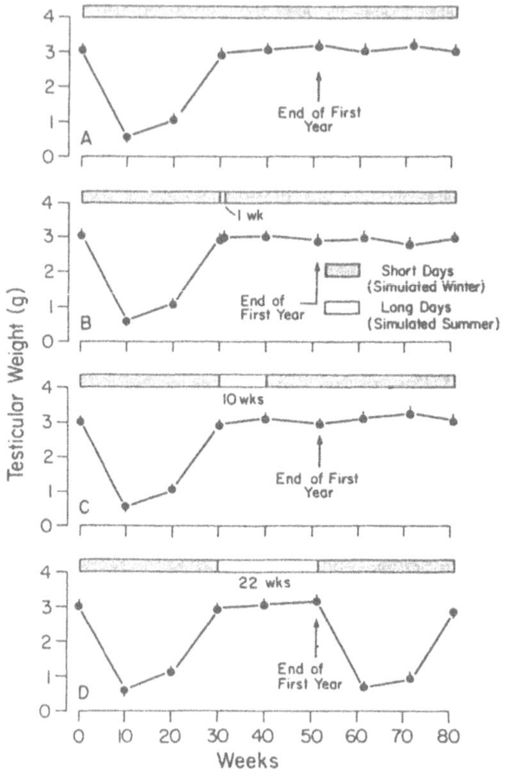

Fig. 4. Refractoriness of the reproductive systems of Syrian hamsters to short days is interrupted if the animals are exposed to long day conditions for 12 or more weeks. Short periods of long day exposure, e.g., 1 or 10 weeks (middle two panels), does not interrupt the refractory period to darkness.

and, as a result, the removal of this organ completely eliminates seasonal variations in sexual competence (Reiter, 1980b). In effect, pinealectomy allows hamsters to be continual as opposed to seasonal breeders even when they are exposed to natural environmental conditions.

CONCLUDING REMARKS

For the survival of the species, seasonal fluctuations in reproductive competence are absolutely essential for animals subjected to the vissitudes of nature. It is essential that the young be born at the time of the year which is most compatible with their survival; this is the spring. To ensure spring delivery of the young, hamsters as well as many other species have evolved systems for limiting their reproductive capability to very restricted portions of the year. To do this the photosensitive species rely on the regularly changing daylengths to signal their pineal gland which, in turn, has the capability of inducing regression of hypothalamo-pituitary-gonadal axis. Thus, the pineal is an absolutely necessary intermediary between the environment and the reproductive system. Without it, hamsters do not respond to the prevailing photoperiodic conditions and, as a result, they are continual, as opposed to seasonal, breeders.

REFERENCES

Berndtson, W. W., and Desjardins, C., 1974, Circulating LH and FSH levels and testicular function in hamsters during light deprivation and subsequent photoperiod stimulation, Endocrinology, 95: 195.

Bex, F., Bartke, A., Goldman, B. D., and Dalterio, S., 1978, Prolactin, growth hormone, luteinizing hormone receptors, and seasonal changes in testicular activity in the golden hamster, Endocrinology, 103: 2069.

Brainard, G. C., Petterborg, L. J., Richardson, B. A., and Reiter, R. J., 1983, Pineal melatonin in Syrian hamsters: Circadian and seasonal rhythms in animals maintained under laboratory and natural conditions, Neuroendocrinology, in press.

Chen, H. J., and Reiter, R. J., 1980, The combinaton of twice daily luteinizing homone-releasing factor administration and renal pituitary homografts restores normal reproductive organ size in male hamsters with pineal-mediated gonadal atrophy, Endocrinology, 106: 1382.

Brown, G. M., Tsui, H. W., Niles, L. P., and Grota, L. J., 1981, Gonadal effects of the pineal gland, in: "Pineal Function", C. D. Matthews and R. F. Seamark, ed., Elsevier/North Holland, Amsterdam.

Elliott, J. A., Stetson, M. H., and Menaker, M., 1972, Regulation of testis function in golden hamsters. A circadian clock measures photoperiodic time, Science 178: 771.

Elliott, J., 1976, Circadian rhythms and photoperiodic time measurement in mammals, Fed. Proc. Amer. Soc. Exp. Biol., 35: 2339.

Eskes, G. A., and Zucker, I., 1978, Photoperiod regulation of the hamster testis: Dependence on circadian rhythms. Proc. Nat. Acad. Sci., 75: 1034.

Gaston, S., and Menaker, M., 1967, Photoperiodic control of hamster testis, Science, 158: 925.

Goldman, B., and Brown, S., 1979, Sex differences in serum LH and FSH patterns in hamsters exposed to short photoperiod. J. Steroid Biochem., 11: 531.

Hall, V., and Goldman, B. D., 1980, Effects of gonadal steroid hormones on hibernation in the Turkish hamster (*Mesocricetus brandti*), J. Comp. Physiol., 135: 107.

Hoffman, R. A., and Melvin, H., 1974, Gonadal responses of hamsters to interrupted dark periods, Biol. Reprod., 10: 19.

Hoffman, R. A., and Reiter, R. J., 1965, Pineal gland: Influence on gonads of male hamsters, Science, 148: 1609.

Hoffman, R. A., and Reiter, R. J., 1966, Responses of some endocrine organs of female hamsters to pinealectomy and light, Life Sci., 5: 1147.

Matthews, M. J., Benson, B., and Richardson, D. L., 1978, Partial maintenance of testes and accessory organs in blinded hamsters by homoplastic anterior pituitary grafts or exogenous prolactin, Life Sci., 23: 1131.

Reiter, R. J., 1967, The effect of pineal grafts, pinealectomy and denervation of the pineal on the reproductive organs of the male hamster, Neuroendocrinology, 2: 138.

Reiter, R. J., 1968, Changes in the reproductive organs of cold-exposed and light-deprived female hamsters (*Mesocricetus auratus*), J. Reprod. Fertil., 16: 217.

Reiter, R. J., 1969, Pineal-gonadal relationships in male rodents, in: "Progress in Endocrinology", C. Gual, ed., Excerpta Medica, Amsterdam.

Reiter, R. J., 1973, Pineal control of a seasonal reproductive rhythm in male golden hamsters exposed to natural daylight and temperature, Endocrinology, 92: 423.

Reiter, R. J., 1973/74, Influence of pinealectomy on the breeding capacity of hamsters maintained under natural photoperiodic and temperature conditions. Neuroendocrinology, 13: 366.

Reiter, R. J., 1974, Circannual reproductive rhythms related to photoperiod and pineal function: A review, Chronobiologia, 1: 365.

Reiter, R. J., 1975a, The pineal gland and seasonal reproductive adjustments, Int. J. Biometeorol., 19: 282.

Reiter, R. J., 1975b, Evidence for refractoriness of the pituitary-gonadal axis to the pineal gland in golden hamsters and its possible implications in annual reproductive rhythms, Anat. Rec., 173: 365.

Reiter, R. J., 1975c, Exogenous and endogenous control of the annual reproductive cycle in the male golden hamster: Participation of the pineal gland, J. Exp. Zool., 191: 111.

Reiter, R. J., 1980a, The pineal gland: A regulator of regulators, Progr. Psychobiol. Physiol. Psychol., 9: 323.
Reiter, R. J., 1980b, The pineal and its hormones in the control of reproduction in mammals, Endocr. Rev., 1: 109.
Reiter, R. J., 1980c, Reproductive involution in male hamsters exposed to naturally increasing daylenghts after the winter solstice, Proc. Soc. Exp. Biol. Med., 163: 264.
Reiter, R.J., 1981, Seasonal aspects of reproduction in a hibernating rodent: Photoperiodic and pineal effects, in, "Survival in Cold", X. J. Musacchia and L. Jansky, eds., Elsevier/North Holland, Amsterdam.
Reiter, R. J., and Johnson, L. Y., 1974a, Depressant action of the pineal gland on pituitary luteinizing hormone and prolactin in male hamsters, Horm. Res., 5: 311.
Reiter, R. J., and Johnson, L. Y., 1974b, Elevated pituitary LH and depressed pituitary prolactin levels in female hamsters with pineal-induced gonadal atrophy and the effects of chronic treatment with synthetic LRF, Neuroendocrinology, 14: 310.
Reiter, R. J., and Sorrentino, S., 1970, Reproductive effects of the mammalian pineal, Amer. Zool., 10: 247.
Rollag, M. D., Panke, E. S., and Reiter, R. J., 1980, Pineal melatonin content in male hamsters throughout the seasonal reproductive cycle, Proc. Soc. Exp. Biol. Med., 165: 330.
Rudeen, P. K., and Reiter, R. J., 1979, Pineal N-acetyltransferase activity in hamsters maintained in shortened light cycles, J. Endocr. Invest., 2: 19.
Rudeen, P. K., and Reiter, R. J., 1980, Influence of a skeleton photoperiod on reproductive organ atrophy in the male golden hamster, J. Reprod. Fertil., 60: 279.
Stetson, M. H., Tate-Ostroff B., 1981, Hormonal regulation of the annual reproductive cycle of golden hamsters, Gen. Comp. Endocr., 5: 329.
Stetson, M. H., Elliott, J. A., and Menaker, M., 1975, Photoperiodic regulation of hamster testis: Circadian sensitivity to the effects of light. Biol. Reprod., 13: 329.
Stetson, M. H., Matt, K. S., and Watson-Whitmyre, M., 1976, Photoperiodism and reproduction in golden hamsters: Circadian organization and the termination of photorefractoriness, Biol. Reprod., 14: 531.
Tamarkin, L., Hutchinson, J. S., and Goldman, B. D., 1976, Regulation of serum gonadotropins by photoperiod and testicular hormone in the Syrian hamster, Endocrinology, 99: 1528.
Turek, F. W., and Campbell, C. S., 1979, Photoeriodic regulation of neuroendocrine-gonadal activity, Biol. Reprod., 20: 32.
Turek, F. W., Alvis, J. D., Elliott, J. A., and Menaker, M., 1976, Temporal distribution of serum levels of LH and FSH in

adult male golden hamsters exposed to long or short photoperiods, Biol. Reprod, 14: 630.
Turek, F. W., Elliott, J. A., Alvis, J. D., and Menaker, M., 1975, Effect of prolonged exposure to nonstimulatory photoperiods on the activity of the neuroendorine-testicular axis of golden hamsters, Biol. Reprod., 13: 475.

MELATONIN AS THE HORMONE WHICH MEDIATES THE EFFECTS OF THE PINEAL GLAND ON NEUROENDOCRINE-REPRODUCTIVE AXIS OF THE SYRIAN HAMSTER

Russel J. Reiter

Department of Anatomy
The University of Texas
Health Science Center at San Antonio
San Antonio, TX 78284

INTRODUCTION

Melatonin is the most widely investigated pineal compound. Since its discovery in 1958 by Lerner and colleagues, it has been the subject of numerous biochemical and physiological investigations. Its synthesis within the pineal gland has been extensively studied and a great deal is known of the factors which control its production (Smith, 1981). Although its effects on the endocrine system have been investigated in many different experimental paradigms (Cardinali, 1981), its sites of action and mechanisms remain obscure. Despite its prominence in the scientific literature, its specific role as a hormonal envoy of the pineal gland continues to be questioned (Brown et al., 1981a; Pevet et al., 1981a). If not melatonin, the hormonal product of the pineal could prove to be another indole (Reiter and Vaughan, 1977), a polypeptide (Benson, 1977; Benson et al., 1981), or another compound (Ebels, 1979). The following paragraphs summarize what is known of synthesis and physilogical actions of melatonin. The actions of the indole are extremely complex and, at times, appear to be contradictory. Many of these apparent contradictions can be explained by what has come to be known as the down-regulation hypothesis. Since many of the studies have dealt with the actions of melatonin on reproduction in the Syrian hamster, these are data that will be considered in greatest depth herein.

MELATONIN PRODUCTION IN THE SYRIAN HAMSTER PINEAL

Inasmuch as the activity of serotonin N-acetyltransferase (NAT) activity was considered to rate limit melatonin production (Klein and Weller, 1970), studies were initially directed at this enzyme. From these studies it rapidly became apparent that if there was a relationship between the activity of the acetylation of serotonin and the actual melatonin produced in the hamster pineal, then melatonin synthesis was clearly higher at night than during the day (Rudeen et al., 1975; Rudeen and Reiter, 1977).

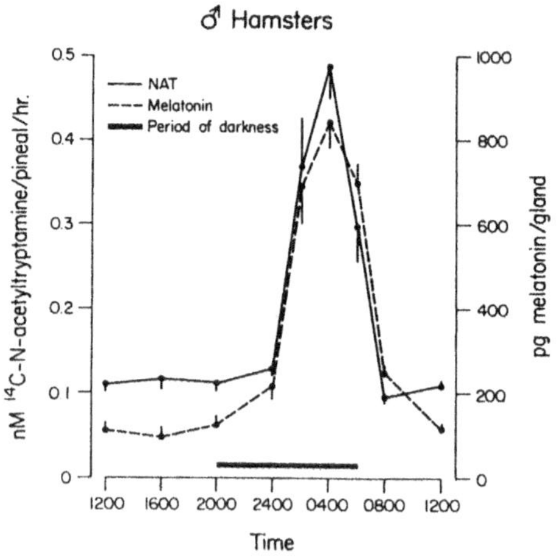

Fig. 1. Pineal N-acetyltransferase (NAT) activity and melatonin content in the pineal of the Syrian hamster throughout a 24 hour period. Characteristically, there is a strong correlation between the activity of the acetylating enzyme and melatonin levels.

Soon thereafter, when a sensitive radioimmunoassay for melatonin became available, a nocturnal rise in pineal melatonin in the Syrian hamster was demonstrated (Panke et al., 1978; 1979; Tamarkin et al., 1979) (Fig. 1). Subsequently, serum levels of melatonin were also reported to be higher at night than during the day in the hamster (Brown et al., 1981b; Reiter, 1982). Of some interest is that photoperiods which are known to suppress the reproductive physiology of hamsters do not change dramatically the apparent production of the compound within the pineal gland (Reiter, 1982).

Besides serotonin being metabolized to melatonin in the pineal, it is also converted to a number of other indoles (Fig. 2). Some of these may have antigonadotrophic actions (Reiter and Vaughan, 1977). Recently, 5-methoxytryptamine has received some interest in this regard (Pevet et al., 1981b).

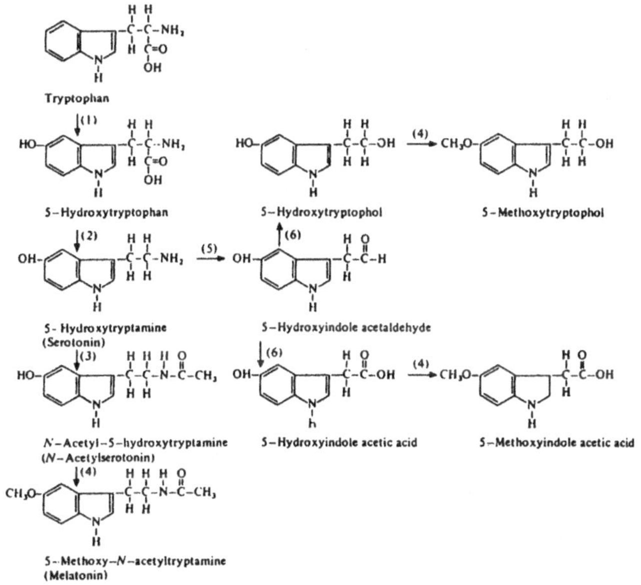

Fig. 2. A summary of tryptophan metabolism in the pineal gland. The enzymes involved include (1) tryptophan hydroxylase, (2), aromatic L-amino acid decarboxylase, (3), N-acetyltransferase, (4), hydroxyindole-O-methyl-transferase, (5), monoamine oxidase. 5-hydroxyindole acetaldehyde (6) is an unstable intermediate and is either oxidized to 5-HIAA or reduced to 5-hydroxy-tryptophol.

ANTIGONADOTROPHIC ACTIONS OF MELATONIN

Although the inhibitory effect of the pineal gland on the reproductive system of the hamster was demonstrated in the mid 1960s (Czyba et al., 2964; Hoffman and Reiter, 1965; 1966), early attempts to illustrate the antigonadotrophic actions of melatonin in this species failed (Reiter, 1974). In these early studies melatonin was administered in a variety of ways (injected daily or subcutaneously implanted); however, the proper experimental protocol for uncovering its gonad-inhibiting action escaped detection. Finally, in 1976 Tamarkin and colleagues finally identified a treatment schedule which inhibited the reproductive system of hamsters to the same degree as did exposure to short days. A chief requirement for the antigaondotrophic action of melatonin in hamsters is its time of administration. In line with this, Tamarkin et al (1976) kept hamsters under light:dark (LD) cycles of 14:10 (lights on daily at 0600 hours) and injected them with 25 µg melatonin at either 1000 hours or 1600 hours daily. Whereas the morning injections, as had been reported previously (Reiter, 1974), completely failed to alter the sexual status of the animals, the same dosage of melatonin given in the afternoon totally suppressed the activity of the neuroendocrine-reproductive axis. Thus, hamsters obviously have a period when they are insensitive to melatonin and a period when they respond to the indole. The changing sensitivity of the system to melatonin may be regulated by melatonin itself (Reiter, et al., 1978; Reiter, 1980a). This aspect of melatonin's action will be discussed later in the present report.

The observations of Tamarkin and co-workers (1976) were readily confirmed in an independent laboratory in which single daily injections of melatonin were given to intact, pinealectomized or pineal denervated hamsters at 1900 hours daily (Reiter et al., 1976). The purpose of this study was to determine whether the ability of afternoon melatonin injections to inhibt reproduction depended upon an intact pineal gland. Indeed, melatonin could inhibit the functioning of the neuro-endocrine-reproductive system only in those hamsters in which the pineal was intact and sympathetically innervated. This implied that the exogenously administered melatonin was either acting directly on the pineal gland to release some other gonad-inhibiting factor or that melatonin administered in the afternoon was synergizing with endogenously produced melatonin (or some other pineal compound) released at night. Judging from more recent investigations, it appears that any procedure which interferes with the nighttime rise in melatonin also curtails the ability of nighttime melatonin injections to suppress the reproductive system (Reiter et al., 1982). The implication of these results is that exogenously administered melatonin probably acts in concert with endogenously produced melatonin to induce regression of the reproductive system. Apparently in this scheme, the two pulses of melatonin must be separated by at least several hours. The need for spacing the two injections is also supported by the findings that melatonin can induce reproductive involution in pinealectomized hamsters provided it is given as widely-spaced daily injections (Tamarkin et al., 1977).

Melatonin, administered by the means described above, is equally as effective as darkness in inhibiting reproduction in the Syrian hamster (Reiter, 1980b) (Fig. 3); likewise, where they have been measured the hormonal changes induced by repeated melatonin injections are similar in timing and magnitude to those caused by reduced photoperiods (Reiter, 1981). This is strong evidence for the conclusion that melatonin is the factor wich mediates the inhibitory effect of the pineal gland on reproduction in this species. Finally, the quantity of melatonin required to inhibit reproduction has been established (Rollag et al., 1980b). It appears that hamsters must receive on the order of 1.6 μg melatonin daily as a subcutaneous injection in order for reproductive collapse to ensue. This is somewhat more than is produced in the hamster pineal gland on a daily basis (Rollag et al., 1980b).

COUNTER ANTIGONADOTROPHIC ACTIONS OF MELATONIN

Not in all cases is melatonin inhibitory to the neuroendocrine-reproductive axis of Syrian hamsters; indeed, in some cases the indole actually *prevents* the suppressive influence of darkness on the reproductive system. When these observations were made they raise serious doubts about the normal role of melatonin in mediating pineal induced gonadal involution. The observations were made in 2 species, the Syrian hamster and the Djungarian hamster (*Phodopus sungorus*), within the same year. In both species melatonin reservoirs, placed subcutaneously, negated the inhibitory effect of short day exposure on reproduction (Hoffmann, 1974; Reiter et al., 1974). Although the minimal effective daily dose of melatonin required to prevent the inhibitory action of short days was never definitively established, it appears that less than 3.6 μg released from a subcutaneous deposit easily negates the suppressive action of darkness and the pineal gland on the neuroendocrine-reproductive axis of hamsters (Reiter et al., 1975).

To this point we had assumed that darkness mediated reproductive collapse in the Syrian hamster was a consequence of melatonin synthesis and release from the pineal gland. However, the results summarized above revealed just the opposite, namely, that melatonin, continuously available, prevented the pineal gland from inducing regression of the testes or ovaries of hamsters exposed to short days (Fig. 3). These results seem to mean that either melatonin was not the pineal antigonadotrophic factor or that a continual uniform level of melatonin, as opposed to a daily nocturnal rise, somehow interfered with the normal mechanisms governing darkness induced reproductive collapse. Also, the interesting possibility was raised that subcutaneous melatonin deposits may even prevent daily afternoon injections of

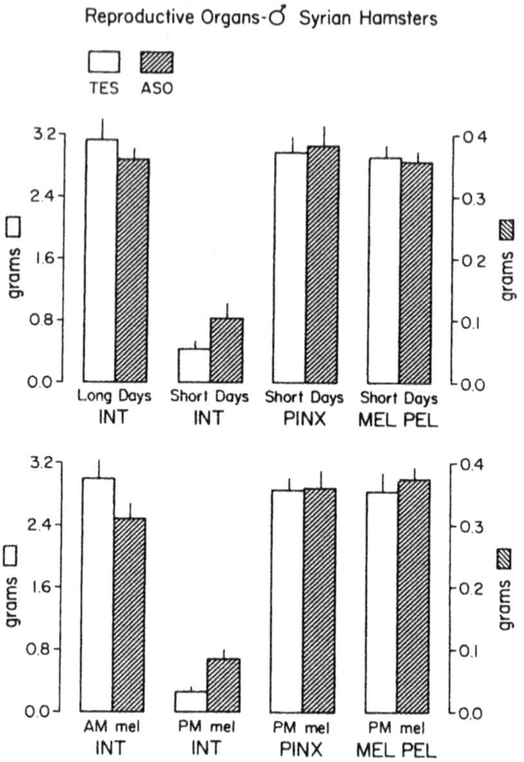

Fig. 3. Testes (TES) and accessory sex organ (ASO) weights in male hamsters kept in either short or long days or given morning (AM mel) or afternoon (PM mel) melatonin injections. Both pinealectomy (PINX) or subcutaneous melatonin pellets (MEL PEL) prevent the action of short days or afternoon melatonin inejctions. INT = intact animals.

melatonin from causing gonadal regression, i.e., perhaps melatonin could inhibit its own action. When tested this was, indeed, found to be the case (Reiter et al., 1977). Thus, if long day exposed male hamsters received a subcutaneous reserve of melatonin (in beeswax) and, additionally, were injected with 25 μg of melatonin each afternoon the reproductive system remained highly functional.

Contrary to the initial impression, these collective results are strongly supportive of the idea that melatonin is the pineal antigonadotrophic factor in the Syrian hamster for the following reasons. Obviously, properly administered melatonin causes gonadal regression in this species (Tamarkin et al., 1976; Reiter et al., 1976); on the hand, continually available melatonin prevents darkness induced gonadal regression (Hoffmann, 1974; Reiter et al., 1974). Most interestingly, melatonin deposits under the skin also negate the action of afternoon melatonin injections (Reiter et al., 1977). This implies that gonadal regression due to darkness and to afternoon melatonin injections are caused by the same compound, i.e., melatonin. Continually available melatonin apparently interferes with the ability of cyclic melatonin (either endogenous from the pineal or exogenous following an injection) to modify the reproductive system. Presumably, melatonin may have these apparently paradoxical effects by virtue of the fact that it alters the sensitivity of its own receptors. This has come to be known as the down regulation hypothesis and is discussed below.

DOWN REGULATION HYPOTHESIS OF MELATONIN'S ACTION IN THE SYRIAN HAMSTER

The concept that melatonin regulates the sensitivity of its own recepetors was first proposed in 1978 by Reiter and colleagues and has been further elaborated in recent years (Reiter, 1980a; Cardinali, 1981). The theory relies on the idea that when melatonin acts on its receptors it temporarily desensitizes (or down regulates) them to further melatonin for a period of time. The theory explains most of the known actions of the indole in terms of the reproductive system of the Syrian hamster. These interactions are summarized in figure 4. When hamsters are kept under long day conditions the endogenous melatonin peak occurs late in the dark phase; when the indole is released it theoretically acts on and down regulates its receptors. As a result, early in the light phase the melatonin receptors are in a desensitized state (Fig. 4, panel A). This explains why daily melatonin injections in the morning are incapable of inducing reproductive collapse. By late in the light period the melatonin receptors have regenerated their sensitivity, i.e., they are no longer down regulated, and daily injections of the indole at this time leads to reproductive inhibition (Fig. 4, panel B). This theory also explains why continually available melatonin prevents both dark-induced and afternoon melatonin injection-induced gonadal regression. When melatonin is persistently available from a subcutaneous reservoir, its receptors are always in a down regulated state (Fig. 4, panel C). Down regulation of the melatonin receptors could, in fact, be a consequence of a change in sensitivity (Reiter, 1980a) or an alteration in their number (Vacas and Cardinali, 1970; Cardinali, 1981). Regardless, the net effect is the same.

There have been a number of physiological tests of the down regulation theory. We predicted that whereas large doses of

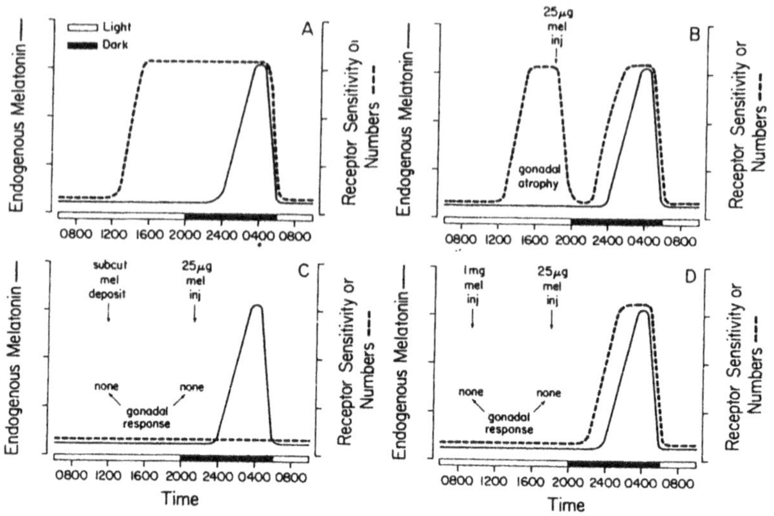

Fig. 4. Summary of the down regulation theory of melatonin action. See text for details.

melatonin, given in the morning, would not induce gonadal regression (since receptors were down regulated), they would prevent the afternoon injection of the indole from causing reproductive collapse since they would prolong the desensitization of the melatonin receptors (Fig. 4, panel D). This is indeed the case. When 1 mg or less of melatonin was injected at 1100 hours (lights on at 0600 hours), it prevented the inhibitory action of smaller doses of melatonin at 1700 hours on reproduction (Chen et al., 1980; Richardson et al., 1981). Presumably, melatonin in the morning prolonged the densensitized state of the melatonin receptors and prevented the later injections of the indole from causing an effect.

Finally, in further tests of the theory it has been found that very large doses of melatonin even when they are given in the afternoon *prevent* darkness-induced gonadal atrophy (Turek and Pappas, 1980; Richardson et al., 1982). Likewise, very large doses of melatonin given in the afternoon to hamsters maintained under long days does not cause gonadal atrophy (Chen, 1981). Presumably, the large bolus of the indole desensitizes the receptors much like does melatonin from a subcutaneous reserve. This further indicates that darkness mediated shut down of the reproductive system in the hamster is due to endogenously produced melatonin. It is important to emphasize that the down regulation theory of melatonin's action applies, at this point, to the Syrian hamster. In other species melatonin may be completely incapable of altering the sensitivity of its own receptors.

ROLE OF MELATONIN IN SEASONAL REPRODUCTION

In photosensitive rodent species, the light:dark cycle acting by way of the pineal gland is clearly involved in determining seasonal variations in reproductive competence (Reiter, 1980a; 1982; Stetson and Tate-Ostroff, 1981). The specific hormone responsible for regulating the seasonal reproductive cycle may be melatonin. Certainly, it causes many of the same changes as does dark exposure (Tamarkin et al., 1976; Trakulrugnsi, et al., 1979). Also, when administered by means of the proper paradigm it can actually duplicate the annual cycle of reproduction in the laboratory. As noted above, afternoon injections of melatonin eventually cause a total collapse of the gonads. After the reproductive organs are in an atrophic state, either short day exposure (Reiter, 1973) or continued melatonin injections (Tamarkin et al., 1976) maintain the reproductive system in a quiescent state for a period of time. Furthermore, if either short day exposure (Reiter, 1972; Stetson et al., 1976) or daily afternoon melatonin administration (Tamarkin et al., 1976) are continued, the neuroendocrine-gonadal axis endogenously regenerates to a functionally mature state. Thus, the reproductive system is capable of becoming refractory to both short days and to melatonin. Once the gonads have recrudesced they remain refractory to both darkness (Reiter, 1975) and melatonin (Bittman, 1978; Reiter et al., 1979) for a period of time. Finally, the refractoriness to both short daylengths and melatonin administration is interrupted by the maintenance of hamsters under long day conditions for a prolonged period (Reiter, 1972; Reiter et al., 1979). The interactions of the photoperiodic changes and melatonin administration on the reproductive system of the Syrian hamster are compared in figure 5.

The similarities between darkness induced and melatonin-mediated gonadal regression as well as the development of refractoriness to both conditions makes it seem likely that they are both caused by the same hormone, i.e., melatonin. Considering that the animals become refractory to melatonin, the amount of melatonin produced by the pineal gland does not have to change throughout the year; rather, the responsiveness of the animal to melatonin changes. When examined, it was found that the pineal gland produces ample amounts of melatonin during all stages of the annual reproductive cycle (Rollag et al., 1980a; Brainard et al., 1983) although under natural photoperiodic

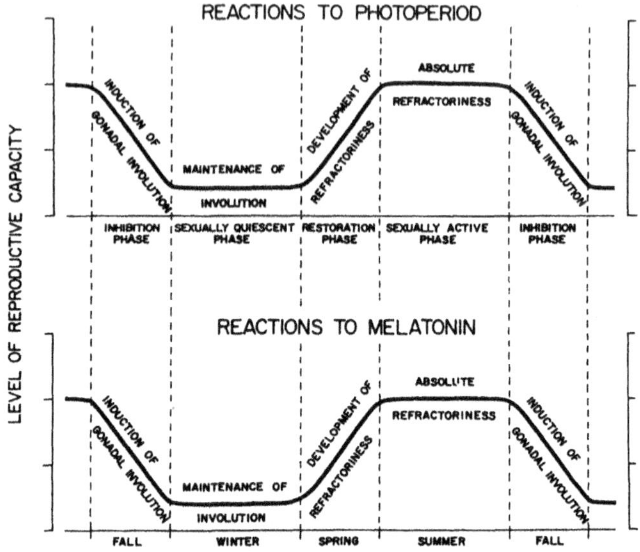

Fig. 5. Summary of the actions of photoperiod or melatonin treatment on seasonal reproduction in hamsters. When properly administered, melatonin duplicates the effects of seasonally changing photoperiods.

conditions there may be slightly less melatonin produced during a given day in the summer than in the winter (Brainard et al., 1983).

CONCLUDING REMARKS

Although compounds other than melatonin have been recovered from pineal tissue, none have been shown to produce the dramatic reproductive effects that melatonin does. The changes induced by the daily afternoon injections of melatonin into Syrian hamsters are similar to those produced by short day exposure. On the other hand, the continual availability of melatonin from a subcutaneous reserve actually counteracts the inhibitory effect of dark exposure on reproduction as well as negating the inhibitory influence of daily afternoon melatonin injections on gonadal physiology. Apparently, melatonin has the capability of desensitizing its own receptors in the Syrian hamster. There is now strong evidence that melatonin may mediate the seasonal changes in reproduction which hamsters experience when they are maintained under natural photoperiods throughout the year.

REFERENCES

Benson, B., 1977, Current status of pineal peptides, Neuroendocrinology 24, 241.

Benson, B., Larsen, B. R., and Findell, P. R., 1981, Melatonin and other pineal products, in: "Melatonin-Current Status and Perspectives", N. Birau and W. Schloot, eds., Pergamon, New York.

Bittman, E., 1978, Hamster refractoriness: Role of insensitivity of pineal target tissues, Science, 202: 648.

Brainard, G. C., Petterborg, L. J., Richardson, B. A., and Reiter, R. J., 1983, Pineal melatonin in Syrian hamsters: Circadian and seasonal rhythms in animals maintained under laboratory and natural conditions, Neuroendocrinology, in press.

Brown, G. M., Tsui, H. W., Niles, L. P., and Grota, L. J., 1981a, Gonadal effects of the pineal gland, in: Pineal Function, C. D. Matthews and R. F. Seamark, eds., Elsevier/North Holland, Amsterdam.

Brown, G. M., Grota, L., and Niles, L., 1981b, Melatonin: Origin, control of circadian rhythm and site of action, in: "Melatonin - Current Status and Perspectives", N. Birau nd W. Schloot, eds., Pergamon, New York.

Cardinali, D. P., 1981, Melatonin. A mammalian pineal hormone, Endocr. Rev., 1: 327.

Chen, H. J., 1981, Melatonin: Failure of pharmacological doses to induce testicular atrophy in the male golden hamster, Life Sci., 28: 767.

Chen, H. J., Brainard, G. C., and Reiter, R. J., 1980, Melatonin given in the morning prevents the suppressive action on the reproductive system of melatonin given in the afternoon, Neuroendocrinology, 31: 129.

Czyba, J. C. Girod, C., et Durand, N., 1964, Sur l'antagonisme épiphyso-hypophysaire et les variations saisonnières de la spermatogénèse chez le hamster doré (*Mesocricetus auratus*), C. R. Soc. Biol., 158: 742.

Ebels, I., 1979, A chemical study of some biologically active pineal fractions, Progr. Brain Res., 52, 309.

Hoffman, R. A., and Reiter, R. J., 1965, Pineal gland: Influence on gonads of male hamsters, Science, 142: 1609.

Hoffman, R. A., and Reiter, R. J., 1966, Repsonse of some endocrine organs of female hamsters to pinealectomy and light, Life Sci., 5: 1147.

Hoffmann, K., 1974, Testicular involution in short photoperiods inhibited by melatonin, Naturwissenschaften, 61: 364.

Klein, D. C., and Weller, J. C., 1970, Indole metabolism in the pineal gland: A circadian rhythm in N-acetyltransferase, Science, 169: 1093.

Lerner, A. B., Case, J. D., Takahashi, Y, Lee, T. H., and Mori, W., 1958, Isolation of melatonin, the pineal gland factor that lightens melanocytes, J. Amer. Chem. Soc., 80, 2587.

Panke, E. S., Reiter, R. J., Rollag, M. D., and Panke, T. W., 1978, Pineal serotonin N-acetyltransferase activity and melatonin concentrations in prepubertal and adult Syrian hamsters exposed to short daily photoperiods, Endocr. Res. Commun., 5: 311.

Panke, E. S., Rollag, M. D., and Reiter, R. J., 1979, Pineal melatonin concentrations in the Syrian hamster, Endocrinology, 104, 194.

Pevet, P., Balemans, M. G. M., Legerstee, W. C., and Vivien-Roels, B., 1981a, Circadian rhythmicity in the activity of HIOMT in the formation of melatonin and of 5-methoxy-tryptophol in the retina, Harderian gland and pineal of the male hamster, in "Melatonin - Current Status and Perspectives", N. Birau and W. Schloot, eds., Pergamon, New York.

Pevet, P., Haldar-Misra, C., and Ôcal, T., 1981b, Effect of 5-methoxytryptamine and 5-methoxytryptophol on the reproductive system of the male golden hamster. J. Neural Transmis., 51: 303.

Reiter, R. J., 1972, Evidence for refractoriness of the pituitary-gonadal axis to the pineal gland in golden hamsters and its possible implications in annual reproductive rhythms, Anat. Rec., 1973: 365.

Reiter, R. J., 1973, Pineal control of a seasonal reproductive rhythm in male golden hamsters exposed to natural daylight and temperature, Endocrinology, 92: 423.

Reiter, R. J., 1974, Pineal regulation of hypotalamicopituitary axis: Gonadotrophins, in "Handbook of Physiology, Endocrinology IV, pt. 2", E. Knobil and W. Sawyer, eds., American Physiological Society, Washington.

Reiter, R. J., 1975, Exogenous and endogenous control of the annual reproductive cycle in the male golden hamster: Participation of the pineal gland, J. Exp. Zool., 191: 111.

Reiter R. J., 1980a, The pineal and its hormones in the control of reproduction in mammals, Endocr. Rev., 1: 109.

Reiter, R. J., 1980b, The pineal gland: A regulator or regulators, Progr. Psychiobiol. Physiol. Psychol., 9: 323.

Reiter, R. J., 1981, Reproductive effects of the pineal gland and pineal indoles in the Syrian hamster and the albino rat, in: "The Pineal Gland. Vol. II. Reproductive Effects", R. J. Reiter, ed., CRC Press, Boca Raton.

Reiter, R. J., 1982, Neuroendocrine effects of the pineal gland and melatonin, in: "Frontiers in Neuroendocrinology, Vol. 7", W. F. Ganong and L. Martini, eds., Raven, New York.

Reiter, R. J., and Vaughan, M. K., 1977, Pineal antigonadotrophic substances: Polypeptides and indoles, Life Sci., 21, 159.

Reiter, R. J., Vaughan, M. K., Blask, D. E., and Johnson, L. Y., 1974, Melatonin: Its inhibition of pineal antigonadotrophic activity in male hamsters, Science, 185: 1169.

Reiter, R. J., Vaughan, M. K., and Waring, P. J., 1975, Studies on the minimal dosage of melatonin required to inhibit pineal antigonadotrophic activity in male golden hamsters, Horm. Res., 6: 258.

Reiter, R. J., Blask, D. E., Johnson, L. Y., Rudeen, P. K., Vaughan, M. K., and Waring, P. J., 1976, Melatonin inhibition of reproduction in the male hamster: Its dependency on time of day of administration and on an intact and sympathetically innervated pineal gland, Neuroendocrinology, 22: 107.

Reiter, R. J., Rudeen, P. K., Sackman, J. W., Vaughan, M. K., Johnson, L. Y., and Little, J. C., 1977, Subcutaneous melatonin implants inhibit reproductive atrophy in male hamsters induced by daily melatonin injections, Endocr. Res. Commun., 4: 35.

Reiter, R. J., Rollag, M. D., Panke, E. S., and Banks, A. F., 1978, Melatonin: Reproductive effects, J. Neural Transmis., Suppl. 13: 209.

Reiter, R. J., Petterborg, L. J., and Philo, R. C., 1979, Refractoriness to the antigonadotrophic effects of melatonin in male hamsters and its interruption by exposure of the animals to long daily photoperiods, Life Sci., 25: 1571.

Reiter, R. J., Hurlbut, E. C., King, T. S., Richardson, B. A., Vaughan, M. K., and Kosub, K. Y., 1982, A 15-minute light pulse during darkness prevents the antigonadotrophic action of afternoon melatonin injections in the male hamster, Int. J. Biometeorol., in press.

Richardson, B. A., Vaughan, M. K., Brainard, G. C., Huerter, J. J., de los Santos, R., and Reiter, R. J., 1981, Influence of morning melatonin injections on te antigonadotrophic effects of afternoon melatonin administration in male and female hamsters, Neuroendocrinology, 33: 112.

Richardson, B. A., King, T. S., Petterborg, L. J., Vaughan, M. K., and Reiter, R. J., 1982, The influence of large doses of melatonin administered in the afternoon on the neuroendocrine-reproductive system of female Syrian hamsters maintained either in a long or short photoperiod, Biomed. Res., 3: 61.

Rollag, M. D., Panke, E. S., and Reiter, R. J., 1980a, Pineal melatonin content in male hamsters throughout the seasonal reproductive cycle, Proc. Soc. Exp. Biol. Med., 165: 330.

Rollag, M. D., Panke, E. S., Trakulrungsi, W. K., Trakulrungsi, C., and Reiter, R. J., 1980b, Quantification of daily melatonin synthesis in the hamster pineal gland, Endocrinology, 106: 231.

Rudeen, P. K., and Reiter, R. J., 1977, Effect of shortened photoperiods on pineal serotonin N-acetyltransferase activity and rhythmicity, J. Interdisc. Cycle Res., 8: 47.

Rudeen, P. K., Reiter, R. J., and Vaughan, M. K., 1975, Pineal serotonin-N-acetyltransferase in four mammalian species, Neurosci. Letters, 1: 225.

Smith, J. A., 1981, The biochemistry and pharmacology of melatonin, in: "Melatonin - Current Satus and perspectives", N. Birau and W. Schloot, ed., Pergamon, New York.

Stetson, M. H., and Tate-Ostroff, B., 1981, Hormonal regulation of the annual reproductive cycle of golden hamsters, Gen. Comp. Endocr., 45, 329.

Stetson, M. H., Matt, K. S., and Watson-Whitmyre, M., 1976, Photoperiodism and reproduction in golden hamsters: Circadian organization and termination of photorefractoriness, Biol. Reprod., 14: 531.

Tamarkin, L., Westrom, W. K., Hamill, A. I., and Goldman, B. D., 1976, Effect of melatonin on the reproductive systems of male and female Syrian hamsters: A diurnal rhythm in sensitivity to melatonin, Endocrinology, 99: 1534.

Tamarkin, L., Hollister, C. W., Lefebvre, N. G., and Goldman, B. D., 1977, Melatonin induction of gonadal quiescence in pinealectomized Syrian hamsters, Science, 198: 953.

Tamarkin, L., Reppert, S. M., and Klein, D. C., 1979, Regulation of pineal melatonin in the Syrian hamster, Endocrinology, 104: 385.

Trakulrungsi, C., Reiter, R. J., Trakulrungsi, W. K., Vaughan, M. K., and Waring-Ellis, P. J., 1979, Interaction of daily injections and subcutaneous reservoirs of melatonin on the reproductive physiology of female Syrian hamsters, Acta Endocr., 91: 59.

Turek, F. W., and Pappas, P., 1980, Daily melatonin injections inhibit short day-induced testicular regression in hamsters, Experientia, 36: 1346.

Vacas, M. I., and Cardinali, D. P., 1979, Diurnal changes in melatonin binding sites of hamster and rat brains. Correlations with neuroendocrine responsiveness to melatonin, Neurosci. Letters, 15: 259.

THE 5-METHOXYINDOLES DIFFERENT FROM MELATONIN: THEIR EFFECTS ON THE SEXUAL AXIS

Paul Pévet[1,2]

1) The Netherlands Institute for Brain Research Amsterdam, and 2) Dept. of Anatomy and Embryology, University of Amsterdam, The Netherlands

From the time of Lerner's discovery of melatonin (Lerner et al., 1958) to the present, much pineal research has been centered on the study of the effects of melatonin, and to date this 5-methoxyindole is considered by many authors as the pineal hormone responsible for inhibiting the reproductive system (details and references in Reiter, 1980) or the pineal factor conveying the photoperiodic message (see details and references in Hoffmann, 1981.

Melatonin, however, is only one of the numerous indoleamines synthesized in the pineal gland. In the gland, indeed the amino-acid tryptophan is converted to 5-hydroxytryptophan by hydroxylation (Lovenberg et al., 1967; Jequier et al., 1969). Decarboxylation of this indole derivative (Lovenberg et al., 1962; Snijder and Axelrod, 1964) leads to the formation of 5-hydroxytryptamine (serotonin). Serotonin, then, can be acetylated to N-acetylserotonin (Weissbach et al., 1960), oxidized to 5-hydroxyindole-3-acetic acid (Lerner and Case, 1960), or metabolized to 5-hydroxytryptophol (McIsaac and Page, 1959). The 5-hydroxyindoles, 5-hydroxytryptophan, 5-hydroxytryptamine (serotonin), 5-hydroxy-indole-3-acetic acid, 5-hydroxytryptophol and N-acetyl-5-hydroxytryptamine (N-acetylserotonin) can be methylated by hydroxyindole-O-methyl transferase (HIOMT) (Axelrod and Weissbach, 1960, 1961; Axelrod and Lauber, 1968; Cardinali and Wurtman, 1972). This means that at least 5 different methylated products, 5-methoxytryptophan, 5-methoxytryptamine, 5-methoxyindole-3-acetic acid, 5-methoxytryptophol and melatonin, can be formed in the pineal (for more details, see Balemans, 1979, 1981), the list probably being not complete since O-acetyl-5-methoxytryptophol might also be present in the pineal (Smith et al., 1979). All these 5-methoxyindoles and 5-hydroxyindoles are potential active factors.

In the present review we will examine the physiological properties of the 5-methoxyindoles only, and we will concentrate our analysis on their effect on the sexual axis.

I 5-METHOXYTRYPTOPHOL

5-Methoxytryptophol (5-MTL) has been first identified in bovine pineal by McIsaac et al. (1965) and later demonstrated to be present in numerous other species (Table 1). Contrary to melatonin (MEL), however, its synthesis occurs during the light phase of the light:dark cycle, at least in the hamster (Pévet et al., 1980). As Carter et al. (1979) observed in the rat that during the light period the plasmatic concentration is significantly higher than during the dark period, it would be tempting to correlate these two observations and to conclude that the circulating 5-MTL originated from the pineal. However, in our opinion, to date we cannot raise this conclusion, not only because it is difficult to compare results obtained in rat and hamster, but especially because it is well-known that 5-MTL, like MEL, is also synthesized in different other organs such as, for example, the retina and the Harderian gland (Balemans et al., 1981, 1982; Pévet et al., 1980, 1981a).

Looking at the literature, it appears that 5-MTL, like MEL, is capable of modifying sexual development and reproduction, especially in birds and mammals. In juvenile chicken, for example, the gonadal growth is stimulated by 5-MTL and MEL, whereas in maturing and adult animals both indolic compounds show an inhibitory effect, that of 5-MTL however being more specific than that of MEL (Balemans, 1972; Balemans et al., 1977a,b). A similar stimulatory or progonadotropic effect of 5-MTL during early post-hatching stages of development has been described in the quail and, interestingly, in this bird the effect appears to be sex-, age- and photoperiod-dependent. 5-MTL, indeed, was found to stimulate ovarian growth in young females kept in 16L/8D or 12L/12D but not in 8L/16D, and did not have any effect in males (see details in Alexander, 1969).

Table 1. Pineal 5-methoxytryptophol (5-MTL) content in different mammals

F: fluorometry, GCMS, gas chromatography-mass spectrometry, WT: wet tissue

Species	Concentration	Technique used	References
Cow	1260 ± 150 pmol/g WT	GCMS	Beck et al., 1981
Dog	170 ± 10 ng/g WT	F	Miller and Maickel, 1970
Pig	15.1 ± 2.3 pmol/g WT	GCMS	Beck et al., 1981
Rat	480 ± 90 ng/g WT 154 pg / pineal gland	F GCMS	Miller and Maickel, 1970 Carter et al., 1979
Sheep	370 ± 90 pmol/g WT	GCMS	Beck et al., 1981

Considering mammals, it appears also that very often 5-MTL presents the same effects as MEL. Both compounds, for example, inhibit in mouse ovarian and uterus hypertrophy induced by an injection of HCG or PMSG (Vaughan et al., 1967), inhibit the compensatory ovarian hypertrophy occurring after unilateral gonadectomy (Vaughan et al., 1972), decrease ovarian weight, and inhibit estrus in the rat. In these last two experiments 5-MTL was found to be even more effective than MEL (McIsaac et al., 1964). This is also true when the effects of both compounds implanted in the median eminence or in the reticular formation on the pituitary LH content of castrated male rats are considered. 5-MTL induced indeed a decrease larger than that obtained with MEL (Fraschini et al., 1970). Administration of melatonin (Debeljuk et al., 1970) or 5-MTL (Talbot and Reiter, 1973/1974) induces a similar decrease in the rat pituitary FSH level., but when implanted in the median eminence only 5-MTL reduces FSH secretion in rats (Fraschini, 1969; Fraschini and Martini, 1970). When injected intracisternally into young female rats, both compounds delay similarly the time of vaginal opening (Fraschini et al., 1971), They also delay light-induced estrus in the ferret, but here 5-MTL is less effective than MEL (Herbert, 1971). 5-MTL is also known to block copulation-induced release of gonadotropin in adult rats

Table 2. Pineal 5-methoxytryptamine (5-MT) content in different mammals

F: fluorometry, GCMS: gas chromatography-mass spectrometry, WT: wet tissue

Species	Concentration	Technique used	References
Cow	2.63 pmol/g WT	GCMS	Beck et al., 1981
	117 ± 48 pmol/g WT	GCMS	Bosin et al., 1979
Baboon	2400 ng/g WT	F	Prozialeck et al., 1978
Dog	740 ng/g WT	F	Prozialeck et al., 1978
	740 ± 40 ng/g WT	F	Miller and Maickel, 1970
Man	0-152 pmol/g WT	GCMS	Bosin and Beck, 1979
	12.8 ± 4.9 ng/g WT	F	Prozialeck et al., 1978
Pig	26.4 ± 3.9 pmol/g WT	GCMS	Beck et al., 1981
	228 ± 119 pmol/g WT	GCMS	Bosin et al., 1979
Rat			
(Wistar)			
	0.035 pmol/g	GCMS	Beck et al., 1981
	3010 ± 1700 ng/g WT	F	Prozialeck et al., 1978
	3700 ± 300 ng/g WT	F	Miller and Maickel, 1970
	4200 ± 900 ng/g WT	GCMS	Cattabeni et al., 1972
	2000 ng/g WT	F	Maickel and Miller, 1968
(Sprague-Dawley)			
	0.028 ± 0.005 pmol/g WT	GCMS	Beck et al., 1981
Sheep	1.78 ± 0.53 pmol/g WT	GCMS	Beck et al., 1981
	545 ± 180 pmol/g WT	GCMS	Bosin et al., 1979

(Farrel et al., 1968) and, when regularly injected, to inhibit spermatogenesis (Mas and Oakin, 1977).

5-MTL appears thus to be a very active principle with clear antigonadotropic properties. Surprisingly, however, and although some of these findings (e.g. McIsaac et al., 1964) are contemporary with those made with melatonin, the study of the endocrine properties of 5-MTL has been neglected by most pinealogists. The reason for this disinterest, which is now considered to be due to the fact that 5-MTL has not been shown to be capable of inhibiting the hypothalamo-hypophyseo-gonadal axis of the hamster (Sackman et al., 1977), is not clear and especially illogical when we remember that it was only in 1976 that Tamarkin et al. clearly demonstrated that MEL is able to induce gonadal atrophy in the hamster. Moreover, 5-MTL is known to have a strong effect on the sexual activity of the hamster. Indeed, Reiter et al. (1975) have reported that weekly implantation of 5-MTL in male hamsters exposed to a light/dark cycle of 1:23 prevented the short photoperiodically induced depression of pituitary LH and prolactin, and testicular and accessory sex organ regression. This effect which, depending of the author, is called a progonadal or a counterantigonadal effect, is - in our opinion - very important. Indeed, if we consider with Hoffmann (1981) that the pineal is not an antigonadotropic organ but a gland involved in conveying the effect of photoperiodic stimuli upon the neuroendocrine axis, or with Pévet et al. (1981c) and Pévet and Haldar-Misra (1982a) that all 5-methoxyindoles synthesized by the pineal as well as by other organs such as the retina and the Harderian gland are implicated in a system by which the pineal and some other organs such as the brain are able to perceive, to differentiate and to integrate environmental information (photoperiod, temperature, etc.), the results obtained by Reiter et al. (1975) could be interpreted by the fact that, when constantly released, 5-MTL prevents the detection by the animal of the short photoperiod. This would mean that 5-MTL, like MEL or 5-methoxytryptamine (see next subheading), is deeply implicated in the control of seasonal reproductive activity, at least in the hamster.

II 5-METHOXYTRYPTAMINE

5-Methoxytryptamine (5-MT) is known to be present and synthesized in the pineal gland (Table 2) (Balemans et al., 1980a,b, 1982; Pévet et al., 1981a) and also, like MEL, in the Harderian gland, intestine and retina (Prozialeck et al., 1978; Pévet et al., 1981a; Balemans et al., 1982). Comparing the results obtained, for example, by the same team and with the same technique, it appears that the concentration of 5-MT in the pineal is higher than that of MEL, at least in the rat (Cattabeni et al., 1972; Maickel and Miller, 1968), a result which is probably due to the higher 5-MT synthesis, as was demonstrated in mole, hamster and Spalax pineals (Balemans et al., 1980b, 1982; Pévet et al., 1981a). As 5-MT has been detected in human plasma (Franzen and Gross, 1965; Prozialeck et al., 1978) — and with a higher value during the night (Hooper et al., 1981) — in cerebrospinal fluid (Prozialeck et al., 1978) and urine (Prozialeck et al., 1978; Haddox and Saslaw, 1963), and as the transformation of a small fraction of circulating melatonin into 5-MT (see below) cannot explain the large amount of 5-MT found in the blood (Hooper et al., 1981), it is very probable that 5-MT, like MEL, is released from the pineal and/or other organs into the general circulation.

In the past few years 5-MT has attracted a great deal of interest, mainly because *(1)* it is a potential precursor in the biosynthesis of the psychomimetic agent 5-methoxy-N,N-dimethyltryptamine (Holmstedt and Lindgren, 1967) — a pathway shown to be present in the human pineal (Guchhait, 1976) — *(2)* it possesses psychomimetic properties of its own (De Montigny and Aghajanian (1977), and *(3)* it mimics the action of serotonin on brainstem neurons (Bradley and Briggs, 1974). At the level of the sexual axis the relatively small amount of investigations has permitted to clearly demonstrate that 5-MT has a profound influence on sexual development and reproduction. In the quail, for example, 5-MT influences gonadal maturation and activity. This effect, however, is highly age-, sex- and photoperiod-dependent (Alexander, 1969). Low doses of 5-MT (10 μg/day) stimulate the rate of sexual maturation in juvenile females kept in 16L:8D but not in 12L/12D or 8L:16D. Large doses (100 μg and 500 μg) are less effective (Alexander, 1968). In adult females 5-MT inhibits ovarian growth when they are exposed to a photoperiod of 12L:12D but not when exposed to 8L:16D (Alexander, 1969). 5-MT was also found to prevent delayed maturation induced by prolactin treatment, indicating so that prolactin and 5-MT exert antagonistic effects on a common mechanism controlling gonadal secretion, at least in the female quail (Alexander, 1969). In male quails 5-MT does not seem to stimulate sexual maturation but was found to lower testicular weight at 45 but not at 35 days of age (Alexander, 1969). Concerning mammals, 5-MT appears to be also very potent. In young mice, for example, it inhibits ovarian growth induced by injection of gonadotropin (Vaughan et al., 1976). 5-MT, like MEL, is also known to suppress the LHRH-induced release of LH from neonatal rat anterior pituitary glands in organ culture. In this test, however, MEL appeared to be more effective (Martin et al., 1977). When injected daily in the late afternoon in hamsters kept under long photoperiod (14L:10D), 5-MT, like MEL (Tamarkin et al., 1976), induces a complete gonadal atrophy similar to that observed in hamsters maintained in short light cycles or in darkness (Pévet et al., 1981b).

In contrast to MEL which, under the same experimental conditions, needs the presence of the pineal to be effective (Reiter et al., 1976), 5-MT acts even in pinealectomized animals. However, in such operated animals the regression of the testes is less when compared with that observed in intact hamsters (Pévet et al., 1981c). It thus appears that 5-MT - althogh in itself causing a slight antigonadotropic effect – is, like MEL, more effective if the pineal is present. Moreover, like for MEL (Tamarkin et al., 1976), the reproductive system of the male hamster shows a diurnal rhythm in sensitivity to 5-MT. Indeed, under the same experimental condition, when injected in the morning, 5-MT does not have any effect on testicular activity (Pévet et al., 1981c). Subcutaneous deposits of 5-MT in male hamsters prevent the reproductive quiescence which is the consequence of natural or of experimentally shortened daily light conditions (Pévet and Haldar-Misra, 1982a). Yet, this result is similar to that obtained with MEL (Hoffman, 1974; Reiter et al., 1974). This is also the case when 5-MT is injected daily in the morning in animals kept under short photoperiod (10L:14D). Under these conditions 5-MTL, like MEL (Turek and Pappas, 1980), prevents at least partly the short-photoperiodically induced gonadal atrophy (Pévet and Haldar-Misra, 1982b).

Taken together, these results clearly demonstrate that 5-MT is a physiologically active compound which, if administered properly, can duplicate most of the effects of the photoperiod on the reproductive axis of the male hamster. In the hamster as well as in the rat 5-MT gives apparently the same physiological response as does MEL. This raises thus the question of whether these effects of 5-MT and MEL are independent ones and are due to *(1)* a direct or indirect transformation of 5-MT into MEL, or, at the contrary, *(2)* to a transformation of MEL into 5-MT.

As explained by Balemans et al. (1980a), and although such metabolic pathways have never been demonstrated, the possibility that exogenously administered 5-MT is acetylated by N-acetyltransferase and, thus, transformed into MEL in the pineal or in other organs such as the retina or the Harderian gland, cannot be excluded. The physiological properties of 5-MT would thus be indirectly due to MEL. Such a transformation of 5-MT into MEL, for example, would explain why the antigonadotropic effect of 5-MT injected late in the afternoon can be prevented by morning injections of large doses of MEL but not of 5-MT (Pévet and Haldar-Misra, 1982c). Indeed, as the N-acetyltransferase activity is activated only during the dark period of the light/dark cycle (Klein, 1979), the 5-MT injected in the morning could not be transformed into MEL. However, such a transformation would not permit to explain why morning injections of 5-MT prevent short-photoperiodically induced gonadal atrophy (Pévet and Haldar-Misra, 1982b). For this last-mentioned experiment a possible 5-MT-induced increase of MEL synthesis, demonstrated to exist in rat (Trentini et al., 1982), could perhaps be considered.

5-MT is produced by a direct methylation of serotonin In the rat, Fiske and Huppert (1968) have demonstrated that injection of MEL provokes an increase in pineal serotonin content. The effects of MEL could therefore be due to the increase of 5-MT indirectly provoked by the MEL-induced increase of serotonin production. This could explain for example why, when injected late in the afternoon into hamsters kept under long photoperiod, 5-MT but not MEL is effective in pinealectomized hamsters. It has been demonstrated *in vitro* (Rogawsky et al., 1979) as well as *in vivo* (Beck and Jonsson, 1981) that a small amount of circulating MEL (0.3 - 0.8% *in vitro*) is deacetylated and thus transformed into 5-MT. Considering now that other organs containing aryl acylamidase, the enzyme most probably responsibe of the deacetylation (Beck and Jonsson, 1981), such as the kidney (Krebs et al., 1947; Nimmo-Smith, 1970; Hoagland and Graf, 1971) might also be implicated in such a phenomenon, the possibility that the effect of exogenously administered MEL is in fact due to 5-MT has to be seriously considered. For example, the observation that three daily injections of MEL are effective in pinealectomized hamsters could be due to the fact that so the concentration of 5-MT formed by deacetylation of MEL would be enough to induce gonadal atrophy. However, the deacetylation of the circulating MEL does not explain why MEL injected in the morning in the male hamster (Pévet and Haldar-Misra, 1982c) but not 5-MT prevents gonadal atrophy induced by 5-MT injected in the late afternoon.

It is very probable — at least, this is our present opinion — that both compounds have their own effects, but that these effects are physiologically dependent of each other. Moreover, it is not possible to exclude that, depending of the time of injection and the experimental conditions, one of the 5-methoxyindoles, or part of it, in transformed into the other, and *vice versa*, such a phenomenon being perhaps an integral part of the physiological response.

III 5-METHOXYINDOLE-3-ACETIC ACID

Like other 5-methoxyindoles, 5-methoxyindole-3-acetic acid (5-MIAA) is present (table 3) and synthesized in the pineal gland (Axelrod and Weissbach, 1961; Pévet et al., 1981a; Balemans et al., 1982) and in several other organs such as the retina (Pévet et al., 1981a; Balemans et al., 1982) and the Harderian gland (Balemens et al., 1982). Such a synthesis, when compared to that of MEL, appears not neglectable, and for example Lerner et al. (1960) found that the amount of 5-MIAA in the pineal is 10 times greater than that of MEL. Beck et al. (1981) also found a higher concentration of 5-MIAA than of MEL in the cow and sheep, but not in the pig.

Table 3. Pineal 5-methoxyindole-3-acetic acid (5-MIAA) content in different mammals.

F: fluorometry; GCMS: gas chromatography-mass spectrometry, WT: wet tissue

Species	Concentration	Technique used	References
Cow	17600 ± 1500 pmol/g WT	GCMS	Beck et al., 1981
	2000 ng/g WT	-	Lerner et al., 1960
Dog	80 ng/g WT *	F	Miller and Maickel, 1970
Pig	406 ± 117 pmol/g WT	GCMS	Beck et al., 1981
Rat	1000 ng/g WT	F	Quay, 1964
	170 ng/g WT *	F	Miller and Maickel, 1970
Sheep	22800 ± 5400 pmol/g WT	GCMS	Beck et al., 1981

* concentration of 5-HIAA/5-MIAA

It is very probable that, like the other 5-methoxyindoles, 5-MIAA is released from the pineal and/or other organs into the general circulation. However, this is difficult to determine since the 5-MIAA in the animal is also formed by other pathways. 5-MIAA, indeed, is a metabolite of practically all 5-methoxyindoles (Delvigs et al., 1965), and almost all 5-methoxyindoles, when administered exogenously, are more or less converted to 5-MIAA. For example, when administered into rat or human, 5-MTL is metabolized in approximately 2 hours. The major metabolite is 5-MIAA and has been found by isotope dilution technique to represent 93.3% in rat (Delvigs et al., 1965) and 53-73% in human (Hoskins and Pollit, 1978) of the administered 5-MTL. Similarly it has been demonstrated that most of the 5-MT injected or formed by deacetylation of MEL is rapidly deaminated by MAO and converted to 5-MIAA (Rogawski et al., 1979; Kopin et al., 1961; Kveder and McIsaac, 1961). Using radioactive 5-MT, Kveder and McIsaac (1961) have demonstrated that 96,8% of it was transformed into 5-MIAA. Concerning MEL, the major pathway for metabolism is hydroxylation in position 6, followed by sulphate or glucuronide conjugation. However, 0.5 - 2% of the exogenously administered MEL is also transformed into 5-MIAA (Kveder and McIsaac, 1961; Kopin et al., 1961.

5-MIAA appears thus to be a metabolite of the three 5-methoxyindoles known to have a clear effect on the sexual axis (see before). This raises immediately the question of whether the observed effects of MEL and especially of 5-MTL and 5-MT could not be indirectly due to 5-MIAA. Apparently, this is not the case. Indeed, 5-MIAA injected late in the afternoon into hamsters kept under long photoperiod is not able, contrary to MEL and 5-MT, to induce testicular atrophy (Pévet, unpublished result). It would also be possible to ask whether a compound which is an end-product of 5-MTL, 5-MT and partly of MEL metabolism, excreted into urine (Delvigs et al., 1965; Kopin et al., 1961; Kverder and McIsaac, 1961) could not be simply considered as a product of degradation without physiological effects. This will explain why 5-MIAA was found to have no effect in the *in vitro* rat pituitary response to LHRH (Martin et al., 1977) and on the ovarian hypertrophy occurring after unilateral gonadectomy in mice (Vaughan et al., 1972). However, as 5-MIAA is synthesized, like other 5-methoxyindoles, in many organs (see before) it cannot, in our opinion, simply be considered as a product of the degradation of the other 5-methoxyindoles. This opinion is supported by the fact that in young mice 5-MIAA inhibits the growth of ovaries, induced by injection of gonadotropin (Vaughan et al., 1976).

IV 5-METHOXYTRYPTOPHAN

To our knowledge, Balemans et al. (1978) were the first to mention 5-methoxytryptophan (5-MTP) as a 5-methoxyindole possibly formed in the pineal gland. Later it has been demonstrated that this compound was indeed synthesized in the pineal (hamster: Balemans et al., 1982; mole: Pévet et al., 1981a; parakeet: Balemans et al., 1981, and rat: Balemans et al., 1980a) as well as in the retina (mole, hamster) and the Harderian gland (hamster)

(Pévet et al., 1981a; Balemans et al., 1982). From all 5-methoxyindoles, 5-MTP appears also to be the one of which the synthesis is the most important. In the mole, for example, the pineal synthesizes 50 to 100 times more 5-MTP than MEL (Pévet et al., 1981a) while in the hamster 4 to 10 times more 5-MTP than MEL is produced by the Harderian gland (Balemans et al., 1982). Moreover, looking at the annual variations of the capacity of the pineal, retina and Harderian gland to synthesize 5-methoxyindoles in hamsters kept under natural climate conditions, 5-MTP appears to be the compound which presents the highest annual variations. It is thus very probable, at least in our opinion, that 5-MTP is a 5-methoxyindole of great physiological importance. Unfortunately, as the properties of 5-MTP have practically never been studied, no experimental data support this opinion. For example, 5-MTP injected in the late afternoon or morning (Pévet et al., 1981b) or continuously administered (Pévet, unpublished data) into hamsters kept under long photoperiod does not have any effect on testicular activity. This, however, does not mean that 5-MTP has no biological effect. It simply means that in the experimental model used 5-MTP has no effect on the parameter studied, the testes of the hamster.

CONCLUSION

From the experiments described in this review it is evident that some 5-methoxyindoles different from MEL greatly influence sexual development and reproduction, at least in birds and mammals. Therefore, MEL is clearly not the only indolic compound which has to be considered when pineal function is studied.

It has been suggested by Pévet and Haldar-Misra (1982 a) that the 5-methoxyindoles would be implicated in a system enabling the pineal and some other organs such as the brain to perceive, to differentiate and to integrate environmental information such as photoperiod, temperature, food, etc. In response the pineal, then, would synthesize and release proteic/peptidic hormone(s) which would have an effect on the function of the reproductive system.

The question whether all 5-methoxyindoles are implicated in the transfer of all information or whether each of the 5-methoxyindoles is specialized for one or more of these parameters, has now to be answered. To date the experimental data are so rare that it is difficult to do so. However, it has been demonstrated that both MEL and 5-MT are deeply implicated in the conveyance of photoperiodic information. Moreover, as 5-MTP is synthesized in the retina, Harderian gland and pineal of hamsters kept under natural environmental conditions only during winter, it would be possible that 5-MTP is implicated in conveying information on temperature. This, however, remains to be demonstrated.

REFERENCES

Alexander, B., 1969, The synthesis and effects of pineal gland 5-methoxyindoles in relation to sexual maturation in the Japanese quail, *Coturnix coturnix*, japonica. Ph.D. Thesis, Northwestern University, Evanston, Ill., U.S.A.

Axelrod, J., and Lauber, J.K., 1968, Hydroxyindole-O-methyltransferase in several avian species, Biochem. Pharmacol., 17:828.

Axelrod, J., and Weissbach, H., 1960, Enzymatic O-methylation of N-acetylserotonin to melatonin, Science, 131:1312.

Axelrod, J., and Weissbach, H., 1961, Purification and properties of hydroxyindole-O-methyltransferase activity in the rat pineal gland by environmental lighting, J. Biol. Chem., 236:211.

Balemans, M.G.M., 1972, Age-dependent effects of 5-methoxytryptophol and melatonin on testes and comb growth of the white leghorn (*Gallus domesticus*, L.), J. Neural Transm., 33:179.

Balemans, M.G.M., Bary, F.A.M., Legerstee, W.C., and Van Benthem, J., 1980a, Seasonal variations in HIOMT activity during the night in the pineal gland of 21-day old male Wistar rats, J. Neural Transm., 49:107.

Balemans, M.G.M., Collin, J.P., Legerstee, W.C., and Juillard, M.T., 1981, Preliminary investigations on the circadian rhythmicity of the methylation of 5-hydroxyindoles in the pineal gland of the parakeet, Biol. Cell., 42:167.

Balemans, M.G.M., Noordegraaf, E.M., Bary, F.A.M. and Van Berlo, M.F., 1978, Estimation of the methylating capacity of the pineal gland. With special reference to indole metabolism, Experientia 34:887.

Balemans, M.G.M., Pévet, P., Legerstee, W.C., and Nevo, E., 1980b, Melatonin and 5-methoxytryptophol synthesis in the pineal, the retina and the Harderian gland of the mole-rat (*Spalax ehrenbergi*, Nehring) and in the pineal of the mouse "eyeless", J. Neural Transm., 49:225.

Balemans,, M.G.M., Pévet, P., Van Benthem, J., Haldar-Misra, C., Smith, I., and Hendriks, H., 1982, Seasonal rhythmicity in the capacity of HIOMT to synthesize different 5-methoxyindoles in the pineal, the retina and the Harderian gland of the golden hamster *(Mesocricetus auratus)* (submitted).

Balemans, M.G.M., Van der Veerdonk, F.C.G., and Van de Kamer, J.C., 1977a, Effect of injecting 5-methoxyindoles on testicular weight of white leghorn cockerels, J. Neural Transm., 41:47.

Balemans, M.G.M., Van der Veerdonk, F.C.G., and Van de Kamer, J.C., 1977b, The influence of 5-methoxytryptophol, a pineal compound, on comb growth, ovarian weight, follicular growth and egg production of juvenile and maturing adult white leghorn hens. J. Neural Transm., 41:37.

Beck, O., and Jonsson, G., 1981, In vivo formation of 5-methoxytryptamine from melatonin in rat, J. Neurochem., 36:2013.

Beck, O., Jonsson, G., and Lundman, A., 1981, 5-Methoxyindoles in pineal gland of cow, pig, sheep and rat. Arch. Pharmacol., 318:49.

Bosin, T.R., and Beck, O., 1979, 5-Methoxytryptamine in the human pineal gland: identification and quantification by mass fragmentography, J. Neurochem., 32:1853.

Bosin, T.R., Jonsson, G., and Beck, O., 1979, On the occurrence of 5-methoxytryptamine in brain, Brain Res., 173:79.

Bradley, P.B., and Briggs, L., 1974, Further studies on the mode of action of psychomimetic drugs: antagonism of the excitatory actions of 5-hydroxytryptamine by methylated derivatives of tryptamine, Brit. J. Pharmacol., 50:345.

Cardinali, D.P., and Wurtman, R.J., 1972, Hydroxyindole-O-methyltransferase in rat pineal, retina and Harderian gland, Endocrinology, 91:247.

Carter, S.J., Laud, C.A., Smith, I., Leone, R.M., Silman, R.E., Hooper, R.J.L., Larson-Carter, D.L., Finnie, M.D.A., and Mullen, P.E., 1979, 5-Methoxytryptophol in rat pineal glands and other tissues, in: "The Pineal Gland of Vertebrates including Man", J.Ariëns Kappers and P.Pévet, eds., Prog. Brain Res., 52, Elsevier, Amsterdam.

Cattabeni, F., Koslow, S.H., and Costa, E., 1972, Gas chromatographic-mass spectrometric assay of four indole alkylamines of rat pineal. Science, 178:166.

Debeljuk, L., Feder, V.M., and Paulucci, D.A., 1970, Effect of treatment with melatonin on the pituitary-testicular axis of the male rat, J. Reprod. Fertil., 21:363.

Delvigs, P., McIsaac, W.M., and Taborsky, R.G., 1965, The metabolism of 5-methoxytryptophol, J. biol. Chem., 240:348.

De Montigny, C., and Aghajanian, G.K., 1977, Preferential action of 5-methoxytryptamine and 5-methoxydimethyltryptamine on presynaptic serotonin receptors; a comparative iontophoretic study with LSD and serotonin, Neuropharmacology, 15:811.

Farrel, G., Powers, D., and Otani, T., 1968, Inhibition of ovulation in the rabbit: seasonal variation and the effects of indoles, Endocrinology, 83:599.

Fiske, V.M., and Huppert, L.C., 1968, Melatonin action on pineal varies with photoperiod, Science, 162:279.

Franzen, F., and Gross, H., 1965, Tryptamine, N,N-Dimethyltryptamine, N,N-Dimethyl-5-hydroxytryptamine and 5-methoxytryptamine in human blood and urine, Nature 206:1052.

Fraschini, F. 1969, The pineal gland and the control of LH and FSH secretion, in: "Progress in Endocrinology" C.Gual, ed., Excerpta Medica, Amsterdam.

Fraschini, F., and Martini, L., 1970, Rhythmic phenomena and pineal principles, in: "The Hypothalamus", L. Martini and F.Fraschini, eds., Academic Press, New York, London.

Fraschini, F., Collu, R., and Martini, L., 1971, Mechanisms of inhibitory action of pineal principles on gonadotropin secretion, in: "The Pineal Gland", G.E. W. Wolstenholm and J.Knight, eds., Churchill, New York.

Guchhait, R.B., 1976, Biogenesis of 5-methoxy-N,N-dimethyltryptamine in human pineal gland, J. Neurochem., 26:187.

Haddox, C.H., and Saslaw, M.S., 1963, Urinary 5-methoxytryptamine in patients with rheumatic fever, J. clin. Invest. 42:435.

Herbert, J., 1971, The role of the pineal gland in the control by light of the reproductive cycle of the ferret, in: "The Pineal Gland", G.E.W.Wolstenholm and J.Knight, eds., Churchill Livingston, Edinburgh, U.K.

Hoagland, R.E., and Graf, G., 1971, Nitroacetanilides as chromogenic substrate for assaying de-acetylating activity: The isolation and partial purification of aryl acylamidases from erepsin and tulip, Enzymologia, 41:313.

Hoffmann, K., 1974, Testicular involution in short photoperiods inhibited by melatonin, Naturwissensch., 61: 364.

Hoffmann, K., 1981, Photoperiodic function of the mammalian pineal organ, in: "The Pineal Organ, Photobiology - Biochronometry - Endocrinology", A.Oksche and P.Pévet, eds., Elsevier/North-Holland, Amsterdam.

Holmstedt, B., and Lindgren, J.E., 1967, Chemical constituents and pharmacology of south american snuffs, in: "Ethnopharmacologic Search for Psychiatric Drugs", D.H.Efron, ed., U.S.Public Health Service, Washington.

Hooper, R.J.L., Silman, R.E., Leone, R.M., and Young, I. M., 1981, The development of a plasma assay for 5-methoxytryptamine using gas chromatography-mass spectrometry, in: "Pineal Function", C.D.Matthews and R.F.Seamark, eds., Elsevier/North-Holland.

Hoskins, J.A., and Pollit, R.J., 1978, The determination of 5-methoxyindole-3-acetic acid in human urine by mass fragmentography, J. Chromatography, 145:285.

Jequier, E., Robinson, D.S., Lovenberg, W., and Sjoerdsma, A., 1969, Further studies on tryptophan hydroxylase in rat brainstem and beef pineal. Biochem. Pharmacol., 18:1071.

Klein, D.C., 1979, Circadian rhythms in the pineal gland, in: "Endocrine Rhythms", D.T.Krieger, ed., Raven Press, New York.

Kopin, I.J., Pare, C.M.B., Axelrod, J., and Weissbach, H., 1961, The fate of melatonin in animals, J. Biol. Chem., 236:3072.

Krebs, H.A., Sykes, W.O., and Bartley, W.C., 1947, Acetylation and deacetylation of the p-amino group of sulphonamide drugs in animal tissues, Biochem. J., 41:622.

Kveder, S., and McIsaac, W.M., 1961, The metabolism of melatonin (N-acetyl-5-methoxytryptamine) and 5-methoxytryptamine, J. Biol. Chem., 236:3214.

Lerner, A.B., and Case, J.D., 1960, Melatonin, Fed. Proc., 19:590.

Lerner, A.B., Case, J.D., Takahashi, Y., Lee, T.H., and Mori, W., 1958, Isolation of melatonin, the pineal gland factor that lightens melanocytes, J. Amer. chem. Soc., 80:2587.

Lerner, A.B., Case, J.D., and Takahashi, Y., 1960, Isolation of melatonin and 5-methoxyindole-3-acetic acid from bovine pineal glands, J. Biol. Chem., 235:1992.

Lerner, A.B., Case, J.D., Biermann, K., Anthony, R.V., and Krivis, A., 1959, Isolation of 5-methoxyindole-3-acetic acid from bovine pineal glands, J. Amer. chem. Soc., 81:5264.

Lovenberg, W., Weissbach, H., and Udenfriend, A., 1962, Aromatic l-amino acid decarboxylase, J. Biol. Chem., 237:89.

Lovenberg, A., Jequier, E., and Sjoerdsma, A., 1967, Tryptophan hydroxylation. Measurement in pineal gland, brainstem and carcinoid tumor, Science, 155:217.

Maickel, R.P., and Miller, F.P., 1968, The fluorometric

determination of indolealkylamines in brain and pineal gland, Adv. Pharmacol., 6a:71.
Martin, J.M., Engel, J.N., and Klein, D.C., 1977, Inhibition of the *in vitro* pituitary response to luteinizing hormone-releasing hormone by melatonin, serotonin and 5-methoxytryptamine, Endocrinology, 100:675.
Mas, M., and Oaknin, S., 1977, Effects of pineal methoxyindoles on male sex behaviour and spermatogenesis, Pineal Gland Symposium, Abstract 18, Israel.
McIssac, W.M., and Page, I., 1959, The metabolism of serotonin (5-hydroxytryptamine), J. biol. Chem., 234:858
McIssac, W.M., Farrell, G., Taborsky, R.G., and Taylor, A.N., 1965, Indole compounds: isolation from pineal tissue, Science, 148:102.
McIsaac, W.M., Taborsky, R.J., and Farrell, G., 1964, 5-Methoxytryptophol: effect on estrus and ovarian weight, Science, 145:63.
Miller, F.P., and Maickel, R.P., 1970, Fluorometric determination of indole derivatives, Life Sci., 9:747.
Nimmo-Smith, R.H., 1970, Aromatic N-deacetylation by chick-kidney mitochondria, Biochem. J., 75:284.
Pévet, P., and Haldar-Misra, C., 1982a, Effect of 5-methoxytryptamine on the testicular atrophy induced by experimental or natural short photoperiods in the golden hamster, J. Neural Transm., in press.
Pévet, P., and Haldar-Misra, C., 1982b, Daily 5-methoxytryptamine injections inhibit short-day induced testicular atrophy in golden hamsters, J. Neural Transm in press.
Pévet, P., and Haldar-Misra, C., 1982c, Morning injection of large doses of melatonin, but not of 5-methoxytryptamine, prevent in the hamster the antigonadotropic effect of 5-methoxytryptamine administered late in the afternoon, J. Neural Transm., in press.
Pévet, P., Balemans, M.G.M., Bary, F.A.M., and Noordegraaf, E.M., 1978, The pineal gland of the mole (*Talpa europaea*, L.). V. Activity of hydroxy-indole-O-methyltransferase (HIOMT) in the formation of melatonin/5-hydroxytryptophol in the eyes and the pineal gland, Ann. Biol. anim. Biochem. Biophys., 18:259.
Pévet, P., Balemans, M.G.M., Legerstee, W.C., and Vivien-Roels, B., 1980, Circadian rhythmicity of the activity of hydroxyindole-O-methyltransferase (HIOMT) in the formation of melatonin and 5-methoxytryptophol in the pineal, retina and Harderian gland of the golden hamster, J. Neural Transm., 19:229.
Pévet, P., Balemans, M.G.M., and De Reuver, G.F., 1981a, The pineal gland of the mole (*Talpa europaea*, L.). VII. Activity of hydroxyindole-O-methyltransferase (HIOMT) in the formation of 5-methoxytryptophan, 5-

methoxytryptamine, 5-methoxyindole-3-acetic acid, 5-methoxytryptophol and melatonin in the eyes and the pineal, J. Neural Transm., 51:271.

Pévet, P., Haldar-Misra, C., and Öcal, T., 1981b, Effect of 5-methoxytryptophan and 5-methoxytryptamine on the reproductive system of the male golden hamster, J. Neural Transm., 51:303.

Pévet, P., Haldar-Misra, C., and Öcal, T., 1981c, The independency of an intact pineal gland of the inhibition by 5-methoxytryptamine of the reproductive organ in the male hamster, J. Neural Transm., 52:95.

Prozialeck, W.C., Boehme, D.H., and Vogel, W.H., 1978, The fluorometric determination of 5-methoxytryptamine in mammalian tissues and fluids, J. Neurochem., 30:1471.

Quay, W.B., 1964, Circadian and estrous rhythms in pineal melatonin and 5-hydroxyindole-3-acetic acid, Proc. Soc. Exp. Biol. Med., 115:710.

Reiter, R.J., 1980, The pineal and its hormones in the control of reproduction in mammals, Endocr. Rev., 1:109.

Reiter, R.J., Blask, D.E., Johnson, L.Y., Rudeen, P.K., Vaughan, M.K., and Waring, P.J., 1976, Melatonin inhibition of reproduction in the male hamster: its dependency on time of day of administration and on an intact and sympathetically innervated pineal gland, Neuroendocrinol., 22:107.

Reiter, R.J., Vaughan, M.K., Blask, D.E., and Johnson, L.Y., 1974, Melatonin: its inhibition of pineal antigonadotrophic activity in male hamsters, Science, 185:1169.

Reiter, R.J., Vaughan, M.K., Blask, D.E., and Johnson, L. Y., 1975, Pineal methoxyindoles: new evidence concerning their function in the control of pineal mediated changes in the reproductive physiology of male golden hamsters, Endocrinology, 96:206.

Rogawski, M.A., Roth, R.H., and Aghajanian, G.K., 1979, Melatonin deacetylation to 5-methoxytryptamine by liver but not brain aryl acylamidase. J. Neurochem., 32:1219.

Sackman, J.W., Little, J.C., Rudeen, P.K., Waring, P.J., and Reiter, R.J., 1977, The effects of pineal indoles given late in the light period on reproductive organs and pituitary prolactin levels in male golden hamsters, Horm. Res., 8:84.

Smith, I., Larson-Carter, D.L., Laud, C.A., Leone, R.M., Silman, R.E., Carter, S.J., Francis, P., Mullen, P. E., Hooper, R.J.L., and Finnie, M.D.A., 1979, O-Acetyl-5-methoxytryptophol - Tentative identification in pineal glands, in: "The Pineal Gland of Vertebrates including Man", J.Ariëns Kappers and P. Pevet, eds., Prog. Brain Res., 52, Elsevier/North-Holland, Amsterdam.

Snyder, S.H., and Axelrod, J., 1964, A sensitive assay for 5-hydroxytryptophan decarboxylase, Biochem. Pharmacol., 13:805.

Talbot, J.A., and Reiter, R.J., 1973/1974, Influence of melatonin, 5-methoxytryptophol and pinealectomy on pituitary and plasma gonadotropin and prolactin levels in castrated adult male rats, Neuroendocrinology, 13:164.

Tamarkin, L., Westerom, W.K., Hamill, A.I., and Goldman, B.D., 1976, Effect of melatonin on the reproductive system of male and female Syrian hamsters: A diurnal rhythm in sensitivity of melatonin, Endocrinology, 99 99:1534.

Trentini, G.P., De Gaetani, G.F., Criscuolo, M., Balemans, M.G.M., Vaessen, H.M.B., and Smith, I., 1982, The effect of melatonin and other indole derivatives in maintaining ovulation in rats kept in continuous light and the influence of these indoles on HIOMT activity in the pineal gland, J. Neural Transm., in press.

Turek, F.W., and Pappas, P., 1980, Daily melatonin injections inhibit short-day-induced testicular regression in hamsters, Experientia, 36:1426.

Vaughan, M.K., Reiter, R.J., Vaughan, G.M., Bigelow, L., and Altschule, M.D., 1972, Inhibition of compensatory ovarian hypertrophy in the mouse and vole: a comparison of Altschules pineal extract, pineal indoles, vasopressin and oxytocin, Gen. comp. Endocr., 18:372.

Vaughan, M.K., Vaughan, G.M., and Reiter, R.J., 1976, Inhibition of human chorionic gonadotrophin-induced hypertrophy of the ovaries and uterus of immature mice by some pineal indoles, 6-hydroxy-melatonin and arginine vasotocin, J. Endocr., 68:397.

Weissbach, H., Redfield, B.G., and Axelrod, J., 1960, Biosynthesis of melatonin. Enzymic conversion of serotonin to N-acetylserotonin. Biochim. biophys. Acta, 43:352.

Acknowledgement: The author wishes to thank Miss J.Sels for skillful secretarial aid.

ACTIONS OF THE PINEAL GLAND AND MELATONIN ON THE SECRETION OF CEREBROSPINAL FLUID

W. B. Quay

Neuroendocrine Laboratory, Department of Anatomy
University of Texas Medical Branch
Galveston, Texas, U.S.A.

INTRODUCTION

Association of pineal function with flow or content of the cerebrospinal fluid (CSF) is a very old concept. In western civilization it came into prominent view with the writings of Herophilos of Alexandria (325-280 BC). However, then and through many subsequent centuries, this functional association was based essentially on philosophical speculations, which were contested and eventually discredited, bringing down with them many of the notions derived from, and dependent upon, them. This long and faltering early phase of "pinealology" corresponds to the first of the three periods in pineal research as summarized by Kappers (1979).

The present, and third, period in the historical evolution of pineal research has given us new vistas of the true functional relations of the pineal. One of these research vistas that I believe to hold especial importance concerns major functional interrelations between the pineal and the central nervous system (CNS). I believe that present evidence suggests that on both the pineal and the nervous system sides of these interactions, both neural and humoral mechanisms are involved. By neural mechanisms we mean innervations. By humoral mechanisms we mean chemical mediations at the tissue humor or parahormone level as well as at the truly hormonal level. The hormonal level of chemical mediation has the more prominent status in current concepts of mammalian pineal physiology and in the design and interpretation of experiments. In comparison, the range of probable mechanisms of mediation involving innervations and local humoral mechanisms is, I believe, understated. Having said this, we must admit, however, that the physiological and pharmacological actions we will be discussing are known chiefly in terms of a hormonal paradigm of pineal action. Furthermore, on the pinealopetal side, aside from sympathetic innervation, relatively little is known about how information concerning brain and/or CSF characteristics and composition may be transmitted to and received by pinealocytes.

It is now clear that the pineal and its hormones have multiple sites of action and kinds of effects within the CNS. References are available in the review of Trentini et al. (1979), and in the present volume those by Cardinali and Mess. My report here is limited to considerations of evidence of pineal actions on mechanisms governing the secretion and composition of CSF, a hitherto relatively neglected subject. The possible physiological impact, however, of pineal actions on secretion and composition of CSF may extend far beyond the particular primary target cells involved in these actions. This is because controls of neuronal excitability and transport mechanisms are affected significantly by small changes in electrolyte concentrations and in intracranial pressure, through the media of CSF and related compartments within the brain and its tissue coverings. CSF represents the extra- and intercellular fluid of the CNS, and is known and studied chiefly as the fluid within the internal ventricular system and the external arachnoidal covering and cisternal systems of the CNS and its meninges.

My concern with a possible pineal role in regulation of the electrolyte concentrations of the brain started with the obtaining of unexpected experimental results (Quay, 1965). These showed that in pinealectomized rats, as compared with sham-operated and non-operated sibling control animals, a sodium-deficient diet led to a significantly lower potassium content in some brain regions (Quay, 1965). Furthermore, continuous light produced the same effect as pinealectomy on brain potassium concentrations in animals on a sodium-deficient diet. But brain potassium concentrations were not lower in pinealectomized rats following a potassium-deficient diet. These and other results from experiments and available published information led to the hypothesis that in the evolution of the pineal and the astrocyte system of the brain in tetrapods, the pineal acquired a role in regulation of electrolyte concentration possibly by acting on the selective transport and metabolic activities of astrocytes (Quay, 1965, 1969, 1970b). In studies on temporal relations of mitotic activity of astrocytes following a standardized forebrain lesion, there was no generalized effect of pinealectomy. But a difference was seen in the phase relations of astrocyte mitoses in the amygdaloid nuclei in pinealectomized animals (Quay, 1974b). Pinealectomy also failed to modify significantly brain water content after two kinds of phase-shifts in photoperiod, and at two times in relation to the timing of the daily photoperiod (Quay, 1970a). It has been difficult to study quantitatively and with sensitivity the responses of astrocytes since they are usually dispersed among other cell types within CNS tissue. However, with technical advances in the use of potassium-sensitive microelectrodes, autoradiography and immunocytochemistry, important research advances now appear more likely in this area.

The hypothesis that the mammalian and human pinealocyte responds to changes in intracranial pressure is currently supported by only circumstantial and indirect evidence. This hypothesis is, however, worthy of note because it points out one category of relationship on the afferent side of a pineal role in the control of secretion and composition of CSF. In early and detailed work on human pineal pathology Walter (1923) observed that hypertrophy of pinealocytes was a concomitant of a diverse array of chronic pathological or disease conditions which were either causes of, or associated with, increased intracranial pressure. A reevaluation (Quay, 1974a) of a set of necropsy data on pineal enzyme activities in man (Otani et al., 1968) also can be interpreted as relevant to this hypothesis.

Our present report and review focuses primarily on recent experimental evidence suggesting that the mammalian pineal through the agency of at least one of its hormones, melatonin, has a regulatory role in relation to secretory or transport, and metabolic, activities of the ependymal epithelial cells of the choroid plexuses of the lateral ventricles of the brain.

STIMULATORY EFFECTS OF MELATONIN ON CHOROIDAL EPITHELIAL CELLS

Recently we have obtained ultrastructural and morphometric evidence for stimulatory actions of melatonin on choroidal epithelial cells in male Golden Hamsters (*Mesocricetus auratus*)(Decker and Quay, in press). The Golden Hamster is a species in which a greater role for humoral control of choroid plexuses might be expected on the basis of the observation by Lindvall et al. (1977) that of ten studied species, the Golden Hamster had the least evidence of a cholinergic nerve supply to its choroid plexuses. It is to be appreciated, however, that other kinds of innervating fibers are not ruled out, and that there are differences between the choroid plexuses of the different ventricles in terms of their characteristics of growth (Quay, 1972), metabolism, composition (Quay, 1966) and innervation (Lindvall et al., 1977).

Animals and Procedures

Male Golden Hamsters were injected daily for 28 consecutive days starting at 4 to 5 weeks of age, with vehicle alone (0.1 ml 0.9% NaCl 30% ethanol), or 25 μg melatonin (Sigma), or 2500 μg melatonin in 0.1 ml of vehicle. The animals were kept in a LD 14:10 photoperiod and received their injections in dorsal skin folds on the body between 13:00 and 13:45 (= $L_{11.00}$ to $L_{11.75}$). They were killed by cervical dislocation on the afternoon of the day following the last injection day. Choroid plexuses (CPs) were removed and fixed within 5 minutes of the time of death. Each of the 16 animals provided 3 CP samples: (1) CPs of ventricles I and II, the lateral ventricles, pooled; (2) CP of ventricle III, including its posterior portion which extends into the suprapineal recess (= dorsal sac); and (3) CP of ventricle IV. Primary fixation was in 2% glutaraldehyde 2% paraformaldehyde in 0.1 M pH 7.3 sodium cacodylate buffer, first at room temperature and then overnight at 5° C. The CPs subsequently were rinsed in cacodylate buffer and post-fixed in 1% osmium tetroxide 1.5% potassium ferricyanide in 0.1 M sodium cacodylate (Langford and Coggeshall, 1980). Staining en bloc with uranyl acetate, dehydration through ethanol solutions and passage through propylene oxide preceded embedding in an Epon/Araldite mixture (Mollenhauer, 1964). Thick (1 µm) sections stained with toluidine blue were inspected and suitable areas were thin-sectioned and stained with uranyl acetate and lead citrate (Venable and Coggeshall, 1965).

Thin sections were examined with a Philips EM 300 and the first ten "whole" cells per each animal's regional (ventricular) sample were photographed. For the purposes of this study, "whole" cells were those choroidal ependymal cells that showed within the particular section field their nuclei and both the apical (ventricular) and basal plasma membranes. For morphometry, electron photomicrographs were taken at tap 1 (x 3,160) and later enlarged (x 2.8) giving a final magnification of x 8,850.

Ten choroidal ependymal cells from each of the three ventricular CP regions (I+II, III and IV) were studied from each of the animals. Thirteen different kinds of counts or measurements were made. These were not converted to volumes or volume densities since sampling was biased in favor of cells sectioned in the vertical system of planes with respect to the plane of the ependymal epithelial layer. Therefore, the significance of the resulting measurements resides chiefly in their comparative value for the purpose of testing for kind(s) and magnitude(s) of effect(s) of melatonin. They may not accurately represent true absolute values of component volumes within the cells.

Observations

The ultrastructural characteristics of the choroidal ependymal cells in our Golden Hamsters conformed to what has been described for these cells in other species (Westergaard, 1970; Davis and Milhorat, 1975; Agnew et al., 1980; Koshiba et al., 1980; Allen et al., 1981). However, our chief interest was in those features that relate to functional activities and that demonstrated significant differences following daily injections of melatonin.

Significant differences were found in four features of the cells suggesting stimulatory actions by melatonin: (1) increase in cytoplasmic volume (Table 1), (2) increase in nuclear volume (Table 1),

Table 1. Effects of Melatonin and Ventricular Origin on Mean Areal Measurements of Choroidal Ependymal Epithelial Cells in Male Golden Hamsters (data of Decker and Quay, in press)

Measurements (μm^2) Melatonin µg/day	Ventricles I + II $\bar{x} \pm SE$	P^a	III $\bar{x} \pm SE$	IV $\bar{x} \pm SE$
Total Cell Area				
0	74.6±3.6	<.001	75.1±4.5	71.8±3.8
25	78.6±1.8	<.001	78.0±4.6	86.2±5.9
2500	115.6±8.4	*	82.9±5.0	78.0±4.0
Cell Area Minus Nucleus & Blebs				
0	58.7±3.8	<.001	60.5±4.0	55.5±3.1
25	60.0±2.6	<.001	63.3±4.5	66.4±7.5
2500	88.4±7.8	*	65.6±3.8	61.3±3.3
Nuclear Area				
0	15.5±1.5		14.8±2.2	13.7±1.1
25	15.4±0.9	<.05	13.5±1.1	15.9±0.9
2500	20.4±1.8	*	16.3±1.9	15.3±0.9

[a] P = probability based on Scheffé tests following ANOVA, in comparisons with starred (*) value in same column and kind of measurement.

(3) increase in mitochondrial area per cell (Table 2), and (4) increase in length of apical (ventricular) microvilli (Table 3). All of these effects observed at the ultrastructural level were either more notable in, or were restricted to, the choroidal ependyma of the lateral ventricles as compared with those of ventricles III and IV (Tables 1 through 3).

Table 2. Effects of Melatonin and Ventricular Origin on Mean Mitochondrial Area Per Cell, Mitochondrial Fraction of the Cytoplasm and Area Per Mitochondrion, in Choroidal Ependymal Epithelial Cells in Male Golden Hamsters (data of Decker and Quay, in press)

Measurements	Ventricles			
Melatonin μg/day	I + II $\bar{x} \pm$ SE	P^a	III $\bar{x} \pm$ SE	IV $\bar{x} \pm$ SE
Mitochondrial Area Per Cell (μm^2)				
0	7.34±0.42	*	7.57±0.57	8.65±0.76
25	9.23±0.43	<.05	8.76±0.59	9.28±0.42
2500	9.97±0.50	<.005	9.07±0.57	9.97±0.43
Mitochondrial Fraction of Cytoplasm (1 = 100%)				
0	.1251±.0068	<.025	.1250±.0073	.1559±.0072
25	.1539±.0066	*	.1383±.0083	.1398±.0082
2500	.1128±.0049	<.001	.1382±.0121	.1626±.0090
Area Per Mitochondrion (μm^2)				
0	.1308±.0134		.1099±.0190	.1317±.0058
25	.1417±.0086		.1312±.0173	.1469±.0153
2500	.1346±.0076		.1413±.0095	.1520±.0173

[a] P = probability based on Scheffé tests following ANOVA, in comparisons with starred (*) value in same column and kind of measurement.

Table 3. Effects of Melatonin and Ventricular Origin on Mean Length of Microvilli on the Ventricular Apices of Choroidal Ependymal Epithelial Cells in Male Golden Hamsters (data of Decker and Quay, in press)

Melatonin	Ventricles			
μg/day	I + II $\bar{x} \pm$ SE	P^a	III $\bar{x} \pm$ SE	IV $\bar{x} \pm$ SE
0	827.2±40.6	<.001[b,c]	1052.4±41.7	1012.8±34.1
25	1026.0±30.8	<.001[c]	1053.5±50.5	1027.1±31.9
2500	1245.7±30.8		1146.8±91.2	1113.9±46.1

[a] P = as in Table 2, with comparisons with 25 (b) and 2500 μg (c).

DISCUSSION

The cellular changes produced by melatonin injections may represent two general categories of cellular response. The increase in cytoplasmic and nuclear areas or volumes (Table 1) appears to indicate a kind of cellular hydrops. This interpretation is based on the fact that this swelling of the choroidal ependymal cells occurred only in animals receiving the very high (2500 μg/day) dose of melatonin, while the other cellular responses to melatonin had a graded increase from 0 to 25 to 2500 μg dose levels. Nevertheless, the swelling of the choroidal ependymal cells with the high dose serves to demonstrate regional differences in the sensitivity of the choroidal ependymal cells to melatonin. The results in the top and middle thirds of Table 1 show that the choroidal ependymal cells of the lateral ventricles increased about 50% both in total cell area and in cell area minus nucleus and blebs, while there was no such increase in the comparable cells of ventricles III and IV after the high dose of melatonin.

The second category of cellular response to melatonin here, is of a graded nature and may be more confidently interpreted as representing cellular stimulation by melatonin. The increase in mitochondrial area per cell (top third of Table 2) appears to represent an increased number of mitochondria per cell and not an increase or change in average size of the mitochondria (bottom third of Table 2). The low value for mitochondrial fraction in choroidal ependymal cells of the lateral ventricles after high melatonin dosage (middle third of Table 2) can be attributed at least largely to the aforementioned swelling of these cells. Visual comparisons of the cells suggest that it is chiefly in the apical cytoplasm, that toward the ventricle, that the increase in number of mitochondria occurred.

Mean length of apical microvilli also showed a graded response to melatonin. This too was notable only in the choroidal ependymal cells of the lateral ventricles. On the basis of circumstantial evidence this response appears to be interpretable as representing a stimulatory action of melatonin. Although the morphological evidence is incapable of providing directional determinations of changes in fluid movement across the choroidal epithelium, the 24 (25 μg melatonin) to 50% (2500 μg melatonin) increase in mean length of the microvilli suggests increased capacity for such flow or transport. The recent demonstration of ouabain-sensitive Na^+,K^+-ATPase in these apical microvilli (Masuzawa et al., 1981) indicates the involvement of these structures at this particular location in the secretion of CSF and in ion transport. The driving force for CSF secretion has been believed for many years to be the active transport of Na^+ across the choroid plexus cells. According to a model (Wright, 1972) of how this system works, Na^+ enters the choroidal epithelial cell from the serosal solution in relation to an electrochemical gradient, and is actively secreted into the CSF across the brush border (= apical microvilli) by a ouabain-sensitive Na+, K+ exchange pump. The location of this pump has been demonstrated also by means of autoradiography of the binding of ouabain from solutions of very low concentration (Quinton et al., 1973). Much remains to be done of course, in order to determine: (1) the extent to which this pump and its actions and linkages are influenced by pinealectomy and/or melatonin, (2) the generality among different species and choroid plexuses of such pineal or melatonin effects, and (3) the contribution of such effects to physiological and pathological regulation of CSF volume and pressure.

Since melatonin in this study was administered systemically, it appears likely that the accessibility of melatonin to the choroid plexuses of the different ventricles was equivalent. It may be functionally significant that the choroidal ependyma at the start of CSF's pathway of flow (ventricles I + II) was affected by melatonin while those situated close to the pineal (ventricle III, including the suprapineal recess) and "downstream" (ventricle IV) were not. This suggests that at least in the Golden Hamster, if endogenous melatonin has a role in influencing the CPs of the lateral ventricles in their secretion of CSF, the effective route of access can be vascular and systemic rather than necessarily either intraventricular or by a special local vascular route or portal system starting in the pineal. Melatonin has been found in CSF and blood of man (I. Smith et al., 1976; Arendt et al., 1977; Vaughan et al., 1978; Brown et al., 1979; J. A. Smith et al., 1979; Tan and Khoo, 1981) and other mammals (Withyachumnarnkul and Knigge, 1980; Perlow et al., 1981). CSF levels of melatonin are generally lower than blood levels, and several other lines of evidence have been raised for the conclusion that at least in man and rat the major route of pineal release of melatonin is into the bloodstream rather than the CSF (Vaughan et al., 1978; Brown et al., 1979; Withyachumnarnkul and Knigge, 1980; Tan and Khoo, 1981). After intravenous injection of ^{3}H-melatonin in rats, melatonin is detectable in CSF within 1 minute and reaches peak levels in CSF and choroid plexuses at 5 minutes (Mess and Trentini, 1974). However, in the bullfrog melatonin can be taken up from CSF by the choroidal epithelium by an active transport process located at the apical brush border (Smulders and Wright, 1980).

Increased intracranial pressure (ICP) either from increased CSF pressure or increased vascular perfusion pressure is generally thought to lead to a lowering of the rate of secretion of CSF by the choroid plexuses (Sahar, 1972b). However, elevation of ICP does not appear to influence the rate of production of CSF if the cerebral vascular perfusion pressure is maintained above a particular level, such as 70 mm of Hg in cats (Weiss and Wertman, 1978). These and other findings serve as caveats for anyone tempted to take a simplistic view of the control of either ICP or rate of formation of CSF. In relation to the interpretation of our results with melatonin, some partial satisfaction may accrue from the general belief that no bulk absorption of CSF occurs via the choroid plexuses (Sahar, 1972a), and the finding that the potassium concentration of CSF secreted by the choroid plexuses is actively regulated by them whether the primary alteration in potassium levels occurs in plasma or CSF (Husted and Reed, 1976).

Among important questions yet to be addressed are the following: (1) Are the choroidal ependymal effects of melatonin general in relation to different species? (2) Are they, like many of melatonin's effects, dependent upon critical timing within circadian and related biorhythms? (3) What, if any, kinds of stimuli relating to ICP and/or CP function serve in the elicitation of melatonin release for regulation of CP cells?

CONCLUSIONS

In one species, the Golden Hamster, melatonin has been shown here to affect quantitative characteristics of the choroidal ependymal cells of the lateral ventricles, but not those of ventricles III and IV. Those affects that are increased with dose level appear to be stimulatory in nature, and to be consistent with the hypothesis that among other roles for melatonin as a circulating hormone is one relating it to stimulation of secretion of CSF and possibly of related activities linked to regulation of transport of particular ions, such as K^+, through the choroidal ependymal epithelium.

These findings possibly have relevance to melatonin's actions in humans, and provide a basis for caution in experimental administration of large doses of melatonin to human subjects. They may also provide the basis for testing humoral and additional pharmacological approaches to control of CSF secretion and of some types of dangerous increase in intracranial pressure. Melatonin, other pineal 5-methoxyindoles, and their analogues merit study with regard to these possible relationships and therapeutic problem areas.

REFERENCES

Agnew, W. F., Yuen, T. G. H., and Achtyl, T. R., 1980, Ultrastructural observations suggesting apocrine secretion in the choroid plexus: A comparative study, Neurological Research, 1: 313.

Allen, D. J., Highison, G. J., Werneck, H., and Gentry, G., 1981, Morphological specializations of the ventricular surface of the choroidal epithelium and associated epiplexus cells, in: Eleventh International Congress of Anatomy: Advances in the Morphology of Cells and Tissues, p. 11, Alan R. Liss, New York.

Arendt, J., Wetterberg, L., Heyden, T., Sizonenko, and Paunier, L.,

1977, Radioimmunoassay of melatonin: Human serum and cerebrospinal fluid, Hormone Res., 8:65.
Brown, G. M., Young, S. N., Gauthier, S., Tsui, H., and Grota, L. J., 1979, Melatonin an human cerebrospinal fluid in daytime: Its origin and variation with age, Life Sci., 25:929.
Davis, D. A., and Milhorat, T. H., 1975, The blood-brain barrier of the rat choroid plexus, Anat. Rec., 181:779.
Decker, J. F., and Quay, W. B., in press, Stimulatory effects of melatonin on ependymal epithelium of choroid plexuses in Golden Hamsters, J. Neural Transmission.
Husted, R. F., and Reed, D. J., 1976, Regulation of cerebrospinal fluid potassium by the cat choroid plexus, J. Physiol., 259: 213.
Kappers, J. A., 1979, Short history of pineal discovery and research, Prog. Brain Res., 52:3.
Koshiba, K., Oshima, H., and Uematsu, H., 1980, Fine structure of the choroid plexus in pigeon by transmission electron microscopy, J. Electron Microscopy, 29:129.
Langford, L. A., and Coggeshall, R. E., 1980, The use of potassium ferricyanide in neural fixation, Anat. Rec., 197:297.
Lindvall, M., Edvinsson, L., and Owman, Ch., 1977, Histochemical, ultrastructural and functional evidence for a neurogenic control of cerebrospinal fluid production from the choroid plexus, Acta Physiol. Scand., Suppl. 452:77.
Masuzawa, T., Saito, T., and Sato, F., 1981, Cytochemical study on enzyme activity associated with cerebrospinal fluid secretion in the choroid plexus and ventricular ependyma, Brain Res., 222:309.
Mess, B., and Trentini, G. P., 1974, ^{3}H-Melatonin level in cerebrospinal fluid and choroid plexus following intravenous administration of the labelled compound, Acta Physiol. Acad. Sci. Hung., 45:225.
Mollenhauer, H. H., 1964, Plastic embedding mixtures for use in electron microscopy, Stain Technol., 39:111.
Otani, T., Györkey, F., and Farrell, G., 1968, Enzymes of the human pineal body, J. Clin. Endocrinol. Metab., 28:349.
Perlow, M. J., Reppert, S. M., Boyar, R. M., and Klein, D. C., 1981, Daily rhythms in cortisol and melatonin in primate cerebrospinal fluid. Effects of constant light and dark, Neuroendocrinology, 32:193.
Quay, W. B., 1965, Experimental evidence for pineal participation in homeostasis of brain composition, Prog. Brain Res., 10:646.
Quay, W. B., 1966, Regional differences in metabolism and composition of choroid plexuses, Brain Res., 2:378.
Quay, W. B., 1969, The role of the pineal gland in environmental adaptation, in: "Physiology and Pathology of Adaptation Mechanisms," E. Bajusz, ed., p. 508, Pergamon Press, Oxford and New York.

Quay, W. B., 1970a, Experimental studies of brain water content and its 24-hour rhythmicity, Rassegna Neurologia Vegetativa, 24:1.
Quay, W. B., 1970b, Endocrine effects of the mammalian pineal, Am. Zool., 10:237.
Quay, W. B., 1972, Regional and quantitative differences in the post-weaning development of choroid plexuses in the rat brain, Brain Res., 36:37.
Quay, W. B., 1974a, "Pineal Chemistry in Cellular and Physiological Mechanisms," Charles C Thomas, Springfield.
Quay, W. B., 1974b, Temporal mitotic patterns around a brain lesion: Cellular and regional asynchronisms and an effect of pinealectomy, Chronobiologia, 1:237.
Quinton, P. M., Wright, E. M., and Tormey, J. M., 1973, Localization of sodium pumps in the choroid plexus epithelium. J. Cell Biol., 58:724.
Sahar, A., 1972a, Choroidal origin of cerebrospinal fluid, Israel J. Med. Sci., 8:594.
Sahar, A., 1972b, The effect of pressure on the production of cerebrospinal fluid by the choroid plexus, J. Neurol. Sci., 16:49.
Smith, I., Mullen, P. E., Silman, R. E., Snedden, W., and Wilson, B. W., 1976, Absolute identification of melatonin in human plasma and cerebrospinal fluid, Nature, 260:718.
Smith, J. A., Barnes, J. L., and Mee, T. J., 1979, The effect of neuroleptic drugs on serum and cerebrospinal fluid melatonin concentrations in psychiatric subjects, J. Pharm. Pharmacol., 31:246.
Smulders, A. P., and Wright, E. M., 1980, Role of choroid plexus in transport of melatonin between blood and brain, Brain Res., 191:555.
Tan, C. H., and Khoo, J. C. M., 1981, Melatonin concentrations in human serum, ventricular and lumbar cerebrospinal fluids as an index of the secretory pathway of the pineal gland, Hormone Res., 14:224.
Trentini, G. P., Mess, B., De Gaetani, C. F., and Ruzsás, C., 1979, Pineal-brain relationship, Prog. Brain Res., 52: 341.
Vaughan, G. M., McDonald, S. D., Jordan, R. M., Allen, J. P., Bohmfalk, G. L., Abou-Samra, M., and Story, J. L., 1978, Melatonin concentration in human blood and cerebrospinal fluid: Relationship to stress, J. Clin. Endocrinol. Metab., 47:220.
Venable, J. H., and Coggeshall, R., 1965, A simplified lead citrate stain for use in electron microscopy, J. Cell Biol., 25:407.
Walter, F. K., 1923, Weitere Untersuchungen zur Pathologie und Physiologie der Zirbeldrüse, Z. Ges. Neurol. Psychiatrie, 83:411.
Weiss, M. H., and Wertman, N., 1978, Modulation of CSF production by alterations in cerebral perfusion pressure, Arch. Neurol., 35:527.

Westergaard, E., 1970, "The Lateral Cerebral Ventricles and Their Ventricular Walls," Andelsbogtrykkeriet, Odense.

Withyachumnarnkul, B., and Knigge, K. M., 1980, Melatonin concentration in cerebrospinal fluid, peripheral plasma and plasma of the confluens sinuum of the rat, Neuroendocrinology, 30:382.

Wright, E. M., 1972, Mechanisms of ion transport across the choroid plexus, J. Physiol., 226:545.

PINEAL-HYPOTHALAMIC INTERACTIONS: POSSIBLE ROLE OF THE MONO-AMINERGIC NEURON SYSTEM

B. Mess[1], G.P. Trentini[2], C. Ruzsás[1], and C.F. De Gaetani[2]

1. Department of Anatomy, Univ. Med. School Pécs Hungary
2. Department of Pathological Anatomy, Univ. of Modena, Italy

I. INTRODUCTION

Reviewing the program for this training course, it is easy to realize that many of the lectures are somehow related to the topics of the present paper. (See lectures of Cardinali, Demaine, Moore, Quay). Therefore, it is difficult to find the proper boundary between the different aspect of similar research problems. In this view, the general morphological, physiological and biochemical changes in the central nervous system, due to the action of the pineal hormones, will be touched upon only very briefly. Therefore, I would like to concentrate rather on the problem of the site and mechanism of action of pineal hormones by which this gland influences the function of the neuroendocrine system. I will attempt to summarize the data indicating a central nervous point of attack of pineal indoles that influence the secretion of the hypothalamic releasing hormones.

II. THE ROUTE OF PINEAL SECRETION.

In the introductory lecture of the Ciba Foundation's Symposium on the pineal gland, held in London, Ariens-Kappers (1971) devoted a special chapter entitled "Mode of secretion of pineal products". This chapter thoroughly discussed the possibility, whether the pineal gland secretes its products into the capillary network of the organ, like a classical endocrine gland, or if these bioactive principles (hormones) are secreted directly into the cerebrospinal fluid (CSF). In this lecture, and later in the general discussion of the entire symposium, Ariens-Kappers strongly stressed that "... all light and electron microscopical data point to the extrusion of the pineal products into the general blood circulation. ... pharmacological findings are, in itself, no reason for accepting that, under physiological circumstances, the pineal secretory products are extruded into the cerebrospinal fluid". This view was supported by the reeent findings of Withyachumnarnkul and Knigge (1980). They found that the melatonin concentration was significantly higher in the blood of the confluens sinuum than in the peripheral blood. CSF taken from the cysterna magna had an even lower melatonin concentration. The melatonin level was increased in the blood but not in the CSF in the dark-night period. Besides these morphologically and physiologically well-founded statements, some physiological (Fraschini et al., 1971), pharmacological (Martini and Fioretti, 1971), anatomical (Sheridan et al., 1969) or ultrastructural (Hewing, 1980) data suggest the possibility of a direct CSF route for pineal secretions.

Rather than belabor the pros and cons concerning this matter, I would like to briefly summarize our investigations performed in the early seventies to approach this puzzling problem.

The most direct evidence would be to compare the indoleamine (e.g. melatonin) level of the general circulation, of the effluent blood of the cannulated pineal veins and of the cerebrospinal fluid, as had been suggested by the speaker at the above mentioned pineal symposium in London (Mess 1971). Because of technical reasons we had to omit the cannulation of the pineal veins and an alternate method was applied. ^{3}H-Melatonin was injected intravenously into adult male rats, and the presence of labelled melatonin was parallelly determined in the peripheral blood, CSF, choroid plexus, hypothalamus, midbrain and cerebral cortex at different intervals after ^{3}H-melatonin injection. In another set of experiments, the tadpole skin-lightening effect was used for the comparative bioassay of melatonin concentration of the blood, CSF and choroid plexus of cats.

Table 1. summarizes the results. Only 2 minutes after its i.v. injection, labeled melatonin appeared in the CSF in high concentrations. A significant isotope accumulation was detected in the CSF even after 2 hours. On the other hand, the choroid plexus showed 2-6 fold higher melatonin content than either hypothalamic or mesencephalic tissue. To verify that melatonin passes from the blood stream to the CSF in a biologically active form, a bioassay (Bagnara, 1963) was performed. As demonstrated in Table 2., biologically active melatonin appears in the CSF and in the choroid plexus in high concentrations 1 and 2 hours after i.v. injection of purified melatonin. - These data lead to the conclusion that melatonin is secreted into the capillaries of the pineal gland, and then concentrated by the choroid plexus and secreted into the CSF. It might explain the controversy detailed in the introduction. The pineal secretes its hormones, like all true endocrine glands, into the capillary network of the gland, but being concentrated in the CSF, they might act on the central nervous system through the CSF.

Table 1

Levels of ^{3}H-melatonin in blood, CSF and different areas of the brain in the albino rat (Dose of melatonin=40μCi/rat)

Minutes after melatonin injection	CPM/ml of blood and CSF, and CPM/mg brain and choroid plexus				
	Blood serum	CSF	Choroid plexus°	Hypothalamus	Mesencephalon
1	43230	9950	63,0	21,5	18,7
2	29838	10698	54,2	17,3	19,0
5	37878	20969	66,0	11,4	9,4
10	31093	10119	52,0	11,2	11,4
20	22123	8800	33,7	5,0	4,5
30	16252	5388	34,1	4,7	4,8
60	8046	3801	18,4	4,2	3,4
120	2273	1000	19,8	3,2	1,6
180	2497	1636	9,9	2,1	2,4
360	1624	890	21,4	1,4	1,8
720	873	793	17,8	2,0	2,0

°The statistical difference between the uptake of the choroid plexus and of the hypothalamus and mesencephalon was equally highly significant $p < 0,001$.

III. ACTIONS OF PINEAL HORMONES ON THE CENTRAL NERVOUS SYSTEM

Our results, summarized above, as well as the data of Cardinali et al., (1973, 1978), clearly indicate that either i.v. or intracysternally injected melatonin will be stored by the hypothalamus and mesencephalon in considerably higher concentrations than by the cerebral or cerebellar cortex. The

Table 2

Melatonin level in CSF, blood serum and choroid plexus of melatonin injected cats as assayed by Xenopus laevis skin melanophore test.

Experimental group	Melanophore index				
	CSF	Serum	Choroid plexus	Melatonin 1 µg/ml	Untreated tadpoles
Untreated control cats	1,79	4,73	3,8 ± 0,30	2,17	4,73
1 hour after 500 µg melatonin injection	1,80	2,20[x]	2,00[xx]	2,30	4,80
Untreated control cats	1,74 ± 0,23	3,96 ± 0,34	3,59 ± 0,30	2,35	4,31
1 hour after 1000 ug melatonin injection	1,29[xx] ± 0,06	1,95[x] ± 0,15	1,22[x] ± 0,12	1,76	4,53
Untreated control cats	3,31 ± 0,27	4,81 ± 0,08	2,57 ± 0,30	1,68	4,51
2 hours after 1000 µg melatonin injection	1,83[x] ± 0,12	4,30[xx] ± 0,13	1,60[xx] ± 0,06	1,94	4,67

x $p < 0{,}01$ compared to the corresponding untreated control cats

xx $p < 0{,}05$ compared to the corresponding untreated control cats melatonin content of different regions of the human brain, determined by radioimmunoassay, was always many-fold (about 50 times) lower than that found in the pineal . No measurable HIOMT activity has been shown in brain tissue indicating that melatonin, which is present in the brain, is not synthetized in situ (Kopp et al., 1980). Since Niles et al. (1979) detected melatonin receptors in the brain, it seems to be very probable that melatonin is taken up and stored by specific melatonin receptors within the brain tissue. The trapped pineal hormones significantly influence the function and metabolism of the brain tissue. As mentioned in the introduction, only the major points of these latter aspects of pineal-hypothalamic interactions will be mentioned briefly.

Pinealectomy significantly increases the MAO activity of the pituitary and less dramatically of the hypothalamus, whereas melatonin decreases MAO activity in the same organs. It was suggested that MAO is a terget enzyme for melatonin (Urry and Ellis, 1975). Data, indicating the role of species differences in the reaction of the hypothalamus to melatonin under in vitro conditions, were reported by Frehn et al. (1974). Δ^4-reductase activity of the hypothalamus was decreased in the hamster and increased in the rat when incubated with melatonin (10^{-6} mol/l). Other enzymes of the brain also are influenced by the pineal gland: Barbanti and Trentini (1965) reported that the ablation of the superior cervical sympathetic ganglia induced a marked increase in the acetylcholinesterase activity of the magnocellular hypothalamic neurosecretory nuclei. Pinealectomy, on the other hand, depressed the thiamine diphosphat-phosphohydrolase activity of the same hypothalamic nuclei. This latter affect was reversed by melatonin or by 5-methoxytryptophol (De Vries, 1972).

Not only were different enzymatic activities of the brain influenced by pineal hormones, but the protein synthesis of the hypothalamus and hypophysis, demonstrated by incorporation of ^{3}H leucine into protein, also was dramatically depressed by melatonin pretreatment (Orsi et al., 1973). However, the protein synthetic rate of the pineal, or that of HIOMT activity, were not influenced by melatonin. The investigators considered these results to provide further experimental evidence for the role of the hypothalamo-hypophyseal system as a target organ for pineal melatonin.

The functional activity of neurons from different brain areas also is generally influenced by pineal indoles. Aside from detailing the well-known data showing that melatonin has a sedating effect of the CNS (Anton Tay, 1974; Marczynski et al., 1964 etc.), the more recent electrophysiologic results of Demaine and Kann (1979 a and b) should be mentioned. They studied the effects of the pineal indoles, melatonin, 5-methoxytryptophol and 5-hydroxytryptophol, on the electrical activity of neurones in the preoptic and anterior hypothalamic areas of rats and hamsters. Upon electrical unit recording of single neurones within these brain areas, melatonin increased, while 5-hydroxytryptophol decreased the rate of neuronal discharge in the majority of the cases. Pineal indoles also modified the inhibitory responses to dopamine and noradrenaline. In evaluating these results, one should take into consideration that only these areas play a major role in the regulation of the hypothalamo-hypophyseal-gonadal axis.

There are still several important effects indicating the action of pineal hormones exerted on the CNS. The main aim of this paper, however, is the discussion of a special aspect of pineal-hypothalamic interplay. In the subsequent part of this talk, I shall try to summarize the findings dealing with the mechanism of pineal-hypothalamic interactions and its integrative role in the regulation of the hypothalamo-hypophyseal neuroendocrine circuit.

IV. EFFECTS OF PINEAL HORMONES AT THE LEVEL OF THE ADENOHYPOPHYSIS

As discussed above, the pineal gland and its hormones exert considerable influence on the central nervous system, mainly at the level of the hypothalamus and mesencephalon. However, the action of the pineal hormones, at the pituitary level also can not be excluded. The *in vivo* experiments, performed in immature (Mieno et al., 1978) and in adult dogs (Yamashita et al., 1978), seem to verify this possibility. The rate of secretion of testicular 17-oxosteroids was significantly enhanced after intracarotid injection of synthetic LHRH both in mature and immature dogs. Melatonin itself did not influence oxosteroid output. However, the effect of exogenous LHRH was markedly diminished by melatonin pretreatment (10 or 100 µg/kg intracarotidally 3 hours before LHRH administration). It was concluded by the authors that melatonin may act directly on the anterior pituitary gland in dogs to inhibit the LHRH-induced release of LH (Fig. 1.).

In addition to melatonin, two other bioactive pineal principles, were reported to have similar effects: 5-methoxytryptophol was less effective (Mieno et al., 1981), while arginine vasotocin (AVT) proved to be as effective as melatonin in the inhibition of the LHRH-induced rise in testicular secretion of 17-oxosteroids (Mieno et al., 1980). It was suggested that 5-methoxytryptophol might act on the same anterior pituitary site as melatonin, whereas AVT might have a different site of action.

This aspect was systematically investigated in rats both under *in vitro* and *in vivo* conditions by Dr. Martin's group. LHRH-induced LH release from pituitary cultures donated by rats 2, 5 and 10 days of age, was significantly diminished by adding

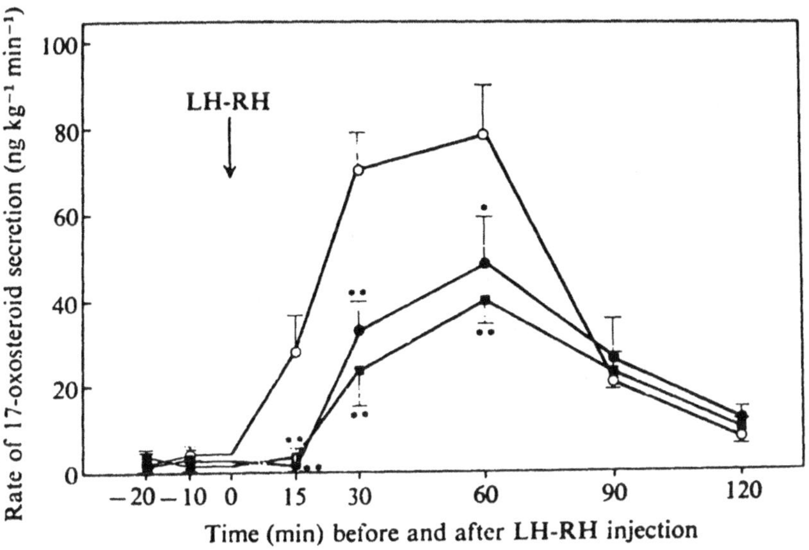

Fig.1. Rate of secretion of 17-oxosteroids by one testis in responses to an intracarotid injection of luteinizing hormone releasing hormone (LH-RH, 5 µg/kg body wt) in dogs given melatonin into the carotid artery 3 h before, expressed as ng produced per kg body wt per min. ■, 100 ug melatonin/kg body wt (seven dogs); ●, 10 µg melatonin/kg (five dogs); ○, melatonin vehicle (six dogs). Each point represents the mean ±S.E.M. *$P<0,05$; **$P<0,01$ compared with the corresponding value when LH-RH was given without pretreatment with melatonin (Yamashita et al., 1978).

melatonin to the medium at 10^{-7}-10^{-10}M concentrations. Serotonin and 5-methoxytryptamine were similarly effective, but other pineal indoles failed to inhibit LH release. However, melatonin, serotonin and 5-methoxytryptamine were ineffective in the inhibiting LH release when pituitary glands from animals 21 and 30 days were incubated (Martin et al., 1977). Comparable results were obtained under in vivo experimental conditions. Serum LH levels were significantly lower in the melatonin-pretreated

LHRH injected male and female neonatal rats than in the solely LHRH-injected animals. Melatonin pretreatment, however, failed to inhibit the LHRH-induced LH release in the older (35-44 day--old) animals (Martin et al., 1980 a). It was concluded that melatonin and serotonin can act directly on the neonatal pituitary gland to suppress LHRH-induced release of LH.

The melatonin metabolite (6-hydroxymelatonin) and a melatonin analog (6 fluoro- and 4-6 difluoromelatonin) also were tested under in vitro conditions. The hydroxy-metabolite of melatonin lost, whereas the fluoro-analog preserved the capability to decrease LH-release at the pituitary level of neonatal rats (Martin et al., 1980 b).

The release of FSH in response to LHRH also was inhibited by adding melatonin to the incubation medium (Martin and Sattler 1979). However, this effect was inversely related to the age of young rats used as pituitary donors (Fig.2.). After 20 days of age, the melatonin sensitivity of the neonatal rat pituitary rapidly disappeared in this model. In contrast, Orts et al. (1980) found that a pineal tripeptide, threonylseryllysine, simultaneously injected intravenously with synthetic LHRH, caused a significant delay of the FSH surge for least 30 minutes after injection in the adult male rat.

The discrepancy experienced between the findings of Yamashita et al. (1978) and of Martin et al. (1977) concerning the age of the animals might easily be due to species differences (dog and rat). The importance of species differences in these types of studies also has been suggested by other investigators. Both in the immature (8 and 12 day of age) and in the adult Syrian hamster, melatonin had no effect at all on the LHRH-induced LH release from the pituitary, equally under in vivo or in vitro experimental conditions (Bacon et al., 1981). Furthermore, Weinberg et al. (1980) also obtained no decrease in LHRH response following melatonin infusion in young human volunteers.

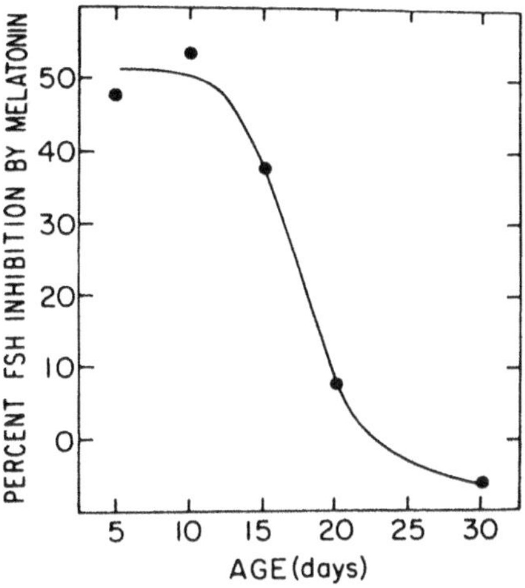

Fig.2. Relationship between animal age and percent maximal inhibition by melatonin of pituitary FSH response to LHRH. The difference in FSH values between groups treated with LHRH alone and groups treated with LHRH and 10^{-8}M melatonin is expressed for each age as a percent of stimulation over controls by LHRH alone (Martin and Sattler, 1979).

Thus, four different mammalian species with completely different pituitary reactions to melatonin treatment is a point to be underscored! One should notice that melatonin does not act at all at the level of the anterior pituitary in two species (hamster and man), or acts at this level only in a very limited perinatal period in the other model (rat), and is fully active only in the dog. It might be assumed therefore that the direct action of melatonin on the adenohypophysial tissue, inhibiting LH release in response to the LHRH might not play an essential no less a general role in the regulatory mechanism exerted by the pineal gland on the neuroendocrine system.

V. PINEAL-CNS-GONADOTROPHIN CIRCUIT

The gonadotrophic hormone system will be used as typical example to try to summarize the mechanism of pineal hypothalamic interactions to provide a framework to understand this pineal-pituitary trophic (e.g. LH) relationships. This will be done for two reasons: firstly, the role of the pineal gland has been most evidently and clearly demonstrated for reproductive functions and secondly, this phenomenon has been of major interest to me and my collegues for more than a decade.

Pinealectomy was found to reastablish ovulation and luteinization in rats subjected to the constant estrous-anovulatory (CEA) syndrome (Fig. 3.). This syndrome can be provoked by either electrolytic lesion of the suprachiasmatic area, or by frontal deafferentation of the hypothalamus (Mess et al., 1971). It subsequently was shown that pinealectomy also provokes ovulation and luteinization in the neonatally-androgenized CEA animals (Ruzsás et al., 1977). Since the anovulatory animals ovulated after pinealectomy, though at irregular intervals (Mess et al., 1973), a preovulatory LH peak was noted (Trentini et al., 1976), and the majority of these rats became pregnant following pinealectomy (Trentini et al., 1973); thus it appears that pinealectomy can provoke an increased LH release in CEA animals.

It is beyond the scope of this paper to discuss all the considerations and evidence which have led us to the assumption that this effect of pinealectomy might be mediated by the

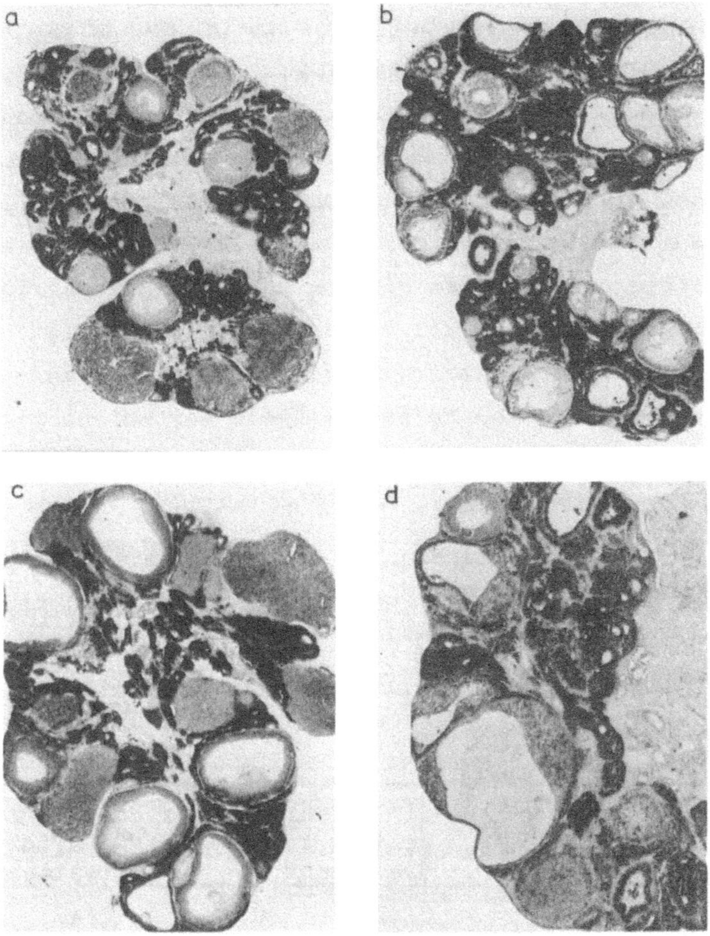

Fig.3. Effect of pinealectomy on the **hypothalamic lesion** induced **polyfollicular ovaries** of **the** rat. a, intact **control; b, polyfollicular** ovary; **c,** reappearance of **corpora lutea** in the **polyfollicular** ovary **following** pinealectomy; d, **marked** thecal **luteinization** in **the polyfollicular ovary following** pinealectomy.

serotoninergic neuron system of the brain stem. This assumption was, however, verified by experiments demonstrating that the ovulation-inducing effect of pinealectomy can be counteracted by daily injections of low doses (1 mg/kg day) of serotonin. Furthermore, the blockade of serotonin biosynthesis by para-chloro-phenylalanine (pCPA) treatment or by maintaining CEA rats on a tryptophan-poor diet (TPD) also provoked ovulation and luteinization in non-pinealectomized CEA animals (Table 3.) (Tima et al., 1973).

Table 3

Effect of pinealectomy, serotonin (5-HT) or parachlorophenylalanine (pCPA) treatment, or keeping the animals on tryptophan-poor diet (TPD, for 30 days) on the occurrence of luteinization in the hypothalamic deafferentation-induced constant estrous anovulatory (CEA) syndrome.

Exp. group	Occurrence of corpora lutea at the end of treatment	%	4 weeks after stopping treatment	%
1. CEA+saline	0/47	0,0	0/47	0,0
2. CEA+pinealectomy	11/17	64,7 x	11/17	64,7 x
3. CEA+pinealectomy+5-HT	2/25	8,0 xx	3/25	12,0 xx
4. CEA+pCPA	11/40	27,5 x	21/40	52,5 x
5. CEA+TPD	11/30	36,6 x	—	—

x $p < 0,01$ vs. group 1.

xx $p < 0,01$ vs. group 2.

All these data have been summarized in detail in a monograph (Mess et al., 1978). Only the basic data were reviewed here, serving as a starting point for our further studies.

It appears curious that pinealectomy was completely ineffective in the third type of the CEA syndrome, i.e. in the light-induced constant-estrous (LCE) animals. In our subsequent and presently ongoing experiments, this latter syndrome proved to be an appropriate model for the investigation of the mechanism of action whereby serotoninergic pathways mediate the effects of the pineal gland on the reproductive system.

The explanation of the above finding that pinealectomy failed to induce ovulation and luteinization in the LCE animals, in contrast to the two other types of CEA syndrome, seemed to be principle. Wurtman et al. (1964) stated that constant light exposure reduces considerably the enzyme activity (HIOMT) of the pineal gland which is necessary for indoleamine production. Therefore, continuous light exposure was called by the Axelrod-Wurtman group as "functional pinealectomy". The surgical extirpation of a functionally eliminated gland cannot provoke considerable effects. The problem, however, still remained unresolved why the removal, or functional inactivation of the pineal gland, elicites LH-release and consequently ovulation in the hypothalamically lesioned, or neonatal androgenization-induced CEA syndrome, and not in the LCE animals? The seemingly contradictory finding of Reiter et al. (1974) provided one probable way for an approach to this question. They reported that under special light-mediated experimental conditions, melatonin inhibits the antigonadotrophic activity of the pineal gland. In other words, under some carefully controlled conditions, melatonin has, instead of the commonly known antigonadotrophic action, a "progonadotrophic" function.

In addition to our own results mentioned above, much data in the literature (for survey see Kamberi, 1974) have stressed the role of the serotoninergic neuron system played in the regulation of gonadotrophin secretion. It should also be taken into account that melatonin and serotonin concentrations of the brain were shown closely correlated (Anton Tay et al., 1968; Cardinali 1975). On the basis of these latter considerations, we tried to approach the problem by utilizing melatonin or serotonin substitution in LCE animals.

Daily subcutaneous injections of 0,1 mg serotonin/rat elicited luteinization in the ovaries of LCE animals. Neither a higher (1,0 mg) nor a lower (0,01 mg) dose of this monoamine had any effect (Hagino et al., 1974) suggesting that just the substitution of the physiological amount of serotonin might play a role in this mechanism.

A single intracisternal, but not subcutaneous injection of 50 µg of melatonin also provoked ovulation in 37,5% of the LCE animals. Even more pronounced was the effect of chronic melatonin administration: daily subcutaneous injection of 2x100 µg of this indoleamine, or its subcutaneous implantation in beeswax pellets, restored the formation of corpora lutea in over 70% of the LCE rats (Table 4.). The same melatonin doses were, however, significantly less effective, when serotoninergic receptors were parallelly blocked by Methiothepin injection, or if the animals were maintained on a tryptophan poor diet (Trentini et al., 1978). These results clearly indicate that the increase in the melatonin level might elicit LH release in the LCE animals on the one hand, and on the other that this effect of melatonin requires functionally intact serotoninergic receptors.

Rats with the fully developed CEA syndrome returned to a normal light-dark rhythm, leads to the reappearance of vaginal cyclicity and ovulation. However, the daily injection of

Table 4

Luteinization provoked by chronic melatonin treatment (for 20 days) in the light-induced constant estrous anovulatory (LCE) animals with or without pinealectomy, and counteraction of this effect by methiothepin treatment or by feeding the animals with tryptophan-poor diet (TPD).

Experimental group	Rats revealing luteiniz. / No. of treated animals	Luteinization %
LCE+solvent injection	0/30	0,0
LCE+Melatonin s.c. (2x100 μg)	18/25	72,0[xx]
LCE+Melatonin implanted in beeswax (3x1 mg)	11/15	73,3[xx]
LCE+Melatonin s.c. (2x100 μg)+ +pinealectomy	11/15	73,3[xx]
LCE+Melatonin s.c. (2x100 μg) + +Methiothepin i.p. (2x10 mg/kg)	6/20	30,0[x]
LCE+Melatonin s.c. (2x100 μg)+TPD	7/25	28,0[x]

x $p < 0{,}01$ vs. melatonin treated groups
xx $p < 0{,}01$ vs. controls

methiothepin (serotonin receptor blocker), or maintaining rats on a tryptophan-poor diet after replacement to a normal light-dark environment, significantly inhibits the restoration of ovulation. These results again suggest a role of the serotoninergic system in this mechanism. The direct proof for this assumption was provided by the experimental group, where LCE animals were kept on TPD after replacement to a light-dark cyclic environment, but were simultaneously daily injected with 50mg/kg serotonin. All of the animals of this latter group ovulated in contrast to the 48% occurrence of ovulation in the light-dark replaced TPD fed animals (Mess et al., 1979) (Table 5.)

Table 5

Ovulation provoked by replacement of light-induced constant estrous anovulatory (LCE) rats into a cyclic light-dark schedule (LD), and the counteraction of this effect by methiothepin treatment or by a tryptophan-poor diet (TPD) (20 days of treatments).

Experimental group	Occurrence of ovulation Number	Percent	Statistical significance
1. LCE rats	0/17	0,0	
2. LCE+LD	30/45	66,6	$p < 0,01$ vs.Gr.1.
3. LCE+LD+Methiothepin (2x10 mg/kg)	2/19	10,5	$p < 0,01$ vs.Gr.2.
4. LCE+LD+TPD	12/25	48,0	$p < 0,05$ vs.Gr.2.
5. LCE+LD+TPD+5-HTP (1 mg/kg)	10/10	100,0	$p < 0,01$ vs.Gr.4.

On the basis of this last series of experiments, we proposed a working hypothesis regarding the role of a "threshold level" of serotonin in the elicitation of ovulation. According to this hypothesis, a critical serotonin level is required to induce or permit the activation of the complicated neuroendocrine mechanism leading to the preovulatory LH-peak. Continuous light exposure suppresses serotonin levels below this threshold level by reducing melatonin biosynthesis in the pineal gland. The replacement of the animals into the physiological LD environment gradually restores serotonin biosynthesis, the brain serotonin concentration reaches the "threshold level" and ovulation can then occur. This working hypothesis is compatible with the findings of Hagino et al. (1974) and Reiter et al. (1974).

Melatonin treatment starting on the first day of continuous light exposure, and injected at the beginning of the dark phase (8 PM), prevents the development of the CEA syndrome (Trentini et al., 1981). This latter finding led us to conclude that melatonin might mimic the dark phase of the nyctohemeral rhythm in the CEA rat, therefore counteracting the effects of continuous light exposure through the serotoninergic neuron system.

Finally, we sought for some morphological proof supporting our view that serotoninergic neuron system might be the most important mediator between the pineal gland and gonadotrophin regulation. Furthermore, we tried to localize the most important structures in the CNS participating in this serotoninergic mechanism.

It is generally accepted that the origin of the ascending serotoninergic pathway terminating in the hypothalamus are the raphe nuclei of the mesencephalon (Ungerstedt, 1971). The serotonin concentration of these nuclei was quantitatively and comparatively determined by Palkovits et al. (1974) and their data strongly suggested the importance of the midbrain raphe nuclei in the serotoninergic mechanism outlined above. The suprachiasmatic nuclear area of the hypothalamus was shown to possess high concentrations and intensive turnover of serotonin, which are closely related to neuroendocrine-regulating properties of these areas (Raisman and Brown Grant, 1977; Héry et al., 1981).

On the basis of these findings, we investigated the effect of electrolytic or of pharmacological (neurotoxin) lesions of these two regions of the CNS that are presumed to play an integrative role in the mediation of pineal effects in the CEA syndrome.

Electrolytic destruction of the median or dorsal raphe nucleus or of the suprachiasmatic area failed to counteract the effect of continuous light (LL) exposure. The occurrence of corpora lutea was equally very low in each lesion-bearing group following LL exposure. Under normal light-dark (LD) environmental circumstances, only suprachiasmatic lesions reduced markedly the occurrence of corpora lutea, as is well known (Hillarp, 1949; Flerkó, 1954). Destruction of the raphe nuclei failed to diminish luteinization in LD animals. However, LL-exposed sham-operated rats injected daily with melatonin revealed luteinization in 75% of the animals. Electrolytic lesion of the suprachiasmatic area and of the median raphe nucleus significantly counteracted the luteinization-inducing effect of melatonin administration in the LL animals. A similar, but less significant effect was elicited by the dorsal raphe lesions (Ruzsás et al., 1981) (Table 6.). These results lend support to the assumption that melatonin influences the reproductive function of the rat by affecting the activity of the median and (less pronouncedly) the dorsal raphe nuclei and of the suprachiasmatic area.

Electrolytic lesion of these regions also might destroy some non-serotoninergic structures belonging to the catecholaminergic or LHRH systems, which could interfere with the effect of lesions of the raphe system.

Pharmacologic destructions of the serotoninergic neurons

Table 6

Effect of electrolytic lesion of the median raphe (MR), dorsal raphe (DR) or of the suprachiasmatic nuclei (SCN) on the occurrence of corpora lutea in LL-exposed rats treated daily with 10 µg melatonin (Mel).

Experimental groups	Occurrence of corpora lutea No of rats	%	Statistical significance
LD			
1. Sham	14/15	93,3	
2. MR	10/10	100,0	
3. DR	16/16	100,0	
4. SCN	1/11	9,1	$p < 0{,}01$ vs.gr.1.
LL			
5. Sham	1/20	5,0	$p < 0{,}01$ vs.gr.1.
6. MR	3/12	25,0	$p < 0{,}01$ vs.gr.2.
7. DR	2/21	9,5	$p < 0{,}01$ vs.gr.3.
8. SCN	1/12	8,3	
LL+Mel			
9. Sham	15/20	75,0	$p < 0{,}01$ vs.gr.1 and 5.
10. MR	1/9	11,1	$p < 0{,}01$ vs.gr.9.
11. DR	10/23	43,5	$p < 0{,}05$ vs.gr.9. and 7.
12.SCN	2/21	9,5	$p < 0{,}01$ vs.gr.9.

by local microinjections of 5-7 dihydroxytryptamine seemed to be more specific than electrolytic lesions. This drug was microinjected with the aid of stereotaxic instrument into the same areas of the CNS as were destroyed by electrolytic lesions in

Table 7

Effect of 5,7-dihydroxytryptamine lesions of the median raphe (MR), dorsal raphe (DR) or of the suprachiasmatic nuclei (SCN) on the occurrence of corpora lutea in LL-exposed rats, treated with melatonin (Mel. 10 µg/day).

Experimental groups	Occurrence of corpora lutea No of rats	%	Statistical significance
LD			
1. Sham	14/15	93,3	
2. MR	10/12	83,3	
3. DR	9/10	90,0	
4. SCN	4/7	57,1	$p < 0,05$ vs.gr.1.
LL			
5. Sham	1/20	5,0	$p < 0,05$ vs.gr.1.
6. MR	1/8	12,5	$p < 0,01$ vs.gr.2.
7. DR	0/9	0,0	$p < 0,01$ vs.gr.3.
8. SCN	0/8	0,0	$p < 0,05$ vs.gr.4.
LL+Mel			
9. Sham	15/20	75,0	$p < 0,01$ vs.gr.1. and 5.
10. MR	10/18	55,0	$p < 0,05$ vs.gr.6.
11. DR	11/22	50,0	$p < 0,05$ vs.gr.7.
12. SCN	5/19	26,3	$p < 0,01$ vs.gr.9.

the previous experiment. Neurotoxic lesions of the raphe nuclei or of the suprachiasmatic area did not influence the development of the CEA syndrome (abolishment of formation of corpora lutea) under LL conditions, similar to the effect of electrolytic lesions. Under LD conditions, only suprachiasmatic lesion influenced luteinization. However, the decreased percentage occurrence of corpora lutea (57,1%) was less pronounced compared to that of electrolytic lesions of the same area (9,1 %). Pharmacologic lesions of all the three investigated brain areas significantly counteracted the luteinization-inducing effect of melatonin treatment in the LL exposed animals. There was no difference between the effect of the dorsal or of the median raphe lesions (occurrence of corpora lutea 50 and 55% respectively), however, these were significantly less effective in counteracting the luteinization inducing effect of melatonin than suprachiasmatic lesion (26,3% occurrence of corpora lutea; Table 7.) , (Ruzsás et al., 1982).

Comparing the results obtained by electrolytic and by neurotoxin intervention, two main differences should be stressed. The neurotoxin lesion of the suprachiasmatic area was less effective in provoking the CEA syndrome under LD conditions than was electrolytic destruction. This might be explained by presuming that the electrolytic lesion destroyed not only serotoninergic terminals but some other structures, probably belonging to the LHRH system. This interpretation is supported by the findings of Mess et al. (1966), Palkovits et al. (1974) and Samson et al. (1980) showing that the suprachiasmatic area possess high LHRH content. Pharmacological lesions of the median raphe nucleus was significantly less effective in counteracting the luteinization-inducing effect of melatonin in LL animals than was the electrolytic damage of the same nucleus. It is conceivable that the electrolytic lesion also destroyed some noradrenergic fibres and terminals in the vicinity of the median raphe nuclei (Roizen and Jacobowitz 1976) (Fig.4.).

The function of the pineal-serotoninergic system-hypothalamic-gonadotrophin circuit could be schematically summarized in the following way: Environmental impulses (e.g. light sti-

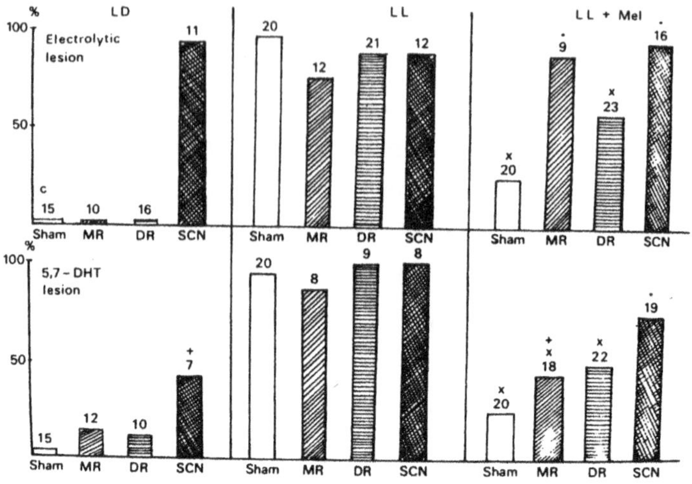

Fig.4. Diagrammatic representation of the occurrence of the constant estrous-anovulatory (CEA) state in rats maintained in normal light-dark cycles (LD) or in continuous light (LL), treated daily with 10 µg melatonin (Mel) and bearing either electrolytic destruction or 5,7-dihydroxytryptamine (5,7-DHT) lesion in the median raphe (MR), dorsal raphe (DR) or in the suprachiasmatic nuclei (SCN).

$p < 0,01$ vs. sham LL+Mel

$p < 0,01$ vs. the corresponding LL and LD

$p < 0,05$ vs. the corresponding electrolytic lesioned group

muli) regulate the secretory activity of the pineal gland; Melatonin acts upon the serotoninergic neurons of the midbrain; Ascending serotoninergic fibres terminate in the hypothalamus, mainly in the suprachiasmatic area; The serotoninergic inputs might regulate LHRH release (Fig.5.).

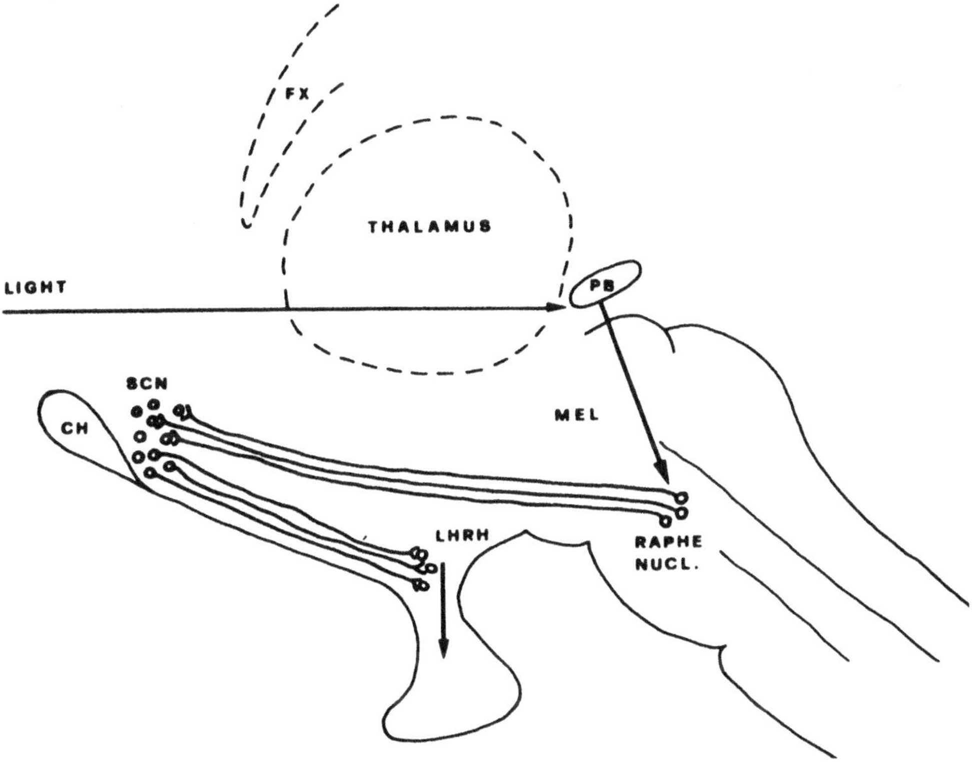

Fig.5. Diagrammatic representation of the postulated mechanism by which light impulses might influence gonadal activity through melatonin secretion.

SUMMARY

It is proposed that the pineal might secrete its hormones towards the capillary system of the gland. Pineal indoles (melatonin) are then concentrated by the choroid plexus into the cerebrospinal fluid. Since the melatonin content of the hypothalamus and midbrain is considerably higher than that of the cerebral cortex or of any other peripheral tissue, these pineal hormones may influence many different functions of the brain.

It is very probable that the pineal hormones influence pituitary activity mainly through the hypothalamus. While a direct pineal-pituitary relationship has not been unequivocally proved, it might have at least minor importance.

The pineal gland acts via the serotoninergic neuron system to influence the hypothalamo-gonadotrophin axis, thus participating in the regulation of ovulation and luteinization. A working hypothesis is presented to explain the mechanism of action of the pineal gland in modulating the function of the reproductive system.

REFERENCES

Anton Tay F., 1974, Melatonin: effects on brain function, Adv. Biochem. Psychopharmacol., 11: 315.

Anton Tay F., Chou C., Anton S., Wurtman R.J., 1968, Brain serotonin concentration: Elevation following intraperitoneal administration of melatonin, Science, 162: 277.

Ariens Kappers J., 1971, The pineal organ: an introduction, in: The pineal gland. A Ciba Foundation Symposium, Wolstenholme G.E.W., Knight J., eds., Churchill Livingstone, Edinburgh and London, p. 3.

Bacon A., Sattler C., Martin J.E., 1981, Melatonin effect on the hamster pituitary response to LHRH, Biol. Reprod. 24: 993.

Bagnara J.T., 1963, The pineal and the body lightening reaction of larval amphibiens, Gener. comp. Endocrin. 3: 86.

Cardinali D.P., Vacas M.J., Boyer E.E., 1978, High affinity of melatonin in bovine medial basal hypothalamus, IRCS Med. Sci. 6: 357.

Barbanti Silva C., Trentini G.P., 1965, Modificazione della neurosecrezione e della attivita acetilcolinesterasica dell' ipotalamo anteriore di ratta albina dopo asportazione del ganglio cervicale superiore, Arch. Vecchi Anat. pat., 45: 807.

Cardinali D.P., 1975, Changes in hypothalamic neurotransmitter uptake following pinealectomy, superior cervical ganglionectomy or melatonin administration to rats, Neuroendocrinology, 19: 91.

Cardinali D.P., Hyyppä M.T., Wurtman R.J., 1973, Fate of intracysternally injected melatonin in the rat brain, Neuroendocrinology, 12: 30.

Demaine C., Kann C.H., 1979a, Modification of the electrical activity of hypothalamic neurons by pineal indoles, Progr. Brain Res., 52: 373.

Demaine C., Kann H.C., 1979b, Hypothalamic neurons as possible target cells for pineal indoles, J. Physiol. (Lond), 291: 49.

De Vries R.A.C., 1972, Influence of pinealectomy on hypothalamic magnocellular neurosecretory activity in the female rat during normal light conditions, light-induced persistent oetrus, and after gonadectomy, Neuroendocrinology, 9: 244.

Flerkó B., 1954, Zur hypothalamischen Steuerung der gonadotrophen Funktion der Hypophyse, Acta Morph. Acad. Sci. Hung., 4: 475.

Fraschini F., Collu R., Martini L., 1971, Mechanism of inhibitory action of pineal principles on gonadotrophin secretion, in: The pineal gland. A Ciba Foundation Symposium, Wolstenholme G.E.W., Knight J., eds., Churchill Livingstone, Edinburgh and London. p. 259.

Frehn J.L., Urry R.L., Ellis L.C., 1974, Effect of melatonin and short photoperiod on Δ 4-reductase activity in liver and hypothalamus of the hamster and the rat, J. Endocr., 60: 507.

Hagino N., Tima L., Flerkó B., 1974, Ovulation inducing capacity of serotonin in persistent oestrous rats, Acta biol. Acad. Sci. Hung., 25: 327.

Héry M., Dusticier G., Faudon M., Barrit M.C., 1980, Kinetic study of serotonin metabolism in the suprachiasmatic nucleus of the rat: neuroendocrine incidence, in: The serotoninergic neuron, Calas A., ed., J. Physiol. (Paris)., p. 497.

Hewing M., 1980, Cerebrospinal fluid-contacting area in the pineal recess of the vole (Microtus agrestis), guinea pig (Cavia cobaya) and rhesus monkey (Macaca mulatta), Cell Tis. Res., 209: 473.

Hillarp N.A., 1949, Studies on the localization of hypothalamic centers controlling the gonadotrophic function of the hypophysis, Acta endocrinol. (Kbh.), 2: 11.

Kamberi I.A., 1974, Catecholaminergic, indoleaminergic and cholinergic pathways and the hypothalamic-hypophysiotrophic neurohormones involved in control of gonadotrophin secretion, in: Endocrinology of Sex, Dörner G., ed., Barth, Leipzig, p. 166.

Kopp N., Chaustrat B., Tappar M., 1980 Evidence for the presence of melatonin in the human brain, Neurosci. Lett. 19: 237.

Marczynski T.J., Yamaguchi N., Ling G.M., Grodinzska L., 1964, Sleep induced by the administration of melatonin (5-methoxy-N-acetyltryptamine) to the hypothalamus in unrestrained cats, Experientia (Basel), 20: 435.

Martin J.E., Engel J.N., Klein D.C., 1977, Inhibition of the in vitro pituitary response to luteinizing hormone-releasing hormone by melatonin, serotonin, and 5-methoxytryptamine, Endocrinology, 100: 675.

Martin J.E., Sattler C., 1979, Developmental loss of the acute inhibitory effect of melatonin on the in vitro pituitary luteinizing hormone and follicle-stimulating hormone responses to luteinizing hormone-releasing hormone, Endocrinology, 105: 1007.

Martin J.E., McKellar S., Klein D.C., 1980a, Melatonin inhibition of the in vivo pituitary response to luteinizing hormone-releasing hormone in the neonatal rat, Neuroendocrinology 31: 13.

Martin J.E., Kirk N.L., Klein D.C., 1980b, Effects of 6-hydroxy-, 6-fluoro-, and 4-6-difluoromelatonin on the in vitro pituitary response to luteinizing hormone-releasing hormone, Endocrinology, 106: 398.

Martini L., Fioretti M.C., 1971, Behavioural effects of pineal principles. in: The pineal gland. A Ciba Foundation Symposium, Wolstenholme G.E.W., and Knight J., eds., Churchill Livingstone, Edinburgh and London, p. 368.

Mess B., Fraschini F., Motta M., Martini L., 1967, The topography of the neurons synthesizing the hypothalamic releasing factors, in: Proc. Second Int. Congr. Hormonal Steroids, Martini L., ed., Excerpta Med. Int. Congr. Ser. 132: 1004.

Mess B., Heizer A., Tóth A., Tima L., 1971, Luteinization induced by pinealectomy in the polyfollicular ovaries of rats bearing anterior hypothalamic lesions, in: The pineal gland. A Ciba Foundation Symposium, Wolstenholme E.G.W., and Knight J., eds., Churchill Livingstone, Edinburgh and London. p. 229.

Mess B., Tima L., Trentini G.P., 1973, The role of pineal principles in ovulation Progr. Brain Res. 39: 251.

Mess B., Trentini G.P., Kovács L., De Gaetani C.F., 1975, Melatonin, cerebrospinal fluid, pineal gland interrelationships, in: Brain-Endocrine Interaction II. The ventricular system in neuroendocrine mechanisms, Knigge K.M., Scott D.E., Kobayashi H., Ishii S., eds., Karger, Basel p. 355.

Mess B., Trentini G.P., Tima L., 1978, Role of the pineal gland in the regulation of ovulation. Stud. Biol. Hung. Vol. 16. Akadémiai Kiadó, Budapest

Mess B., Trentini G.P., Ruzsás Cs., De Gaetani C.F., 1979, Possible role of the brain serotoninergic neuron system in inducing ovulation in the light-induced constant estrous anovulatory syndrome, in: Recent advances in reproduction and regulation of fertility, Talwar G.P., ed., Elsevier, Amsterdam p. 37.

Mieno M., Yamashita E., Iimori M., Yamashita K., 1978, An inhibitory effect of melatonin on the luteinizing hormone releasing activity of luteinizing hormone releasing hormone in immature male dogs, J. Endocr., 78: 283.

Mieno M., Yamashita E., Koba H., Iimori M., Yamashita K., 1980, Combined effects of melatonin and arginine-vasotocin on the canine pituitary response to luteinizing hormone releasing hormone. - IRCS Med. Sci., 8: 458.

Mieno M., Ogawa E., Tanaka R., Iimori M., Yamashita K., 1981, Effect of 5-Methoxytryptophol on the action of luteinizing hormone releasing hormone in the release of luteinizing hormone and its competition with melatonin, IRCS Med. Sci., 9: 87.

Niles L.P., Wong Y.W., Miskra R.K., Brown G.M., 1979, Melatonin receptors in brain, Europ. J. Pharmacol., 55: 219.

Orsi L., Demari J.H., Nagle C.A., Cardinali D.P., Rosner J.M., 1973, Effects of melatonin on the synthesis of proteins by the rat hypothalamus, hypophysis and pineal organ, J. Endocr. 58: 131.

Orts R.J., Bruot B.C., Sartin J.L., 1980, Inhibitory properties of a bovine pineal tripeptide, Threonylseryellisine, on serum Follicle-Stimulating Hormone, Neuroendocrinology, 30: 92.

Palkovits M., Arimura A., Brownstein M., Schally A.V., Saavedra J.M., 1974, Luteinizing Hormone Releasing Hormone (LHRH) content of the hypothalamic nuclei in rat, Endocrinology, 95: 554.

Palkovits M., Brownstein M., Saavedra J.M., 1974, Serotonin

content of the brain stem nuclei in the rat, Brain Res., 80: 237.

Raisman G., Brown Grant K., 1977, The "suprachiasmatic syndrome". Endocrine and behavioural abnormalites following lesions of the suprachiasmatic nuclei in the female rat, Proc. Roy. Soc., 198: 297.

Reiter R.J., Vaughan M.K., Blask D.E., Johnson L.Y., 1974, Melatonin: its inhibition of pineal antigonadotrophic activity in male hamsters, Science, 185: 1169.

Roizen M.F., Jacobowitz D.M., 1976, Studies on the origin of innervation of the noradrenergic area bordering the nucleus raphe dorsalis, Brain Res., 101: 561.

Ruzsás Cs., Trentini G.P., Mess B., 1977, The role of the pineal gland in the regulation of LH release in rats with different types of the anovulatory syndrome, Endokrinologie, 70: 142.

Ruzsás Cs., De Gaetani C.F., Criscuolo M., Mess B., Trentini G.P., 1981, Possible role of the midbrain serotoninergic raphe nuclei in the regulation of ovulation exerted by melatonin in the rat, Neuroendocr. Letters, 3: 331.

Ruzsás Cs., De Gaetani C.F., Criscuolo M., Trentini G.P., 1982, Effect of selective 5,7-dihydroxytryptamine lesions of the midbrain serotoninergic neuron system on ovarian cyclicity maintained by melatonin in the continuous light exposed rat, In press.

Samson W.K., McCann S.M., Chud L., Dudley C.A., Moss R.L., 1980, Intra and extrahypothalamic Luteinizing hormone releasing hormone (LHRH) distribution in the rat with special reference to single neurons responsible to LHRH, Neuroendocrinology, 31: 66.

Sheridan M.N. Reiter R.J., Jacobs J.J., 1969, An intersting anatomical relationship between the hamster pineal gland and the ventricular system of the brain, J. Endocrin., 45: 131.

Tima L., Trentini G.P., Mess B., 1973, Effect of serotonin on ovulation induced by pinealectomy in anovulatory frontal deafferented rats, Neuroendocrinology, 12: 149.

Trentini G.P., Tima L., Mess B., 1973, Effects of bilateral cervical sympathectomy on ovulation and fertility of rats bearing frontal hypothalamic deafferentation and anovulatory syndrome, Hormone Res., 4: 349.

Trentini G.P., De Gaetani C.F., Martini L., Mess B., 1976, Changes in luteinizing hormone levels following pinealectomy and bilateral cervical ganglionectomy in constant oestrous anovulatory rat, Proc. Soc. exp. Biol. Med., 153: 490.

Trentini G.P., Mess B., De Gaetani C.F., Ruzsás Cs., 1978, Effect of melatonin and possible role of the brain serotoninergic system, J. endocr. Invest., 1: 305.

Trentini G.P., Mess B., De Gaetani C.F., Poggioli R., Ferrari P., Di Gregorio C., 1981, The role of melatonin and brain serotoninergic system in the maintenance of rat ovarian cyclicity, in: Pineal function, Matthews C.D., and Seamark R.F., eds., Elsevier, Amsterdam p. 77.

Ungerstedt U., 1971, Stereotaxic mapping of the monoamine pathways in the rat brain, Acta Physiol. Scand., 82: Suppl. 367: 1.

Urry R.L., Ellis L.C., 1975, Monoamine oxydase activity of the hypothalamus and pituitary: Alterations after pinealectomy, changes in photoperiod, or additions of melatonin in vitro, Experientia (Basel)., 31: 891.

Weinberg U., Weitzman E.D., Fukushima D.K., Cancel G.F., Rosenfeld R.S., 1980, Melatonin does not suppress the pituitary luteinizing hormone response to Luteinizing Hormone-Releasing Hormone in Men, J. Clin Endocrinol. Metab., 51: 161.

Withyachumnarnkul B., Knigge K.M., 1980, Melatonin concentration in cerebrospinal fluid, peripheral plasma and plasma of the confluence sinuum of the rat, Neuroendocrinology, 30: 382.

Wurtman R.J., Axelrod J., Fisher J.E., 1964, Melatonin synthesis in the pineal gland. Effect of light mediated by the sympathetic nervous system, Science, 143: 1328.

Yamashita K., Mieno M., Shimizu T., Yamashita E., 1978, Inhibition by melatonin of the pituitary response to luteinizing hormone releasing hormone in vivo, J. Endocr., 76: 487.

HUMORAL INTERRELATIONS OF THE PINEAL GLAND WITH LATERAL EYES AND ORBITAL GLANDS

W. B. Quay

Neuroendocrine Laboratory, Department of Anatomy
University of Texas Medical Branch
Galveston, Texas, U.S.A.

INTRODUCTION

Three kinds of general interrelationships are currently accepted concerning the pineal gland, the lateral eyes and the glands of the orbit: (1) The mammalian pineal gland has evolved from eye-like structures ("median or third eyes") in the pineal complex of lower vertebrates, and its specialized parenchymal cells, the pinealocytes, have residual features of photoreceptor cells (Collin, 1969, 1979; and reviews by Oksche and Kappers in the present volume). (2) The mammalian pineal gland, starting a few days after birth, receives its photic input via the lateral eyes, CNS and sympathetic innervation (Wurtman et al., 1967; Moore et al., 1968; Klein and Moore, 1979). (3) Pineal gland, lateral eye retina, and, to a lesser degree, the Harderian gland of the orbit, share some special biochemical characteristics or capacities. At the present time, the best known and most relevant of these shared characteristics is the presence of the melatonin-forming enzyme, hydroxyindole-*O*-methyltransferase (HIOMT)(Quay, 1965; Quay et al., 1969; Cardinali and Wurtman, 1972; Nagle et al., 1972a, 1972b, 1974; Cardinali et al., 1974; Pévet et al., 1978; Suzuki and Yagi, 1978; Balemans et al., 1980; Pévet et al., 1980a, 1980b, 1981; Ralph, 1980; Vivien-Roels et al., 1981) and its 5-methoxyindole products (Bubenik et al., 1974, 1976a, 1976b; Pang et al., 1977, 1980; Reiter et al., 1981; Yu et al., 1981).

A central theme in the following review is that the synthesis and possible release of melatonin and related indole derivatives by pineal, eye and Harderian gland may form the basis for humoral or hormonal interactions between these organs, or between particular of their constituent cells. An important, and as yet unresolved, question remains about possible tissue sources of circulating melatonin. If the only source of melatonin in the blood of mammals is the pineal gland, as has been thought by many, then melatonin's physiological relations as a hormone are more limited in scope and are less complex for our understanding. However, recent research presents conflicting results and interpretations of whether there are extrapineal sources of hormonal or circulating melatonin (Pang and Ralph, 1975; Lewy et al., 1980; Meyer et al., 1981; Harlow et al., 1981; Yu et al., 1981; Tetsuo et al., 1982). A tentative and conservative conclusion concerning this question would place the burden of proof for extrapineal sources of circulating melatonin on demonstration of chemical specificity in the assay system used. Whatever the ultimate answer to this question may be, one conclusion remains clear: It is possible that endogenous melatonin at levels below detection by current assay methods, and administered or exogenous melatonin, and/or its congeners may possibly participate in humoral or hormonal interactions within and between these three organs of concern here - pineal, eyes and Harderian glands.

Our use of the term "humoral" here, refers to chemical mediator action across intercellular spaces of diverse extent, including those between adjacent cells within the same tissue as well as between cells in different tissues. Possible function of melatonin as such a tissue humor or parahormone can be hypothesized for its actions within its tissues or organs of origin, namely pineal, eyes and Harderian glands of present concern, and possibly certain others, especially in lower vertebrates. Evidence is available, albeit essentially indirect and incomplete, for melatonin functioning as a tissue humor or parahormone within the pineal and eyes, but not as yet within the Harderian gland. The suggestion (Quay, 1974) that the pineal itself may be a first or direct target for melatonin has been supported by experiments in which effects of exogenous melatonin on the pituitary and reproductive organs of rats and hamsters are blocked by prior pinealectomy (Reiter et al., 1976; Sackman et al., 1977; Cardinali et al., 1979; Quay et al., 1982). However, in some as yet undefined circumstances, administered melatonin has antigonadal effects in pinealectomized hamsters (Hoffmann and Küderling, 1977; Tamarkin et al., 1977; Turek, 1977). The pineal's permissive role in the elicitation of some of the effects of administered melatonin may possibly be explained in two ways. Both are consistent with available information, and either one or both may be true: (1) Melatonin receptor and/or other mechanisms within the pineal itself are necessary for melatonin's peripheral actions on some cells in some circumstances. And (2), a chronic or daily conditioning action of the pineal on hypothalamic and/or peripheral receptor mechanisms is necessary for melatonin's actions in some circumstances (Quay et al., 1982). Direct response of Guinea Pig pineal cells to melatonin has been demonstrated recently by recording their electrical activity following microelectrophoretic application of melatonin to them (Semm et al., 1980).

Evidence concerning melatonin's possible functioning within the vertebrate eye is more complex than that concerning its possible actions within the pineal. Therefore, the next major section of this review is devoted to evidence for actions by the pineal or by melatonin within the eye. It should be emphasized that the omission of other pineal products from this discussion is due to the near absence of information about their occurrences rather than to a belief in melatonin's being necessarily the only functionally important pineal product with these functional relations.

The third major section of this review, in an analogous manner, is concerned with evidence for interactions between the Harderian and pineal glands of mammals, the only animals in which such possible relationships have been studied. Here too, much of the information is about melatonin, and little is known about the possible participation of other pineal or Harderian gland products. The interpretation of available experimental results concerning the Harderian gland is most difficult and leaves us with fundamental questions unanswered.

HUMORAL INTERACTIONS INVOLVING THE EYES

Our attention here is directed chiefly to experimental results suggesting actions of the pineal, or of melatonin, on the eyes and their constituent tissues and cells. Although we will assume here that the specific pineal and melatonin actions to be discussed are humoral or hormonal in nature, the possibility remains that some of these actions may either be by way of, or be influenced by, centrifugal (non-sensory) innervation of the eye (see Repérant et al., 1981, for references). This remains an open question. Also remaining outside of our review here are possible interactions with the autonomic innervation of some of the intraocular cells and tissues (Cardinali et al., 1981).

There are several general features of melatonin's relations to the eyes, additional to its synthesis there noted in the Introduction above. First, uptake of administered labeled melatonin by the iris-choroid tissues of the eye is second only to that by the pineal itself, and is therefore greater than that by the usually considered target organs of melatonin (Wurtman et al., 1964; Axelrod and Wurtman, 1966; Table 1). The possable functional significance of this will be noted later in this section, in relation to increased fluid content of hamster eyes following daily injections of melatonin. Secondly, in experiments concerning melatonin's effects on other tissues, melatonin has sometimes been injected "intraocularly" (Pomerantz and Sorrentino, 1973; Pomerantz and Reiter, 1974). It remains to be determined whether in such experiments intraocular effects of melatonin have modified its neuroendocrine or more general systemic effects. Thirdly, if the retina or eye contributes directly to plasma melatonin or related hormone levels, then experiments

Table 1. Tissue Distribution of Radioactivity as ^{3}H-Melatonin or Metabolites One Hour after Intravenous Injection of 100 μc (125 μg) ^{3}H-Melatonin to Cats (Axelrod and Wurtman, 1966)

Tissue	^{3}H-Melatonin Content (mμg/g)
Pineal	25.3
Iris-choroid	6.3
Ovary	5.6
Pituitary	2.9
Sympathetic chain	2.6
Peripheral nerve	2.3
Testis	1.9
Thyroid	1.8
Adrenal	1.5
Kidney	1.3
Liver	0.9
Spleen	0.7
Heart	0.6
Plasma	0.6
Skin	0.5
Brain	0.5
Diaphragm	0.3
Fat	0.2

employing blinding by enucleation or removal of the eyes may possibly have different effects potentially than those in which blinding was by other means.

A related and important topic that we will not review here is the similarity in photic and related responses of retinal and pineal serotonin-*N*-acetyltransferase activity. Most of this research has used chick pineal and retina (Binkley et al., 1979, 1980; Hamm and Menaker, 1980, 1981; Sagar et al., 1982).

The subtopics that will be covered here relate to melatonin's actions, in what are essentially pharmacological studies, on eyes and their parts in vivo. These subtopics can be organized in a variety of ways - all arbitrary. I will follow a plan that is partly by developmental age and partly by functional relationship.

Early Developmental Effects of Indoleamines

Indole biogenic amines are detectable in early developmental stages of both invertebrate and vertebrate animals. They appear to be likely participants in some embryonic inductions and in some developmentally transient mechanisms and phenomena (Buznikov and Manukhin, 1961; Buznikov, 1964; Buznikov et al., 1964, 1966; Quay, 1968; Baker and Quay, 1969; Kostellow and Morrill, 1971). Most studies in this topic area have been concerned chiefly with serotonin (5-HT) and its precursors. However, in the development of the South African Clawed Toad (*Xenopus laevis*) melatonin has been shown to have distinctive stage-related localizations and probable functional relations (Baker et al., 1965; Baker, 1969; Baker and Quay, 1969). In *Xenopus* embryos HIOMT and melatonin are detectable as early as the gastrula stage. But the first major rise in melatonin content occurs just before the time of active swimming (stages 41-42). This leads to a developmental peak in melatonin content, which is located mainly in the eyes (Baker et al., 1965; Baker, 1969; Baker and Quay, 1969). At this developmental stage one role of melatonin within the eyes may be linked to the mediation of retinal photomechanical responses, the subject of the next subsection.

Melatonin and Photomechanical Responses within the Eye

Photomechanical responses of the eye are responses to illumination in which morphological changes or physical movements are seen. When they occur within the retina they are known also as retinomotor responses. They are ancillary to the primary photosensory reactions and responses of the retinal photoreceptor cells, and can occur either within these or other cell types. Some of the more frequently observed photomechanical responses within the eye are: (1) movement of pigment granules within pigment cells, (2) changes in form or length of particular kinds of retinal photoreceptor cells or their parts, and (3) changes in the form and staining reactions of retinal nuclear layers. Early and classical work on this subject included that by Arey (1915, 1916a), Detwiler (1943) and Welsh and Osborne (1937). Later sources are given by Ali (1975, 1981). Although photomechanical responses of the eye are more diverse, higher in amplitude, and more frequently studied in lower vertebrate eyes, they are known also in those of some mammals (Detwiler, 1943; Pang et al., 1978). Two examples of melatonin-activated photomechanical responses will be noted here. It should be kept in mind, however, that the hypothesis of melatonin's mediation or activation of these responses is based upon circumstantial and pharmacological evidence, and is not supported by proof that melatonin is the endogenous mediator.

In *Xenopus laevis* larvae after the developmental stage of the rise in ocular melatonin content, contraction of the cone myoid segment occurs as a retinomotor or photomechanical response to light (Quay and McLeod, 1968). In darkness the lengthening of the cone

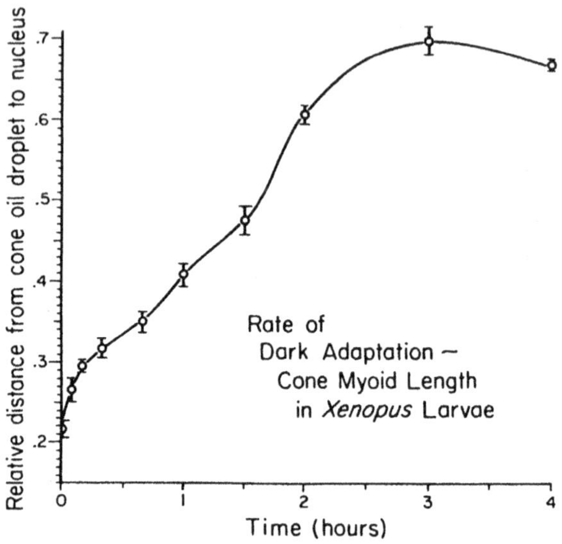

Fig. 1. Course of dark adaptation in cone myoid elongation of larva of *Xenopus laevis*. Means ± standard errors are plotted for groups of sibling larvae in separate glass vials. At times indicated, larvae were killed by the addition of formalin. Distances within the retinas were measured with a calibrate ocular micrometer and mounted and stained tissue sections (Quay, nonpublished data).

myoid is a relatively slow process (Fig. 1). But in either light or a melatonin-containing solution, the cone myoid begins to contract within a few seconds and attains an asymptote in about 45 seconds after exposure (Fig. 2). The near minimal effective concentration o melatonin in the medium to produce a response greater than that of control medium is 5×10^{-10} g/ml (Fig. 2). More information on char- acteristics of these mobile (type 1) cones and technical factors important for their study can be found in a monograph by Saxén (1954 Several circumstances still pose problems for answering the question

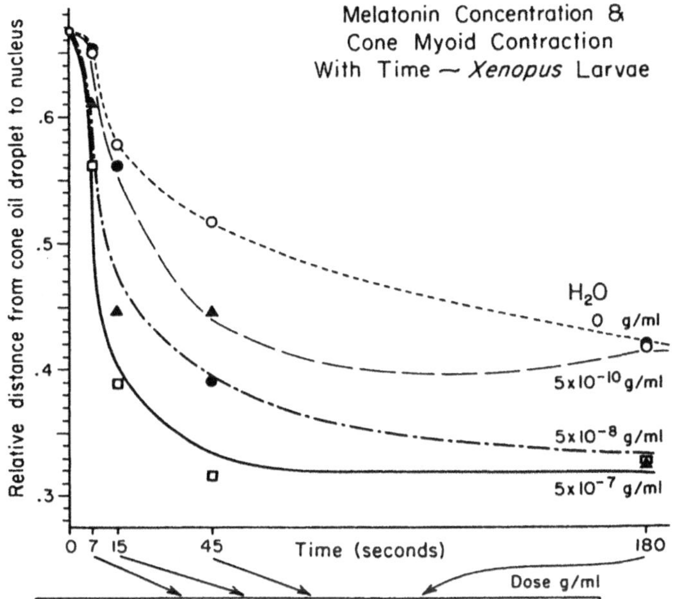

Standard Errors of the Means					Dose g/ml
○	.020	.038	.040	.017	0
●	.026	.042	.029	.024	5×10^{-10}
▲	.029	.026	.026	.026	5×10^{-8}
□	.015	.026	.044	.019	5×10^{-7}

Fig. 2. Comparisons of cone myoid contraction after addition of melatonin or H_2O (control) at "0" time to media containing dark-adapted sibling *Xenopus* larvae. Other information as in legend of Fig. 1.

whether melatonin is the natural mediator of retinomotor responses of larval *Xenopus* cones or whether it is mimicking the action of the natural mediator. Physical agitation of the medium, including that produced by gently adding a "control" solution produces a low level of response (Fig. 2). It might be suggested that this represents a kind of "stress" response mediated by way of efferent fibers to the retina (Arey, 1916b; Kline and Pickering, 1975). But at least in some species retinomotor responses are under local cellular control within the retina (Easter and Macy, 1978). Retinomotor responses in lower vertebrates can be visualized as one category of adaptive response to the peculiarities of photic environment in particular habitats to which the species is adapted (Ali, 1981). Further study of melatonin's possible involvement as a mediator in these responses may well shed light on the physiological evolution of melatonin's roles within the pineal complex of vertebrates.

Melatonin's actions on or in photomechanical responses may also provide a good system for study of melatonin's molecular mechanisms within the cytosol of responding cells. It has been suggested that the mechanism of action of serotonin and some of its congeners, such as melatonin, may involve actions on changes in microtubular or related organelle systems (Quay, 1968). Although melatonin may not be a general inhibitor of microtubules (Poffenbarger and Fuller, 1976), melatonin administration or endogenous release of melatonin can affec at least some cellular mechanisms that depend upon microtubules (Cardinali, 1980).

Photomechanical responses have been demonstrated also to be influenced by melatonin in the eyes of a mammal, the Guinea Pig. In this species light adaptation causes a clumping (lightening) of the contents of the cells of the retinal pigment epithelium and a dispersion (darkening) within the pigment cells of the choroid layer. Dark adaptation leads to opposite effects in both kinds of cells (Pang et al., 1978). Melatonin administration leads to pigment aggregation (lightening) in both (Pang and Yew, 1979).

Melatonin and Light-entrained Disc-shedding by Retinal Rods

Shedding of discs by the rod outer segment and the phagocytosis of these discs by the retinal pigment epithelium follow light-cued circadian rhythms (LaVail, 1976a, 1976b; Goldman et al., 1980). Acute effects of light and darkness on this shedding response in adult frogs are not affected by pinealectomy (Currie et al., 1978). Nor does pinealectomy affect shifts in the disc-shedding rhythm (Goldman, 1982). There appear to be both local (intraretinal) and central (CNS) factors in the regulation of the circadian rhythm in disc-shedding by rat retinal rods (Teirstein et al., 1980). The possibility of an intraretinal contribution by melatonin is still to be tested.

Melatonin's Enhancement of Retinal Damage Induced by Illumination

Prolonged exposure to bright light leads to retinal damage and photoreceptor degeneration in albino rats (O'Steen, 1970; O'Steen et al., 1972, 1974; Reiter and Klein, 1971; Reiter, 1973; Lanum, 1978). This damage is increased when melatonin (100 μg/day) is administered concurrently (Bubenik and Purtill, 1980). In other studies it is reported that the earliest signs of damage by light occur in the retinal pigment epithelium (RPE). The RPE cells seem to be unable to phagocytize outer segment discs and to digest previously ingested ones as a result of changes in the endoplasmic reticulum and the phagolysosomal system (Lai et al., 1980). It would be interesting to determine which if any of these processes are affected by melatonin.

Immunological Relations of Retina with Pinealocytes

Evidence has been presented recently of immunological cross-reactivity between retinal photoreceptors and pinealocytes (Kalsow and Wacker, 1978). A kind of experimental allergic uveitis (EAU) can be elicited in Guinea Pigs by injection of an extract of homologous retinal homogenate (= "S preparation"). During the development of the EAU there is involvement of the pineal gland as well. The multiple changes observed in both tissues suggest that in inflammatory diseases of the retina or pineal, involvement of the other structure should be assayed. There are several interesting implications in these findings that merit investigation. One of these is the possibility that particular populations of pinealocytes might be immunologically identifiable and might also be immunologically deleted by appropriate techniques.

Quantitative Effects of Melatonin on the Eye

Effects of pinealectomy, or of melatonin or other pineal hormones, on weights or volumes of mammalian eyes and their parts have not been reported previously. In recently completed work I have found that melatonin can have significant effects on the eyes of Golden Hamsters. These results are from the same set of animals that was used in study of effects of pinealectomy and melatonin on growth of endocrine and reproductive organs (Quay et al., 1982), and in study of effects of melatonin on choroid plexuses (Decker and Quay, in press). Male Golden Hamsters in an LD 14:10 photoperiod represented five surgical groups: non-operated, sham-pinealectomized, sham-pinealectomized with black plastic shielding over the pineal region of the brain, pinealectomized and pinealectomized with black plastic shielding over the pineal region of the brain. Starting 4 to 5 days after surgery, animals of each of the 5 surgical groups were distributed among 3 injection groups: vehicle (= 0.1 ml 0.9% NaCl 30% ethanol) only, 25 µg melatonin (Sigma) in 0.1 ml vehicle and 2500 µg melatonin in 0.1 ml vehicle. Each animal received a daily subcutaneous injection at $L_{11.00}$ to $L_{11.75}$ in the afternoon of the daily light phase for 28 days.

At autopsy during the afternoon of the day following the last injection, the left eyes were removed and fixed. They were later rinsed, dissected and weighed. None of the surgical procedures had any effect on the weights of the eyes or their parts. However, melatonin injections did have highly significant effects. There was an increase in average weight of intact eyes, and this increase was

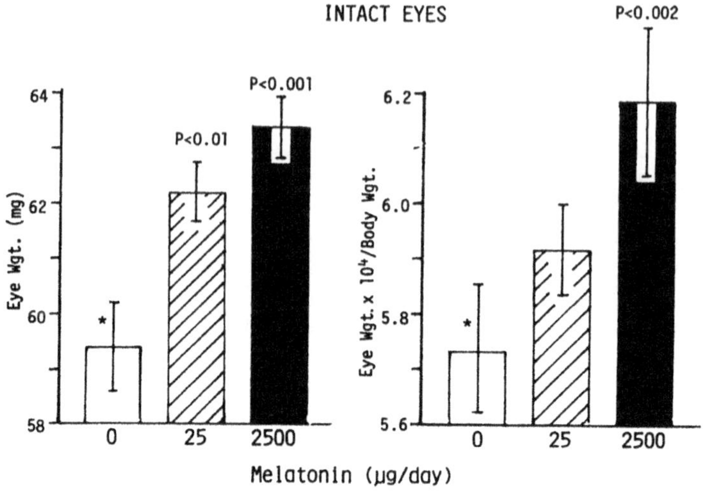

Fig. 3. Effect of 28 consecutive daily injections of melatonin on the weights of intact whole eyes of male Golden Hamsters. Means ± standard errors are shown for absolute weights of left eyes (left panel) and left eye weights in relation to whole animal (body) weights (right panel). Vehicle-injected animals - clear; 25 µg melatonin/day - cross hatched; 2500 µg melatonin/day - solid black. N = 37 or 38 per injection group. * = Groups with which other groups are compared in terms of *P* values based on the Student-Fisher *t*.

greater in relation to daily melatonin dose level (Fig. 3). This was not a consequence of differences in overall body weights, as shown by the equivalence of effects whether absolute or relative eye weights are analyzed (Fig. 3). After the lens and the fluid content of the chambers of the eye were removed, the remaining nonlenticular tissues of the eye showed a lower weight after melatonin (Fig. 4, right panel). In contrast, the fluid in the chambers was increased by melatonin (Fig. 4, left panel). Lens weight was also increased by melatonin, but only with the higher dose level (Fig. 5).

These effects of melatonin on hamster eyes probably represent two kinds of actions. First, there was a chronic stimulation of lens growth by a very high dose of melatonin. This is of interest since most hormones and drugs that are known to affect the lens, are chiefl inhibitory. It will be of interest to determine whether melatonin has any effect on the occurrence or progress of degenerative changes in some kinds of cataract. Second, melatonin's increase of intra-

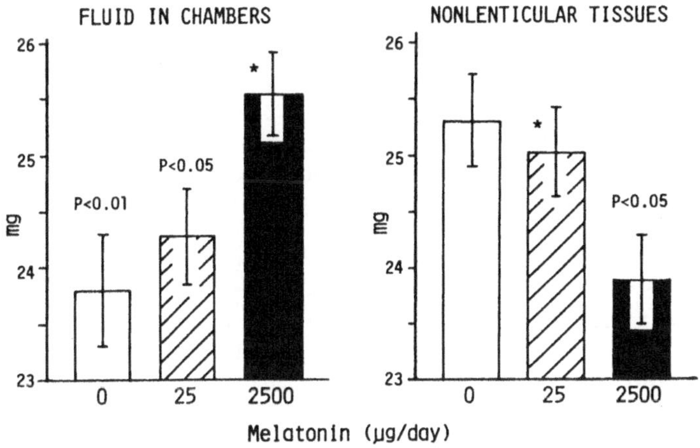

Fig. 4. Effects of 28 consecutive daily injections of melatonin on weights of fluid in chambers and of nonlenticular tissues of left eyes of male Golden Hamsters. Additional information is the same as that given in the legend for Fig. 3.

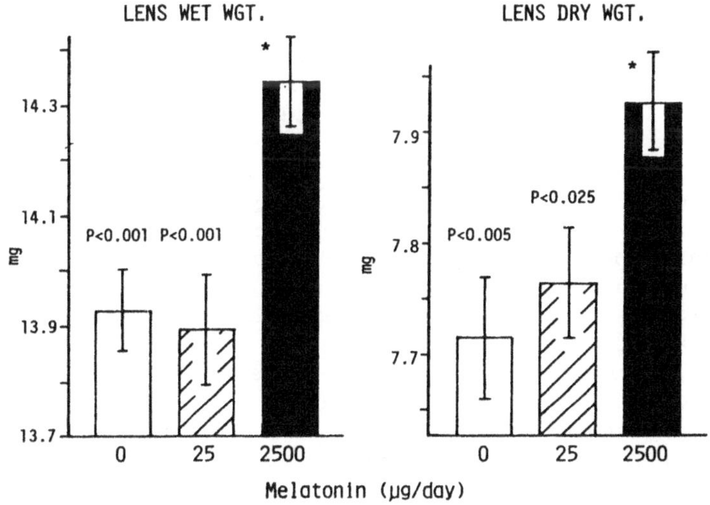

Fig. 5. Effects of 28 consecutive daily injections of melatonin on wet and dry weights of lenses from left eyes of male Golden Hamsters. Additional information is the same as that given in the legend for Fig. 3.

ocular fluid with decreased nonlenticular tissue (Fig. 4) could have resulted through a single mechanism, such as increased intraocular secretion or diminished outflow of aqueous humor, thence possibly leading to increased intraocular pressure (IOP). This effect of melatonin is of particular interest in relation to the possibility that it is relevant to some type(s) of glaucoma, or defects in control of IOP. Whether or not this is true, it is a possibility that should be considered a caveat to those who experimentally administer melatonin to humans.

It may be physiologically significant that the "iris-choroid" has been shown to have an uptake of melatonin second only to that of the pineal (Wurtman et al., 1964; Axelrod and Wurtman, 1966)(Table 1 This suggests that melatonin in the systemic circulation, whether exogenous or derived from the pineal, may have an action on iris-choroid tissue. Since the ciliary epithelium was presumably include in the "iris-choroid" (Table 1) anatomical fraction of the eye, it may have been involved in this melatonin uptake activity. This is worthy of investigation since the eye's aqueous humor is usually thought to be secreted by the ciliary epithelium. Nevertheless, we can not deny that there are alternatives still to be tested for the mechanism of melatonin's actions in the eye. There remains as well the question whether retinal melatonin could possibly reach the ciliary epithelium and thereby modify secretion rate of aqueous humo

More needs to be determined about rhythmic and physiological changes in melatonin formation, release, levels and actions within mammalian and human eyes. Twenty-four-hour rhythmic changes in diverse vertebrate species have been reported in retinal contents of melatonin and of enzymic activities involved in melatonin's biosynthesis (Nagle et al., 1972; Pang et al., 1977; Gern et al., 1978; Joss, 1978). However, in male Golden Hamsters little 24-hour change has been found in retinal HIOMT activity, possibly for technical reasons (Pévet et al., 1980a, 1980b). In rats in a LD 14:10 photoperiod the daily peak in retinal HIOMT activity has been reported to occur near the end of the light phase (Nagle et al., 1972). But retinal melatonin levels in the rat have been reported to be minimal at this time and to peak near the onset of light (Pang et al., 1980) This more closely matches the timing of the daily cycle in retinal *N*-acetylserotonin in this species (Pang et al., 1981), suggesting that serotonin-*N*-acetyltransferase rather than HIOMT activity may be rate-limiting for retinal synthesis of melatonin. This is similar to what has been demonstrated for rat pineal synthesis of melatonin.

IOP in man follows a 24-hour rhythm. Although there is some disagreement or discrepancy concerning both mean pressures and the daily timing of peak pressures, peaks and acrophases are generally attributed to the morning (deVenecia and Davis, 1963; Follmann, 1977; Ferrario et al., 1982), and this is true in the glaucoma population also (Worthen, 1978). This timing, at least superficially, resembles that of the daily melatonin peak in the rat eye. Information on ocular or retinal melatonin rhythms in man is yet to be reported. Nevertheless, it may be of interest in the evaluation of humoral contributions to changes in IOP. This is most important in trying to improve understanding of mechanisms in the etiology, and the pharmacologic approaches for control, of different types of glaucoma.

HUMORAL INTERACTIONS INVOLVING THE HARDERIAN GLANDS

Harderian glands are members of the complex of glands occupying the orbit and adjacent regions, and whose secretions bathe the cornea and conjunctiva. These secretions also can drain to the nasal chambers via the nasolacrimal canal. The relevance of the Harderian glands to pineal physiology and pharmacology stems from four kinds of information: (1) Harderian glands of at least some mammals contain HIOMT, serotonin-*N*-acetyltransferase and melatonin (Cardinali and Wurtman, 1972; Bubenik et al., 1976a, 1976b, 1978; Pang et al., 1977; Brammer et al., 1978; Pévet et al., 1980a, 1980b; Balemans et al., 1980; Reiter et al., 1981). Removal of the Harderian glands in rat and hamster leads to changes in the pineal and reproductive organs (Wetterberg et al., 1970a, 1970b; Reiter and Klein, 1971; Panke et al., 1979). (3) Castration of male hamsters produces Harderian glands of the female type, and studies combining pinealectomy with other procedures have been interpreted as suggesting a pineal involvement in the mediation of such changes (Hoffman, 1971; Clabough and Norvell, 1973). (4) Injection of melatonin modifies the porphyrin content of rat Harderian glands (Joó and Kahán, 1975). (5) Continuous light and blinding each have effects on the Harderian glands and their endocrine relations (Hoffman, 1971; Reiter and Klein, 1971; Reiter, 1973; Clabough and Norvell, 1973; O'Steen et al., 1978).

The microscopic structure of the Harderian glands is that of a merocrine exocrine gland. But the above noted research reports have sometimes interpreted their results to suggest photosensory and/or endocrine activities as well. These still need to be provided with necessary supporting information. Abundant experimental evidence based upon rodents indicates that at least in this animal group, Harderian glands have complex relations with the endocrine system. This is most strongly indicated by: (1) sexual dimorphism, (2) responsiveness to steroid hormones, and (3) contributions to pheromonal mechanisms which in turn may have secondary impacts on diverse aspects of physiology and behavior (Thiessen et al., 1976, 1977; Thiessen, 1977; Kittrell and Thiessen, 1981; Lin and Nadakavukaren, 1981; Shirama et al., 1981). The currently confusing picture of Harderian gland function in mammals results partly from the fact that basic information, concepts and experimental results used and obtained by specialists in one biological or biomedical discipline are either ignored by, or unknown to, specialists in other topics.

Another and more fundamental problem is that the homologies and exact identity or derivation of the Harderian gland, among orbital glands of vertebrates, are not reliably known (Sakai, 1981). The location within a particular part of the orbit, and the gross and microscopic anatomy, of the Harderian glands differ among the major Orders of the Class Mammalia. Biochemical Harderian gland markers, such as lipid and porphyrin contents, have been studied in very few taxa.

Harderian glands in mammals are reportedly relatively largest in lagomorphs and rodents, and to be lacking in primates (Sakai, 1981). This circumstance of itself might be taken as reason enough for considering the Harderian gland of little significance in human biology. However, this conclusion may be premature, since it rests upon faulty and deficient information. This view is offered on the basis of: (1) The homologies and exact identities of the various orbital glands in man and other mammals are poorly known. Therefore, it may be premature to assume that either a homologue or a functional analogue of the Harderian gland does not exist in humankind. (2) Harderian and certain other orbital glands are present and of large size in some marine mammals, including cetaceans and pinnipeds, suggesting that their functional importance is not limited to bathing the exterior surfaces. (3) Endocrine and neuroendocrine studies are either lacking or present conflicting and confusing results for the various named orbital glands. Thus, much more information is needed before confident conclusions can be made about presence or absence of neuroendocrine relations of such glands in man.

CONCLUSIONS

Experimental evidence has been summarized suggesting that some of melatonin's actions may occur in, or by way of, pineal tissue itself, rather than always or necessarily directly upon central, neuroendocrine or peripheral targets.

Melatonin is also both a product of, and a likely chemical mediator within, the lateral eyes of vertebrates including mammals. New findings on melatonin's effects on hamster and Guinea Pig eyes show that at least pharmacologically it can cause: (1) movement of melanin granules in retinal pigment epithelium and choroidal melanocytes, (2) increase in lens material, and (3) increased fluid content of ocular chambers with associated decreased weight of nonlenticular tissues. The latter result suggests that melatonin may be capable of increasing IOP. For this reason its chronic administration at high dosage should not be undertaken without ocular tonometry or equivalent monitoring of IOP.

The high uptake of ^{3}H-melatonin from blood in vivo by the iris-choroid fraction of the eye may possibly be related to the increased fluid content of ocular chambers after systemic melatonin administration. This suggestion follows from the fact that the intraocular fluid with the most rapid turnover, the aqueous humor, is secreted by the ciliary epithelium covering a part of this anatomical fraction of the eye. A hypothesis to explain these findings can be offered - melatonin may have a stimulatory action on the secretory activity of the ciliary epithelium of the eye, leading to increased production of aqueous humor and increased intraocular pressure (IOP) in some conditions or circumstances. This may be analogous to melatonin's stimulatory action on the choroidal ependymal epithelial cells described in another report in this volume. In the latter case, increased secretion of CSF is a possible result. Other kinds of experimental studies are needed, with the objective of testing these and alternative hypotheses.

Harderian glands of some rodents contain melatonin and the enzymes for its biosynthesis. Although results of experiments suggest complex interrelations involving Harderian, pineal and certain of the endocrine glands in rodents, the mechanisms and biological significances of these relationships remain to be demonstrated or clarified.

REFERENCES

Ali, M. A., 1975, Retinomotor responses, in:"Vision in Fishes," M. A. Ali, ed., p. 313, Plenum Press, New York.

Ali, M. A., 1981, Adaptations rétiniennes aux habitats, Rev. Can. Biol., 40:3.

Arey, L. B., 1915, The occurrence and significance of photo-mechanical changes in the vertebrate retina - an historical survey, J. Comp. Neurol., 25:535.

Arey, L. B., 1916a, The movements in the visual cells and retinal pigment of the lower vertebrates, J. Comp. Neurol., 26:121.

Arey, L. B., 1916b, The function of the efferent fibers of the optic nerve of fishes, J. Comp. Neurol., 26:213.

Axelrod, J., and Wurtman, R. J., 1966, The formation, metabolism and some actions of melatonin, a pineal gland substance, in: "Endocrines and the Central Nervous System," Assoc. for Res. in Nerv. and Ment. Dis., 43:200, Williams & Wilkins, Baltimore.

Baker, P. C., 1969, Melatonin levels in developing *Xenopus laevis*, Comp. Biochem. Physiol., 28:1387.

Baker, P. C., and Quay, W. B., 1969, 5-Hydroxytryptamine metabolism in early embryogenesis, and the development of brain and retinal tissues," Brain Res., 12:273.

Baker, P. C., Quay, W. B., and Axelrod, J., 1965, Development of hydroxyindole-O-methyl transferase activity in eye and brain of the amphibian, *Xenopus laevis*, Life Sci., 4:1981.

Balemans, M. G. M., Pévet, P., Legerstee, W. C., and Nevo, E., 1980, Preliminary investigations on melatonin and 5-methoxy-tryptophol synthesis in the pineal, retina and Harderian gland of the Mole Rat and in the pineal of the mouse "Eyeless", J. Neural Transmission, 49:247.

Binkley, S., Hryshchyshyn, M., and Reilly, K., 1979, *N*-Acetyltransferase activity responds to environmental lighting in the eye as well as in the pineal gland, Nature, 281:479.

Binkley, S., Reilly, K. B., and Hryshchyshyn, M., 1980, *N*-Acetyltransferase in the chick retina. I. Circadian rhythms controlled by environmental lighting are similar to those in the pineal gland, J. Comp. Physiol., 139:103.

Brammer, G. L., Yuwiler, A., and Wetterberg, L., 1978, *N*-Acetyltransferase activity of the rat Harderian gland, Biochim. Biophys. Acta, 526:93.

Bubenik, G. A., Brown, G. M., and Grota, L. G., 1976a, Differential localization of N-acetylated indolealkylamines in CNS and the Harderian gland using immunohistology, Brain Res., 118:417.

Bubenik, G. A., Brown, G. M., and Grota, L. J., 1976b, Immunohistochemical localization of melatonin in the rat Harderian gland, J. Histochem. Cytochem., 24:1173.

Bubenik, G. A., Brown, G. M., Uhler, I., and Grota, L. J., 1974, Immunohistological localization of N-acetylindolealkylamines in pineal gland, retina and cerebellum, Brain Res., 81:233.

Bubenik, G. A., and Purtill, R. A., 1980, The role of melatonin and dopamine in retinal physiology, Can. J. Physiol. Pharmacol., 58:1457.

Bubenik, G. A., Purtill, R. A., Brown, G. M., and Grota, L. J., 1978, Melatonin in the retina and the Harderian gland. Ontogeny, diurnal variations and melatonin treatment, Exp. Eye Res., 27: 323.

Buznikov, G. A., 1964, Use of tryptamine derivatives in the study of the role of 5-oxytryptamine (serotonin) during embryonic development of invertebrates, Dokl. Akad. Nauk SSSR, Otd. Biol., 152:1243.

Buznikov, G. A., Chudakova, I. V., and Zvezdina, N. D., 1964, The role of neurohumors in early embryogenesis. I. Serotonin content of developing sea urchin and loach, J. Embryol. Exp. Morph., 12:563.

Buznikov, G. A., and Manukhin, B. N., 1961, Serotonin-like substance in the embryogenesis of some gastropod molluscs, Zh. Obshch. Biol., 22:223.

Buznikov, G. A., Zvezdina, N. D., and Makeeva, R. G., On the possible participation of serotonin and other neurohormones in the regulation of protein biosynthesis (Experiments on sea urchin eggs), Dokl. Akad. Nauk SSSR, Otd. Biol., 166:1252.

Cardinali, D. P., 1980, Molecular biology of melatonin: Assessment of the "microtubule hypothesis of melatonin action," in: "Melatonin: Current Status and Perspectives," N. Birau and W. Schloot, eds., p. 247, Pergamon Press, Oxford and New York.

Cardinali, D. P., Faigón, M. R., Scacchi, P., and Moguilevsky, J., 1979, Failure of melatonin to increase serum prolactin levels in ovariectomized rats subjected to superior cervical ganglionectomy or pinealectomy, J. Endocr., 82:315.

Cardinali, D. P., Nagle, C. A., and Rosner, J. M., 1974, Periodic changes in rat retinal and pineal melatonin synthesis, Acta Physiol. Latino. Amer., 24:91.

Cardinali, D. P., Vacas, M. I., and Gejman, P. V., 1981, The sympathetic superior cervical ganglia as peripheral neuroendocrine centers, J. Neural Transmission, 52:1

Cardinali, D. P., and Wurtman, R. J., 1972, Hydroxyindole-O-methyl transferases in rat pineal, retina and Harderian gland, Endocrinology, 91:247.

Clabough, J. W., and Norvell, J. E., 1973, Effects of castration, blinding, and the pineal gland on the Harderian glands of the male Golden Hamster, Neuroendocrinology, 12:344.

Collin, J.-P., 1969, Contribution à l'étude de l'organe pinéal. De l'épiphyse sensorielle à la glande pinéale: Modalités de transformation et implications fonctionelles, Ann. Sta. Biol. Besse-en-Chandesse, Suppl. 1:1.

Collin, J.-P., Recent advances in pineal cytochemistry. Evidence of the production of indoleamines and proteinaceous substances by rudimentary photoreceptor cells and pinealocytes of Amniota, Prog. Brain Res., 52:271.

Currie, J. R., Hollyfield, J. G., and Rayborn, M. E., 1978, Rod outer segments elongate in constant light: Darkness is required for normal shedding, Vision Res., 18:995.

Decker, J. F., and Quay, W. B., in press, Stimulatory effects of melatonin on ependymal epithelium of choroid plexuses in Golden Hamsters, J. Neural Transmission.

Detwiler, S. R., 1943, "Vertebrate Photoreceptors," Macmillan, New York.

de Venecia, G., and Davis, M. G., 1963, Diurnal variation of intraocular pressure in the normal eye, Arch. Ophthal., 69:752.

Easter, S. S., Jr., and Macy, A., 1978, Local control of retinomotor activity in the fish retina, Vision Res., 18:937.

Ferrario, V. F., Bianchi, R., Giunta, G., and Roveda, L., 1982, Circadian rhythm in human intraocular pressure, Chronobiologia, 9:33.

Follmann, P., 1977, Distribution of normal intraocular pressure. Trans. Ophthal. Soc. U. K., 97:683.

Gern, W. A., Owens, D. W., and Ralph, C. L., 1978, The synthesis of melatonin by the trout retina, J. Exp. Zool., 206:263.

Goldman, A. I., 1982, The pineal gland does not mediate phase shifts in the disc shedding rhythm of the rat retina, Invest. Ophthal. Visual Sci., 22:111.

Goldman, A. I., Teirstein, P. S., and O'Brien, P. J., 1980, The role of ambient lighting in circadian disc shedding in the rod outer segment of the rat retina, Invest. Ophthal. Visual Sci., 19: 1257.

Hamm, H. E., and Menaker, M., 1980, Retinal rhythms in chicks: Circadian variation in melatonin and serotonin *N*-acetyltransferase activity, Proc. Natl. Acad. Sci. USA, 77:4998.

Hamm, H. E., and Menaker, M., 1981, Pineal and retinal serotonin *N*-acetyltransferase activity: Modulation by phosphate, J. Neurochem., 37:1567.

Harlow, H. J., Phillips, J. A., and Ralph, C. L., 1981, Day-night rhythm in plasma melatonin in a mammal lacking a distinct pineal gland, the Nine-banded Armadillo, Gen. Comp. Endocrinol., 45: 212.

Joó, I., and Kahán, A., 1975, The porphyrin content of Harderian glands of rats and the melatonin-melanocyte stimulating hormone-system, Endokrinologie, 65:308.

Joss, J. M. P., 1978, A rhythm in hydroxyindole-*O*-methyltransferase (HIOMT) activity in the scincid lizard, *Lampropholas guichenoti*, Gen. Comp. Endocrinol., 36:521.

Kalsow, C. M., and Wacker, W. B., 1978, Pineal gland involvement in retina-induced experimental allergic uveitis, Invest. Ophthal. Visual Sci., 17:774.

Kittrell, E. M. W., and Thiessen, D. D., 1981, Does the removal of the Harderian gland affect the physiology of the Mongolian Gerbil (*Meriones unguiculatus*)?, Physiol. Psychol., 9:299.

Klein, D. C., and Moore, R. Y., 1979, Pineal *N*-acetyltransferase and hydroxyindole-*O*-methyltransferase: Control by the retinohypothalamic tract and the suprachiasmatic nucleus, Brain Res., 174:245.

Kline, L. W., and Pickering, S. G., 1975, An efferent effect on a rhythmic response in the frog retina, Can. J. Physiol. Pharmacol., 53:816.

Kostellow, A. B., and Morrill, G. A., 1971, Tryptophan induction of the serotonin and kynurenine pathways in the early amphibian embryo, Nature New Biol., 231:119.

Lai, Y.-L., Lug, R., Yao, P. C., Hayasaka, S., and Hayasaka, I., 1980, Studies of the pathogenic mechanisms of light on rat retina, Acta Anat., 107:407.

Lanum, J., 1978, The damaging effects of light on the retina. Empirical findings, theoretical and practical implications, Surv. Ophthal., 22:221.

LaVail, M. M., 1976a, Rod outer segment disc shedding in relation to cyclic lighting, Exp. Eye Res., 23:277.

LaVail, M. M., 1976b, Rod outer segment disc shedding in rat retina: Relationship to cyclic lighting, Science, 194:1071.

Lewy, A. J., Tetsuo, M., Markey, S. P., Goodwin, F. K., and Kopin, I. J., 1980, Pinealectomy abolishes plasma melatonin in the rat, J. Clin. Endocrinol. Metab., 50:204.

Lin, W.-L., and Nadakavukaren, M. J., 1981, Harderian gland lipids of male and female Golden Hamsters, Comp. Biochem. Physiol., 70B:627.

Meyer, A. C., Wassermann, W., Meyer, B. J., Joubert, W. S., Roux, S., and Biagio, R., 1981, Melatonin rhythm in the Chacma Baboon

(*Papio ursinus*) and the effect of pinealectomy and superior cervical ganglionectomy on the rhythm. South African J. Sci., 77:39.

Moore, R. Y., Heller, A., Bhatnager, R. K., Wurtman, R. J., and Axelrod, J., 1968, Central control of the pineal gland: Visual pathways, Arch. Neurol., 18:208.

Nagle, C. A., Cardinali, D. P., and Rosner, J. M., 1972, Light regulation of rat retinal hydroxyindole-*O*-methyl transferase (HIOMT) activity, Endocrinology, 91:423.

Nagle, C. A., Cardinali, D. P., and Rosner, J. M., 1973, Retinal and pineal hydroxyindole-*O*-methyl transferases in the rat: Changes following cervical sympathectomy, pinealectomy or blinding, Endocrinology, 92:1560.

Nagle, C. A., Cardinali, D. P., and Rosner, J. M., 1974, Effects of castration and testosterone administration on pineal and retinal hydroxyindole-*O*-methyl transferases of male rats, Neuroendocrinology, 14:14.

O'Steen, W. K., 1970, Retinal and optic nerve serotonin and retinal degeneration as influenced by photoperiod, Exp. Neurol., 27:194.

O'Steen, W. K., Anderson, K. V., and Shear, C. R., 1974, Photoreceptor degeneration in albino rats: Dependency on age, Invest. Ophthal., 13,334.

O'Steen, W. K., Kraeer, S. L., and Shear, C. R., 1978, Extraocular muscle and Harderian gland degeneration and regeneration after exposure of rats to continuous fluorescent illumination, Invest. Ophthal. Visual Sci., 17:847.

O'Steen, W. K., Shear, C. R., and Anderson, K. V., 1972, Retinal damage after prolonged exposure to visible light. A light and electron microscopic study, Am. J. Anat., 134:5.

Pang, S. F., Brown, G. M., Grota, L. J., Chambers, J. W., and Rodman, R. L., 1977, Determination of *N*-acetylserotonin and melatonin activities in the pineal gland, retina, Harderian gland, brain and serum of rats and chickens, Neuroendocrinology, 23:1.

Pang, S. F., and Ralph, C. L., 1975, Pineal and serum melatonin at midday and midnight following pinealectomy or castration in male rats, J. Exp. Zool., 193:275.

Pang, S. F., and Yew, D. T., 1979, Pigment aggregation by melatonin in the retinal pigment epithelium and choroid of Guinea-pigs, *Cavia porcellus*, Experientia, 35:231.

Pang, S. F., Yew, D. T., and Tsui, H. W., 1978, Photomechanical changes in the retina and choroid of Guinea Pigs, *Cavia porcellus*, Neuroscience Letters, 10:221.

Pang, S. F., Yu, H. S., Suen, H. C., and Brown, G. M., 1980, Melatonin in the retina of rats: A diurnal rhythm, J. Endocr., 87:89.

Pang, S. F., Yu, H. S., and Tang, P. L., 1981, A diurnal rhythm of *N*-acetylserotonin in the retina of rats, Neuroscience Letters, 21:197.

Panke, E. S., Reiter, R. J., and Rollag, M. D., 1979, Effect of removal of the Harderian glands on pineal melatonin concentra-

tions in the Syrian Hamster, Experientia, 35:1405.
Pévet, P., Balemans, M. G. M., Bary, F. A. M., and Noordegraaf, E. M., 1978, The pineal gland of the Mole (*Talpa europaea*, L.). V) Activity of hydroxyindole-*O*-methyl transferase (HIOMT) in the formation of melatonin/5-methoxytryptophol in the eyes and the pineal gland, Ann. Biol. Anim. Bioch. Biophys., 18:259.
Pévet, P., Balemans, M. G. M., Legerstee, W. C., and Vivien-Roels, B., 1980a, Circadian rhythmicity in the activity of HIOMT in the formation of melatonin and of 5-methoxytryptophol in the retina, Harderian gland and pineal of the male hamster, in: "Melatonin: Current Status and Perspectives," N. Birau and W. Schloot, eds., p. 201, Pergamon Press, Oxford and New York.
Pévet, P., Balemans, M. G. M., Legerstee, W. C., and Vivien-Roels, B., 1980b, Circadian rhythmicity of the activity of hydroxyindole-*O*-methyl transferase (HIOMT) in the formation of melatonin and 5-methoxytryptophol in the pineal, retina, and Harderian gland of the Golden Hamster, J. Neural Transmission, 49: 229.
Pévet, P., Balemans, M. G. M., and de Reuver, G. F., 1981, The pineal gland of the Mole (*Talpa europaea* L). VII. Activity of hydroxyindole-*O*-methyltransferase (HIOMT) in the formation of 5-methoxytryptophan, 5-methoxytryptamine, 5-methoxyindole-3-acetic acid, 5-methoxytryptophol and melatonin in the eyes and the pineal gland, J. Neural Transmission, 51:271.
Poffenbarger, M., and Fuller, G. M., 1976, Is melatonin a microtubule inhibitor?, Exp. Cell Res., 103:135.
Pomerantz, G., and Reiter, R. J., 1974, Influence of intraocularly-injected pineal indoles on PMS-induced ovulation in immature rats, Int. J. Fertil., 19:117.
Pomerantz, G., and Sorrentino, S., Jr., 1973, The influence of melatonin administered subcutaneously, intravenously, or intraocularly upon ovulation in the PMS-treated immature rat, Neuroendocrinology, 12:354.
Quay, W. B., 1965, Retinal and pineal hydroxyindole-*O*-methyl transferase activity in vertebrates, Life Sci., 4:983.
Quay, W. B., 1968, Comparative physiology of serotonin and melatonin, Adv. Pharmacol., 6A:283.
Quay, W. B., 1974, "Pineal Chemistry in Cellular and Physiological Mechanisms," Charles C Thomas, Springfield.
Quay, W. B., and McLeod, R. W., 1968, Melatonin and photic stimulation of cone contraction in the retina of larval *Xenopus laevis*, Anat. Rec., 160:491.
Quay, W. B., Payer, A. F., Parkening, T. A., Banerji, T. K., and Collins, T. J., 1982, Melatonin's inhibition of pituitary, adrenal, testicular and accessory gland growth in male Golden Hamsters: Pineal dependence and organ differences with shielding and intracranial surgery, J. Neural Transmission, 53:59.
Quay, W. B., Smart, L. I., and Hafeez, M. A., 1969, Substrate specificity and tissue localization of acetylserotonin methyltransferase in eyes of trout (*Salmo gairdneri*), Comp. Biochem.

Physiol., 28:947.

Ralph, C. L., 1980, Melatonin production by extra-pineal tissues, in: "Melatonin: Current Status and Perspectives," N. Birau and W. Schloot, eds., p. 35, Pergamon Press, Oxford and New York.

Reiter, R. J., 1973, Comparative effects of continual lighting and pinealectomy on the eyes, the Harderian glands and reproduction in pigmented and albino rats, Comp. Biochem. Physiol., 44A:503.

Reiter, R. J., Blask, D. E., Johnson, L. Y., Rudeen, P. K., Vaughan, M. K., and Waring, P. J., 1976, Melatonin inhibition of reproduction in the male hamster: Its dependency upon time of day of administration and on an intact and sympathetically innervated pineal gland, Neuroendocrinology, 22:107.

Reiter, R. J., and Klein, D. C., 1971, Observations on the pineal gland, the Harderian glands, the retina, and the reproductive organs of adult female rats exposed to continuous light, J. Endocr., 51:117.

Reiter, R. J., Richardson, B. A., and Hurlbut, E. C., 1981, Pineal, retinal and Harderian gland melatonin in a diurnal species, the Richardson's Ground Squirrel (*Spermophilus richardsonii*), Neuroscience Letters, 22:285.

Repérant, J., Vesselkin, N. P., Rio, J. P., Ermakova, T. V., Miceli, D., Peyrichoux, and Weidner, C., 1981, La voie visuelle centrifuge n'existe-t-elle que chez les oiseaux?, Rev. Can. Biol., 40: 29.

Sackman, J. W., Little, J. C., Rudeen, P. K., Waring, P. J., and Reiter, R. J., 1977, The effects of pineal indoles given late in the light period on reproductive organs and pituitary prolactin levels in male Golden Hamsters, Hormone Res., 8:84.

Sagar, S. M., Martin, J. B., and Reppert, S. M., 1982, Circadian rhythm of melatonin content in chick retina, Endocrine Soc. Prog. and Abstracts, 990.

Sakai, T., 1981, The mammalian Harderian gland: Morphology, biochemistry, function and phylogeny, Arch. Histol. Japon., 44:299.

Saxén, L., 1954, The development of the visual cells, embryological and physiological investigations on Amphibia, Ann. Acad. Sci. Fenn., Ser. A, IV. Biol., 23:1.

Semm, P., Demaine, C., and Vollrath, L., 1980, The effects of microelectrophoretically applied melatonin, putative transmitters, thyroxine and sex hormones on the electrical activity of pineal cell in the Guinea-pig, in: "Melatonin: Current Status and Perspectives," N. Birau and W. Schloot, eds., p. 129, Pergamon Press, Oxford and New York.

Shirama, K., Furuya, T., Takeo, Y., Shimizu, K., and Maekawa, K., 1981, Influences of some endocrine glands and of hormone replacement on the porphyrins of the Harderian glands of mice, J. Endocr., 91:305.

Suzuki, O., and Yagi, K., 1978, A nonisotopic assay for acetylserotonin methyltransferase, Analyt. Biochem., 88:580.

Tamarkin, L., Hollister, C. W., Lefebvre, N. G., and Goldman, B. D., 1977, Melatonin induction of gonadal quiescence in pinealectomized Syrian Hamsters, Science, 198:953.

Teirstein, P. S., Goldman, A. I., and O'Brien, P. J., 1980, Evidence for both local and central regulation of rat rod outer segment disc shedding, Invest. Ophthal. Visual Sci., 19:1268.

Tetsuo, M., Perlow, M. J., Mishkin, M., and Markey, S. P., 1982, Light exposure reduces and pinealectomy virtually stops urinary excretion of 6-hydroxymelatonin by Rhesus Monkeys, Endocrinology, 110:997.

Thiessen, D. D., 1977, Thermoenergetics and the evolution of pheromone communication, Prog. Psychobiol. Physiol. Psychol., 7:91.

Thiessen, D. D., Clancy, A., and Goodwin, M., 1976, Harderian gland pheromone in the Mongolian Gerbil *Meriones unguiculatus*. J. Chem. Ecol., 2:231.

Thiessen, D. D., Graham, M., Perkins, J., and Marcks, S., 1977, Temperature regulation and social grooming in the Mongolian Gerbil (*Meriones unguiculatus*), Behav. Biol., 19:279.

Turek, F. W., 1977, Antigonadal effect of melatonin in pinealectomized and intact male hamsters, Proc. Soc. Exp. Biol. Med., 155:31.

Vivien-Roels, B., Pévet, P., Dubois, M. P., Arendt, J., and Brown, G. M., 1981, Immunohistochemical evidence for the presence of melatonin in the pineal gland, the retina and the Harderian gland, Cell Tissue Res., 217:105.

Welsh, J. H., and Osborn, C. M., 1937, Diurnal changes in the retina of the catfish, *Ameiurus nebulosus*. J. Comp. Neurol., 66:349.

Wetterberg, L., Geller, E., and Yuwiler, A., 1970a, Harderian gland: An extraretinal photoreceptor influencing the pineal gland in neonatal rats, Science, 167:884.

Wetterberg, L., Yuwiler, A., Ulrich, R., Geller, E., and Wallace, R., 1970b, Harderian gland: Influence on pineal hydroxyindole-*O*-methyltransferase activity in neonatal rats, Science, 170:194.

Worthen, D. M., 1978, Intraocular pressure and its diurnal variation, in: "Glaucoma. Conceptions of a Disease," K. Heilmann and K. T. Richardson, eds., p. 54, Georg Thieme Publ., Stuttgart.

Wurtman, R. J., Axelrod, J., and Potter, L. T., 1964, The uptake of H^3-melatonin in endocrine and nervous tissues and the effects of constant light exposure, J. Pharmacol., 143:314.

Wurtman, R. J., Axelrod, J., Chu, E. W., Heller, A., and Moore, R. Y., 1967, Medial forebrain lesions: Blockade of effects of light on rat gonads and pineal, Endocrinology, 81:509.

Yu, H. S., Pang, S. F., and Tang, P. L., 1981a, Increase in the level of retinal melatonin and persistence of its diurnal rhythm in rats after pinealectomy, J. Endocr., 91:477.

Yu, H. S., Pang, S. F., Tang, P. L., and Brown, G. M., 1981, Persistence of circadian rhytms of melatonin and *N*-acetylserotonin in the serum of rats after pinealectomy, Neuroendocrinology, 32: 262.

MODIFICATION OF HYPOTHALAMIC ELECTRICAL ACTIVITY BY PINEAL INDOLES

C. Demaine

Department of Physiology
Chelsea College
London University
London, England

INTRODUCTION

In a number of mammalian species, the pineal gland has been shown to mediate the inhibitory effects of short photoperiods on the secretion of several pituitary hormones. Although most studies have centred on gonadotropin secretion, in the last twenty five years there have been reports which have provided evidence for pineal-mediated inhibition of the production and release of each of the adenohypophyseal hormones (see review Mess et al., 1979). The Syrian hamster (Mesocricetus auratus) is a highly photosensitive species, in which the pineal can readily be shown to exert a dramatic and potent influence on the release of luteinising hormone (LH), follicle stimulating hormone (FSH) and prolactin (Reiter, 1980). Light restriction, obtained by blinding hamsters or placing them in continuous darkness or short photoperiod, results in reproductive involution. In males, after 50 days of light restriction, the testes are reduced to less than 20% of their original weight and do not produce spermatozoa (Hoffman and Reiter, 1965). These changes are associated with a reduction in the pituitary content of both LH and FSH (Reiter and Johnson, 1974a), and of prolactin (Reiter et al., 1975). Turek et al. (1976) found that serum LH and FSH levels in animals exposed to short photoperiods were lower than in hamsters maintained in a L:D, 14:10 lighting cycle. In light restricted females, the oestrous cycles are interrupted and the uteri become infantile (Johnson and Reiter, 1978). Although the ovaries tend to increase in weight under these conditions, there are pronounced changes in ovarian morphology. There is extensive proliferation of interstitial tissue, a small number of antral follicles and a distinct lack of corpora lutea (Vaughan et al., 1981). The relationship between reproductive collapse and pituitary gonadotropin secretion is more complicated in female hamsters. In fact, LH levels actually appear to rise when animals are blinded (Reiter and Johnson, 1974b). Presumably the occurrence of daily surges in both LH and FSH in fema hamsters exposed to short days (Seegal and Goldman, 1975) underlies the disruption of normal reproductive function.

Evidence for an inhibitory role of the pineal gland on the pituitary-thyroid axis of the hamster has recently been reported (Vriend et al., 1977). These workers found that blinding male hamsters resulted in a depression of plasma thyroxin. The effect was found to be abolished by surgical removal of the pineal gland or by superior cervical ganglionectomy, an operation which interrupts th neural pathways by which photic information from the retina is transmitted to the pineal gland (Moore, 1978). In the rat at least, the pineal gland appears to exert its influence on the thyroid gland via a reduction of pituitary thyrotropin (TSH) output (Niles et al., 1979a). Thus in animals kept under short photoperiods, pinealectomy was associated with a highly significant increase in serum TSH. Similar data on TSH levels in hamsters are not currently available.

Numerous investigators have attempted to determine the pineal factor(s) responsible for the effects of the pineal gland on the neuroendocrine system, with by far the greatest attention being paid to the pineal methoxyindoleamine, melatonin. In the Syrian hamster, daily injections of this indole can induce both gonadal atrophy and thyroid suppression similar to that obtained by light restriction (Vriend and Reiter, 1977; Vriend et al., 1979), although the effect depends on the time of day of administration (Reiter et al., 1976; Tamarkin et al., 1976; Vriend and Reiter, 1977). Despite the large number of investigations of the effects of melatonin, the site(s) of its action on the hypothalamo-hypophyseal-endocrine axes have yet to be firmly established. These sites may include one or a variety of tissues such as the gonad (Ellis, 1972; Cohen et al., 1978), the pituitary gland (Martin and Sattler, 1979) or the central nervous system (CNS) (Fraschini et al., 1968; Trentini et al., 1979; Chen et al., 1980; Reiter et al., 1981). Most recently, the hypothalamus has been regarded as the most likely target tissue (Reiter, 1980). Melatonin is known to be taken up (Anton-Tay and Wurtman, 1969) and specifically bound (Niles et al., 1979b; Vacas and Cardinali, 1979) in the hypothalamus of the laboratory rat. Using fluorescence histochemistry, Bubenik et al. (1976) were able to localise an affinity for melatonin in the hypothalamic region. Fraschini et al. (1971) have shown that implants of various pineal indoles, including melatonin, into the hypothalamus of female rats significantly reduced the level of gonadotropins in the anterior pituitary. In addition, the injection of melatonin into the lateral ventricles of the brain just before the 'critical period' on the day of proestrus, either blocked ovulation or reduced the number of ova shed. Kamberi et al. (1970) were also able to show a significant reduction in the output of pituitary LH following the injection of melatonin into the third ventricle of the brain in male rats. Vriend (1981) has recently arrived at the conclusion that a single CNS neuroendocrine mechanism is involved in both pineal-thyroid and pineal-gonadal effects in the Syrian hamster, since the parallels between the action of melatonin on the two systems are so striking.

POSSIBLE SITES OF ACTION OF PINEAL INDOLES IN THE BRAIN

In order to modify the hypothalamic control of anterior pituitary secretion, pineal hormones must ultimately influence the secretion of releasing hormones into the hypophyseal portal system. At least three possible mechanisms of action may be considered. Firstly, melatonin could induce increased release of an inhibitory transmitter, e.g. serotonin, by acting on neurones with axons terminating in the hypothalamus. The parts of the brain most likely to be target sites for the indoleamine would then be the amygdala, the hippocampus and the midbrain monoaminergic neurone system. Secondly, pineal hormones could increase the sensitivity of steroid (or thyroid) hormone negative feedback sites in the hypothalamus itself, and thirdly, they could themselves have inhibitory 'transmitter-like' or neuromodulatory effects on hypothalamic neurones involved in releasing hormone secretion.

Trentini et al. (1979) have suggested that serotonin actually mimics some of the effects which are known to follow melatonin administration, and that, in general, the action exerted by the pineal gland on CNS activity is similar to that induced by variations in brain serotonin levels. This neurotransmitter has been particularly implicated in the control of pituitary growth hormone secretion (Collu et al. 1972; Smythe and Lazarus, 1973; Smythe et al., 1973) and the effects of pineal melatonin on this neuroendocrine axis may well operate via the serotonergic neuronal system. Similarly, Yates and Herbert (1976) reported that melatonin given to ferrets kept in long photoperiods caused serotonin rhythms, characteristic of animals exposed to short photoperiods, in both the pineal gland and the hypothalamus. The role of serotonin in the hypothalamic control of pituitary gonadotropin, prolactin and TSH secretion is extremely ambiguous. Although many authors have described evidence for an inhibitory effect of increased serotonin turnover on gonadotropin release and ovulation, equal numbers of reports have indicated a separate positive action of the transmitter on the cyclic release of gonadotropins and on prolactin secretion (see review Trentini, 1979). However, electrical stimulation of the medial raphe nucleus was found to be inhibitory to ovulation and led to decreased plasma LH levels (Carrer Taleisnik, 1972). Pazo (1979) showed that melatonin had prediminantly inhibitory effects on the spontaneous neuronal electrical activity of the mesencephalic reticular formation. He suggests that this effect may contribute to the changes in the sleep-wakefulness cycle and anticonvulsant action attributed to the pineal hormone. It is clear that if this is the main influence of melatonin on midbrain serotonergic neurones, then their release of this neurotransmitter would be reduced.

Brain regions usually considered to have a more important influence on the activity of gonadotropin releasing hormone (GnRH) neurones are the amygdala and the hippocampus (Komisurak et al., 1981). Although most studies have shown the amygdala capable of stimulating LH secretion, electrochemical stimulation of the basolateral amygdala in the freely moving rat resulted in a delayed and diminished LH surge (Carrillo et al., 1977). The findings of Brown et al. (1981) indicate that melatonin binds with high affinity in limbic brain areas including the hippocampus. This region has been shown to be predominantly inhibitory with respect to LH secretion, thus electrochemical stimulation results in blockade of the LH surge during proestrus in the freely moving rat (Carillo et al., 1977). However, the effects of manipulations of the hippocampus on FSH secretion are contradictory (Komisurak et al., 1981).

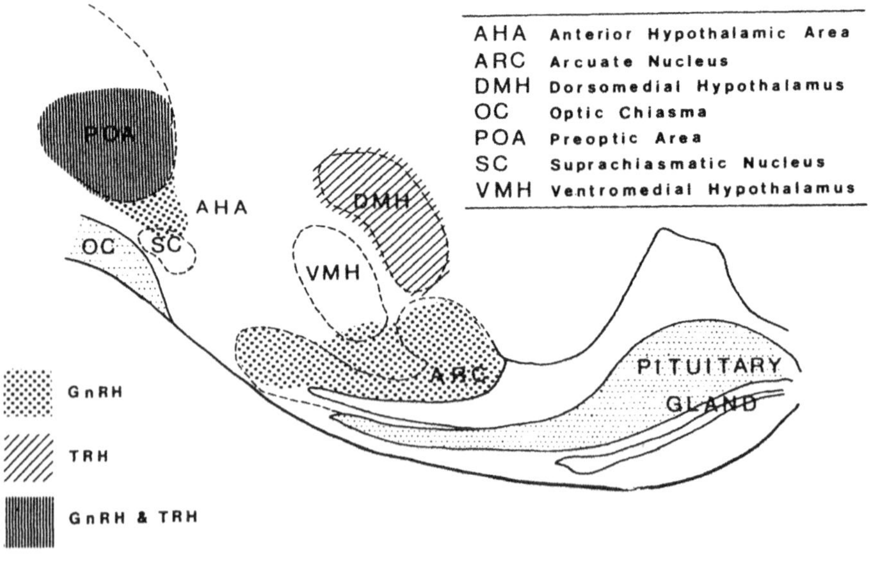

Fig. 1. Localisation of some releasing hormones in the preoptic-hypothalamic region of the brain.

Although these inputs to the hypothalamic releasing hormone neurones must undoubtedly be considered as potential sites for pineal indoles, their influence on the general levels of GnRH release, particularly in males, has not been demonstrated. A more likely site of action for the active principles of the pineal are the hypothalamic regions themselves. In the rat brain, the distribution of the endogenous melatonin has been demonstrated immunohistochemically, particularly in the suprachiasmatic nuclei (SCN) and the ventral region of the anterior hypothalamus (Bubenik et al., 1976).

Fig. 1 shows the localisation of some of the releasing hormone neurones within the hypothalamic nuclei. If pineal indoles act on these cells to alter their sensitivity to gonadal or thyroid hormone negative feedback, or if they themselves can influence neuronal electrical activity, then these areas must be considered as prime target sites. Melatonin has been shown to have significant effects on microtubule protein content of the hypothalamus (Cardinali and Freire, 1975; Cardinali et al., 1975) and to specifically bind in the mediobasal hypothalamus of bovine brain (Cardinali et al., 1979) and in the hypothalamus of the rat (Niles et al., 1979b). An inverse relationship between diurnal variations in pineal melatonin and GnRH content of the preoptic-suprachiasmatic area and of the medial basal hypothalamus has been demonstrated (Brown et al., 1977). The pineal indoleamine content was found to be low throughout the light phase, when GnRH levels were significantly higher while the reverse was observed throughout the dark period. Of the neural pathways influencing GnRH release, attention may be focussed on the preoptic-suprachiasmatic area rather than the mediobasal hypothalamus following experiments in which the latter region was surgically isolated from the rest of the hypothalamus in hamsters (Sorrentino and Reiter, 1971; Reiter and Sorrentino, 1972). These investigators found that this type of surgical isolation prevents the gonadal atrophy that usually follows as a consequence of light restriction. Since the lesions were unlikely to interrupt the pathways which serve to transmit photic information to the pineal gland, it is reasonable to conclude that the site of action of pineal antigonadotropic hormones may lie anterior to the cut. More recently, Reiter et al. (1981) have shown that the inhibitory action of exogenously applied melatonin on reproduction in male pinealectomised hamsters depends on intact neural pathways in the anterior hypothalamus (AHA). The authors suggest that one possible interpretation of this finding is that melatonin may modify the functioning of the hypothalamo-pituitary axis by working on neurones which influence the synthesis and secretion of hypothalamic releasing hormones.

The suprachiasmatic nuclei in the hypothalmus are involved in the generation and entrainment of circadian rhythms (Rusak and Zucker, 1979). These rhythms appear to be entrained to the photoperiod via the action of the pineal gland (Elliott and Goldman, 1981), in a number of mammalian species. Thus the SCN have been considered as possible target sites for pineal indoles (Brown et al., 1981). The suggestion has been endorsed by immunohistological studies which have localised melatonin in the rat SCN (Bubenik et al., 1976) and the finding that the nuclei have the ability to concentrate the indoleamine from the peripheral circulation (Reiter, 1980). Bittman et al. (1979) have recently studied the potential action of melatonin in hamster SCN in relation to the antigonadotropic properties of the pineal indole. Male hamsters were housed under a typical lighting schedule (L:D, 14:10). Groups of animals were pinealectomised and some were subjected to bilateral lesions of the SCN. The animals were injected with melatonin (25μg dose) either once a day or three times daily. Thrice daily melatonin injections caused testicular atrophy within six weeks in the pinealectomised hamsters whether the SCN had been completely lesioned or left intact. In hamsters with intact pineal glands, reproductive collapse in response to thrice daily melatonin injections only occurred in animals with complete SCN lesions. Thus electrolytic destruction of the SCN does not interfere with the ability of melatonin to induce gonadal involution when applied in this manner. Bittman et al. (1979) conclude that the SCN are not essential target tissues for the effects of exogenously administered melatonin on the reproductive system. In fact their results demonstrate that SCN lesions in hamsters with intact pineal glands have the same effects on the sensitivity of the reproductive system to exogenously applied melatonin as do pinealectomy or superior cervical ganglionectomy. The SCN are intimately involved with the regulation of pineal secretory activity since bilateral lesions of these structures abolish the darkness induced peak of pineal N-acetyltransferase activity and therefore the normal production of melatonin in the gland (Moore, 1978).

In the rat, there is considerable evidence for the occurrence of a population of GnRH neurone cell bodies in the medial preoptic area (POA) of the rostral hypothalamus (Fig. 1), immediately overlying the optic chiasma. Not only can GnRH be extracted from this region, but lesions of the axons of these neurones which project to the mediobasal hypothalamus are followed by a decline in the content of GnRH stored in the median eminence as measured by subsequent assays. Such a loss would be consistent with the degeneration of preoptico-tuberal GnRH neurones and loss of the stored GnRH in their axon terminals (McCann, 1981). In the case of thyrotropin releasing hormone (TRH), activity is also found in the POA, particularly in the region of the interstitial nucleus of the stria terminalis. The activity is present in an area extending caudally from this region to the dorsomedial nucleus (DMN) where a higher activity is present (Fig. 1). McCann (1981) postulates that cell bodies of TRH neurones are likely to be present in the POA, the activity of the DMN representing the caudal projection of their axons or the cell bodies of a second population of TRH neurones. Thus a considerable proportion of both GnRH and TRH neuronal cell bodies can be localised in the POA. Moreover, there is also a large amount of evidence to indicate that elements in the POA may also be involved in the generation of circadian rhythms (Rusak and Zucker, 1979), although unlike the SCN, POA neurones do not appear to play a part in the neural pathways by which retinal photic information is transmitted to the pineal gland (Moore, 1978).

POSSIBLE MECHANISMS OF ACTION OF PINEAL HORMONES IN THE ROSTRAL HYPOTHALAMUS

One of the possible mechanisms for the effects of pineal indoles in the hypothalamus is the alteration of the sensitivity of releasing hormone neurones to the negative feedback signals provided by the final products of the various neuroendocrine axes which are influenced by the pineal gland. However, decreases in serum prolactin levels in light restricted male hamsters occur even in castrated animals and prolactin levels are not affected by exogenous androgen. Castration also does not alter serum prolactin concentrations in males kept on a long photoperiod. Thus photoperiod-induced changes in prolactin secretion do not result from alterations in the effectiveness of the feedback signal provided by the levels of circulating testicular hormones in this species (Bex et al., 1978; Elliott and Goldman, 1981). By contrast, in the ewe, Goldman (1981) postulates that seasonal changes in LH secretion are mediated by alterations in sensitivity to the negative feedback effect of oestrogen. Ovariectomised ewes maintain high serum LH levels throughout the year, but when ovariectomised animals were administered a constant dose of oestrodiol, they showed pronounced seasonal fluctuations in serum LH concentrations, with the highest levels coinciding with the normal breeding season (Legan et al., 1977). On the other hand, in the ram, changes in LH secretion occur in castrated animals in relation to alterations in photoperiod (Pelletier and Ortavant, 1976). Moreover, in castrated Soay rams, exposure to long days after a period of short days resulted in a decline in the plasma levels of both LH and FSH despite the fact that the levels were always much above the normal range for the intact animal (Lincoln and Short, 1980). These latter results are consistent with the idea that pineal-mediated effects of the photoperiod occur directly on the hypothalamus regulating the release of GnRH independently of androgen feedback.

To return to the situation in the Syrian hamster, the decrease in serum gonadotropins in light restricted males appears to be at least partly caused by an increased sensitivity of the hypothalamo-pituitary axis to the feedback effects of androgens (Tamarkin et al., 1976; Turek, 1977). However, there is evidence for an androgen-independent suppression of gonadotropins in short photoperiods in the male (Ellis and Turek, 1979), and the pattern of LH secretion in light restricted female hamsters shows little or no change even after combined ovariectomy and adrenalectomy (Bittman and Goldman, 1979).

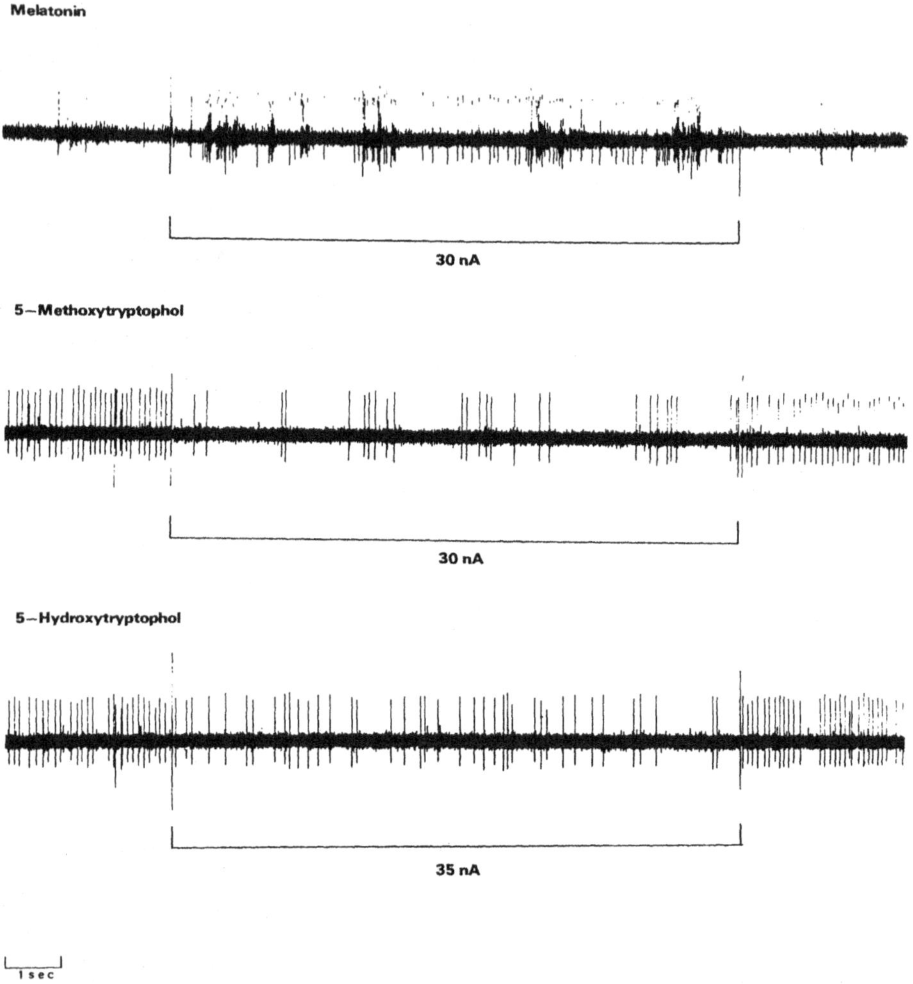

Fig. 2. Responses of rostral hypothalamic neurones to the application of pineal indoles.

Effects of Pineal Indoles on the Electrical Activity of Rostral Hypothalamic Neurones

In a search for an alternative mechanism for the action of the pineal gland on hypothalamic releasing hormone neurones we have shown that the putative pineal hormones, melatonin, 5-methoxytryptophol (5MTPH) and 5-hydroxytryptophol (5HTPH) can influence the rate of discharge of neurones in the rostral hypothalamus in both rats and Syrian hamsters (Demaine and Kann, 1979). By contrast, neurones randomly tested in the cerebellum and cerebral cortex showed their electrical activity to be unaffected by the application of these pineal indoles. Recordings of extracellular unit activity were made by means of a multibarrelled micropipette stereotaxically inserted in either the preoptic or anterior hypothalamic areas (POA/AHA) of males anaesthetised with urethane. In some experiments melatonin was expelled in picolitre quantities from the tip of the micropipette by means of a thermal expansion device (Demaine et al., 1977). Otherwise the drugs were all expelled from the pipette by passing cationic currents of 30-40 nA for periods of 10 seconds.

In this first series of experiments, recordings were made from fifty POA/AHA neurones in hamsters kept under normal animal house conditions. The lighting schedule was 14 hours of light and 10 hours of darkness (i.e. long photoperiods - L:D, 14:10), and recordings made in the first seven hours of the light phase. Under these conditions, melatonin, 5MTPH and 5HTPH all modified the spontaneous electrical activity of some rostral hypothalamic neurones. Fig. 2 shows typical actions of these three indoles on single POA/AHA neurones in the hamster, while Fig. 3 illustrates the proportions of neurone cell bodies in which spontaneous activity was either increased, reduced or unaffected by local application of the three indoles. Melatonin increased the rate of discharge of spontaneously active neurones by between 10 and 187% in twenty-nine out of forty cells tested, while equal numbers of excitatory and inhibitory responses to 5MTPH were observed. By contrast, of the forty-one neurones tested with 5HTPH, thirty-two were inhibited by between 8 and 60%. The remaining neurones were unresponsive to the application of these indoles. When cells were tested with the neurotransmitter, dopamine, in the presence of the pineal indoles, the inhibitory responses to standard doses of the catecholamine were modified. 5HTPH enhanced the inhibitory responses of thirty-four neurones to the application of the transmitter. However, both melatonin and 5MTPH caused a reduction of dopamine-induced inhibition in the majority of cells tested in this way. In seven neurones the inhibitory response to the catecholamine was effectively abolished when melatonin was present.

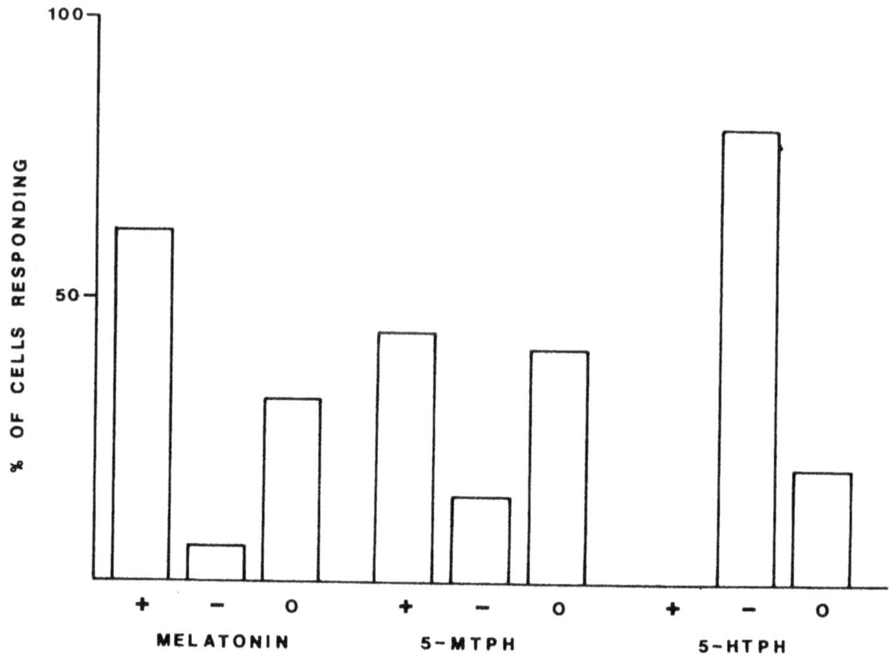

Fig. 3. Responses of rostral hypothalamic neurones to the application of melatonin, 5-methoxytryptophol (5MTPH) and 5HTPH, expressed as percentages of the total numbers of cells tested with each substance.
+, excitation; -, inhibition; O, no response

The influence of Light Restriction on the Electrical Activity of Hypothalamic Neurones which respond to Pineal Indoles

Although striking endocrinological changes mediated by the pineal gland occur in Syrian hamsters exposed to short photoperiods, little is known about the effects of long periods of darkness on the activity of hypothalamic neurones involved in the control of neurohormone release. The sexually quiescent phase in the seasonal reproductive capability of the hamster is known to require an intact, fully innervated, pineal gland (Reiter, 1987; 1980a). Thus during this phase the target site for antigonadotropic pineal hormones must be sensitive at least for a substantial part of each day. Since the overall output of pineal melatonin is not markedly increased during short day exposure (Tamarkin et al., 1979), it is likely that receptor sensitivity may be greater under these conditions than in animals housed in long days.

Having established that, using electrophysiological techniques, pineal indoles may influence the hypothalamo-pituitary neuroendocrine system by acting on neurones in the rostral hypothalamus which may be implicated in either GnRH or TRH release, we went on to study the effect of housing hamsters in short photoperiods (L:D, 2:22) on POA/AHA neurones. Again the electrical responses of these cells to the application of dopamine and the same three pineal indoles were investigated. Male hamsters were housed in short photoperiods for six to eleven weeks, during which complete testicular regression occurred, and the results of pineal indole application compared with those previously described. In complete contrast to animals housed in L:D, 14:10, very few rostral hypothalamic neurones in animals kept in L:D, 2:22 exhibited spontaneous electrical activity, although when active units were found the iontophoretic application of dopamine caused characteristic inhibition. Table 1 gives the proportions of POA/AHA cells responding to or unaffected by application of pineal indoles in the two groups of hamsters. Whereas in the long day hamsters melatonin caused excitation in 70% of the cells tested and never induced inhibition, in the short day animals a considerable proportion of neurones were inhibited by the indole. Similarly, the proportion of inhibitory responses to 5MTPH was substantially greater in the dark-accommodated animals. Excitatory responses to 5HTPH were only recorded in hamsters which had been kept in short photoperiods. Finally, in the hamsters kept in short days, dopamine-induced inhibitory responses appeared to be unaffected by the presence of all three pineal indoles. Thus the results indicate that exposure to short photoperiods caused marked changes in the electrical behaviour of these rostral hypothalamic neurones, particularly in their response to pineal indoles.

Table 1. Responses of rostral hypothalamic neurones, in hamsters housed in long or short photoperiods, to the application of pineal indoles expressed as percentages of the total number of cells tested with each substance

	Melatonin			5MTPH			5HTPH		
	O	-	+	O	-	+	O	-	+
LONG DAY L:D, 14:10	70	0	30	20	20	60	0	40	60
SHORT DAY L:D, 2.22	52	32	16	20	60	20	19	50	31

Key: +, excitation; -, inhibition; O, no response

Effect of Pinealectomy on responses of Rostral Hypothalamic Neurones to Pineal Indoles

Early reports of the ability of melatonin to induce testicular atrophy when administered by daily subcutaneous injection to male Syrian hamsters clearly showed that the effects were markedly different in pinealectomised animals (Reiter et al., 1976; Tamarkin et al., 1976). Both groups of investigators showed that daily late afternoon or evening injections only induced reproductive collapse in hamsters with intact, fully innervated pineal glands. The results were consistent with, although not proof of, the idea that the pineal gland itself may be a target organ, at least for exogenously administered melatonin, as suggested by the biochemical investigations of Freire and Cardinali (1975). However, Tamarkin et al. (1977) were subsequently able to show that thrice daily injections of melatonin, during the night, resulted in testicular regression in both pinealectomised and intact hamsters. Their data clearly show that the action of melatonin in inducing gonadal quiescence need not involve the stimulation or inhibition of the release of another pineal product.

One consistent finding from all investigations of pineal-mediated effects of light restriction in the hamster is that the action of exogenously administered melatonin on the neuroendocrine system is particularly dependent on the time of day of administration (Reiter et al., 1976; Tamarkin et al., 1976; Vriend and Reiter, 1977). Thus in intact hamsters exposed to long photoperiods, even thrice daily injections have been shown to be ineffective in inducing testicular regression when administration is made in the first half of the light phase. In pinealectomised hamsters, Tamarkin and Goldman (1978) found that thrice daily injections resulted in testicular atrophy regardless of whether the injections were given during the day or the night. This finding indicates that the target tissue for melatonin is sensitive throughout the light phase in pinealectomised hamsters. Accordingly, a series of experiments was designed in order to assess the influence of an intact pineal organ on the electrical responses of POA/AHA neurones (Demaine and Kann, 1981). The actions of pineal indoles on nerve cells in hamsters which had undergone pinealectomy were compared with those occurring in animals subjected to the appropriate sham operation. All the animals were housed in long photoperiods (L:D, 14:10) and recordings made during the first half of the light phase of the L:D cycle.

The results of these experiments are summarised in Table 2. Hypothalamic neurones in the pinealectomised hamsters showed marked differences in electrical behaviour from that of cells in sham operated animals. Relatively more units appeared to be spontaneously active in the pinealectomised group, but, in most instances, the activity of a given neurone did not persist for more than a few seconds. Thus the proportion of active cells which could be tested with the drugs was quite small. The neurotransmitter dopamine inhibited the units in both groups of animals. By contrast, marked differences in the responses to the application of the methoxyindoles, melatonin and 5MTPH were observed when recordings were made in pinealectomised hamsters. Whereas all the units responding to melatonin in the sham operated animals showed excitatory responses, only inhibitory responses to the indoleamine were recorded when cells in pinealectomised animals were tested. In a similar way, a significantly greater proportion of units were inhibited by 5MTPH in the pinealectomised animals. There were no significant differences, however, in the effects of 5HTPH application between the two groups of animals.

Table 2. Responses of rostral hypothalamic neurones, in pinealectomised or sham operated hamsters, to the application of pineal indoles expressed as percentages of the total number of cells tested with each substance.

	Melatonin			5MTPH			5HTPH		
	+	-	O	+	-	O	+	-	O
Pinealectomised hamsters	O	72	28	14	48	38	O	53	47
Sham operated hamsters	69	O	31	40	26	34	O	67	33

Key: +, excitation; -, inhibition; O, no response

The results of these experiments are clearly in line with the concept that the action of melatonin on the hypothalamus of intact hamsters is dependent on the activity of the pineal gland. The administration of exogenous melatonin the the light phase is only effective in inducing testicular regression in hamsters in which the pineal has either been removed or deprived of normal neural inputs (Tamarkin et al., 1977; Bittman et al., 1979). The findings embodied in Table 2, therefore, suggest that melatonin inhibition of the electrical acrivity of rostral hypothalamic neurones may reflect the antigonadotropic or anthithyrotropic action of the compound.

Circadian Variation in the Electrical Responses of Rostral Hypothalamic Neurones to Pineal Methoxyindoles

As described in the previous section, the antigonadotropic and antithyrotropic actions of exogenously applied melatonin in Syrian hamsters are dependent on the time of day of administration, suggesting that the target tissue of this putative pineal hormone is less sensitive or responds differently at different times of day. In an effort to further strengthen the hypothesis, that the inhibitory responses of POA/AHA neurones to pineal methoxyindoles may reflect their abilities to suppress hypothalamic releasing hormone neurones, experiments were conducted in which the effects of pineal indoles on rostral hypothalamic neurones were assessed in hamsters kept in long photoperiods, the recordings being additionally made during the dark phase of the L:D cycle (Demaine and Stoughton, 1982)

When recordings made in the dark phase were compared with those obtained in the light, there were significant differences between the proportion of neurones excited or inhibited by the methoxyindoles. The proportions of neurones responding to or unaffected by application of pineal indoles are given in Table 3. The predominant response to melatonin in the L phase was once again excitation, while of the units tested in the D phase the majority were inhibited by the indoleamine. There were similar findings in the responses to 5MTPH, the proportion of cells inhibited being significantly increased. In contrast, neuronal responses to 5HTPH appeared to be unaffected bv the L:D cycle.

These results demonstrate that the responses of neurones which may control releasing hormone secretion, to both melatonin and 5MTPH,

Table 3. Responses of POA/AHA neurones to the application of pineal indoles expressed as percentages of the total number of cells tested with each substance[a]

	Melatonin			5MTPH			5HTPH		
	+	–	O	+	–	O	+	–	O
LIGHT PHASE	70	0	30	24	22	54	0	45	55
DARK PHASE	12	54	34	8	56	36	0	60	40

Key: +, excitation; –, inhibition; O, no response.

[a]Recordings were made either in the light phase or in the dark phase of the L:D cycle.

are clearly dependent on the time of day at which they are applied to the cells.

CONCLUSIONS

In a number of photoperiod-sensitive mammalian species there is now good evidence that the action of the pineal gland is mediated through specific receptive sites located in the hypothalamus. In the Syrian hamster, we have found that under conditions when the exogenous administration of melatonin causes involution of the gonads, the indoleamine appears predominantly to inhibit POA/AHA neurones. These results suggest that the rostral hypothalamus is one of the target sites for pineal methoxyindoles, and that their possible mechanism of action is a direct modification of the rate of electrical discharge of target neurones and consequently of their transmitter (or neurohormone) release. Vacas and Cardinali (1979) have located receptors with high affinity binding of melatonin in the membrane fraction of the hypothalamus of both rats and hamsters. Since melatonin is lipid soluble, however, it seems reasonable to assume that it exerts its effect on electrical activity as it crosses the neuronal membrane without the necessity of combining with membrane receptors. Similar rapid modification of the electrical activity of excitable cells by other lipid soluble substances, including steroids and thyroxin, have been observed (Pfaff, 1981; Semm et al., 1981a; 1981b).

The effects exerted by pineal methoxyindoles on the firing rate of POA/AHA neurones is clearly not always the same, and thus the interactions between the cell membranes of the neurones and these indoles must be altered by other factors. There is some evidence that the pineal gland itself affects the responsiveness of neurones in this region of the brain, perhaps by its endogenous rhythm of methoxyindole secretion. It is possible that the down-regulation hypothesis of the melatonin receptor (Reiter, 1980b) could be tested on these neurones by studying the effects of repeated applications of melatonin onto the same cell over a long period.

ACKNOWLEDGEMENTS

I would like to thank Hazel Kann and Roger Stoughton for providing much of the data presented here, and Diane Newman for secretarial help. Some of the work reported here was financially supported by the Medical Research Council.

REFERENCES

Anton-Tay, F. and Wurtman, R.J., 1961, Regional uptake of ^{3}H-melatonin from blood or cerebrospinal fluid by rat brain, Nature, 221:474.

Bex, F., Bartke, A., Goldman, B.D. and Dalterio, S., 1978, Prolactin, growth hormone, luteinizing hormone receptors and seasonal changes in testicular activity in the golden hamster, Endocrinology, 103:2069.

Bittman, E.L. and Goldman, B.D., 1979, Serum gonadotropin levels in hamsters exposed to short photoperiods: effects of adrenalectomy and ovarectomy, Endocrinology, 83:113.

Bittman, E.L., Goldman, B.D. and Zucker, I., 1979, Testicular responses to melatonin are altered by lesions of the suprachiasmatic nuclei in golden hamsters, Biol. Reprod., 21:67.

Brown, G.H., Pang, S.F., Friend, W., Seggie, J. and Grota, L.J., 1977, Inverse relationship of hypothalamic LHRH to pineal melatonin and N-acetylserotonin, J. Neural Transm., suppl.13.

Brown, G., Grota, L. and Niles, L., 1981, Melatonin: origin, control of circadian rhythm and site of action, Adv. Biosci, 29:193.

Bubenik, G.A., Brown, G.M. and Grota, L.G., 1976, Differential localization of N-acetylated indolealkylamines in CNS and the harderian gland using immunohistology, Brain Res., 118:417.

Cardinali, D.P. and Freire, F., 1975, Melatonin effects on brain. Interaction with microtubule protein, inhibition of fast axoplasmic flow and induction of crystaloid and tubular formations in the hypothalamus, Mol. Cell. Endocr., 2:317.

Cardinali, D.P., Freire, F., Nagle, C.A. and Rosner, J.M., 1975, Effects of environmental lighting, superior cervical ganglionectomy and adrenergic drugs on microtubule protein levels of the rat hypothalamus, Neuroendocrinology, 19:44.

Cardinali, D.P., Vacas, M.I. and Boyer, E.E., 1979, Specific binding of melatonin in bovine brain, Endocrinology, 105:437.

Carrer, H.F. and Taleisnik, S., 1972, Neural pathways associated with the mesencephalic inhibitory influence on gonadotropin secretion, Brain Res., 38:299.

Carrillo, A.J., Rabii, J., Carrer, H.F. and Sawyer, C.H., 1977, Modulation of the proestrous surge of luteinizing hormone by electrochemical stimulation of the amygdala and hippocampus in the unanaesthetized rat, Brain Res., 128:81.

Chen, H.J., Brainard, G.C. and Reiter, R.J., 1980, Melatonin given in the morning prevents the suppressive action on the reproductive system of melatonin given in late afternoon, Neuroendocrinology, 31:129.

Cohen, M., Roselle, D. and Chabner, B., 1978, Evidence for a cytoplasmic melatonin receptor, Nature, 274:894.

Collu, R., Fraschini, F., Visconti, P. and Martini, L., 1972, Adrenergic and serotoninergic control of growth hormone secretion in adult male rats, Endocrinology, 90:1231.

Demaine, C. and Kann, H.C., 1979, Modifications of the electrical activity of hypothalamic neurones by pineal indoles. Progr. in Brain Res., 52:373.

Demaine, C. and Kann, H.C., 1981, Effect of pinealectomy on the electrical responses of hypothalamic neurones to the application of melatonin, Advances in Biosciences, 29:123.

Demaine, C. and Stoughton, R., 1982, Circadian variation in the electrical responses of hypothalamic neurones to pineal indoles, J. Physiol. (Lond.), (in press).

Demaine, C., Forsdyke, M., Kann, H.C. and Wormald, D.I., 1977, A technique for the ejection of solutions from glass micropipettes, based on digitally controlled thermal expansion, J. Physiol. (Lond.), 270:1.

Elliott, J.A. and Goldman, B.D., 1981, Seasonal reproduction, photoperiodism and biological clocks, in:"Neuroendocrinology of Reproduction,"N.T. Adler, ed., Plenum Press, New York.

Ellis, L.C., 1972, Inhibition of rat testicular androgen synthesis in vitro by melatonin and serotonin, Endocrinology, 90:17.

Ellis, G.B. and Turek, F.W., 1979, Time course of the photoperiod-induced change in sensitivity of the hypothalamic-pituitary axis to testosterone feedback in castrated male hamsters, Endocrinology, 104:625.

Fraschini, F., Mess, B. and Martini, L., 1968, Pineal gland, melatonin and the control of leuteinizing hormone secretion, Endocrinology, 82:919.

Fraschini, F., Collu, R. and Martini, L., 1971, Mechanisms of inhibitory action of pineal principles on gonadotropin secretion, in:"The Pineal Gland," Wolstenhulme and Knight, eds., a Ciba Foundation Symposium, Churchill Livingstone, Edinburgh.

Freire, F. and Cardinali, D.P., 1975, Effects of melatonin treatment and environmental lighting on the ultrastructural appearance, melatonin synthesis, norepinephrine turnover and microtubule protein content of the rat pineal gland, J. Neural Transmission, 37:237.

Hoffman, R.A. and Reiter, R.J., 1965, Pineal gland: influence on gonads of male hamsters, Science, 148:1609.

Johnson, L.Y. and Reiter, R.J., 1978, The pineal gland and its effects on mammalian reproduction, Prog. Reprod. Biol., 4:116.

Kamberi, I.A., Mical, R.S. and Porter, J.C., 1970, Effects of anterior pituitary perfusion and intraventricular injection of catecholamines and indoleamines on LH release, Endocrinology, 87:1.

Komisaruk, B.R., Terasawa, E. and Rodriguez-Sierra, J.F., 1981, How the brain mediates ovarian responses to environmental stimuli, in:"Neuroanatomy and Neurophysiology," N.T. Adler, ed., Plenum Press, New York.

Legan, S.J., Karsch, F.J. and Foster, D.L., 1977, The endocrine control of seasonal reproductive function in the ewe: a marked change in response to the negative feedback action of estradiol on luteinizing hormone secretion, Endocrinology, 101:818.

Lincoln, G.A. and Short, R.V., 1980, Seasonal breeding: nature's contraceptive, Recent Progress in Hormone Research, 36:1.
Martin, J.E. and Sattler, C., 1979, Developmental loss of the acute inhibitory effect of melatonin on the in vitro pituitary luteinizing hormone and follicle-stimulating hormone responses to luteinizing hormone-releasing hormone, Endocrinology, 105:1007.
McCann, S., 1981, CNS control of the pituitary: neurochemistry of hypothalamic releasing and inhibitory hormones, in: "Neuroendocrinology of Reproduction," N.T. Adler, ed., Plenum Press, New York.
Mess, B., Trentini, G.P., Ruzsas, C. and Gaetani, C.F., 1979, Some endocrine effects of the pineal gland and melatonin with special reference to reproduction, Progr. in Brain Res., 52:329.
Moore, R.Y., 1978, The innervation of the mammalian pineal gland, Prog. Reprod. Biol., 4:1.
Niles, L.P., Brown, G.M. and Grota, L.J., 1979a, Role of the pineal gland in diurnal endocrine secretion and rhythm regulation, Neuroendocrinology, 29:14.
Niles, L.P., Wong, Y.W., Mishra, R.K. and Brown, G.M., 1979b, Melatonin receptors in brain, Europ. J. Pharmacol., 55:219.
Pazo, J.H., 1979, Effects of melatonin on spontaneous and evoked neuronal activity in the mesencephalic reticular formation, Brain Research Bulletin, 4:725.
Pelletier, J. and Ortavant, R., 1976, Photoperiodic control of LH release in the ram. II.Light-androgens interaction, Acta Endocrinol., 78:442.
Pfaff, D.W., 1981, Electrophysiological effects of steroid hormones in brain tissue, in: "Neuroendocrinology of Reproduction," N.T. Adler, ed., Plenum Press, New York.
Reiter, R.J., 1978, Interaction of photoperiod, pineal and seasonal reproduction as exemplified by findings in the hamster, Prog. Reprod. Biol., 4:169.
Reiter, R.J., 1980a, The pineal and its hormones in the control of reproduction in mammals, Endocrine Reviews, 1:109.
Reiter, R.J., 1980b, "The Pineal," Eden Press, London.
Reiter, R.J. and Johnson, L.Y., 1974a, Depressant action of the pineal on pituitary luteinizing hormone and prolactin in male hamsters, Hormone Res., 5:311.
Reiter, R.J. and Johnson, L.Y., 1974b, Pineal regulation of immunoreactive luteinizing hormone and prolactin in light deprived female hamsters, Fert. Steril., 25:958.
Reiter, R.J. and Sorrentino, S.jr., 1972, Prevention of pineal-mediated reproductive responses in light-deprived hamsters by partial or total isolation of the medial basal hypothalamus, J. Neurovisc. Relat., 32:355.
Reiter, R.J., Vaughan, M.K., Vaughan, G.M., Sorrentino, S.jr. and Donofrio, R.J., 1975, The pineal gland as an organ of

internal secretion, in: "Frontiers of Pineal Physiology," M.D. Altschule, ed., MIT Press, Cambridge, MA.

Reiter, R.J., Blask, D.E., Johnson, L.Y., Rudeen, P.K., Vaughan, M.K. and Waring, P.J., 1976, Melatonin inhibition of reproduction in the male hamster: its dependency on time of day administration and on an intact and sympathetically innervated pineal gland, Neuroendocrinology, 22:107.

Reiter, R.J., Vaughan, M.K., Chen, H.J., Meyer, A.C., Philo, R.C., Dinh, D.T., de los Santos, R. and Guerra, H.C., 1981, Reproductive consequences of melatonin in mammals, Advances in the Biosciences, 29:151.

Rusak, B. and Zucker, I., 1979, Neural regulation of circadian rhythms, Physiol. Rev., 59:449.

Seegal, R.F. and Goldman, B.D., 1975, Effects of photoperiod on cyclicity and serum gonadotrophins in the Syrian hamster, Biol. Reprod., 12:223.

Semm, P., Demaine, C. and Vollrath, L., 1981a, The effects of sex hormones, prolactin and chorionic gonadotrophin on pineal electrical activity in guinea pigs, Cell. and Mol. Neurobiology, 1:259.

Semm, P., Demaine, C. and Vollrath, L., 1981b, Electrical responses of pineal cells to thyroid hormones and parathormone: a microelectrophoretic study, Neuroendocrinology. 33:212.

Smythe, G.A. and Lazarus, L., 1973, Growth hormone regulation by melatonin and serotonin, Nature, 244:230.

Smythe, G.A., Compton, P. and Lazarus, L., 1973, The biogenic amine control of corticosteroid secretion, in: Proc. 16th Annual Meeting Endocrinol. Socl. Australia, Abstr.16:12.

Sorrentino, S.,jr. and Reiter, R.J., 1971, Lack of pineal-induced gonadal regression in dark-exposed or blind hamsters after surgical isolation of the medial-basal hypothalamus, Gen. Comp. Endocr., 17:227.

Tamarkin, L. and Goldman, B., 1978, Effects of melatonin on the reproductive system in intact and pinealectomized hamsters, J. Neural Transm., suppl.13:398.

Tamarkin, L., Hutchison, J.S. and Goldman, B.D., 1976a, Regulation of serum gonadotropins by photoperiod and testicular hormone in the Syrian hamster, Endocrinology, 99:1528.

Tamarkin, L., Westrom, W.K., Hamill, A.I. and Goldman, B.D., 1976b, Effect of melatonin on the reproductive systems of male and female Syrian hamsters: a diurnal rhythm in sensitivity to melatonin, Endocrinology, 99:1534.

Tamarkin, L., Hollister, C.W., Lefebvre, N.G. and Goldman, B.D., 1977, Melatonin induction of gonadal quiescence in pinealectomized Syrian hamsters, Science, 198:953.

Tamarkin, L., Reppert, S.M. and Klein, D.C., 1979, Regulation of pineal melatonin in the Syrian hamster, Endocrinology, 104:358.

Trentini, G.P., Mess, B., de Gaetani, C.F. and Ruzsas, C., 1979 Pineal-brain relationship, Progr. in Brain Res., 52:329

Turek, F.W., 1977, The interaction of the photoperiod and testosterone in regulating serum gonadotropin levels in castrated male hamsters, Endocrinology, 101:1210.

Turek, F.W., Alvis, J.D., Elliot, J.S. and Menaker, M., 1976, Temporal distribution of serum levels of LH and FSH in adult male golden hamsters exposed to long or short days, Biol. Reprod., 14:630.

Vacas, M.I. and Cardinali, D.P., 1979, Diurnal changes in melatonin binding sites of hamster and rat brains. Correlation with neuroendocrine responsiveness to melatonin, Neuroscience Letters, 15:259.

Vaughan, M.K., Herbert, D.C., Brainard, G.C., Johnson, L.Y., Zeagler J.W. and Reiter, R.J., 1981, A comparison of blinding and afternoon melatonin injections on the histology of the reproductive organs. Pineal ultrastructure and gonadotrophin hormone levels in female Syrian hamsters, Advances in Biosciences, 29:65.

Vriend, J., 1981, The pineal and melatonin in the regulation of pituitary-thyroid axis, Life Sciences, 29:1929.

Vriend, J. and Reiter, R.J., 1977, Free thyroxin index in normal melatonin-trested and blind hamsters, Horm. Metab. Res., 9:231.

Vriend, J., Sackman, J.W. and Reiter, R.J., 1977, Effects of blinding pinealectomy and superior cervical ganglionectomy on free thyroxin index of male golden hamsters, Acta Endocrinol., 86:758.

Vriend, J., Reiter, R.J. and Anderson, G.R., 1979, Effects of the pineal and melatonin on the thyroid activity of male golden hamsters, Gen. Comp. Endocrinol., 38:189.

Yates, C.A. and Herbert, J., 1976, Differential circadian rhythms in pineal and hypothalamic 5-HT induced by artificial photoperiods or melatonin, Nature, 262:219.

NEUROBIOLOGICAL INVESTIGATIONS OF THE PINEAL GLAND AND ITS HORMONE MELATONIN

P. Semm

J.W.Goethe University, Dept. of Zoology

Siesmayerstr.70,6000 FRANKFURT, FRG

Since very recently an extensive review of the results of electrophysiological experiments in the pineal gland of rodents and birds has been published(Semm,1981), in this article I will focus on four important neurobiological aspects of the pineal in guinea-pigs,rats and homing pigeons: 1. circadian rhythmicity,as shown by long-term recordings and the use of microelectrophoresis, 2. electrophysiological and morphological studies on the central innervation of the gland 3. the transmitter-like activity of melatonin and its precursors in the pineal and other brain areas and 4. as a major point investigations of the magnetic sensitivity of the pineal, including single unit recordings, biochemical measurements of melatonin synthesis during and following earth-strength magnetic stimulation and the metabolic mapping of brain areas, which are sensitive for magnetic stimuli in the range of the natural field. The latter results may be of special interest for pineal researchers as most of the brain areas exhibiting magnetic sensitivity, as shown by the

(^{14}C)2-deoxyglucose-method, also contain or synthesize N-acetylalkylamines or melatonin respectively.

CIRCADIAN RHYTHMICITY

Long-term recordings from the guinea-pig pineal (Semm and Vollrath, 1980; Semm, 1982) have shown that three types of intrinsic cells can be distinguished: (a) Cells showing constant firing rates over a period of up to 24h. (b) Cells that are highly active during the day and show a low firing rate during the night (light activated cells). Artificial darkness during the day depresses the activity of these cells gradually, but it seems that this is an effective stimulus only until 16.00 h (Semm, 1982). Furthermore, artificial illumination during the night had no measurable effect. (c) Cells that exhibit low activity during the day and enhanced activity during the night (darkness activated cells). Artificial light given for 1 min during the night leads, in contrast to the effect of artificial darkness given to light-activated cells during the day, to an abrupt decrease of activity followed by a gradual renewed increase in firing rates. Artificial darkness during the day had no measurable effect on these cells. Both the light-and darkness-activated cells showed an oscillatory pattern in maintained activity, which closely followed season-dependent differences in day-and night-lengths. Darkness-activated cells were discussed as being involved in melatonin secretion (Semm and Vollrath, 1979) and in measuring different night-lengths, while light-activated cells may serve to measure day lengths. Both these cells together may represent a complex time-keeping device in which continuously oscillating mechanisms may play an important role. The constantly firing cells, which show no oscillating spontaneous activity, could serve

as a reference principle, related to the amount of activation of the two other cell types (see also Fig.1).
It could be shown that some of the constantly firing cells are under the influence of the CNS (Semm, 1982), whilst the other two cell types cease firing following sympathectomy.
When long-term recordings in the pineal gland were made from animals kept under continuous illumination for more than six weeks, no darkness activated cells could be demonstrated. Light-activated cells exhibited an altered rhythmicity when compared with cells from animals kept under natural lighting conditions.
Using the technique of microelectrophoresis and thus applying melatonin, noradrenaline and acetylcholine during the day to single pinealocytes, most of the cells were excited by melatonin and noradrenaline, the remaining cells being inhibited or showing no response. Acetylcholine caused in most of the cells either an inhibition or no response.
When the responses to melatonin and noradrenaline during the day were compared with those occuring at night, the numbers of excitatory/inhibitory effects were markedly different. The majority of responses that were recorded after 19.00 hours were inhibitory. No rhythmicity could be observed in the responses to the application of acetylcholine. Cells that could be excited by electrical stimulation of the lateral habenular nuclei showed also an excitation to the application of acetylcholine (Semm et al., 1981b).
In another study (Semm et al., 1981c), the responses to thyroxine showed a similar circadian pattern as observed with melatonin, calcitonin exhibiting a predominantly inhibitory and parathormone a predominantly excitatory action on the activity of pinealocytes.
Surprisingly, also the administration of a ß-blocking agent

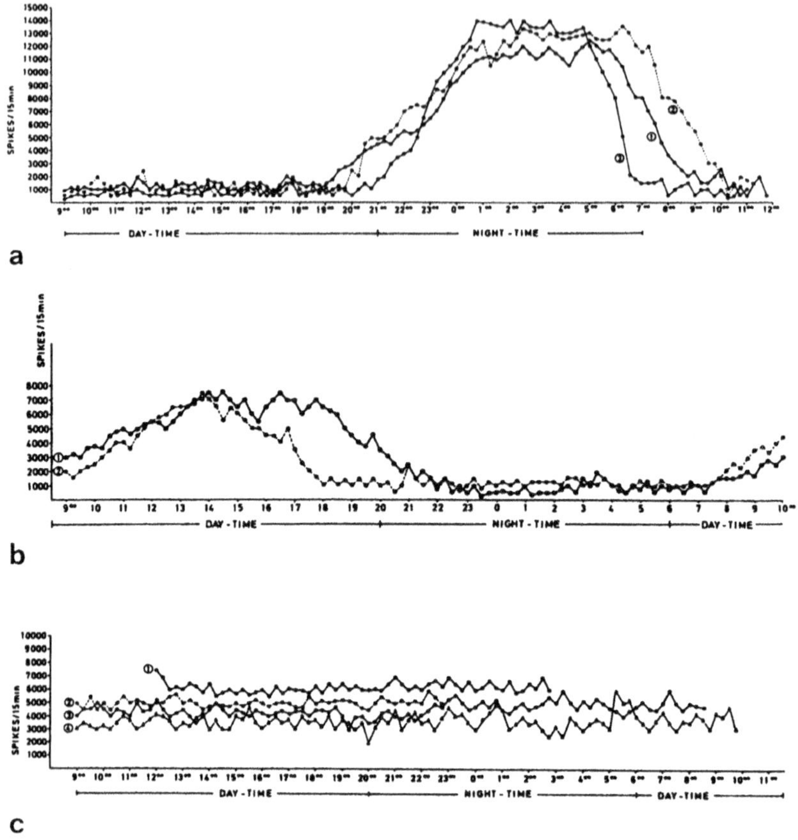

Fig. 1. Long-term recordings from guinea-pig pineal cell showing three different patterns of activity.
a. Three darkness-activated cells, showing low electrical activity during the day and higher activity during the night.
Note the close correspondence between the length of the night at different seasons(1, autumn; 2, winter; 3, summer)and the nocturnal activation.
b. Two light-activated cells, the activity of whic is high during the day and diminished during nighttime.
c. Four constantly firing cells; note the absence circadian rhythmicity.

(propranolol) elicited a circadian rhythmicity in the responsiveness of single pineal cells. The majority of responses that were recorded during daytime were excitatory, whilst most of the cells responded with an inhibition during the night. This implicates that a higher intrinsic activity of pinealocytes is depressed by the low sympathicotonus during daytime (Semm, unpublished observation). However, the direct proof for this assumption has to come from recordings of cultured pinealocytes lacking associations with nervous elements.
Using the same technique (Semm et al., 1981d), oestrone application in pregnant animals lead to an increase in firing frequency of most of the cells, while in males the majority of cells were inhibited. After administration of testosterone in females the pineal cells were usually inhibited, while in males the cells were often excited. Whereas in males a large majority of cells responded with excitation between 14.00 and 22.00 hours, in the morning the predominant effect was one of inhibition.

CENTRAL INNERVATION

A number of neuroanatomical and electrophysiological reports (see Ueck, 1979 and Semm, 1981 for the pertinent literature) have demonstrated a clear neural connection between the brain and the pineal gland. Some of the neural pathways are pinealofugal, while others are clearly pinealopetal. Some recent findings from our laboratory may be briefly discussed here.
In order to evaluate the possible influence of central structures on the electrical activity of single pineal cells, in the rat stimulation electrodes were inserted into the hippocampal formation. In the guinea-pig, the Nc. paraventricularis and the inferior and superior colliculi

were stimulated electrically(Reuss and Semm,1982). Stimulation of the hippocampus elicited a gradual augmentation of spontaneous electrical activity in some of the rat pineal cells. This excitation usually lasted for some minutes following stimulation(Fig.2). All stimulated pinea cells met the criteria of orthodromic stimulation, when cc pared with usual central neuron properties. The latencies of these cells were variable and in the range from 4.0 tc 6.0 msec. In contrast to earlier findings in sympathetically influenced cells(Semm and Vollrath,1979a,b) and similar to other centrally influenced cells(Semm et al., 1981a) no particular pattern in the firing of these cells was detected. As projections from the hippocampus to the lateral habenular nucleus have been studied both anatomically and electrophysiologically(literature reviewed in Mok and Mogenson,1974), it may well be, that information from the hippocampus reaches the pineal via the habenular nuclei. Furthermore, as both the habenular nuclei and the hippocampus contain melatonin(Brown et al.,1981), it can be speculated, that the pineal may influence these structures via melatonin and that in turn the melatonin-induced alterations are fed back via nervous pathways.

Electrical stimulation of the Nc.paraventricularis in the guinea-pig elicited both orthodromic and antidromi responses in the pineal gland, thus confirming again that both efferent and afferent elements are present in the gland.

Recently central fibres coming probably from and/or via the lateral habenular nucleus could be identified as being neurosecretory in nature in the guinea-pig(Schneide et al.,1981).Moreover, a direct projection from the Nc. paraventricularis to the posterior part of the pineal gland could be demonstrated, using the horseradish peroxidase method(Korf and Wagner,1980).

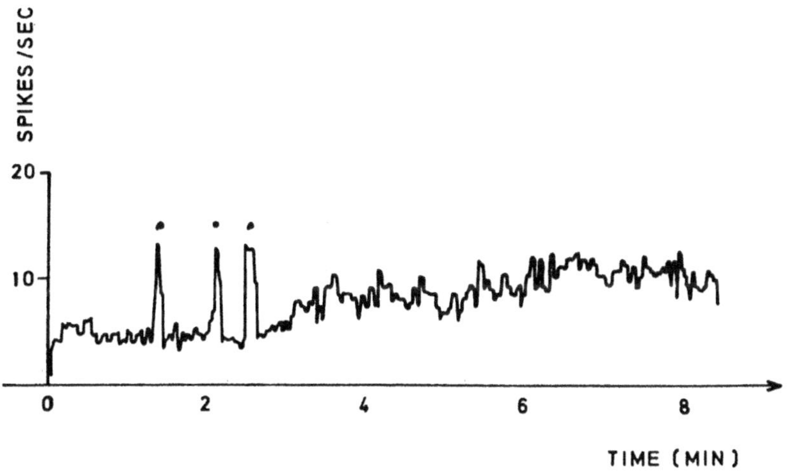

Fig. 2. Peri-stimulus time histogram of a rat pineal cell, the activity of which was gradually augmented following a train of electrical stimuli applied to the hippocampus (indicated by black circles)

Another central route of innervation via the posterior commissure is indicated by pineal responses following stimulation of the superior and inferior colliculi. Again, both orthodromic and antidromic responses could be obtained. Since the posterior commissure is known as a structure, which receives fibres from the colliculi and the tegmentum and short-term photic and acoustic stimuli can influence the electrical activity of some pineal cells in the rat(McClung and Dafny,1975) and in the guinea-pig(Semm, 1978), this second central innervation appears likely. However, in the rat the superior colliculus, the pretectum and the midbrain central gray receive inputs from the lateral habenular nuclei(Mok and Mogenson,1974). Further electrophysiological studies have to be undertaken to show whether some of these structures can also be influenced by the pineal gland.

MELATONIN AS A TRANSMITTER IN THE PINEAL, VISUAL SYSTEM AND CEREBELLUM

It has been hypothetized that one of the target organs for melatonin and/or 5-methoxyindoles could be the pineal gland itself, where indoles could possibly function as chemical messengers(for review see Vollrath, 1981). It is therefore not suprising that in addition to the previously demonstrated neurophysiological action of melatonin(Semm et al.,1981b) 5-methoxytryptophol and also 5-hydroxytryptophol influence the electrical activity of pineal cells(Semm and Vollrath,1982).
5-Methoxytryptophol caused on average a similar distribution of excitatory and inhibitory responses of the cells recorded, as did melatonin. However, only 29% of the single cells showed the same reactions to melatonin and the methoxyindole when applied successively. It is therefore not unlikely that both substances exhibit on most of the pineal cells a different intrinsic action, possibly influencing intracellularly or extracellularly the formation, storage, and/or release of different peptidergic or other effector compounds(see Quay,1974; Benson,1977).

Very recently it could be demonstrated that the delta sleep-inducing peptide(DSIP), a peptide which induces slow-wave EEG(sleep)(for review see Schoenenberger et al.,1978), is highly concentrated in the pineal after i.v. injection when compared with 9 other different regions in the brain (Graf et al.,1981a). Injections of DSIP at different times of the day(7.00 and 18.00 h) produced significant changes in the 24-h rhythmicity of 5-HT, dopamine and noradrenaline in the rat brain and of prolactin, growth hormone and corticosterone levels(Graf et al.,1981b). This strongly suggests (a) that DSIP acts on a higher level of coordination as a "programmer" for the circadian rhythmicity, the sleep inducing property appearing as a subset and (b) as melatonin shows a similar overall effect, that there is a functional relationship between the two substances. Microelectrophoretical studies have been made to determine the interaction of DSIP and melatonin on the single cell level in the pineal and the cerebellum(Semm and Reuss, 1982).

Recordings in the visual system of the guinea-pig reveal that melatonin and 5-methoxytryptophol can alter the spontaneous and light evoked or depressed activity in the retina, optic chiasma and geniculate body, but not in the superior colliclus(Semm and Vollrath, 1982). Suprisingly the different cell classes in the visual system responded identically with an augmentation of activity to the application of the substances, the methoxyindole exhibiting in most of the cases a more pronounced effect than melatonin. Since melatonin plays a role in the pigment aggregation of the retinal pigment epithelium in the guinea-pig(Pang and Yew, 1979) it may together with its excitatory action in the retinal ganglion cells act as a substance which makes the visual cells more sensitive to light.

In the cerebellum of the homing pigeon and the guinea-pig, melatonin and two of its precursors caused clear alterations in the electrical activity of Purkinje cells and other cell types(granule cells, Golgi cells).
In the pigeon, an equal number of Purkinje cells was inhibited and excited during day time by melatonin and 5-methoxytryptophol, whilst the responses to 5-hydroxytryptophol were predominantly inhibitory. The responses to the indoles varied significantly depending on whether the cel[l]s were tested during the day or at night($p>0.001$). 90% of the cells were inhibited by noradrenaline independently of the time of recording.
In the guinea-pig, most of the Purkinje cells were inhibited by the application of melatonin and 5-methoxytryptophol, while 5-hydroxytryptophol caused no response in most of the cells. In this species no significant circadian changes in sensitivity to the application of the indoles could be observed(Fig.3).
Purkinje cell responses to microelectrophoretically applied indoles were examined before and during noradrenaline aplication. This substance was found to augment indole excitation and inhibition in all the cases investigated, supporting the hypothesis that noradrenaline and thereby the indoles act on postsynaptic processes to increase the responsiveness of the Purkinje cells to afferent inputs.
Most of the other cerebellar units in the pigeon were inhibited by melatonin and 5-methoxytryptophol, whilst 5-hydroxytryptophol exhibited an excitation in most of the cells investigated. In the guinea-pig the responses of these cells were more complicated. Melatonin caused an excitation in most of the units, while the predominant response caused by 5-methoxytryptophol was one of inhibition. Only 30% of these units responded to 5-hydroxytryptophol, the remaining cells showing no measurable response.

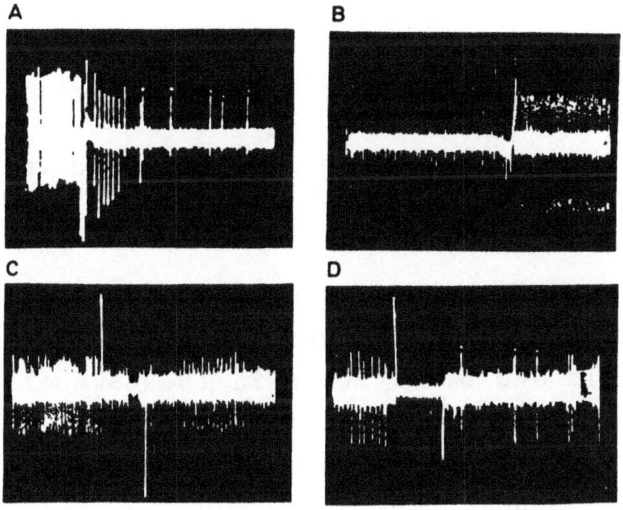

Fig.3 Purkinje cell responses to the application of melatonin(A;guinea-pig), 5-methoxytryptophol(B;pigeon), 5-hydroxytryptophol(C;pigeon) and noradrenaline (D;guinea-pig).

A:the left part of the picture shows spontaneous activity, which is strongly inhibited during application of melatonin. The downward deflection of the oscilloscope beam indicates the current artifact caused by electrophoresis. The discharge of the simple spikes is completely inhibited, while the complex spike is apparently not influenced (indicated by white circles).

B:The left part shows spontaneous activity, which is clearly augmented by 5-methoxytryptophol.

C:spontaneous activity of another Purkinje cell, which is inhibited by 5-hydroxytryptophol. During application time(indicated by the two current artifacts) background activity remains unaffected.

D:Purkinje cell electrical activity is completely inhibited by the application of noradrenaline. Following application time simple spike activity remains depressed, complex spike activity shows ongoing activity.
Calibrations A:0.5sec 2mV;B:0.5sec 2mV;C:0.5sec 1mV;D:0.5sec 1mV.

It is apparent from these studies that(1) pineal indoles can bring about direct changes in the firing frequency of Purkinje cells and other cells in the cerebellum of two different species and (2) that their action and thereby their influence on locomotor activity may be different in the two species investigated and (3) that these substances although primarily regarded as hormones represent new types of neurotransmitters or neuromodulators.
The results of the present study, showing a predominantly inhibitory action of melatonin on Purkinje cell activity at night, when levels of circulating melatonin are higher than during the day, implicate that melatonin causes a disinhibition of the inhibitory influences of the cerebellum on other brain areas, which are involved in the regulation of locomotor activity. Whether or not the result of this influence is a depression of overall locomotor activity remains to be elucidated(Semm and Vollrath,1982)

NEUROBIOLOGICAL INVESTIGATIONS ON THE MAGNETIC SENSITIVITY OF THE PINEAL GLAND IN RODENTS AND PIGEONS

In the past 15 years growing evidence has accumulated that the geomagnetic field can influence the behavior and orientation of a variety of organisms from bacteria through vertebrates, including man (for review see Wiltschko,

1980). From this body of experimental work it seem important to note, that most behavioral responses generally fall into two categories: either they involve magnetic compass orientation or, in addition, imply some form of physiological sensitivity to weak fluctuations in natural magnetic intensity.

Some of the characteristics of the bird's magnetic compass, its dependence on a rather narrow intensity range centred around the actual intensity of the natural field, its plasticity and its working as an inclination compass are well documented(Wiltschko,1980). On the other hand, there is a number of pathological phenomena, the temporal occurence of which can be correlated with geomagnetic activity(Gerisch and Becker,1979;Malin and Srivastava, 1979).

These results imply the existence of some sort of sensory receptors for sensing and transducing the magnetic field to the central nervous system. However, it is not known, where in vertebrates the magnetoreceptor is located. As the initial orientation of individual birds can be altered by changes in the direction of the magnetic field around their heads, it was assumed that the magnetic sensitive system is located in the head(Walcott, 1977).

From a theoretical point of view the pineal gland was considered as being a part of the postulated magnetic system(Semm et al.,1980), and testing this hypothesis by means of electrophysiological, biochemical and autoradiographic methods, it could be shown, that this structure, probably beneath others, is indeed involved in sensing magnetic variations in the natural environment.

This chapter summarizes the very new findings on the magnetic sensitivity of the pineal gland of rodents and pigeons and of other parts of the pigeon's brain.

1. Electrophysiology

a Guinea pig

About 20% of single cells in the guinea-pig pineal gland showed a marked effect to earth strength magnetic stimulation (Semm et al., 1980). Spontaneous electrical activity during daytime could be diminuished by more than 50% by an induced field, generated by Helmholtz-coils (the vertical component was added to the natural field), and restored when the magnetic field was inverted. Both, the latency of depression and renewed activation, was in the range of about 2 min. After turning off the inverted magnetic field the cells showed their normal spontaneous activity which could be depressed again with a low magnetic stimulus. Some cells responded in the range of msec, but they were infrequently found.

In addition, brain regions in the neighbourhood of the pineal gland were tested under identical experimental conditions. A comparable response (activation of the spontaneous activity) could be obtained in the molecular layer of the cerebellum, a finding which could be made valid indirectly later by means of the deoxyglucose method.

Some of the magnetical sensitive pineal cells in the guinea-pig showed an oscillatory electrical activity with a pattern characteristic for that of "darkness-activated cells." Thus, it was suggested that these cells may combine the abilities of time measurement, sensing variations in the natural magnetic field and in lighting conditions, and the synthesis of hormones, probably melatonin (Semm, 1982).

From this point of view it appears likely that these cells belong to the magnetic system which is sensitive to the daily fluctuations of the magnetic field rather than being involved in orientation within the magnetic field.

Because the daily heating and cooling of the atmosphere displaces the jet streams north and south, a more or less regular circadian variation of the magnetic field is observed on the ground. This furthermore implies that this rhythmicity of the magnetic field may have a "Zeitgeber" function, a hypothesis, which could be confirmed in hamsters (Brown and Scow, 1978) and in birds (Bliss and Heppner, 1976). On the other hand, the normal fluctuations in the earth's magnetic field can influence pigeon orientation (Keeton et al., 1974). Since most of the magnetic fluctuations in these behavioral experiments were less than 70 gamma, it was concluded that the sensitivity of pigeons to magnetic cues probably approaches that already demonstrated for insects.

b Homing pigeon

As to our knowledge nothing is known about effects of magnetic fields on the behavior of guinea-pigs, we turned to homing pigeons, which are known to use the natural magnetic field for directional information, especially under total overcast (Walcott, 1977).
As only a few recordings had been made from the pigeon pineal gland (Oksche et al., 1969) we checked the properties of the pigeon's pinealocytes, especially their sensitivity to light, given either only to the pineal following enucleation or to both the eyes and the pineal (Semm and Vollrath, 1982). Immediately after penetration of the pineal capsule action potentials could be observed exhibiting a wide range of firing rates from 2 to more than 90 actionpotentials/sec. Even after enucleation some pineal cells responded to light, mostly in a gradual manner (Fig, 4).

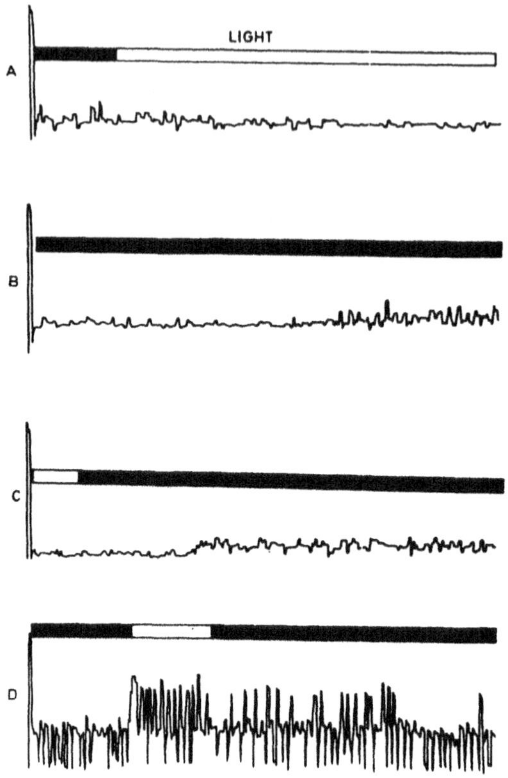

Fig.4 Single unit recordings in the pineal of enucleated pigeons (Averager Histograms, 1sec/bin)
A: Spontaneous electrical activity of a pineal cell, the activity of which is depressed by light and renewed gradually by artificial darkness (B).
C: Response of another pineal cell, which is gradually excited by darkness.
D: Activity of another pineal cell, the reaction of which starts immediately with the presentation of a light stimulus. The augmentation of activity during the light stimulus is followed by an attenuated activity for some seconds.

The electrical activity of about 30% of cells in the upper part of the pigeon's pineal gland could be altered by changing the strength of the local magnetic field and the direction of its vertical or horizontal component, using two pairs of Helmholtz coils(Semm et al.,1982).
In contrast to the most cells in the guinea-pig, such cells responded to a rapid and/or a gradual change in the magnetic field with either excitation or inhibition of electrical activity with a latency in the range of msec.
A considerable number of cells showed responses to slow or fast changes in the magnetic field, but the reactions could not be repeated within a time of 30 min. although recordings had a good quality and the experimental situation was unchanged. It may well be that these cells are due to unknown reasons refractory to a second magnetic stimulus within a certain period.
At present it is yet not possible to decide whether the magnetical sensitive cells react primarily to changes in the direction or to changes in the intensity of the magnetic field. As some of the cells responded as well to a complete rapid inversion of one component(i.e. a different direction but the same intensity as in the natural magnetic field) as to a gradual change(i.e. a distinct range of a different direction and an altered intensity), it may be possible that some of the cells are only direction selective and others can encode both direction and intensity changes by different amounts of excitation or inhibition of activity.
Comparing the electrophysiological results under magnetic stimulation in the guinea-pig and the pigeon one may have the impression that pigeon pineal cells are more sensitive and complicated in response to alterations in the magnetic field. However, the participation of these elements in magnetic orientation remains to be elucidated.

2. Biochemistry

The magnetic sensitivity of single pineal cells poses the question whether the secretory activity of the gland and thereby the synthesis of melatonin is also affected by an artificial magnetic field in the strength of that of the earth.
To test this hypothesis, the effects of a magnetic field on pineal serotonin-N-acetyltransferase(NAT) activity and the melatonin content in male rats were investigated (Welker et al.1982a+b).
Experimental inversion of the horizontal component of the natural magnetic field, performed at different time points during the night, led to a significant decrease of both parameters investigated, although the animals were free moving within a "globus," consisting of three Helmholtz-coils in north-south, east-west and vertical direction (Semm, 1982). During day-time, when the secretory activity of the gland is normally low, this effect was less conspicuous. Inversion of the horizontal component was followed by a reduced pineal melatonin formation, for about 2 hours, when the magnetic stimulation was performed at 24.00 hours. When the experiments were carried out at 22.00 hours, a clear depression was still noticeable after 2 hours.
In this context it is relevant to note that exposure of rats to 1 min of light during the night leads to a striking decrease of pineal melatonin, the synthesis of which is activated again, when the light puls is given at 22.00 hours. After midnight, in the second half of the scotophase a further increase is no longer detectable (Illnerova et al., 1979). As this effect is discussed as indicating a circadian rhythmicity in photosensitivity, a similar circadian rhythmicity in magnetosensitivity may also be

present in the pineal, thus giving further evidence for a functional relationship between the 24 h cycle and the magnetic field.

After 24 hours exposure to the inverted horizontal component, return to the natural condition was followed by a renewed clear depression of pineal NAT activity and melatonin content, indicating that the change of the magnetic field is the stimulus. Changing the inclination of the local magnetic field also decreased the secretory activity of the rat pineal gland.

In pigeons a similar response of the nocturnal melatonin synthesis in the pineal gland could be demonstrated, but due to unknown reasons, the results were not as clear as in rats. There were much more individual differences in the experimental group, whilst the melatonin values of the controls were in the normal range.

In conclusion, as the alteration in the ambient magnetic field rather than a movement of the head of a free moving animal causes a depression of melatonin synthesis, it appears likely that there is a compass system in the brain calibrated either on the polarity or the intensity of the natural field. In turn, after staying for 24 hours in the inverted magnetic field, the compass system is adaptet and calibrated to the new polarity.

Very recently also the melatonin synthesis in the retina of the rat could be demonstrated as being sensitive to earth-strength magnetic stimuli (Cremer-Bartels and Krause, pers. communication).

These results pose the question whether the pineal gland itself is sensitive to changes in the magnetic field or whether these effects are neuronally mediated, either by the sympathetic nerve fibres or the central commissural fibres. Since the sympathetic nervous system as a whole is sensitive to magnetic alterations (Sakharova, 1976), the sympathetic innervation of the pineal and perhaps also of the retina may be responsible for the magnetic sensitivity. A further explanation could be that the magnetic field is capable of influencing the enzymes themselves, which are involved in melatonin synthesis. It was pointed out by Beischer(1963) that hydrogen nuclei and other cell constituents briefly precess with frequencies according to their mechanical and magnetic moments when the body turns about in the earth's magnetic field. It was suggested, that such an interaction may provide living matter with spatial cues.

In general, what effect the magnetically induced short term decrease in melatonin synthesis has for the functioning of the organism as a whole remains to be determined. From the current status of knowledge it may be speculated, that also other hormones which are functionally coupled to melatonin(prolactin, growth hormone, sex hormones, thyroxine) and other brain regions, which either contain melatonin or are affected by the hormone, are sensitive to changes in the magnetic environment. Possibly this concept explains the phenomena which have been attributed to extremely low intensity magnetic fields(Busby, 1968; Gerisch and Becker, 1979; Ossenkopp and Nobrega, 1979).

Some observations indicate that interactions between organisms and the magnetic field have, during evolution, become a necessity for developing and maintaining normal functional integrity of higher organisms, the magnetic field possibly serving as a governing force that controls the rate of cellular growth and proliferation (Ossenkopp et al., 1972; Beck, unpublished observations). The relative importance of the pineal gland and its hormone melatonin in overall orientation will be tested in pinealectomized and/or melatonin treated pigeons and migratory birds under magnetic and non magnetic conditions.

It may well be that the pineal, independently from its magnetic sensitivity, plays a role in the time compensated sun compass orientation, and that in turn the time compensation of this compass can be affected by magnetic fields, which influence melatonin secretion. In comparison, when tiger salamanders are trained to select a special direction (Adler and Taylor, 1980), they cannot orient themselves under continuous light or darkness, but when melatonin, but not serotonin and thyroxin are given, the animals were again capable of orientation, but move in a 90° clockwise shifted direction.

3. Deoxyglucose-technique

Since the electrophysiological and biochemical studies have made it probable that the pineal gland may be part of the postulated magneto-sensitive system, an attempt was made to find further areas of the brain related to detection and transformation of magnetic stimuli from the natural environment, using the (^{14}C)2-deoxyglucose technique. This method has been shown to provide auto radiographic maps of regional glucose consumption under several experimental conditions, in particular during sensory stimulation (Sokoloff et al., 1977; Streit et al., 1980).
It could be shown that the functional uptake of deoxyglucose induced by magnetic field (inversion of the horizontal component) takes place in different brain regions in the pigeon (Mai and Semm, 1982). Those regions exhibiting extensive glucose uptake as compared with controls included the molecular layer of the cerebellum, the optic tectum, the parvocellular component of the isthmic nuclei, the intercollicular nuclei, the trapezoid body, the habenular nuclei and the pineal gland.
Even in experiments during daytime and at night after

presentation of light the pineal was well labelled, a fact which speaks in favor for specific magnetic sensitivity within the gland, since secretory activity is low during daytime and depressed by light during daytime. The labelling of the habenular nuclei, however, can also be explained by their involvement in the transmission of light to the pineal, as demonstrated earlier in the quail (Herbuté and Baylé, 1977).

In view of the fact, that the rat and pigeon pineal gland melatonin synthesis is influenced by magnetic fields, it appears meaningful, that some brain areas, which contain and/or synthesize melatonin (cerebellum, visual system, brain stem) are also responding to magnetic stimuli, as shown by the incorporation of (^{14}C)deoxyglucoe. Since electrophysiological studies have shown that these areas probably contain receptors for melatonin (Semm and Vollrath, 1982a+b; Pazo, 1979), it may well be that a magnetically induced alteration in blood melatonin concentration influences these areas.

Specific magnetic sensitivity in the optic tectum in animals stimulated in darkness during the night supports the proposal of Leask (1978) that magnetic field detection could take place in the retina, mediated by rhodopsin. In this context it is interesting to note, that (a) retinal ganglion cells can respond to magnetic stimuli by a specific alteration in their by light induced activity (Lövsund et al., 1981), (b) that in humans the scotopic critical flickerfusion frequency shows a tendency to diminuish gradually during exposure to a magnetic field with reduced intensity (Beischer et al., 1967) and (c) that behavioral studies have made it likely (Wiltschko and Wiltschko, 1981) that light is an important prerequisite for sampling magnetic information during an outward journey of young unexperienced pigeons. Furthermore, rhodopsin is also present

in the pineal gland(Deguchi,1981) and may have contributed to the magnetic sensitivity in this area.
However, the direct proof for these assumptions supporting the proposal of Leask (1978) has to come from experiments in total darkness showing that for example the spontaneous activity of single cells in magnetic sensitive areas cannot be influenced by a magnetic field
In general, I believe that all these results have opened a new door for the understanding of the complex orientation behavior of vertebrates and that the multiple implications of geomagnetism on behavior can now be investigated also in the brain itself. We have done only the first steps in this direction and we are full of hope, that the combination of neurobiological and behavioral methods will allow further insight into the old enigma of brain-magnetic field interrelationships.

REFERENCES

Adler,K.,and Taylor,D.H.,1980,Melatonin and thyroxine: influence on compass orientation in salamanders, J.comp.Physiol.136:235

Beck,W.,1982,pers.communication

Beischer,D.E.,1971,The null magnetic field as a reference for the study of geomagnetic directional effects in animals and man.Ann.New York Acad.Sci.188:324

Beischer,D.E.,Knepton,J.C.,and Kembro,D.V.,1967,Letter Rept.,NASA order No.R-39,U.S.Naval Aerospace Medical institute,Pensacola,Fla.(cited in Busby,D.E., 1968,see ibid.)

Benson,B.,1977,Current status of pineal peptides. Neuroendocrin.24:241

Brown,G.,Grota,L.,Bubenik,G.,Niles,L.,and Tsui,H.,1981, Physiologic regulation of melatonin,Adv.in Biosci., 29:95.

Brown,F.A.,Jr.,and Scow,K.M.,1978,Magnetic induction of a circadian cycle in hamsters,J.interdiscipl.Cycle Res.,9:137

Bliss,V.L.,and Heppner,F.H.,1976,Circadian activity rhyth influenced by near zero magnetic field,Nature 261: 411.

Busby,D.E.,1968,Space biomagnetics,Space Life Sci.,1:23.

Cremer-Bartels,G.,and Krause,K.(pers.communication).

Deguchi,T.,1981,Rhodopsin-like photosensitivity of isolated chicken pineal gland.Nature,290:702.

Gerisch,W.,and Becker,G.,1979,Geomagnetobiologisch bedingter Zusammenhang zwischen der Fraßaktivität von Termiten und der Zahl der Sterbefälle.Bundesanstalt für Materialprüfung(Berlin),Forschungsbericht 62:1.

Graf,M.,Christen,H.,Tobler,H.J.,Baumann,J.B.,and Schoenenberger,G.A.,1981,DSIP a circadian'programming' substance?,Experientia 37:624

Graf,M.,Lorez,H.P.,Gillesen,D.,Tobler,H.J.,and Schoenenberger,G.A.,1981,Distribution and specific binding of ^{3}H-DSIP.,Experientia 37:625.

Herbute,J.,and Bayle,J.D.,1977,Suppression of pineal multiunit response to flash after habenular lesion in quail,Am.J.Physiol.,231:136

Illnerova,H.,Vanecek,J.,Krecek,J.,and Wetterberg,L.,1979, Effect of one minute light exposure to light at night on rat pineal serotonin N-acetyltransferase and melatonin,J.Neurochem.,32:673

Keeton,W.T.,Timothy,S.L.,and Windsor,D.M.,1974,Normal fluctuations in the earth's magnetic field influenc pigeon orientation,J.comp.Physiol.,95:95.

Korf,H.-W.,and Wagner,V.,1980,Evidence for a nervous connection between the brain and the pineal organ in the guinea-pig,Cell.Tiss.Res.,209:505

Leask,M.J.M.,1977,A physikochemical mechanism for magnetic field detection by migratory birds and homing pigeons, Nature,267:144.

Lövsund,P.,Nilsson,S.E.G.,and Öberg,P.A.,1981,Influence on frog retina of alternating magnetic fields with special reference to ganglion cell activity,Med.& Comput.,19:679.

Mai,J.K.,and Semm,P.,1982,Anatomy of magnetic communication syste :Metabolic mapping of magnetically sensitive areas of the pigeon's brain by means of the autoradiographic (^{14}C)deoxyglucose technique (submitted).

Malin,S.R.C.,and Srivastava,B.J.,1979,Correlation between heart attacks and magnetic activity.Nature,277:646.

McClung,R.,and Dafny,N.,1975,Neurophysiological properties of the pineal body.II.Single unit recording,Life Sci.,16:621.

Mok,A.C.S.,and Mogenson,G.J.,1974,Effects of electrical stimulation of the lateral hypothalamus,hippocampus,amygdala and olfactory bulb on unit activity of the lateral habenular nucleus of the rat,Brain Res.,77:417.

Ossenkopp,K.-P.,Koltek,W.T.,and Persinger,M.A.,1972,Prenatal exposure to an extremely low frequency-low intensity rotating magnetic field and increase in thyroid and testicle weight in rats, Develop Psychobiol.,5:275.

Ossenkopp,K.-P.,and Nobrega,J.N.,1978,Significant relationship between perinatal geomagnetic field activity and anxiety levels in females,Arch.Met.Geoph.Biokl. Ser.B.,27:75.

Pang,S.F.,and Yew,D.T.,1979,Pigment aggregation by melatonin in the retinal pigment epithelium and choroid of guinea-pigs,Experientia,35:231

Pazo,J.H.,1979,Effects of melatonin on spontaneous and evoked neuronal activity in the mesencephalic reticular formation,Brain Res.Bull.,4:725

Quay,W.B.,1974,Pineal chemistry,Springfield,III.:Charles C.Thomas.

Reuss,S.,and Semm,P.,1982,Electrophysiological investigations on the sympathetic and central innervation of the mammalian pineal gland,Abstracts of the 75th Symposium of the German Zoological Society,Hannover, p.99.

Sakharova,S.A.,1977,Reactions of the central and peripheral mediator links of the sympatho-adrenal system to a single exposure to an alternating magnetic field,Biol.Nauki,9:35.

Schneider,T.,Semm,P.,and Vollrath,L.,1981,Ultrastructural observations on the central innervation of the guinea-pig pineal gland,Cell Tiss.Res.,220:41

Schoenenberger,G.A.,Maier,P.F.,Tobler,H.J.,Wilson,K.,and Monnier,M.,1978,The delta EEG(Sleep)-inducing peptide(DSIP)XI.Amino-acid analysis,Sequence,Synthesis and Activity of the Nonapeptide,Pflügers Arch.,376:119.

Semm,P.,1978,Electrophysiological and morphological aspects of the guinea-pig epiphysis cerebri,J.neural Transm.,Suppl.,13:394.

Semm,P.,1981,Electrophysiological aspects of the mammalian pineal gland,in:The pineal organ:photobiology-biochronometry-endocrinology,Oksche,A.,and Pevet, P.,eds.,pp.81-96,Elsevier/North Holland

Semm,P.,1982,Electrophysiology of the mammalian pineal gland:evidence for rhythmical and non rhythmical elements and for magnetic influence on electrical activity, in:Structure and Function of the vertebrate circadian system,Aschoff,J.,ed.,Berlin-Heidelberg-New York:Springer(in press).

Semm,P.,and Reuss,S.,1982,Interactions of melatonin and delta sleep-inducing peptide(DSIP) in the pineal gland and cerebellum(in preparation).

Semm,P.,and Vollrath,L.,1979,Electrophysiology of the guinea-pig pineal organ:sympathetic influence and different reactions to light and darkness,Prog. Brain Res.,52:107.

Semm,P.,and Vollrath,L.,1980,Electrophysiological evidence for circadian rhythmicity in a mammalian pineal organ,J.neural Transm.,47:181.

Semm,P.,and Vollrath,L.,1982a,Alterations in the spontaneous activity of cells in the guinea-pig pineal gland and visual system produced by pineal indoles, J.neural Transm.53(in press).

Semm,P.and Vollrath,L.,Electrical responses of homing pigeon and guinea-pig Purkinje cells to pineal indoles applied by microelectrophoresis,Exp.Brain Res.(submitted)

Semm,P.,and Vollrath,L.,1982c,Spontaneous electrical activity in the pineal gland of the pigeon(in preparation).

Semm,P.,Demaine,C.,and Vollrath,L.,1981b,Electrical responses of pineal cells to melatonin and putative transmitters:evidence for circadian changes in sensitivity, Exp.Brain Res.,43:361.

Semm,P.,Demaine,C.,and Vollrath,L.,1981c,Electrical responses to thyroid hormones and parathormone.A microelectrophoretic study,Neuroendocrin.,33:212

Semm,P.,Demaine,C.,and Vollrath,L.,1981d,The effects of sex hormones,prolactin and chorionic gonadotropin on pineal cell electrical activity in guinea-pigs, Cell.mol.Neurobiol.,1:259.

Semm,P.,Schneider,T.,and Vollrath,L.,1980,The effects of an Earth-strength magnetic field on the electrical activity of pineal cells,Nature,288:607.

Semm,P.,Schneider,T.,and Vollrath,L.,1981a,Morphological and electrophysiological evidence for habenular influence on the guinea-pig pineal gland,J.neural Transm.,50:247.

Semm,P.,Schneider,T.,Vollrath,L.,and Wiltschko,W.,1982, Magnetic sensitive pineal cells in pigeons,Proc. in Life Sci:Avian Navigation, Papi,F.and Wallraff,H. eds.,pp.329-337, Springer Verlag.

Sokoloff,L.,Reivich,M.,Kennedy,C.,Des Rosiers,M.H.,Patlak C.S.,Petitgrew,K.D.,Sakurada,O.,and Shinohara,M., 1977,The (^{14}C)deoxyglucose method for the measurement of local cerebral glucose utilization:theory, procedure,and normal values in the conscious and anesthetized albino rat,J.Neurochem.,28:897.

Streit,P:,Burkhalter,M.,Stella,M.,and Cuénod,M.,1980, Patterns of activity in pigeon brain's visual relays as revealed by the (^{14}C)2-deoxyglucose method,Neurosci.,5:1053.

Ueck,M.,1979,Innervation of the vertebrate pineal,Prog. Brain Res.,52:45.

Vollrath,L.,1981,The pineal organ,in:Handbuch der mikroskopischen Anatomie des Menschen,Oksche,A.,and Vollrath,L.,eds.,Vol.6,part 7,Berlin-Heidelberg-New York:Springer Verlag.

Walcott,C.,1977,Magnetic fields and the orientation of homing pigeons under sun,J.exp.Biol.,70:105.

Welker,H.A.,Semm,P.,and Vollrath,1982a,Effects of an artificial magnetic field on the secretory activity of the rat pineal gland, Symposium of the German Endocrinological Society, Salzburg, Abstract 139, p. 122.

Welker, H.A., Semm, P., Willig, R.P., Wiltschko, W., and Vollrath, L., 1982b, Effects of an artificial magnetic field on serotonin N-acetyltransferase activity and melatonin content of the rat pineal gland (in preparation).

Wiltschko, W., 1980, The earth's magnetic field and bird orientation, Trends in Neurosci., 3:140.

Wiltschko, W., and Wiltschko, R., 1981, Disorientation of inexperienced young pigeons after transportation in total darkness, Nature, 291:433.

Addendum:

Oksche, A., Morita, Y., and Vaupel-von Harnack, M., 1969, Zur Feinstruktur und Funktion des Pinealorgans der Taube (Columba livia), Z. Zellforsch., 102:1.

INFLUENCE OF THE PINEAL GLAND ON TUMOR GROWTH IN MAMMALIANS: A REAPPRAISAL FROM A BIOCHEMICAL POINT OF VIEW

Maria E. Ferioli, Giuseppe Scalabrino, and Franco Fraschini

Institute of General Pathology and C.N.R. Center for Research in Cell Pathology (M.E.F., G.S.) and Institute of Pharmacology, Chair of Chemotherapy (F.F.), University of Milan, 20133 MILANO, Italy

> In nature infinite book of secrecy
> A little I can read
> W. Shakespeare, Antony and Cleopatra, Act I, Scene 2.

INTRODUCTION

The regulation of the growth of mammalian tumors by different hormones remains an intriguing mystery. The claim that the pineal gland is involved in controlling neoplastic growth is one segment of that large area.

Experimental and clinical reports summarized by Bartsch and Bartsch (1981) and by Lapin and Ebels (1981) indicate that there is some link between cancer development and pineal function, although the data so far available are often contradictory. The link was detected when the effects on neoplastic growth of pinealectomy and/or administration of melatonin were studied. This approach has been complemented and supplemented by identification of other active pineal substances with anti-tumor activity and by the study of histological changes in the pineal glands of humans and animals that have died of cancer.

It has been demonstrated that pinealectomy accelerates the growth of transplanted melanoma in hamsters (Das Gupta and Terz, 1967a,b), of transplanted Walker 256 carcinoma in rats (Rodin, 1963; Barone and Das Gupta, 1970) and of transplanted Yoshida sarcoma in rats (Lapin, 1975; Lapin and Frowein, 1981). Similarly, removal of the pineal gland enhanced the induction of mammary adenocarcinoma in the rat by 7,12-dimethylbenz(α)-anthracene (DMBA), particularly when the carcinogen was administered at low doses (Tamarkin et al., 1981). But, when DMBA was administered at high doses, there was little if any difference in the occurrence of mammary neoplasms in pinealectomized and sham-pinealectomized rats (Lapin, 1978; Aubert et al., 1980).

The administration of the pineal indole hormone melatonin to pinealectomized hamsters abolished the accelerating effect of pinealectomy on the growth of melanoma implants (El-Domeiri and Das Gupta, 1973). Analogously, tumor development was slowed down and tumor incidence decreased by exogenous melatonin administration to female rats in which mammary adenocarcinomas were being induced by DMBA (Aubert et al., 1980; Tamarkin et al., 1981). The only opposite finding was reported by Hamilton (1969). Furthermore, melatonin was shown to reduce the incidence of methylcholanthrene-induced sarcoma in mice (Lapin and Ebels, 1976). Interestingly enough, it was demonstrated by Bartsch and Bartsch (1981) that the inhibitory effect of melatonin on the growth of fibrosarcoma ascites cells or of Ehrlich ascites cells injected into the peritoneal cavities of mice was dependent on the photoperiod and the time of day of administration. Melatonin administration also slowed down the development of Yoshida sarcoma cells transplanted into rats (Lapin and Ebels, 1976; Lapin and Frowein, 1981), but for this type of tumor, too, an opposite finding has been reported (Huxley and Tapp, 1972a). The inhibitory effect of melatonin on tumor growth was also demonstrated on a transplantable leukemia in mice (Buswell, 1975) and a transplantable mammary tumor in female rats (Karmali et al., 1978). Unexpectedly, melatonin reduced the survival time of mice bearing Lewis lung carcinoma (Lapin and Ebels, 1976).

Melatonin does not have any in vitro anti-tumor activity on KB cells (Bindoni et al., 1976), but there is some substance in sheep pineal gland that has antimitotic activity on that cell strain. It has been partially purified (Bindoni et al., 1976). The bovine pineal gland also contains a polypeptide substance with anti-tumor activity on different transplantable mouse tumors and chemically induced rat mammary adenocarcinomas (Dilman et al., 1979).

From this survey of the literature, it is quite clear that so far no studies using the reliable putative biochemical markers of neoplastic growth have been done to evaluate whether or not pinealectomy accelerates neoplastic development or whether or not exogeneous melatonin slows down tumor growth. A large body of experimental evidence demonstrates that the physiologically occurring di- and poly-amines accumulate during all cell growth processes, including growth of experimental animal tumors and human tumors (Scalabrino and Ferioli, 1981). Polyamines appear to be tightly linked to biochemical processes crucial for cell growth and proliferation, although the precise function(s) of these cationic substances are not fully understood. L-Ornithine decarboxylase (L-ornithine carboxy-lyase, EC 4.1.1.17) (ODC) catalyzes the conversion of ornithine to putrescine and is widely reported to be the rate-limiting enzyme in the biosynthetic pathway of polyamines. Besides being the first enzyme in the polyamine biosynthetic pathway, ODC is easily inducibile and to a considerable degree by a great variety of stimuli. Its half-life is the shortest known for any enzyme and, last but not least, it carries out the only known reaction for putrescine biosynthesis in mammalian cells (Scalabrino and Ferioli, 1981). Moreover, induction of ODC activity is an early and marked event when eukaryotic cells are stimulated to grow, in all systems studied to date (Scalabrino and Ferioli, 1981). With this in mind, we monitored the levels of ODC activity sequentially in the livers of pinealectomized rats and of sham-pinealectomized rats, both in the process of chemical hepatocarcinogenesis induced by 3′-methyl-4-(dimethylamino)-azobenzene (3′-Me--DAB). Additionally, we compared the levels of ODC activity in the mammary glands of pinealectomized and of sham-pinealectomized rats, both in the process of DMBA chemical carcinogenesis, since we hypothesize that the influence, if any, of the epiphysis on neoplastic growth and development might be more easily detected in a neoplasm whose growth is under hormonal control.

We did not consider it necessary to perform any histological or biochemical studies of the pineal glands of our rats with tumors, since the relevance of that would be marginal and inconclusive. Very few studies have included the weights and the morphology of the pineal glands of human who died with cancer. The pineal weights were heavier than those in patients who died of nonmalignant diseases in two studies (Rodin and Overall, 1967; Tapp, 1980a), but just the opposite has also been reported (Tapp and Blumenfield, 1970). Pineal morphological abnormalities (Hajdu et al., 1972) and calcification (Cohen et al., 1978) have also been described. In experimental animals, the lipid content of pineal glands from rats with chemically induced fibrosarcomas was significantly greater than that of glands from the controls (Huxley and Tapp, 1972b; Tapp, 1980b).

MATERIALS AND METHODS

Chemicals

L-[1-^{14}C] Ornithine monohydrochloride (SA 58 mCi/-mmol) was purchased from the Radiochemical Centre (Amersham, U.K.). The volatile $^{14}CO_2$ contamination of labeled L-ornithine was removed before use (Jänne and Williams-Ashman, 1971). The source of 3'-methyl-DAB was Koch Light (Colnbrook, U.K.) and that of DMBA was Sigma, Chemical Co., St. Louis, Mo., U.S.A.. All of the other chemicals were the highest grades commercially available.

Animals

Adult rats of both sexes (Sprague-Dawley strain) were obtained from Charles River (Calco, Italy). All rats were rigidly standardized, starting 12 days before pinealectomy or sham-pinealectomy and continuing until they were killed. They were maintained in a climate-controlled room (23°C), artificially illuminated with a light-dark cycle of 12:12 h daily (lights on, 06:00-18:00 h). During both the habituation period and experiment, the nocturnal feeding habit of the rats was reinforced by removal of their food when the lights came on and its replacement just before the lights were turned off. The composition of the semisynthetic basal diet for the controls was essentially that reported by Sneider and Potter (1969). The diet oncogenic for the liver was identical with the basal diet except that 0.06% 3'-Me-DAB was added and this was given to male rats only. 8.0 Mg of DMBA in peanut oil was given by i.g. intubation to 60-day-old female rats, which had been pinealectomized 35 days earlier. Pinealectomy was performed under Nembutal anesthesia by the method of

Hoffman and Reiter (1965). In sham-pinealectomized rats, the forceps were introduced into the vicinity of the pineal gland without injuring the gland or neighboring vascular or brain tissues. Each animal was checked for successful pinealectomy at zooscopy. The operations were always performed between 08:00-11:00. The overall mortality of rats, on either of the oncogenic treatments, was 5%. In order to minimize the diurnal changes of ODC activity levels in the organs of the rats, the rats were always killed between 14:00 and 16:00 h.

Preparation of Organ Extracts

At fixed times the rats were killed by decapitation. The livers were quickly removed from the rats on the diet containing 3'-Me-DAB and were processed individually. The mammary glands were quickly removed from the rats treated with DMBA and were also processed individually. The homogenizing medium was that previously described by Scalabrino et al. (1979). The livers were homogenized in 2 vol and the mammary glands in 4 vol of medium. The 20,000 x g supernatant fraction was used immediately for assay of the ODC activity.

Enzyme Assay

The ODC activity was assayed essentially as previously described by Scalabrino et al. (1979). The assays were always done in duplicate. The concentration of L-ornithine (labeled plus unlabeled) used in the routine assay was saturating (final concentration, 2 mM).

Protein Determination

Protein content was estimated, with bovine serum albumin as the standard, as described by Geiger and Bessman (1972).

RESULTS AND DISCUSSION

Table 1 shows the levels of ODC activity in livers of the control male rats, both sham-pinealectomized and pinealectomized, fed the basal diet and of rats, both sham-pinealectomized and pinealectomized, fed the diet for hepatocarcinogenesis. It appears that pinealectomy does not modify the levels of hepatic ODC activity in

Table 1. ODC Activity in Livers of Sham-pinealectomized or Pinealectomized male Rats during 3'-Me-DAB Hepatocarcinogenesis

Rats	Diet	Time of diet (months)	O D C Activity (pmol $^{14}CO_2$/mg protein/30 min)
Sham-PLX	Basal	1	70.4 ± 12.15 (6)
PLX	Basal	1	65.9 ± 9.84 (6)
Sham-PLX	3'-Me-DAB	1	194.4 ± 25.16 (6)
PLX	3'-Me-DAB	1	195.3 ± 23.42 (6)
Sham-PLX	Basal	3	16.6 ± 2.26 (5)
PLX	Basal	3	11.8 ± 1.15 (5)
Sham-PLX	3'-Me-DAB	3	47.6 ± 3.72 (5)
PLX	3'-Me-DAB	3	38.0 ± 2.34 (5)

The results are expressed as mean values ± E.S. Number of animals in parentheses. Sham-PLX=Sham-pinealectomized rats. PLX=Pinealectomized rats.

rats fed the basal diet throughout the experimentation period, since sham-pinealectomized and pinealectomized animals nearly always had the same liver enzyme activity. Furthermore, again from Table 1, it appears that pinealectomy does not accelerate hepatocarcinogenesis when the ODC activity is the biochemical parameter for evaluating tumor growth. In fact, the increase in hepatic ODC activity, which was observed at the end of the first month, caused by the carcinogen (Scalabrino et al., 1978), is of the same order of magnitude in both sham-pinealectomized and in pinealectomized rats. Similarly, at the end of the third month, the hepatic levels of ODC activity in pinealectomized and in sham-pinealectomized rats did not significantly differ. In keeping with an earlier study (Scalabrino et al., 1978, at this time we observed a marked fall in hepatic levels of ODC activity.

Table 2 shows the levels of ODC activity in the mammary glands of intact untreated rats and of sham-pinealectomized and pinealectomized rats 2 months after i.g. intubation of a low dosage of DMBA. This low dosage of DMBA was chosen because Tamarkin et al., (1981) demonstrated that pinealectomy accelerated tumor development

Table 2. ODC Activity in Mammary Glands of Sham-pinealectomized or Pinealectomized Female Rats Treated with DMBA

Rats	Time after i.g. DMBA intubation (months)	O D C Activity (pmol $^{14}CO_2$/mg protein/30 min)
Intact	0	2.1 ± 0.94 (4)
Sham-PLX	2	5.7 ± 0.71 (4)
PLX	2	22.8 ± 2.15 (4)

The results are expressed as mean values ± E.S. Number of animals in parentheses. Sham-PLX=Sham-pinealectomized rats. PLX=Pinealectomized rats.

in mammary glands only when DMBA was given at a low dosage. From Table 2 it appears that the treatment with DMBA enhances ODC levels, as other carcinogens do in their target tissues. High ODC activity was found by Andersson et al., (1976) in fully developed rat mammary tumors induced by DMBA. But, what is more interesting, is that the enhancement of ODC activity induced by DMBA 2 months after i.g. intubation was much greater in pinealectomized rats than in sham-pinealectomized rats. This demonstrates that the removal of pineal gland facilitated tumor development in mammary glands. This conclusion is in agreement with Tamarkin et al., (1981), but must be taken with caution, since our investigation is still in progress. However, we can state unequivocally at this time that the pineal gland does not at all influence the development and growth of hepatomas induced chemically in the rat.

REFERENCES

Andersson, A. C., Henningsson, S., Lundell, L., Rosengren, E., and Sundler, F., 1976, Diamines and Polyamines in DMBA-Induced Breast Carcinoma Containing Mast Cells Resistant to Compound 48/80, Agents and Actions, 6:577.

Aubert, C., Janiaud, P., and Lecalvez, J., 1980, Effect of Pinealectomy and Melatonin on Mammary Tumor Growth in Sprague-Dawley Rats Under Different Conditions of Lighting, J. Neural Transm.,

47:121.
Barone, R. M., and Das Gupta, T. K., 1970, Role of Pinealectomy on Walker 256 Carcinoma in Rats, J. Surg. Oncol., 2:313.
Bartsch, H., and Bartsch, C., 1981, Effect of Melatonin on Experimental Tumors Under Different Photoperiods and Times of Administration, J. Neural Transm., 52:269.
Bindoni, M. Jutisz, M., and Ribot, G., 1976, Characterization and Partial Purification of a Substance in the Pineal Gland Which Inhibits Cell Multiplication In Vitro, Biochim. Biophys. Acta, 437:577.
Buswell, R. S., 1975, The Pineal and Neoplasia, Lancet, i:34.
Cohen, M., Lippman, M., and Chabner, B., 1978, Role of Pineal Gland in Aetiology and Treatment of Breast Cancer, Lancet, ii:814.
Das Gupta, T. K., and Terz, J., 1967a, Influence of Pineal Body on Melanoma of Hamsters, Nature, 213:1038.
Das Gupta, T. K., and Terz, J., 1967b, Influence of Pineal Gland on the Growth and Spread of Melanoma in the Hamster, Cancer Res., 27:1306.
Dilman, V. M., Anisimov, V. N., Ostroumova, M. N., Morosov, V. G., Khavinson, V. K., and Azarova, M. A., 1979, Study of the Anti-Tumor Effect of Polypeptide Pineal Extract, Oncology, 36:274.
El-Domeiri, A. A. H., and Das Gupta, T. K., 1973, Reversal by Melatonin of the Effect of Pinealectomy on Tumor Growth, Cancer Res., 33:2830.
Geiger, P. J., and Bessman, S. P., 1972, Protein Determination by Lowry's Method in the Presence of Sulfhydryl Reagents, Anal. Biochem., 49:467.
Hajdu, S. I., Porro, R. S., Lieberman, P. H., and Foote, F. W., Jr., 1972, Degeneration of the Pineal Gland of Patients with Cancer, Cancer, 29:706.
Hamilton, T., 1969, Influence of Environmental Light and Melatonin Upon Mammary Tumour Induction, Brit. J. Surg., 56:764.
Hoffman, R. A., and Reiter, R. J., 1965, Rapid Pinealectomy in Hamsters and Other Small Rodents, Anat. Rec., 153:19.
Huxley, M., and Tapp, E., 1972a, Effects of Biogenic Amines on the Growth of Rat Tumors, Life Sci., 11 (Part II):19.
Huxley, M., and Tapp, E., 1972b, The Histochemical Measurements of the Lipid Content of the Pineal Gland in Rats Suffering from Malignancy, Brain Res., 37:123.

Jänne, J., and Williams-Ashman, H. G., 1971, On the Purification of Ornithine Decarboxylase from Rat Ventral Prostate and Effects of Thiol Compounds on the Enzyme, J. Biol. Chem., 246:1725.
Karmali, R. A., Horrobin, D. F., and Ghayur, T., 1978, Role of Pineal Gland in Aetiology and Treatment of Breast Cancer, Lancet, ii:1001.
Lapin, V., 1975, The Pineal and Neoplasia, Lancet, i:341.
Lapin, V., 1978, Effects of Reserpine on the Incidence of 9,10-Dimethyl-1,2-benzanthracene-induced Tumors in Pinealectomised and Thymectomised Rats, Oncology, 35:132.
Lapin, V., and Ebels, I., 1976, Effects of Some Low Molecular Weight Sheep Pineal Fractions and Melatonin on Different Tumors in Rats and Mice, Oncology, 33:110.
Lapin, V., and Ebels, I., 1981, The Role of the Pineal Gland in Neuroendocrine Control Mechanisms of Neoplastic Growth, J. Neural Transm., 50:275.
Lapin, V., and Frowein, A., 1981, Effects of Growing Tumours on Pineal Melatonin Levels in Male Rats, J. Neural Transm., 52:123.
Rodin, A. E., 1963, The Growth and Spread of Walker 256 Carcinoma in Pinealectomized Rats, Cancer Res., 23:1545.
Rodin, A. E., and Overall, J., 1967, Statistical Relationships of Weight of the Human Pineal to Age and Malignancy, Cancer, 20:1203.
Scalabrino G., and Ferioli, M. E., 1981, Polyamines in Mammalian Tumors, Part I, Adv. Cancer Res., 35:151.
Scalabrino, G., Ferioli, M. E., Nebuloni, R., and Fraschini, F., 1979, Effects of Pinealectomy on the Circadian Rhythms of the Activities of Polyamine Biosynthetic Decarboxylases and Tyrosine Aminotransferase in Different Organs of the Rat, Endocrinology, 104:377.
Scalabrino, G., Pösö, H., Hölttä, E., Hannonen, P., Kallio, A., and Jänne, J., 1978, Synthesis and Accumulation of Polyamines in Rat Liver During Chemical Carcinogenesis, Int. J. Cancer, 21:239.
Sneider, T. W., and Potter, V. R., 1969, Deoxycytidylate Deaminase and Related Enzymes of Thymidine Triphosphate Metabolism in Hepatomas and Precancerous Rat Liver, Adv. Enzyme Regul., 7:375.
Tamarkin, L., Cohen, M., Roselle, D., Reichert, C., Lippman, M., and Chabner, B., 1981, Melatonin Inhibition and Pinealectomy Enhancement of 7,12-Dimethylbenz(a)anthracene-induced Mammary Tumors in the Rat, Cancer Res., 41:4432.

Tapp, E., and Blumfield, M., 1970, The Weight of the Pineal Gland in Malignancy, Brit. J. Cancer, 24:67.

Tapp, E., 1980a, The Human Pineal Gland in Malignancy, J. Neural Transm., 48:119.

Tapp, E., 1980b, Pineal Gland in Rats Suffering from Malignancy, J. Neural Transm., 48:131.

INTERACTIONS BETWEEN PINEAL AND NONREPRODUCTIVE ENDOCRINE GLANDS

B. Mess

Department of Anatomy, Univ. Med. School, Pécs
Hungary

INTRODUCTION

It has not been too long a time that the endocrine nature of the pineal gland has been doubted. In the light of the research work performed world-wide in the last few decades, it has become evident and generally accepted that the pineal gland forms an integrate part of the endocrine system. However, the role of the pineal does not fit simply into the general scheme of the hypothalamo-pituitary-peripheral target organ axis.

The most widely investigated and most firmly substantiated endocrine function of the pineal gland is related to the reproductive system. Since, in the last few years, I have had several opportunities to summarize the possible function and significance of the pineal-reproductive interaction (Mess et al., 1978; 1979; 1980; 1981), and that Dr. Reiter spoke yesterday in detail on this subject, I shall omit discussing this problem. Furthermore, it is also my task to summarize the question of pineal-hypothalamic interactions, and consequently the relationship between the pineal and the hypothalamo-pituitary unit. Therefore, in the present paper I shall restrict my subject to the problem generally termed as "interglandular interplays", i.e. to the direct or indirect functional interrelationships between the pineal and the peripheral endocrine glands.

Since pineal-endocrine organ interglandular interplay is a fairly complicated and ambiguous question, I do not wish to increase the confusion by dealing with different phylogenetic orders of animals. Therefore, I shall restrict my talk mainly to mammals, with an occasional reference to birds.

II. PINEAL-THYROID INTERRELATIONSHIPS

The investigation of the functional relationship between the pineal and thyroid gland, and vice versa, has not been a popular subject of pineal research. However, a short monograph (Peschke, 1980-81) and a short, comprehensive chapter (Vollrath 1981) have appeared in the last year dealing with this subject.

The first sentence of Vollrath's survey states "...pineal usually exerts an inhibitory influence on thyroid function, though some studies have pointed to stimulatory effects". May I add that several reports (Yamada, 1961; Mess, 1969; Naber et al., 1969; Rowe et al., 1970; Brammer et al., 1979; Mattila, 1981) have indicated no effect of the pineal gland on the different parameters of thyroid activity that were investigated. A similar dubious position was stated by Reiter et al., (1975): "Ambivalent results concerning the connections between the pineal gland and the thyroid are characteristic".

Omitting chronological order and the older data, we shall focus on the literature of recent pineal research and to the data obtained with the more sophisticated and reliable newer methods of thyroid investigation.

The possible explanations for the contradictions, consistently found in the pineal-thyroid literature, were summarized by Peschke (1980) as follows:

1., Differences in the species of experimental animals, and within the species, sex differences might also be responsible.

2., Age differences of the employed animals

3., Circadian and circannual rhythms, and divergent lighting schedules used by the different investigators

4., Duration of the experiments; different intervals between experimental manipulation and observation of the results

5., Further factors, difficult to standardize, such as: ambient temperature, constituents of the diet, olfactory and acoustic stimuli, different kinds of stress situations etc.

One can easily agree with the arguments of Peschke (1980). To provide only a few characteristic examples supporting each argument, Singh and Turner (1972) found that the thyroid secretion rate (TSH) responded with a maximal decrease after the injection of a lower dose of melatonin (50 μg/100 g body weight) in the female hamster, while a double dose of melatonin was required to obtain the same degree of inhibition in the male rat. According to these data, the TSH-thyroid system of the hamster is more sensitive to melatonin than that of the rat. At the same time, the sex difference also could be taken into account in this experiment. Wurtman (1971) discussed in greater detail the role of species differences in his summary of the pineal colloquium of the Ciba Foundation in London. He pointed out that some species of animals show greater responses to light/dark effects or to melatonin than others, e.g. the hamster showed much greater changes in gonad weight in response to environmental lighting and to removal of the pineal than the

Table 1. Factors influencing pineal reactivity

Factors	Exp. results	References
Species difference	Hamsters are more sensitive to melatonin than rats	Singh and Turner, 1972
	Hamsters are more sensitive to light changes than rats	Wurtman, 1971., Hoffmann 1975
Sex difference	Female animals are more sensitive to melatonin than males	Singh and Turner, 1972
Age or duration of exp.	Pineal regulation on TSH secretion is more marked in immature animals than in adults	Panda and Turner, 1968
	Pineal exerts short-lived inhibition on the thyroid	Relkin, 1972
	Pineal has relatively long-term inhibition on thyroid	Vriend et al., 1977
Ambient temper.	Pineal N-acetyl-transferase activity is highest et 22°C ambient temperature	Ulrich et al., 1974
	Pineal reactivity is affected by cold	Miline et al., 1970
Acoustic stimuli	Acustic stress results in involution of the pineal in the hibernating bat	Miline et al., 1969
Olfactory stimuli	Experimental anosmia inhibits compensatory hypertrophy of different organs	Reiter, 1969; 1972

albino rats (Hoffmann and Reiter, 1965; Hoffmann 1975).

Investigating changes in blood and pituitary TSH concentrations of melatonin-treated rats of different ages, Panda and Turner (1968) concluded that the regulatory effect of the hypophyseal-pineal axis is more marked in immature animals than in the adults. On the other hand, Relkin (1972) demonstrated that pinealectomy performed on the 21st day of life of male rats and which were exposed to constant light or constant darkness, changes in thyroid activity (plasma and pituitary TSH level, protein bound iodine (PBI) blood level) were characteristic when investigated on the 25th day of life, but all these parameters were unchanged on the 28th day of life. These findings might be due to age differences, but more probably to the difference in the duration of the experimental period. The author interpretes his results by stating: "the pineal gland appears to be capable of exerting a significant short-lived inhibitory influence over the pituitary-thyroid axis of prepuberal male rats ..." Vriend et al., (1977) found that blinding of young male hamsters led to a depression of the plasma free thyroxine index after an 8 week period, an effect that was reversed either by pinealectomy or by removal of the ganglion cervicale superius. However, no significant effects were observed 2 weeks after treatment. These latter data, in contrast to Relkin's results, seem to indicate that the pineal gland has a relatively chronic inhibitory effect on thyroid function.

The transference of environmental lighting impulses towards the endocrine system is one of the most obviously proved functions of the pineal gland. Since this organ is an important regulator of different biorhythms (circadian, circannual) and termed "a biological clock" by Axelrod and Wurtmann (1966), this position will be discussed later in detail.

Changes in ambient temperature have been proposed to influence pineal activity. According to the investigations of Ulrich et al. (1974), intact suckling rats exposed to constant dark at 7, 22 and 34°C ambient temperature, produces a significantly higher activity of pineal N-acetyl transferase in the group kept at 22°C than in any other group. It was concluded that light and temperature both regulate pineal function probably by different mechanisms. Similarly,Miline et al. (1970) showed that pineal gland behaviour is severly affected by cold. Other factors, such as olfactory and acoustic stimuli, also were demonstrated to influence pineal function. Vigorous acoustic stimuli, applied twice daily each time for 1 hour, caused characteristic involution of the pineal gland in the hibernating bat (Miline et al., 1969). Olfactory impulses also play an important role in the regulation of pineal function. As it was shown by Reiter (1972), experimental anosmia considerably influences the compensatory hypertrophy of different organs (gonads, adrenals, kidneys). This effect of olfactory bulbectomy was proved to be mediated by the pineal gland,thus demonstrating clearcut relationship between pineal function and olfaction (Reiter, 1969). According to my knowledge, no direct data have yet to prove the existence of an olfactory-pineal-thyroid circuit. Based on the analogy to the above mentioned olfactory-pineal-reproductive interrelationship, this former possibility is a consideration. The role of different stress situations and the pineal-adrenal relationships will be the next point of this report.

As was briefly shown, all factors, mentioned by Peschke (1981), or their different combinations might be responsible for the modifications of the experimental results obtained concerning the role of the pineal gland in the regulation of thyroid function.

If one critically surveys the recent data reported in the literature dealing with this subject (besides the differences in the above mentioned factors), an even more important common feature becomes evident. Positive results demonstrating that the pineal can influence thyroid function have been obtained, when basic thyroid function was not the object of investigation, but rather was focused on finer modifications, under more sophisticated experimental conditions. This character of pineal function was formulated in respect to pineal-gonadal interplay in the last paragraph of our monograph (Mess et al., 1978) as follows: "This point stresses the "modulator" role of the pineal gland plays in the regulation of gonadotrophin secretion, and the importance of employing an appropriate experimental model to be able to demonstrate the true physiological nature of the pineal." I am convinced that this "modulator" character of pineal function is true also in its influence exerted on the TSH-thyroid axis, I would like to provide support for this assumption by presenting selected data reported by outstanding pinealologists in the last decade.

^{131}I uptake was significantly elevated after pinealectomy plus bilateral enucleation, but not following simple pinealectomy (Seibel and Schweisthal, 1973) indicating that the pineal modifies the light-induced changes in thyroid function rather than influencing the basic mechanism of iodide uptake.

The elegant experiment of Relkin (1978) furnished convincing evidence that the pineal mediates the changes in lighting conditions that influences not only the reproductive system, but also pituitary TSH secretion. It was shown (Table 2.) that pinealectomy or constant illumination decreased pituitary TSH and increased serum TSH and T_4 levels. On the contrary, constant darkness roughly decreased all three of these parameters. This latter effect, however, could be counteracted by pinealectomy. - Since animals of each experimental group equally responded to exogenous TRH with TSH release, the conclusion was drawn

Table 2. Changes in pituitary and serum TSH levels in the constant dark exposed, intact and pinealectomized rat (Relkin, 1978)

Exp. Group	Number of animals	Pit. TSH mU/mg	Plasma TSH mU/100 ml	PBI µg/100 ml
Diurnal Sh.	13	47,8±5,4	37±5,5	4,7±0,2
Diurnal Px.	12	15,6±3,7	54±6,2	7,1±0,3
Dark Sh.	12	17,5±3,0	13±2,7	2,9±0,1
Dark Px.	11	22,0±3,9	51±6,5	7,6±0,2

that the inhibitory effect of the pineal is exerted at the level of hypothalamic secretion of TRH. - On surveying these results, all the values are moderately significant, and are shifted around control values. The most pronounced effect of pinealectomy was observed in the dark-exposed group. - The experiments of Niles et al. (1979) also suggest a role of the pineal gland in the light-regulated diurnal secretion of TSH. Fig.1. demonstrates that nearly constant darkness (1L:23D) does not alter plasma levels of TSH (in contrast to Relkin's data, 1978) compared to those of controls maintained under normal (12L:12D) lighting conditions. Under this short photoperiod, pinealectomy increased the average level and shifted the peak of TSH to 6 AM. It was concluded that the pineal may have a role in entraining the TSH rhythm to environmental lighting.

In various instances, it is not only the complicated experimental schedule, but also the nature of the investigated parameter, representing finer variations in thyroid activity, that favour the concept of a "modulator" role for the pineal gland. The thyroid hormone secretion rate (TSR) (Singh and

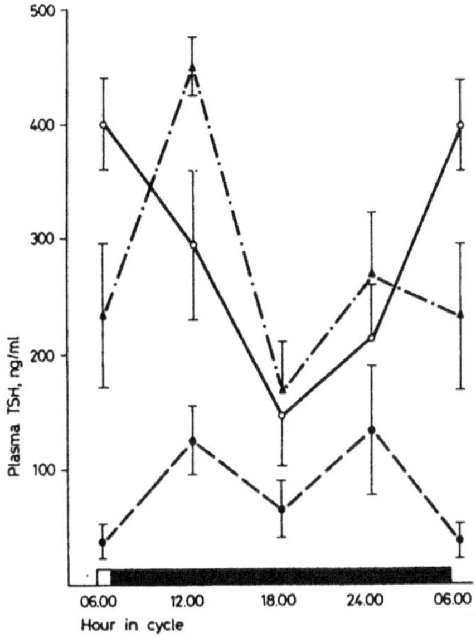

Fig.1. The effects of pinealectomy or immunization against melatonin and NAS on the diurnal plasma TSH rhythm of animals exposed to short daily photoperiods. ●--● =Control 1L:23D; ○—○ =pinealectomized 1L:23D; ▲-·-▲ =NAS-immunized 1L:23D. (Niles et al., 1979).

Turner,1972) and the free thyroxine index (Vriend et al., 1977) both are slowly-reacting indicators of thyroid activity. The free thyroxin index shows a slight but significant decrease after exposure to continuous darkness for an 8 week period. Pinealectomy, or superior cervical ganglionectomy, reverse this effect. TSR was also slightly depressed by daily melatonin treatment continued for over a week. If one takes into account that even these slow reacting parameters of thyroid function show only moderate changes, and very often only under sophisticated experimental conditions, the primary or "vital" role of the pineal gland upon thyroid function can be refuted. The same results verify, however, the "modulatory" role of the pineal gland with respect to its regulation of the TRH-TSH-thyroid system.

Another widely discussed question in this field is the mechanism of action of melatonin by which it influences thyroid activity. This latter question will be mentioned here only very briefly. Baschieri et al. (1963) were the first to describe the inhibitory role of melatonin on thyroid function. They showed that melatonin prevents thiouracil-induced hyperplasia, and causes reduced thyroidal ^{131}I-uptake. The conclusion was drawn that melatonin acts either on TSH secretion or directly on the thyroid gland. A controversial finding has been reported by Thiéblot et al. (1966) who showed that administration of melatonin produced marked hyperactivity of the thyroid as evidenced by histological changes.

This controversy could not be resolved, since subsequent data demonstrated a "goitrogenic-like" effect of melatonin. Panda and Turner (1968) observed an increase in blood TSH levels and a decrease in pituitary TSH content with concomitant hyperplasia and hypertrophy of the thyroid gland equally in the melatonin- and in the goitrogen(tapazole) treated rat. They concluded that "melatonin is similar in its action to that of goitrogen (antithyroid agent) and in some way inhibits thyroid hormone synthesis". Similar findings and conclusions were reported later by De Fronzo and Roth (1972). A proportional increase in thyroid DNA and RNA content with significantly increased ^{131}I-uptake and decreased PBI level were observed in chronically (four weeks) melatonin-treated rat. These findings led to the conclusion that melatonin has a true goitrogenic effect, which occurs at a point beyond the thyroid trapping mechanism. The results of Singh and Turner (1972), showing a decreased TSR in melatonin-treated rats and hamsters, also appear to support this assumption.

However, there are findings which contradict the proposed goitrogenic effect of melatonin. Seibel and Schweisthal (1973) found a significant decline in thyroidal ^{131}I-uptake after melatonin treatment. They stressed that "our results are in essential agreement with Baschieri et al. (1963) ...". Investigating an other characteristic parameter, TSH blood levels also were decreased in melatonin-treated rats compared to saline treated controls. The authors concluded that melatonin might act at the hypothalamic level instead of the proposed peripheral level.

This evident controversy hardly can be explained simply by the different species or ages of animals, or by the differences in the experimental design. The possibility cannot be excluded that melatonin might actually have a direct peripheral action on the thyroid gland thus inhibiting the biochemical mechanisms of thyroid hormone secretion. At the same time, melatonin might act-analogous to that observed for the gonadotrophin system - also at the level of the central nervous system, influencing the function of the serotoninergic neurons of the brain stem and inhibiting TRH release.

III. PINEAL-ADRENAL INTERRELATIONSHIPS

In 1959 when Lerner described the structure of melatonin, Farrel (1959) also proposed the existence of a pineal hormone, adrenoglomerulotropin. No wonder that much enthusiasm was generated by these findings towards research into pineal--adrenal interrelationships. Many investigators, including ourselves (Fraschini et al., 1968), have shown that pinealectomy provokes an increase in the activity of the adrenal gland. Changes in several different parameters of adrenal function were then reported following pineal gland interventions. Lommer (1966) found that pineal extracts inhibit β-hydroxylation within the adrenal cortex, thereby inhibiting production of corticoids. Similarly, Trentini et al. (1965) described altered enzymatic activities and lipid content in the adrenal cortex following pinealectomy. A decreased corticosterone production also was reported following administration of pineal principles (Gromova et al., 1967). These collective findings suggest a peripheral action of pineal hormones exerted directly on adrenocortical tissue. - On the other hand, data indicating a central effect of the pineal gland also appeared during this period. An enhanced ACTH secretion was reported by Jouan (1963) following ablation of the pineal gland. Motta et al. (1971) found significantly decreased plasma corticosterone levels after intraventricular injection of different pineal indoles both under resting conditions and in cold-exposed animals. In contrast, an enhanced aldosterone secretion was reported by Giordano and Balestreri (1963) under *in vitro* conditions in which adrenal tissue was incubated with pineal extract. Kinson and Singer (1967) stated that pinealectomy increases the basal secretion rate of aldosterone, but does not influence the adrenal response to dietary sodium deficiency. The most important data of this period have been summarized in our previous review articles (Mess, 1968; Mess et al., 1979).

Publications dealing with pineal-adrenal interrelationships have become relatively scarce in recent years, indicating that this subject is considered to have only a minor importance among the many functions of the pineal gland.

Partly negative findings were reported in the last decade.

De Proposto and Hurley (1971) found no appreciable change in the weight of the adrenal gland following intraventricular injection of melatonin in rats. Similarly, no weight changes were obtained in pinealectomized, blinded, or blinded and pinealectomized hamsters. Exogenous melatonin also failed to influence adrenal weight in this species (Seibel and Schweisthal 1973). Even under more sophisticated experimental conditions, Vermes and Telegdy (1979) reported negative results. Pinealectomy or superior cervical sympathetic ganglionectomy neither altered the hypothalamic 5-hydroxytryptamine content, nor the diurnal adrenal rhythm or the adrenal stress response.

Some discrepancies are evident even among the positive results, obtained with more advanced techniques. Ogle and Kitay (1977) stated that melatonin (but not aqueous pineal extract) enhances α-reductase activity in the adrenal cortex. This effect also was elicited in hypophysectomized animals. This indicates that melatonin acts directly on the adrenocortical cells augmenting steroid hormone production. In contrast, melatonin was shown (De Fronzo and Roth, 1972) to inhibit, while pinealectomy to stimulate, adrenal activity.

However, the main trend in research of pineal-adrenal relations is the investigation of either the rhythmic adrenal secretion influenced by the pineal gland, or of the finer details of the regulatory mechanism of the adrenal cortex.

Thurley et al. (1975) presented data which indicated that the diurnal rhythm of cortisol production was abolished by pinealectomy. Accordingly, Barrel and Lapwood (1978) found that pinealectomy inhibits the morning rise in cortisol secretion in the ram. They concluded that the pineal mediates not only the long (seasonal), but also the short (daily) changes in photoperiods affecting the neuroendocrine system. - Not only is the rhythmic secretion of adrenocortical steroid influenced by the pineal gland, but changes in the rhythm of the adrenomedullary system also were reported. Kachi et al. (1979; 1980) distinguished two types of synaptic vesicles in the chromaffin cells using quantitative electronmicroscopy. The number of the small light vesicles showed two peaks in normal animals, in the middle of both the light- and dark phase of the day. Pinealectomy abolishes the peak in the dark phase. The so-called big granular vesicles showed the same two peaks, however, to a considerably less pronounced degree. Pinealectomy diminished the smaller peak at the begining of the dark phase. These investigators concluded that both the adrenalin and noradrenalin secretory rhythm is influenced by the pineal gland, however, in a somewhat different way.

Regarding the influence of the pineal gland on the stress--inducing activation of the adrenocortical system, interesting data were reported by Devečerski (1979) and Miline (1980). On the basis of histological and histochemical investigations, they stated that reactive changes in the adrenal cortex under stress conditions are more pronounced in pinealectomized rats than in unoperated animals. It was concluded that pinealectomized animals are more sensitive to stress caused by hyperthermia. - Even more sophisticated experimental models were applied to approach the finer details of the mechanism of the pineal-adrenal interplay. Vaughan et al. (1980) reported that blinding and anosmia, termed "sensory deprivation", caused a moderate increase in ACTH secretion, which was reversed by pinealectomy. The unilateral adrenalectomy-induced increase in ACTH blood levels, however, was not influenced by pinealectomy alone, whereas blinding and anosmia+pinealectomy elicited more than a two-fold increase, compared either to sham operated, to simply pinealectomized or to simply blinded anosmic rats (Fig.2.).

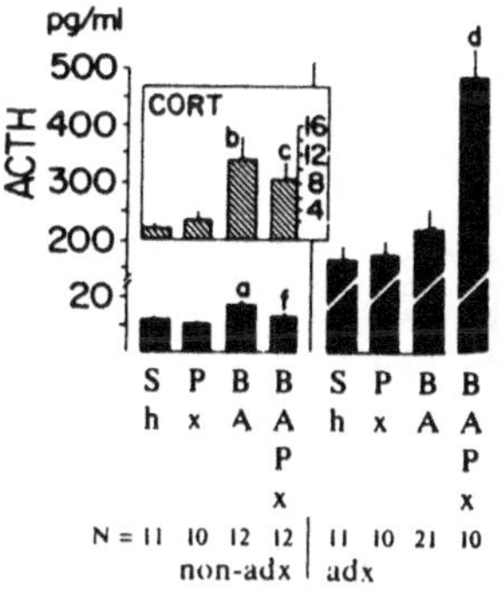

Fig.2. Influence of sham-pinealectomy (SH), pinealectomy (Px), blinding-anosmia (BA) combined BAPx, and bilateral adrenalectomy (adx) on plasma ACTH (mean$\pm$SE). The number of rats per group (n) is indicated below the group symbols near the bottom of the figure. The inset in the ACTH graph, labelled CORT, represents corticosterone concentrations in μg/dl. Significance levels are indicated above the bars: a, $p < 0{,}001$ vs Sh; b, $p < 0{,}01$ vs Sh; c, $p < 0{,}05$ vs Sh; d, $p < 0{,}001$ vs BA; f, $p < 0{,}05$ vs BA, comparisons being between groups with like adrenal status (non-adx or adx) (Vaughan et al., 1980).

Compensatory adrenal hypertrophy following unilateral adrenalectomy was not inhibited by pinealectomy. Removal of the pineal gland only also failed to influence the dopamine β-hydroxylase (DBH) activity of the adrenal cortex. However, pinealectomy combined with unilateral adrenalectomy significantly inhibited the unilateral adrenalectomy-induced rise in this enzyme activity of the remaining and hypertrophied adrenal gland (Banjeri and Quay, 1981).

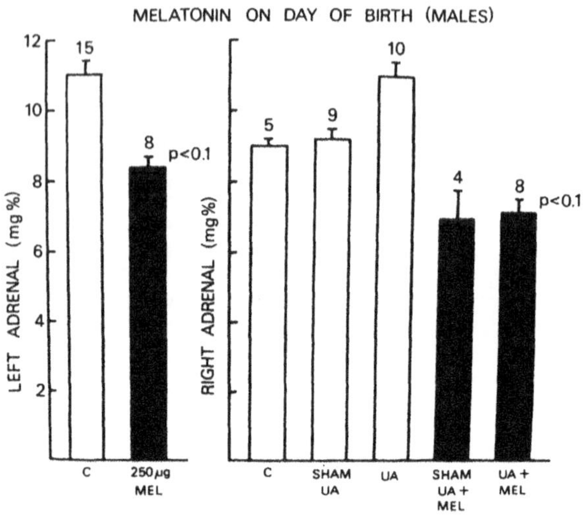

Fig.3. Weight of the left adrenal (mg/100 g body weight) removed at the time of unilateral adrenalectomy is statistically reduced in male mice receiving a single subcutaneous injection of melatonin at the time of birth (stippled bar on left). The weight of the right adrenal (mg/100 g body weight), removed 5 days later, is statistically reduced both in sham and unilaterally adrenalectomized male mice receiving melatonin on the day of birth (stippled bars on right). Compensatory adrenal hypertrophy did not occur in these melatonin-treated males. Vertical lines indicate standard errors. Parentheses signify the number of animals in each group (Vaughan and Vaughan, 1974).

The pineal hormones also appear to influence the differentiating adrenal and gonadal/central neural control mechanisms during the neonatal critical period. A single subcutaneous injection of melatonin into newborn rats or mice prevented the compensatory adrenal- and ovarian hypertrophy following unilateral extirpation of the corresponding endocrine organ during the adult stages (Vaughan and Vaughan, 1974) (Fig.3.).

In summary, similar conclusions can be drawn concerning the role and mechanism of pineal influences exerted upon the adrenal gland, as have been proposed with respect to the pineal--thyroid interplay. The pineal might act as a "modulator" rather than as a basic regulator in influencing the function and responses of the adrenal gland.

IV. INFLUENCE OF THE PINEAL GLAND ON OTHER TROPHIC HORMONES OR ENDOCRINE ORGANS

Sporadic and less firmly-proved data have appeared in the literature suggesting a regulatory role of the pineal gland on other endocrine functions. These reports will be mentioned here only very briefly.

Prolactin (PRL) belongs, in a wider sense, to the gonadotrophins but will be mentioned here. Kuew-Hsiung Lu and Meites (1973) reported that brain serotonin, its precursors (tryptophan and 5-hydroxytryptamine) and melatonin may have a role in stimulating PRL release. Pinealectomy was found to abolish the early morning rise of plasma PRL levels in the male rat; superior cervical ganglionectomy had a similar effect (Ronnekleiv and McCann, 1973). Shino et al. (1974) have shown that pinealectomy prevents the blinding and anosmia-induced rise in serum PRL levels. They concluded that the pineal gland inhibits the secretion of prolactin inhibitory factor (PIF), thereby promoting PRL release. In contrast, Blask and Reiter (1978) proposed that the pineal gland exerts its regulatory influence on PRL release via the proposed hypothalamic prolactin releasing factor (PRF). In an in vitro system, medial-basal hypothalamic extracts of pinealectomized or ganglionectomized and blinded anosmic animals provoked a significantly higher degree of PRL release than that of the solely blinded anosmic rats.

Acidic extracts of sheep pineal (protein fractions) inhibited the release of PRL in pituitary monolayer culture (Demoulin et al., 1977). This finding demonstrates that protein substances of the pineal gland might play an inhibitory role on PRL release. On the other hand, these data suggest a direct action of pineal hormones on adenohypophyseal cells. According to the results of Pavel et al. (1975), the effect of pinealectomy exerted on PRL release can be counteracted by minute amounts of arginine vasotocin (AVT). These data also suggest that hormones other than the indoleamines might play a role in the mechanism whereby pineal gland influences PRL secretion. Vaughan et al. (1976) proposed AVT to be a "potential pineal PRL releasing factor". In agreement with these latter findings, but in contrast to the data of Kuew-Hsiung Lu and Meites (1973), it was shown that melatonin and its derivatives were without effect on PRL release from bovine pituitary tissue culture. Serotonin strongly inhibited, while AVT stimulated, PRL output (Padmanabhan et al., 1979).

It would be difficult to formulate a uniform picture regarding the mechanism of pineal-PRL interplay. The only conclusion that can be drawn with reasonable certainity is that the pineal gland, in contrast to the other two gonadotrophins (FSH and LH), stimulates PRL secretion.

Even less data are available with respect to the role of the pineal gland in the regulation of Growth Hormone (GH) secretion. Melatonin was shown to block, while serotonin to stimulate, GH release in the rat (Collu et al., 1972; Smythe and Lazarus, 1973 a). Furthermore, the stimulating effect of seroto-

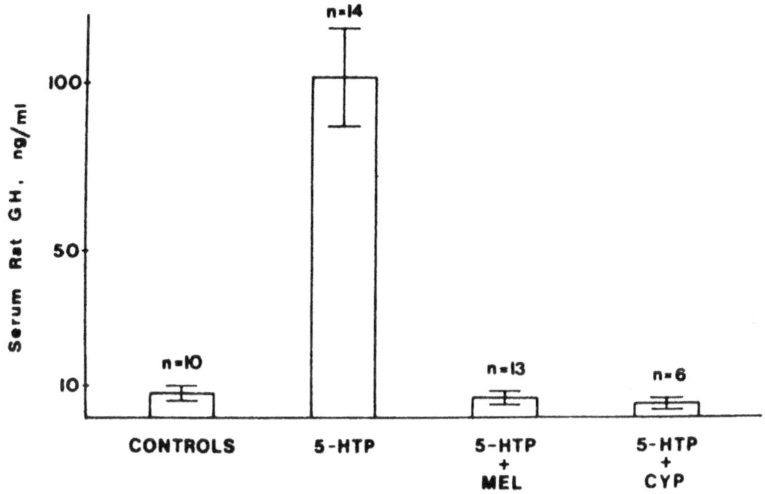

Fig.4. Changes in GH blood levels of the rat following simple serotonin or simultaneous serotonin and melatonin treatment (Smythe and Lazarus, 1973 b).

nin could be prevented by simultaneous melatonin treatment (Smythe and Lazarus, 1973 b) (Fig.4.).
In man, however, melatonin treatment had an opposite effect in which serum GH levels were elevated (Smythe and Lazarus, 1974).

The effect of pinealectomy on GH secretion was investigated mainly in experiments that combined pinealectomy with environmental light-changes, food restriction or sensory deprivation. Osman et al. (1972) described subnormal body growth under conditions of constant darkness, but only in animals with an intact pineal gland. Since the delay in vaginal opening did not show parellelism with the retardation of body growth under the different experimental circumstances, these investigators concluded that the anti-gonadotrophic and anti-growth properties of the pineal gland are separate physiological entities. In the same year, Relkin (1972) reported that constant darkness suppresses GH blood levels, but that effect of light deprivation was counteracted by pinealectomy. It was concluded that the darkness-induced increase in pineal activity inhibits the secretion of GH releasing factor resulting in a decreased release of pituitary GH. Similar results were obtained by Sorrentino et al. (1971; 1972); blinding and anosmia provoked marked decrease in body growth and GH levels. These phenomena only occur in the presence of the pineal gland (Fig.5.). These findings were later corroborated by Shino et al. (1974).

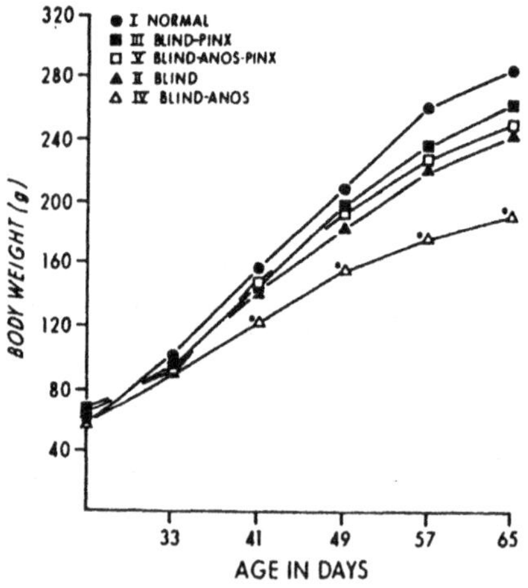

Fig.5. Growth rates of rats as evaluated by body weights. Asterisks indicate groups different from respective pinealectomized group ($p < 0{,}01$) (Sorrentino et al., 1971).

The role of the pineal gland in the maintenance of the diurnal rhythm of GH secretion also was studied. Pinealectomy significantly reduced GH release of rats during the light period, but did not significantly reduce further GH release during the night (Ronnekleiv and McCann, 1978). In contrast, Niles et al. (1979) found no difference between the plasma GH levels of rats kept in a normal (12 L:12 D) or a short photoperiod (1 L:23 D) environment. Neither pinealectomy, nor immunization against melatonin, influenced the rhythmic pattern of GH release. A discussion of the possible explanations for this discrepancy is beyond the scope of this presentation.

Finally, an interesting but somewhat vague concept should be mentioned. On the basis of light- and electronmicroscopic investigations, Csaba and Baráth (1974) demonstrated that the parafollicular cells of the thyroid gland increase both in their number and granulosity following pinealectomy. Pinealectomy also was shown to decrease blood sugar levels under resting conditions. However, the alloxan-induced increase of the blood sugar level was significantly more dramatic in the pinealectomized than in the unoperated animals (Csaba and Baráth, 1971). Neonatal pinealectomy provoked a wasting disease and tetany in the majority of the chickens suggesting that the parathyroid gland also might be regulated by the pineal (Csaba et al., 1972); furthermore, they suggested the existence of a negative feed back mechanism. Adrenaline, noradrenaline and insulin (hormones, which generally are not regulated by pituitary trophic hormones) strongly increased the serotonin fluorescence in pineal tissue culture. It was presumed that the inhibitory effect of the three administered hormones on the conversion or elimination of serotonin could play a role in a negative feed back mechanism (Csaba and Bernád, 1972).

On the basis of these and of a series of other investigations, it was proposed that the pineal gland might regulate those endocrine organs which are of ecto- and endodermal origin (thyroid, parathyroid, adrenomedulla, Langerhans islets), and those of mesodermal origin (adrenocortical tissue, gonads) that are under pituitary influence (Csaba and Bernád, 1972; Csaba and Baráth 1971).

It sufficies to state that the most extensively investigated, confirmed and generally accepted endocrine effect of the pineal gland is its regulatory influence on the reproductive system. Since the gonads, and especially their endocrine targets, are derivatives of the coeloma (mesoderma), the concept of Csaba and associates is without foundation.

V. PINEAL REACTIONS TO CHANGES IN THE ACTIVITY OF THE ENDOCRINE ORGANS

The main task of this talk was to demonstrate the regulatory influence of the pineal gland on the function of the different endocrine systems. The reverse aspect, i.e. the survey of the morphological and functional changes in the pineal gland observed following increased or decreased hormone secretion of one or the other endocrine organs, will be outlined here only very superficially.

Hypophysectomy causes a marked decrease in pineal volume and an atrophy of the pinealocytes (Ito and Matsushima, 1968). All ultrastructural signs of inactivation in the pinealocytes also were shown following hypophysectomy (Satodate et al., 1970; Karasek, 1971). HIOMT activity also decreased in the pineal of hypophysectomized animals (Urry et al. 1976).

Adrenalectomy provoked an increase in the number of pinealocytic processes with a concomitant increase in HIOMT activity (Deussen Schmitter et al., 1976). On the other hand, increased pineal activity was registered following thyroxine administration. In *in vitro* studies, Nir and Hirschmann (1978) found an increase in the melatonin and N-acetyl serotonin content of pineal tissue after adding thyroxin (but not TSH) to the organ culture. They proposed the existence of a direct positive feedback between the thyroid and the pineal gland.

It would exceed the possibilities of the present paper to enter into a discussion of the fascinating question of how pineal gland responded to its own secretory products. Thus, melatonin might enhance the release of AVT (Pavel, 1973), and conversely, AVT might decrease melatonin secretion (Sratin et al., 1978). Benson (1977) proposed in his comprehensive paper that melatonin might exert its antigonadotrophic activity by influencing the synthesis and/or the release of the as yet unknown "pineal antigonadotrophins (PAG)" of purported peptidic character.

SUMMARY

The survey of numerous, and sometimes contradictory data, have led to the conclusion that the pineal gland modulates rather than basically influences the main neuroendocrine circuits regulating thyroid and adrenocortical activities. Mainly, changes in external environmental conditions (photic, osmic or thermal stimuli) are mediated by the pineal gland to the releasing hormone(s)-pituitary trophin(s)-target gland(s) primary regulatory system.

Pineal influences exerted upon the secretion a PRL and GH also are briefly surveyed and the relationship between the pineal and endocrine glands of ecto- and endodermal origin is briefly discussed. Finally, data dealing with the effects of different hormones upon the pineal gland are summarized.

REFERENCES

Axelrod J., Wurtman R.J., 1966, The pineal gland: a biological clock, in: Symposium international sur la neuro-endocrinologie. Problémes actuels d' endocrinologie et de nutrition, Serie No.10: Kotz H.P., ed., 201. Expansion, Paris.

Banerji T.K., Quay W.B., 1981, Compensatory increase in adrenal weight and Dopamine- β -hydroxylase activity after unilateral adrenalectomy: Experimental evidence for a pineal-dependent contribution, Proc. Soc. exp. Biol. Med. 167: 514.

Barrel G.U., Lapwood K.R., 1978, Effects of pinealectomy of rams on secretory profiles of luteinizing hormone, testosterone, prolactin and cortisol, Neuroendocrinology, 27: 216.

Baschieri L., De Luca F., Cramarossa L., Martino C., Oliverio A., Neri M., 1963, Modifications of thyroid activity by melatonin, Experientia (Basel), 19: 12.

Benson B., 1977, Current status of pineal peptides. - Neuroendocrinology, 24: 241.

Blask D.E., Reiter R.J., 1978, Pineal removal or denervation: effects of hypothalamic PRF activity in the rat, Mol. Cell. Endocr. 11: 243.

Brammer G.L., Morley J.E., Geller E., Yuwiler A., Hershman J.M., 1979, Hypothalamus-pituitary-thyroid axis interactions with pineal gland in the rat, Amer. J. Physiol.,236: E416.

Csaba G., Baráth P., 1971, Are Langerhans'Islets influenced by the pineal body?, Experientia (Basel), 27: 962.

Csaba G., Bernád I., 1972, In vitro model of feedback mechanism between different endocrine glands and rat pineal in tissue culture, - Acta biol. Acad. Sci. Hung., 23: 91.

Csaba G., Rados I., Wohlmuth E., 1972, Wasting disease and tetany following neonatal pinealectomy, Acta med. Acad. Sci. Hung., 29: 231.

Csaba G., Baráth P., 1974, The effect of pinealectomy on the parafollicular cells of the rat thyroid gland, Acta Anat. (Basel). 88: 137.

Collu R., Fraschini F., Visconti P., Martini L., 1972, Adrenergic and serotoninergic control of growth hormone secretion in adult male rats, - Endocrinology, 90: 1231.

De Fronzo R.A., Roth W.D., 1972, Evidence for the existence of a pineal-adrenal and a pineal thyroid axis, Acta Endocrinol. (Kbh), 70: 31.

Demoulin A., Hudson B., Legros J.J., Franchimont P., 1977, Influence d'un extrait de glandes pinéales ovines sur la libération de prolactine in vitro, Compt. rend. Soc. Biol. (Paris) 171: 1134.

De Proposto N., Hurley J., 1971, Effects of injecting melatonin and its precursors into the lateral cerebral ventricles on selected organs in rats, J. Endocr., 49: 545.

Deussen-Schmitter M., Garweg G., Schwabedal P.E., Wartenberg H., 1976, Simultaneous changes of the perivascular contact area and HIOMT activity in the pineal organ after bilateral adrenalectomy in the rat, Anat. Embryol., 149: 297.

Devečerski V., 1979, Histological and histochemical changes in the adrenal cortex of epiphysectomized rats due to deep hypothermia, Progr. Brain Res.,52: 387.

Farrel G., 1959, Glomerulotropic activity of an acetone extract of pineal tissue, Endocrinology, 65: 239.

Fraschini F., Mess B., Martini L., 1968, Pineal gland, melatonin and the control of luteinizing hormone secretion, Endocrinology., 83: 919.

Giordano G., Balestreri R., 1963, La glande épiphysaire dans la régulation de lá biosynthése de l' aldosterone, Ann. Endocrinol. (Paris), 24: 331.

Gromowa A.E., Kraus M., Krecek J., 1967, Effect of pineal

substances on production of corticosterone and aldosterone by rat adrenal gland, Gen. comp. Endocr., 9: 455.

Hoffmann K., 1975, On the function of the pineal in photoperiodic mammals, Acta Endocr.(Kbh)., Suppl. 193: 16.

Hoffmann R.A., Reiter R.J., 1965, Rapid pinealectomy in hamsters and other small rodents, Anat. Rec., 153: 19.

Ito T., Matsushima S., 1968, Effects of gonadectomy and hypophysectomy on the pineal body in the mouse: a quantitative morphological study, Anat. Rec., 162: 479.

Jouan P., 1963, Epiphyse, 5-Hydroxytryptamine et corticoidogenése in vitro, Ann. Endocr. (Paris), 24: 365.

Kachi T., Banerji T.K., Quay W.B., 1979, Daily rhythmic changes in synaptic vesicle contents of nerve endings on adrenomedullary adrenaline cells, and their modification by pinealectomy and sham operations, Neuroendocrinology, 28: 201.

Kachi T., Banerji T.K., Quay W.B., 1980, Circadian and ultradian changes in synaptic vesicle numbers in nerve endings on adrenomedullary noradrenaline cells, and their modifications by pinealectomy and sham operations, Neuroendocrinology, 30: 291.

Karasek M., 1971, The enfluence of hypophysectomy on the ultrastructure of the pineal gland in white rats. Preliminary investigations, Acta med. pol., 12: 153.

Kinson G.A., Singer B., 1967, The pineal gland and the adrenal response to sodium deficiency in the rat, Neuroendocrinology, 2: 283.

Kuew-Hsiung Lu, Meites J., 1973, Effects of serotonin precursors and melatonin on serum prolactin release in rats, Endocrinology, 93: 153.

Lerner A.B., Case J.D., Heinzcelmann R.V., 1959, Structure of melatonin, J. Amer. Chem. Soc., 81: 6084.

Lommer D., 1966, Inhibtion of corticosteroid-11-hydroxylation by an extract of pineal body, Experientia, 22: 122.

Mattila J., Männistö P.T., 1981, Studies on the role of the pineal gland in the regulation of TSH secretion in post-puberal male rats, Hormone Res, 14: 24.

Mess B., 1969, Endocrine and neurochemical aspects of pineal function, Int. Rev. Neurobiol. 11: 171.

Mess B., Trentini G.P., Tima L., 1978, Role of the pineal gland in the regulation of ovulation. Studia Biologica Hung., Vol. 16, Akadémiai Kiadó, Budapest

Mess B., Trentini G.P., Ruzsás Cs., De Gaetani C.F., 1979, Some endocrine effects of the pineal gland and melatonin with special reference to reproduction. Progr. Brain Res., 52: 329

Mess B., Trentini G.P., 1980, Interaction between the pineal and other endocrine glands. in: Progress in Psychoneuro-endocrinology, Brambilla F., Racagni G., De Wied D., eds., Elsevier, Amsterdam, p. 439.

Mess B., De Gaetani C.F., Ruzsás Cs., Trentini G.P., 1981, Possible mechanisms by which the pineal gland influences reproductive function. - in: The pineal organ: Photobiology--Biochronometry-Endocrinology, Oksche A. and Pévet P., eds., Elsevier/North Holland Biomedical Press, Amsterdam p. 303.

Miline R., Devečerski V., Kristič R., 1969, Effects des stimuli auditifts sur la glande pineale de la chauve-souris en hibernation, Acta anat., Suppl. 56: 293.

Miline R., Sijacki N., Kristič R., 1970, Pineal gland behaviour as affected by cold, Hormones (Basel), 1: 321.

Miline R., 1980, The role of pineal gland in stress, J. Neural Transm., 47: 191.

Motta M., Schiaffini O., Piva F., Martini L., 1971, Pineal principles and the control of adrenocorticotropin secretion. -

in: The pineal gland. A Ciba Foundation Symposium, Wolstenholme G.E.W., and Knight J., eds., Churchill Livingstone, Edinburgh and London p. 279.

Naber S.P., Goldman M., Peaslee M.H., 1969, The effect of melatonin on the iodine-concentrating mechanism of the thyroid gland in young adult rats, Proc. Soc. Dak. Acad. Sci. 48: 44.

Niles L.P., Brown G.M., Grota L.J., 1979, Role of the pineal gland in diurnal endocrine secretion rhythm regulation, Neuroendocrinology, 29: 14.

Nir I., Hirschmann N., 1978, The effect of thyroid hormones on rat pineal indoleamine metabolism in vitro, J. Neural. Transm. 42: 117.

Ogle T.F., Kitay J.I., 1977, Effects of melatonin and aqueous pineal extract on pineal secretion of reduced steroid metabolites in female rats, Neuroendocrinology, 23: 113.

Osman P., Welschen R.W., Moll J., 1972, Anti-gonadotrophic and anti-growth effects of the pineal gland in immature female rats, Neuroendocrinology 10: 121.

Padmanabhan V., Convey E.M., Tucker A.H., 1979, Pineal compounds after prolactin release from bovine pituitary cells, Proc. Soc. exp. Biol. Med.,160: 340.

Panda J.N., Turner C.W., 1968, The role of melatonin in the regulation of thyrotrophin secretion, Acta endocrinol. (Kbh). 57: 363.

Pavel S., 1973, Arginine vasotocin release into cerebrospinal fluid of cats induced by melatonin, Nature (Lond), 246: 183.

Pavel S., Calb M., Georgescu M., 1975, Reversal of the effects of pinealectomy on the pituitary prolactin content in mice by very low concentrations of vasotocin injected into the third cerebral ventricle, J. Endocr., 66: 289.

Peschke E., 1980-1981, Morphologische, physiologische und

statistische Untersuchungen an der männlichen Wistar-Ratte zum Problem eines möglichen functionellen Connexus: Epiphysis cerebri - Schilddrüse. Parts I-VI. (Part I; Zool. Jb. Anat., 109: 398. Part II; Zool. Jb. Anat., 104: 466. Part III; Zool. Jb. Anat., 105: 77. Part IV; Zool. J . Anat., 105: 147. Part V: Zool. Jb. Anat., 105: 297. Part VI; Zool. Jb. Anat., 105: 320.

Reiter R., 1969, Relationship among the pineal gland, the photoperiod, olfaction and reproduction in female rats. - J. Reprod. Fertil., 18: 152.

Reiter R.J., 1972, Compensatory growth of ovaries, adrenal glands and kidneys in blinded, anosmic rats. - Experientia (Basel), 28: 1492.

Reiter R.J., Sorrentino S., Donofrio J.R., 1975, The pineal gland as an organ of internal secretion. in: Frontiers of pineal physiology, Altschule M.D., ed., MIT Press, Cambridge, Mass-London.

Relkin R., 1972, Effects of pinealectomy and constant light and darkness on thyrotropin levels in the pituitary gland and plasma of the rats, Neuroendocrinology, 10: 46.

Relkin R., 1972, Effects of pinealectomy, constant light and darkness on growth hormone levels in the pituitary and plasma of the rat, J. Endocr., 53: 289.

Relkin R., 1978, Use of melatonin and synthetic TRH to determine site of pineal inhibition of TSH secretion, Neuroendocrinology, 25: 310.

Ronnekliev O.K., McCann S.M., 1973, The effect of pinealectomy on plasma gonadotropins and prolactin in the rat, The Physiologists, 16: 436.

Ronnekliev O.K., McCann S.M., 1978, Growth hormone release in conscious pinealectomized and sham-operated male rats, Endocrinology, 102: 1694.

Rowe J.W., Richert J.R., Klein D.C., Reichlin S., 1970, Relation of the pineal gland and environmental lighting to thyroid function in the rat, Neuroendocrinology, 6: 247.

Sartin J.L., Bruot B.C., Orts R.J., 1978, Interaction of arginine vasotocin and norepinephrine upon pineal indoleamine synthesis in vitro, Mol. Cell. Endocr., 11: 7.

Satodate R., Hsieh K.S., Ota M., 1970, Morphological changes in the pineal gland of the albino rat by hypophysectomy and ovariectomy, Experientia (Basel), 26: 638.

Seibel H.R., Schweisthal M.R., 1973; Relationship between the pineal gland, other endocrine glands and reproductive organs of single and parabiosed golden hamsters, Acta Endocrinol. (Kbh)., 74: 434.

Shino M., Arimura A., Rennels E.G., 1974, Effects of blinding, olfactory bulbectomy, and pinealectomy on prolactin and growth hormone cells of the rat, with special reference to ultrastructure, Amer. J. Anat., 139: 191.

Singh D.V., Turner C.W., 1972, Effect of melatonin upon thyroid hormone secretion rate in female hamsters and male rats. Acta Endocrinol. (Kbh). 6 : 35.

Smythe G.A., Lazarus L., 1973a, Growth hormone regulation by melatonin and serotonin, Nature (Lond), 244: 230.

Smythe G.A., Lazarus L., 1973b, Blockade of serotonin induced GH secretion in rats by melatonin and cyproheptadine, Horm. Metab. Res., 5: 227.

Smythe G.A., Lazarus L., 1974, Growth hormone responses to melatonin in man, Science, 184: 1373.

Sorrentino S.Jr., Reiter R.J., Schalch D.S., 1971, Pineal regulation of growth hormone synthesis and release in blinded and blinded anosmic male rats, Neuroendocrinology, 7: 210.

Sorrentino S.Jr., Reiter R.J., Schalch D.S., 1971, Role of the pineal gland in growth restraint of adult male rats by light and smell deprivation, Neuroendocrinology, 8: 116.

Thiéblot L., Berthelay J., Blaise S., 1966, Effect de la melatonine chez le rat male et female II. Action au niveau de la thyroide, Ann. Endocr.(Paris)., 27: 69.

Thurley D.C., Gibb M., Russel A., 1975, Effect of pinealectomy on diurnal cortisol patterns in sheep, N.Z. med. J., 81: 490.

Trentini G.P., Barbanti Silva G., Vassanelli P., Botticelli A., 1965, Modificazione istoenzimatiche del corticosurrene di ratto consequenti ad epiphisectomia (studio istochimico), Bol. Soc. ital. Biol. sper., 40: 967.

Ulrich R., Yuwiler A., Wetterberg L., Klein D., 1974, Effects of light and temperature on the pineal gland in suckling rats, Neuroendocrinology, 13: 255.

Urry R.L., Dougherty K.A., Frehn J.L., Ellis L.C., 1976, Factors other than light affecting the pineal gland: Hypophysectomy, testosterone, dihydrotestosterone, estradiol, cryptorchism and stress, Amer. Zool., 16: 79.

Vaughan M.K., Vaughan G.M., 1974, Neonatal melatonin administration and subsequent functioning of adult gonadal and adrenal neuroendocrine axes, Endocrinol. exp., 8: 261.

Vaughan M.K., Blask D.E., Vaughan G.M., Reiter R.J., 1976, Dose-dependent prolactin releasing activity of arginine vasotocin in intact and pinealectomized estrogen-progesterone treated adult male rats, Endocrinology, 99: 1319.

Vermes I., Telegdy G., 1979, Correlation between the pineal and brain 5-hydroxytryptamine rhythms and adrenal functions in male rats, Progr. Brain Res., 52: 471.

Vollrath L., 1981, The pineal organ. Handbuch der mikroskopischen Anatomie des Menschen VI/7, Springer-Verlag, Berlin, Heidelberg, New York.

Vriend J., Sackman J.W., Reiter R.J., 1977, Effects of

blinding, pinealectomy and superior cervical ganglionectomy on free thyroxine index of male golden hamsters, Acta Endocr. (Kbh)., 86: 758.

Wurtman R.J., 1971, Summary of symposium. in: The pineal gland. A Ciba Foundation Symposium, Wolstenholme G.E.W. and Knight J., eds., Churchill Livingstone, Edinburgh and London, p. 379.

Yamada T., 1961, The effect of electrical ablation of the nuclei habenulae, pineal body and subcommissural organ on endocrine function, with special reference to thyroid function, Endocrinology, 69: 706.

THE PINEAL GLAND AND SEXUAL FUNCTION IN MAN

Merrily Poth, Sadayoshi Higa, and Sanford Markey

Department of Pediatrics, USUHS, and Laboratory Clinical Clinical Science, NIMH
Bethesda, MD 20814

INTRODUCTION

The pineal gland has potent effects on the pituitary - gonadal system in many different species (1). In some situations it seems clear that these effects are mediated by melatonin secreted by the pineal gland (2).

Humans share several aspects of this system, that is, they have a pineal gland which episodically secretes melatonin in a 24 hour "circadian rhythm," and they have a pituitary-gonadal system, immature at birth, which after sexual motivation exhibits clear rhythmicity in the female although not in the male. However, at every point of potential action and interaction of the neuroendocrine and endocrine systems there is question and even active controversy as to whether or not the pineal gland and specifically melatonin has any effect on human reproductive function.

The major problem in dissecting out the function and relevance of the pineal in man has been the relative inacessibility of the gland and its products in situ. In this chapter we will not consider the question of pineal factors or hormones other than melatonin but will concentrate on melatonin as the active pineal hormone. We will extrapolate to man from the biochemistry and pharmacology of rat and hamster pineal glands which have been well studied. We will assume that in man a rhythm-entraining stimulus communicates with the pineal by a pathway including the superior cervical ganglion and involving β-adrenergic neurotransmission, that specific circadian enzyme change occur and that these enzyme changes are reflected in a circadian rhythm of melatonin production.

This chapter will restate some of the critical questions regarding the function of the pineal in humans, review the evidence, circumstantial and direct from the literature and from our own data, and attempt to formulate a kind of progress, (or lack of progress) report of the area. It will be obvious that this research is in its infancy and that much remains to be done.

METHODOLOGY

Most studies of melatonin in humans have involved the assay of melatonin by radioimmunoassay or by gas chromatography and mass spectrography of serum samples. Ideally this entails nearly constant or integrated blood sampling techniques and a sensitive and highly specific method for melatonin assay. The radioimmunoassays used appear to be sensitive enough for the purpose in that day/night differences are seen clearly using these assays. However, some controversy exists regarding the specificity of the substances measured (3). In addition, obvious problems exist in measuring frequent (30-90 min) serum melatonin samples in a humans, and particularly children. This mitigates against long-term sequential studies.

Melatonin in blood is rapidly metabolized to 6-hydroxymelatonin and excreted in the urine, mostly as the glucuronide or sulfate conjugates (4). We have developed a sensitive, very specific and highly reproducible assay for these compounds in urine (5) and we have used this assay to study human pineal function.

Twenty four hour urines were collected from subjects under normal conditions of work, acitivity, diet, and sleep. The urine was refrigerated (4°C) during the collection period and measured aliquots were frozen at - 20°C until analyzed.

The 6-hydroxymelatonin content of the urine was determined by gas chromatography negative chemical ionization mass spectrometry. Briefly, the assay utilizes tetradeuterated 6-hydroxymelatonin sulfate as an internal standard added to a 3 ml urine aliquot, followed by enzymatic (glusulase) hydrolysis, extraction of free 6-hydroxymelatonin into dichloromethane, reaction to form a stable t-butyldimethylsilylpentafluoropropionyl derivative, column chromatography on silica gel, and quantitation by selected ion recording to two specific and intense fragment ions (endogenous m/z 470 and 450; internal standard m/z 473,453).

The use of this assay allows us to monitor the daily output of melatonin from the pineal gland in both children and adults during normal life activities and to correlate their pineal output with other endocrine events of their life.

We make some important underlying assumptions in interpreting our data. Some of these have been proven and some have been taken for granted and await direct proof in the future.

The assumption is made that the 6-hydroxymelatonin measured in human urine comes from the pineal gland. Pinealectomized humans are not available for study. However, in the Rhesus monkey we measured urinary 6-hydroxymelatonin excretion before and after pinealectomy. This data shows clearly that in the absence of pineal gland >96% of the urinary 6-hydroxymelatonin rapidly disappears (6).

Next, it would be optimal if the urinary excretion of 6-hydroxymelatonin reflected the rhythmic secretion of pineal melatonin allowing both quantitative and temporal measurement of pineal activity. This is shown to be the case in studies of normal adult males where 70-95% of total 24 hour urinary 6-hydroxymelatonin is excreted in the period of time between midnight an 8 am (7). In Rhesus monkey, exposure to constant light abolished the urinary 6-hydroxymelatonin rhythm (6). Dr. Lewey discusses the effects of light on pineal rhythms in man much more fully in his chapter.

Patients with defective autonomic nervous systems (patients with Shy-Drager Syndrome or idiopathic orthostatic hypotension) did not exhibit normal circadian patterns of 6-hydroxymelatonin excretion (8). This implies that an intact sympathetic nervous system is necessaary for normal pineal rhythmicity (7).

The last major assumption upon which clear interpretation of this data depends is that hepatic metabolism and urinary excretion of melatonin and its 6-hydroxymetabolites does not vary significantly with the other parameters under study. Studies have not been performed to confirm this. We assume that these factors remain relatively constant and do not lead to significant changes in urinary 6-hydroxymelatonin. Because of these unknown factors which might be expected to lead to small changes in excretions, we do not consider changes of less than 30% to be of any significance in analysis of our data.

SPECIFIC QUESTIONS REGARDING PINEAL FUNCTION IN HUMANS

Development of melatonin rhythms in children:

It is known that infants synchronize to external stimuli and exhibit circadian rhythms of temperature and cortisol excretion by two weeks of age (9). However, it is not known at what age the pineal has rhythmic function and what effects this pineal rhythm has on other physiologic parameters in children.

We studied 24 hour urinary total 6-hydroxymelatonin excretion in 101 normal children between the ages of 3 and 17 years. We collected urines in 3 aliquots; the first (Aliquot I) from bedtime through the first morning void, the second (Aliquot II) from then through 3 pm and the last (Aliquot III) from 3 pm to bedtime. In most of the children two consecutive 24 hour timed excretions periods were measured.

The results are seen in Figures I & II.

The total amount of 6-hydroxymelatonin excreted was not different for children as a group when compared to adults. All normal children studied had clear circadian excretion of 6-hydroxymelatonin, with 75-90% of that excretion in period I and <10% of excretion during period III. Again, children did not differ from adults in this rhythm.

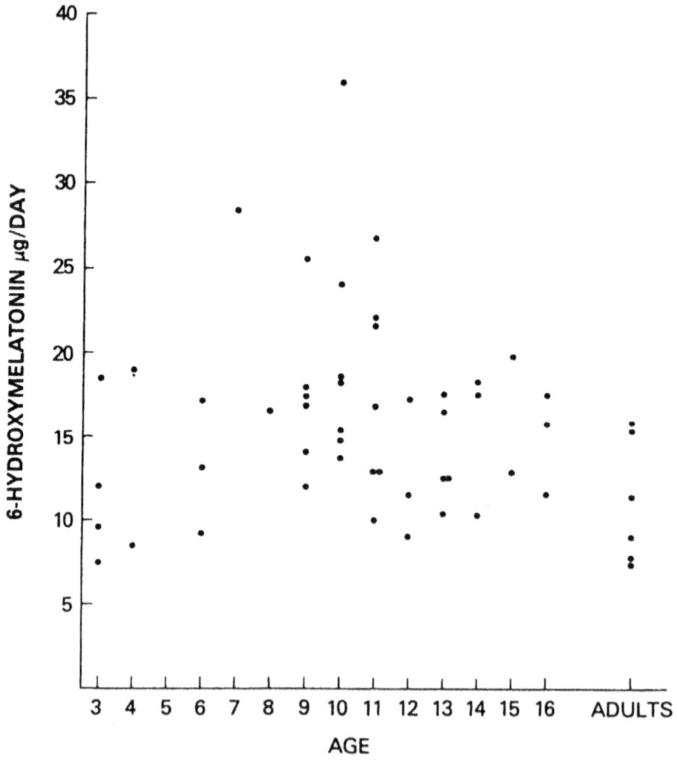

Fig. I Total amount of 6-hydroxymelatonin excreted in urine in 24 hours by normal girls and women.

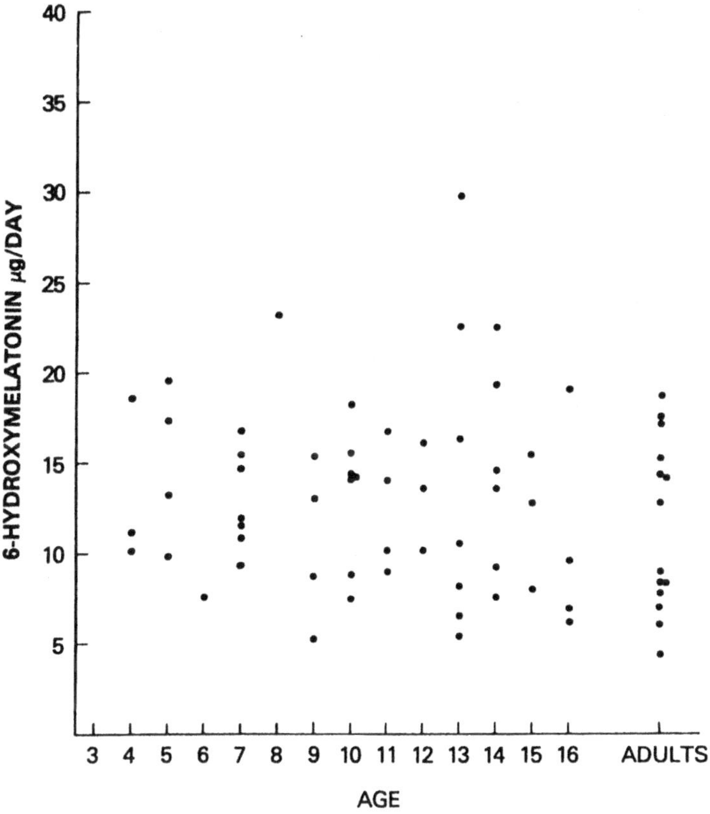

Fig. II Total amount of 6-hydroxymelatonin excreted in urine in 24 hours by normal boys and men.

QUESTION OF INTERACTION OF MELATONIN WITH PUBERTY

There is considerable data both positive and negative in the literature regarding the possibility of pineal puberty interaction in children. Tumors in the area of the pineal gland have been associated with both acceleration and delay of puberty in children (10). This has been explained by hypothesizing that some of these tumors secrete melatonin and hence can suppress gonadal maturation and result in delayed puberty. Other tumors or lesions results in destruction of the normally inhibitory pineal gland and by releasing this suppression result in precocious sexual development. However, no data is available to correlate changes in melatonin secretion with either early or late puberty and pineal lesions. In a recent study Silman assayed serum melatonin in late morning or early afternoon in normal school children (11). He found a marked decline in serum melatonin in boys in Tanner II stage of puberty and hypothesized that this decrease in the inhibiting substance, melatonin allowed puberty to progress. Another study by Tamarkin et al., (12) used integrated 24 hour blood sampling techniques and assayed serum melatonin by RIA and failed to confirm that pre pubescent boys had higher serum melatonin levels.

We studied this question in normal children using two consecutive 24 hour urines assayed for 6-hydroxymelatonin. Each child had a complete history and physical examination, including pubertal staging. This was done for boys using Tanner criteria for public hair and estimation of testicular volume with Prader beads as reference. Girls were staged according to Tanner criteria for breast development and for pubic hair (13).

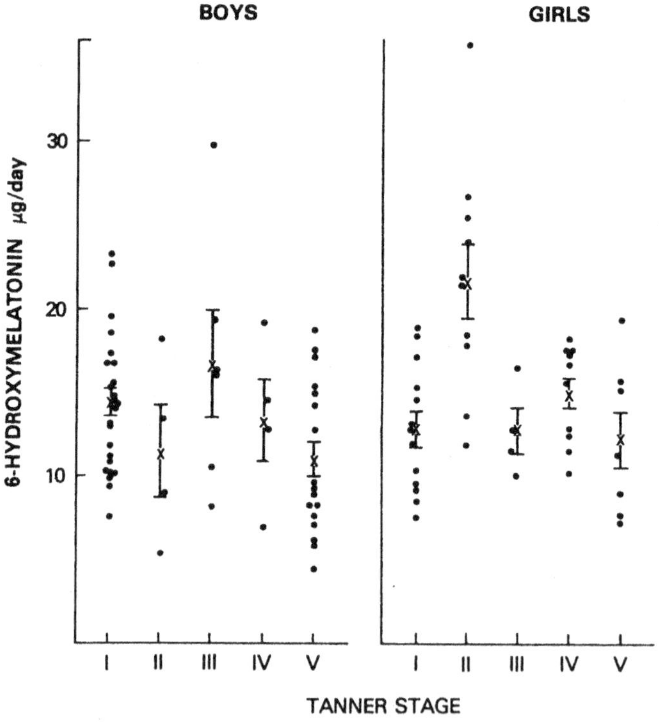

Fig. III 6-hydroxymelatonin excretion in normal children during puberty. Boys were staged by Tanner criteria for pubic hair and girls for breast development.

No change was seen in amount of 6-hydroxymelatonin excreted at any stage in pubertal development in the normal boys (Data is shown in Fig. III).

Surprisingly, girls who were identified as Tanner Stage II for breast development were higher when compared with boys at any stage or girls at any other stage of development. The girls who were at Tanner Stage II of puberty by breast configuration were at various stages of puberty by pubic hair determination, that is they had Tanner Stage I, II or III pubic hair. Pubic hair is a manifestation of adrenarche in girls and is not mediated by pituitary gonadotropins.

INTERACTION OF MENSTRUAL CYCLE AND PINEAL GLAND

There are several reports of measurement of pineal function during the normal menstrual cycle in women. Some of these using plasma melatonin measured by RIA have reported changes coincidental with the cycle (14).

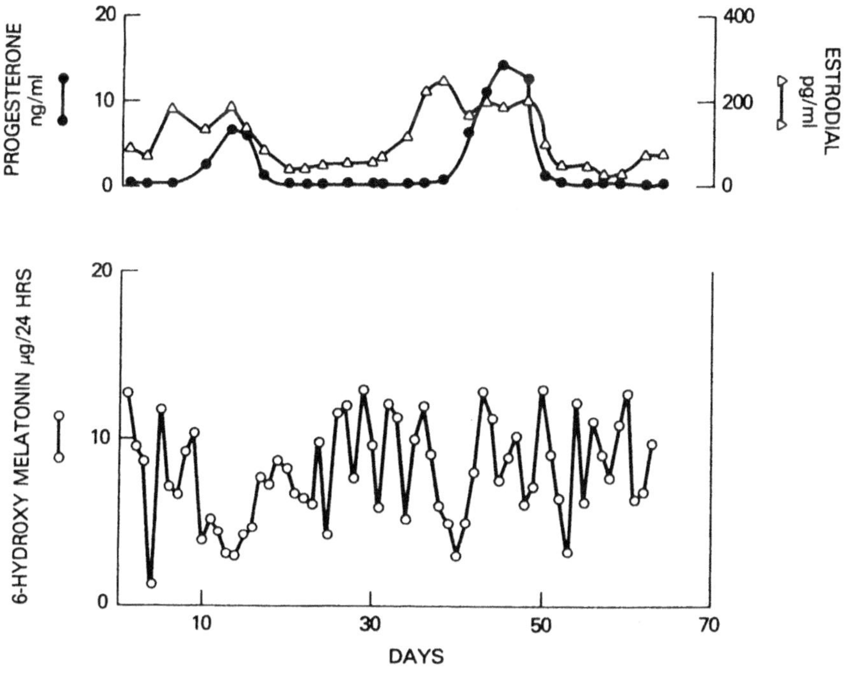

Fig. IV & V 6-hydroxymelatonin excretion during menstrual cycle in normal women, urinary 6-hydroxymelatonin excretion was measured in 24 hour urine for 65 consecutive days. Serum estradiol and progesterone were assayed three times per week.

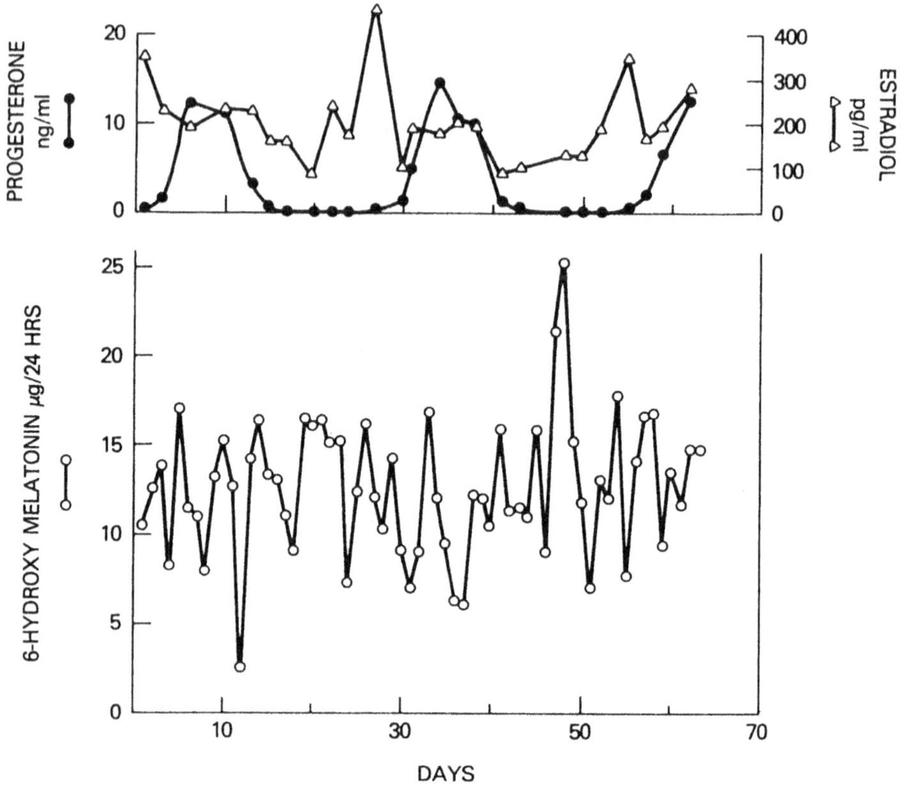

Fig. IV & V (continued)

We are studying women during normal menstrual cycles. Each women is asked to collect 24 hour urines on 65 consecutive days. Results of 2 women followed through 2 consecutive normal cycles are shown in Figs. IV & V. Each had normal ovulatory cycles as measured by morning temperatures and by serum estrogen and progesterone levels measured three times per week. Completeness of 24 hour collection was verified by measurement of urinary creatinine. The women showed marked day to day variation in the amount of 6-hydroxymelatonin excreted. A specific relationship between the components and hormonal changes of the menstrual cycle can not be ascertained from the data, but a 3-4 day rhythm appears to exist in each of these women.

This marked variability in daily 6-hydroxymelatonin excretion in cycling women was in marked contrast to the relatively constant daily excretion rates seen in the prepubertal and early pubertal girls. These girls were studied on 2 consecutive 24 hours - every 2 months in a 14 month time period, data is shown in Fig. VI, note that variability is only in the order of 10-20%.

As a fourth "control" for the normally cycling women we are studying young adult men. Each collects 15 daily 24 hour urine samples which are assayed for 6-hydroxymelatonin. It is clear that there is less variability in 6-hydroxymelatonin in these men than seen in the women (data not shown).

In conclusion we are using a sensitive, specific and highly reproducible gas - chromatographic - mass spectrometric assay for the measurement of the major metabolites of melatonin in human urine. This method allows the non-invasive study of pineal func-

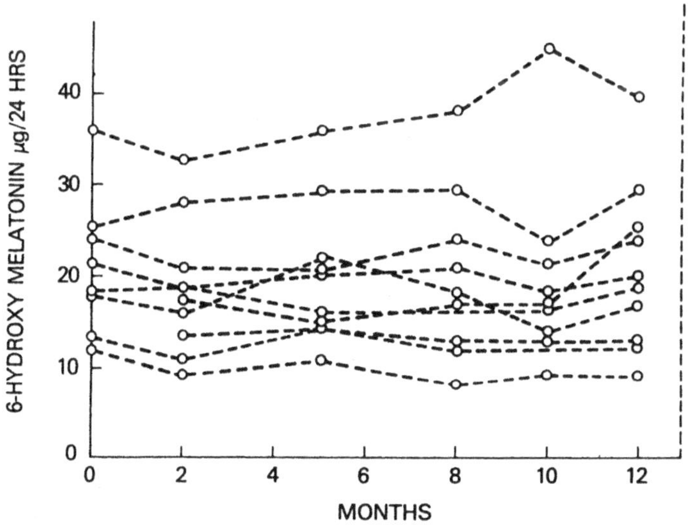

Fig. VI Longitudinal study of 6-hydroxymelatonin excretion in preovulatory girls. Each point is mean of excretion in two consecutive 24 hour urines. Each line is individual girl.

tion in humans. Quantitation of the absolute amount of 6-hydroxymelatonin excreted as well as temporal relationships in the time of excretion are able to be determine on both adults and children during their normal daily activities.

Using this assay we find that children at least by age three, have normal circadian rhythms of pineal function. We find no changes in pineal secretion of melatonin in boys during sexual maturation. Girls in breast Tanner Stage II of puberty excrete greater amounts of 6-hydroxymelatonin during the normal night time peak time with no change in the normally low day time excretion rate. This increase is seen when compared to girls at any other stage of puberty and with boys of any age.

Women monitored during normal ovulatory cycles show marked day to day variation in the amount of 6-hydroxymelatonin excreted compared to preovulatory girls and to males. It is not clear what specific relationships hold between pineal, pituitary and various function in these women. Greater numbers of women are being studied and results will be subjected to more sophisticated statistical treatment to try to ascertain the specific relationships between the hormonal events of the menstrual cycle and pineal activity.

REFERENCES

1. R.J. Reiter, The pineal and its hormones in the control of reproduction in mammals, Endocr Rev 1:109-131 (1980).
2. L. Tamarkin, C.W. Hollister, N.G. Lefebvre, and B.D. Goldman, Melatonin induction of gonadal quiescence in pinealectomized Syrian hamsters, Science 198:935-936 (1977).
3. S.P. Markey, Quantitative mass spectrometry, Biomedical Mass Spectrometry 8:426-430 (1981).
4. I.J. Kopin, C.M.B. Pare, J. Axelrod, and H. Weissbach, The fate of melatonin in animals, J Biol Chem 236:3072-3075 (1961).
5. M. Tetsuo, S.P. Markey, R.W. Colburn, and I.J. Kopin, Quantitative analysis of 6-hydroxymelatonin in human urine by gas chromatography-negative chemical ionization mass spectrometry, Analyt Biochem 109:208-215 (1981).
6. M. Tetsuo, M.J. Perlow, M. Mishkin, and S.P. Markey, Light exposure reduces and pinealectomy virtually stop urinary excretion of 6-hydroxymelatonin by Rhesus monkey, Endocrinology 110:997-1001 (1982).
7. M. Tetsuo, S.P. Markey, R.W. Colburn, and I.J. Kopin, Measurement of 6-hydroxymelatonin in human urine and its diurnal variations, Life Sciences 27:105-109 (1980).
8. M. Tetsuo, R.J. Polinsky, S.P. Markey, and I.J. Kopin, Urinary 6-hydroxymelatonin excretion in patients with orthostatic hypotension, J Clin Endocr Metab 53:607-610 (1981).

9. R.C. Franks, Diurnal variation of plasma 17-hydroxycorticosteroids in children, J Clin Endocr Metab 27:75-78 (1967).
10. J.I. Kitay, Pineal lesions and precocious puberty: a review, J Clin Endocr Metab 14:622-625 (1954).
11. R.E. Silman, R.M. Leone, R.M. Hooper, and M.A. Preece, Melatonin, the pineal gland and human puberty, Nature, Lond 282:301-303 (1979).
12. J.R.L. Ehrenkranz, L. Tamarkin, F. Comite, R.E. Johnsonbaugh, D.E. Bybee, D.L. Lorianx, and G.B. Cutler, Jr., Daily rhythm of plasma melatonin in normal and precocious puberty. J Clin Endocr Metab 55:307-310 (1982).
13. J.M. Tanner, "Foetus into Man," Harvard University Press, Cambridge (1978).
14. L. Wetterburg, J. Arendt, L. Pamnier, P.C. Sizenenko, W. VanDonselaar and T. Heyden, Human serum melatonin changes during the menstrual cycle, J Clin Endocr Metab 42:752-764 (1976).

MELATONIN SECRETION -- A BIOLOGICAL MARKER FOR HUMAN PINEAL ADRENERGIC FUNCTION

Alfred J. Lewy

Departments of Psychiatry, Pharmacology and Ophthalmology
Oregon Health Sciences University
Portland, Oregon, U.S.A.

INTRODUCTION

Spanning the last 20 years, hundreds of anatomical and pharmacological studies have indicated that melatonin secretion by the mammalian pineal gland is an ideal "biological marker" for adrenergic function, the endogenous circadian pacemaker, and the effects of light. Nighttime beta-adrenergic stimulation of melatonin synthesis provides the theoretical basis for using the pineal as a model of adrenergic function. The biochemical steps resulting in melatonin synthesis appear to be intricately coordinated.

In recent years, several controversial areas have been clarified; many of the remaining inconsistencies are in the process of being resolved. One such area, for example, had been the apparent inverse day/night activity rhythm of pineal cyclic AMP; however, it is now known that pineal cyclic AMP levels are greatest during the night (Mikuni et al., 1981).

Another controversial area concerns extrapineal melatonin. There are two issues here. One is whether or not melatonin is synthesized in other organs than the pineal. The other is whether or not plasma melatonin is derived from such sources, if they exist. We have addressed the second issue, using the highly accurate and sensitive gas chromatographic-negative chemical ionization mass spectrometric (neg CI GCMS) technique for measuring plasma melatonin.

GCMS ASSAYS

There seems to be a growing consensus that the neg CI GCMS assay is the most accurate of the human plasma melatonin assays (Arendt, 1981; Rollag, 1981). However, GCMS assays require expensive instrumentation that is difficult to maintain. Therefore, a principle use of this assay is for validation of other assays and for answering certain questions that require a high degree of specificity.

To compare an assay to the neg CI GCMS assay, plasma levels of melatonin should be 2-10 pg/ml during the day and 25-100 pg/ml at night. As with all assays, the lower the values, the more specific the assay. However, the dynamic range is probably more informative: with the mass spectral assay there is a ten to fifty-fold increase from day to night. Values that are similar to those of the GCMS assay at night but higher than the GCMS values during the day probably indicate the presence of a consistent level of a contaminant that appears to be relatively greater during the day than at night. Ideally, values should be corrected for recovery, since artifactually lower results could occur due to low or variable recovery. (The values of the neg CI GCMS assay are corrected for recovery which is therefore 100 percent.)

GCMS is considered to be theoretically highly specific, since compounds are identified by molecular weight as well as by retention time as they are quantified. However, not all GCMS assays are optimally specific. A deuterated internal standard should be used. The deuterated internal standard is chemically similar to endogenous melatonin and melatonin of the external standard curve, except that it has a slightly higher molecular weight, permitting separate identification and quantification by the mass spectrometer. When added to the plasma at the initiation of the assay, the deuterated internal standard coextracts, coderivatizes, cochromatographs, and cofragments with the nondeuterated (endogenous) melatonin. Measurement of two fragment ions at a characteristic intensity ratio is the mass spectral "fingerprint" for the parent molecule. Retention times are known exactly for each injection by comparing the retention time of the plasma sample to that of the internal standard. Interfering substances can be visually noted, because they are characterized by shoulders or humps preceding or following the melatonin peak. A recovery is calculated for each sample: thus each point of the external standard curve and each plasma sample is corrected for any losses that might occur during the assay.

Adequate sensitivity has been a problem for GCMS. The neg CI GCMS assay (Lewy and Markey, 1978) has the requisite sensitivity for measuring plasma melatonin (minimal detectable concentration <1 pg/ml) and is among the most sensitive of the human plasma melatonin assays (Rollag, 1981). Since it uses a deuterated internal standard, this assay also has a high degree of specificity and precision.

Measurement of urinary 6-hydroxymelatonin sulfate has also been performed with mass spectral assays using deuterated internal standards (Fellenberg et al., 1980; Tetsuo et al., 1980). The day/night dynamic range in the concentration of this urinary metabolite is similar to that of plasma melatonin as measured by the neg CI GCMS assay. Extremely low levels of 6-hydroxymelatonin are excreted during the day, indicating that marked daytime pulsatile secretion of melatonin does not occur (this is also in agreement with the plasma melatonin data as measured by the neg CI GCMS assay, because it is rare that plasma levels above 10 pg/ml are found in healthy sighted humans during the day). The urinary 6-hydroxymelatonin assay is the method of choice for measuring amplitude of secretion. Measurement of plasma melatonin requires many time points, not only for determination of onset, maximum, or offset of secretion, but also for description of a curve of values under which an area can be calculated, whereas urinary excretion of 6-hydroxymelatonin usually involves only a few data points for each subject. The plasma melatonin assay is the best suited for evaluating timing of secretion. Urine is difficult to collect in shorter than six-hour intervals; consequently, only a gross estimate of the day/night rhythm can be obtained using 6-hydroxymelatonin.

EXTRAPINEAL MELATONIN?

After pinealectomy, rats have very little melatonin during the day or night, or after administration of isoproterenol as measured by the neg CI GCMS assay (minimal detectable concentration <1 pg/ml) (Lewy et al., 1980) (see Figure 1). Markey's group has recently confirmed these findings measuring 6-hydroxymelatonin in the rat (Markey and Buell, 1982) and in the monkey (Tetsuo et al., 1982). Pelham found similar results in the chicken using a bioassay (Pelham et al., 1972; Pelham, 1975). Arendt showed that only after complete pinealectomy could sheep plasma melatonin be abolished (Arendt et al., 1980). It is not known as yet exactly how low are the "true" levels of melatonin and 6-hydroxymelatonin after pinealectomy, because these levels are close to, or less than, the minimal detectable concentration of the most sensitive assays. Measurement of substantial amounts of melatonin in the circulation after pinealectomy most likely indicates an assay specificity problem. Reduction of melatonin at night in humans after administration of propranolol (Vaughan et al., 1976; Hanssen et al, 1977, author's unpublished data) suggests that most of human plasma melatonin is derived from the pineal gland, because it is unlikely that other sources of melatonin would be under beta-adrenergic regulation. We (Lewy and Neuwelt, in preparation[a]) have recently used the neg CI

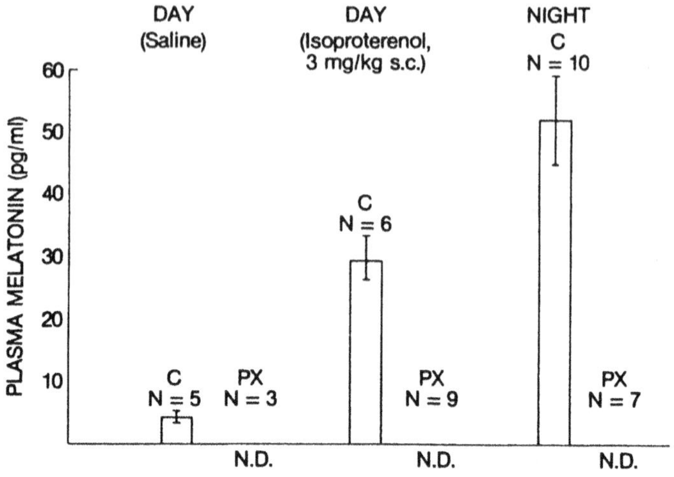

Fig. 1. Sham-operated control (C) rats showed expected low daytime and high nighttime plasma melatonin values. Rats injected with 3 mg/kg l-isoproterenol-d-bitartrate during the day had substantial melatonin levels. Pinealectomized (PX) rats had nondetectable (N.D.) plasma melatonin under all three conditions. From Lewy et al., 1980.

GCMS assay to measure plasma melatonin in a patient whose malignant pineal gland was surgically removed: a normal day/night pattern was present prior to extirpation, whereas no melatonin could be detected following pinealectomy.

Whether or not there exists extrapineal synthesis of melatonin is still controversial. Future studies should be carried out in pinealectomized animals to rule out the pineal as a source for melatonin. Identification of melatonin by immunohistological techniques using antibodies whose specificity has not been absolutely assured leaves unanswered the question of the true identity of "immunoreactive melatonin" in extrapineal tissue. Demonstration of melatonin production in tissue culture after adding serotonin is not adequate proof of extrapineal production in vivo: identification of extrapineal N-acetyltransferase and HIOMT activity should also include demonstration of serotonin in those organs at adequate concentrations. If extrapineal synthesis of melatonin exists, it is unlikely that these sites contribute significantly to plasma or urine levels.

Abolition of plasma melatonin after pinealectomy validates the use of circulating levels for estimating pineal secretion of this hormone and provides the basis for further work with plasma melatonin as a biological marker for adrenergic activity as well as for the circadian system.

ADRENERGIC REGULATION OF MELATONIN PRODUCTION

The pineal is unique in its sympathetic innervation. Most endocrine glands are regulated by blood-born substances usually of pituitary origin. (Some exocrine glands are innervated by peripheral postganglionic neurons.) The adrenal medulla (which is considered to be similar to a postganglionic sympathetic neuron, because it synthesizes and secretes catecholamines) is innervated by preganglionic fibers from splanchic nerves. The pineal is innervated by postganglionic neurons from the superior cervical ganglia. What is also remarkable (and has important implications) is that these neurons do not respond to stress (Klein and Parfitt, 1976; Parfitt and Klein, 1977). The "fight or flight" response typical of the sympathetic nervous system does not include the sympathetic nerves of the pineal (which are discretely regulated by an endogenous circadian pacemaker).

The pineal is also unusual in that melatonin production is not compensatorily regulated by parasympathetic nerves. Cholinergic control (Wartman et al., 1979) is in doubt, because choline acetyltransferase appears not to be present in the rat pineal (Schrier and Klein, 1974). Similarly, reports of parasympathetic innervation of the rabbit pineal (Romijn,1973) are confounded by the fact that choline acetyltransferase is also not present in the pineal of this species (Schrier and Klein, 1974). Acetyl cholinesterase has apparently been measured in the sympathetic fibers innervating the pineal (Eranko et al., 1970a; Eranko et al., 1970b; Manocha, 1970; Trueman and Herbert, 1970) and ganglionectomy (Machado and Lemos, 1971) and 6-hydroxydopamine (Snyder et al., 1965) has apparently diminished it, but again, this work must be evaluated in the light that there may be no choline acetyltransferase present in the pineal (Schrier and Klein, 1974).

There are a few studies, however, that suggest at least in some species the existence of efferent innervation to the pineal whose function is as yet unknown (David and Herbert, 1973; Nielsen and Moller, 1975; Moller, 1978; Ronnekleiv et al., 1980), but at this point it seems safe to assume that neuroregulation of melatonin production is exclusively by sympathetic nerves (Ariens Kappers, 1960; Ariens Kappers, 1979; Ariens Kappers et al., 1979;). The pineal apparently also has presynaptic alpha-2 autoadrenoceptors (Pelayo et al., 1977). (Postsynaptic pineal alpha-adrenergic receptors appear to regulate phosopholipid metabolism in the pineal (Nijjar et al., 1980).

Since the pineal is innervated by peripheral postganglionic sympathetic neurons that appear to be exclusively involved in the regulation of melatonin production, this gland has advantages as an experimental model for evaluation of adrenergic function. It is particularly useful, because synthesis of melatonin occurs at night, and because stress (which increases the levels of circulating catecholamines) does not influence melatonin production due to highly active uptake mechanisms (Klein and Parfitt, 1976; Parfitt and Klein, 1977). Catechol-O-methyltransferase activity is greatest during the day, which perhaps affords another mechanism that minimizes daytime production of melatonin (Backstrom and Wetterberg, 1972). (Since melatonin production is not affected by stress, measurement of melatonin is not a "marker" for stress.) It is also important that the pineal is a peripheral, and not a central, organ (its beta-adrenergic receptors are outside of the blood-brain barrier).

In animal experiments the pineal gland facilitates several ways in which to evaluate adrenergic function. Since the nerves of the pineal are quiescent during the day (and firing rate can be suppressed at any time by light), neuronal firing can be reduced without the use of drugs. During the day postsynaptic function can be selectively evaluated (because the presynaptic nerves are quiescent), by administration of an agonist, such as isoproterenol (see Figure 2, top). Uptake mechanisms can also be evaluated by comparing the melatonin produced in response to isoproterenol (which is not taken up presynaptically) to exogenous administration of norepinephrine (which is). During the night the sympathetic nerves release endogenous norepinephrine that stimulates melatonin production. Measurement of nighttime melatonin secretion evaluates both pre- and postsynaptic function (see Figure 2, bottom).

A hypothetical experiment would be as follows. Suppose a drug caused a 50% reduction of melatonin production at night and a 50% decrease in the response to isoproterenol during the day. This would indicate that the drug had primarily postsynaptic activity and relatively little presynaptic activity. On the other hand, suppose the drug caused only 10% reduction in response to isoproterenol during the day but a 50% reduction at night. Then the drug probably had primarily presynaptic effects.

HUMAN STUDIES

As in all mammals studied to date, human melatonin appears to be under adrenergic control. Patients with transection of the cervical

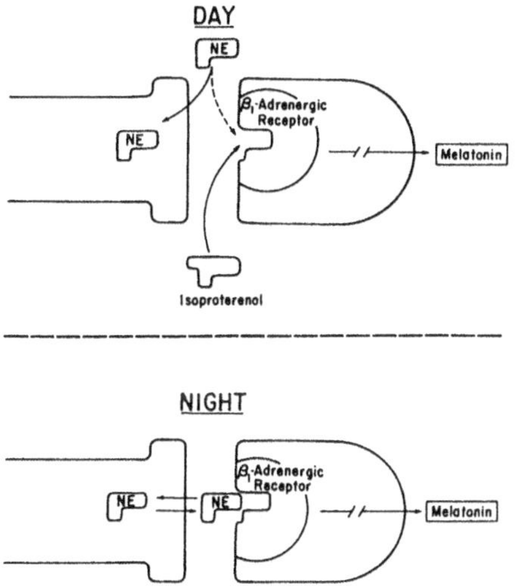

Fig. 2. Postsynaptic function can be selectively evaluted by. measuring melatonin production in response to administration of a beta-adrenergic agonist during the day (top); total pre- and postsynaptic adrenergic activity can be evaluated by measuring nighttime melatonin production stimulated by endogenous release of norepinephrine (NE) from the pineal's sympathetic neurons (bottom).

spinal cord have disturbed melatonin secretory rhythms (Kneisley et al., 1978). Patients with Shy-Drager Syndrome (multiple system atrophy) and idiopathic orthostatic hypotention (diseases which are manifested by central and peripheral sympathetic lesions) have reduced levels of plasma melatonin (Vaughan et al., 1979) and urinary 6-hydroxymelatonin (Tetsuo et al., 1981). These data suggest that the human pineal is regulated by sympathetic adrenergic neurons.

Vaughan originally demonstrated in one healthy subject (Vaughan et al., 1976) that propranolol blocks nighttime melatonin secretion. This was later confirmed in a psychiatric patient (Hanssen et al., 1977) and in several human subjects (author's unpublished data). Propranolol, in doses of 120-140 mg po administered at night, can completely reduce melatonin secretion to daytime levels or below (author's unpublished data). This drug may therefore be used to help evaluate adrenergic function of the human pineal in vivo. The amount of propranolol that reduces melatonin levels may be related to the functional number of beta-adrenergic receptors in the pineal. Propranolol may also have a use in the treatment of melatonin hypersecretion. Administration of propranolol also affords the opportunity to understand the relationship between melatonin secretion and secretion of other endocrine substances. After reducing melatonin secretion with propranolol, changes in the levels of other hormones could then be assessed. On a subsequent night, administration of the same dose of propranolol in addition to an infusion of physiologic levels of melatonin would control for any beta-adrenergic and/or nonspecific effects of propranolol; thus hormonal changes due to the reduction of melatonin levels could be isolated.

Clonidine also reduces human melatonin secretion (Lewy et al., in preparation[b]). This alpha-adrenergic agonist is primarily selective for presynaptic alpha-2 adrenergic autoreceptors but stimulates postsynaptic alpha-1 and alpha-2 adrenergic receptors as well.

Clonidine passes through the blood-brain barrier and has been used to evaluate alpha-adrenergic receptors in a variety of disorders, including affective disorders (Siever et al., 1981). Clonidine's effects have been evaluated by measuring growth hormone or blood pressure. However, these physiological variables are regulated by more than one neurotransmitter. Also, compensatory autonomic mechanisms can confound the cardiovascular effects of clonidine. Consequently, measurement of melatonin secretion in response to clonidine may provide a more specific index of alpha-adrenergic receptor function.

One hour after intravenous administration of clonidine, melatonin levels are reduced in a linear dose-response relationship (Lewy et al., in preparation[b]). Oral clonidine is also effective (this may be a more practical route of administration). After oral administration there is a lag time and a longer duration of action. It may also be possible to measure urinary 6-hydroxymelatonin sulfate in an overnight collection in order to quantify melatonin excretion.

Measurement of nighttime melatonin is the simplest way of evaluating pineal adrenergic activity. In a preliminary study, we measured plasma melatonin in five manic-depressive patients, in both manic and depressed states (Lewy et al., in preparation[c]). Variables that might confound the results, such as age, sex, drug regimen, or time of year, were controlled, in that patients were compared to themselves; they were on the same drug regimen and were studied at the same time of year in each mood state. Melatonin levels were increased in mania compared to depression, which is compatible with increased adrenergic activity in mania compared to depression.

Wetterberg et al. (1979) also found that unipolar patients have decreased melatonin levels when depressed, compared to when they are not depressed. Jimerson et al. (1977) studied bipolar depressed patients and found no difference in urinary melatonin; however, they compared patients to healthy subjects. Since it is not known if phase of the menstrual cycle, age, sex, and time of year, affect melatonin secretion, it is recommended that these variables be appropriately controlled. Either the area under the curve of plasma melatonin levels or an overnight (or 24-hour) urinary 6-hydroxymelatonin sulfate level are the optimal methods for measuring pineal adrenergic activity.

Although the human pineal is probably regulated by sympathetic adrenergic neurons, stimulation of these beta-adrenergic receptors during the day in humans has been difficult, because the dose of agonist that would be necessary would cause untoward cardiovascular effects. At least one hour of continuous stimulation is necessary to provide sufficient time for the induction of N-acetyltransferase. This problem may have been avoided in a recent study of volunteers who exercised for a length of time on a bicycle, resulting in a two-fold increase in melatonin levels (Carr et al., 1981). This increase, however, is considerably less than the response seen in experimental animals with isoproterenol. Assay specificity may have contributed to this apparent difference in response: using the neg CI GCMS assay (Lewy and Markey, 1978), doses of isoproterenol, tyramine, l-dopa, and dopamine (which caused a 50% increase in heart rate for one hour) were insufficient to raise melatonin levels significantly (author's unpublished data). Negative results in attempting to stimulate melatonin production during the day in some species (such as man) is probably a problem of dosage.

SUMMARY

The amplitude of plasma melatonin levels or urinary 6-hydroxymelatonin sulfate levels appears to be a useful biological marker for adrenergic activity of the pineal gland in humans. Perhaps in the future, we will develop the appropriate methods to be able to stimulate human melatonin production with a safe dose of a beta-adrenergic agonist. Until then, reduction of nighttime secretion with clonidine or propranolol remain the best tools for specific evaluation of presynaptic and postsynaptic adrenergic activity in the pineal gland.

REFERENCES

Arendt, J., 1981, Current status of assay methods of melatonin, in: Melatonin -- Current Status and Perspectives, Advances in the Biosciences, vol. 29, N. Birau, and W. Schloot, eds., Pergamon Press, New York, pp. 3-7.

Arendt, J., Forbes, J. M., Brown, W. B., and Marston, A., 1980, Effect of pinealectomy on immunnoassayable melatonin in sheep, J. Endocrinol. 85:1P-2P.

Ariens Kappers, J., 1960, The development, topographical relations and innervation of the epiphysis cerebri in the albino rat, Z. Zellforsch. Mikrosk. Anat. 52:163-215.

Ariens Kappers, J., 1979, Short history of pineal discovery and research, in: The Pineal Gland of Vertebrates Including Man -- Progress in Brain Research, vol. 52, J. Ariens Kappers, and P. Pevet, eds., Elsevier North-Holland Biomedical Press, Amsterdam, pp. 3-22.

Ariens Kappers, J., Smith, A. R., and DeVries, R. A. C., 1979, The mammalian pineal gland and its control of hypothalamic activity, in: The Pineal Gland of Vertebrates Including Man -- Progress in Brain Research, vol. 52, J. Ariens Kappers, and P. Pevet, eds., Elsevier North-Holland Biomedical Press, Amsterdam, pp. 149-174.

Backstrom, M., and Wetterberg, L., 1972, Catechol-O-methyl-transferase, histamine-N-methyltransferase and methanol forming enzyme in the rat pineal gland, Life Sci. 11:293-299.

Carr, D. R., Reppert, S. M., Bullen, B., Skrinar, G., Beitins, I., Arnold M., Rosenblatt, M., Martin, J. B., and McArthur, J. W., 1981, Plasma melatonin increases during exercise in women, J. Clin. Endocrinol. Metab. 53:224-225.

David, G. F. X., and Herbert, J., 1973, Experimental evidence for a synaptic connection between habenula and pineal ganglion in the ferret, Brain Res. 64:327-343.

Eranko O, Rechardt, L., Eranko, L., and Cunningham, A., 1970, Acetylcholinesterase (AChE) activity in the sympathetic nerve fibres of the pineal body (PB) of the rat, Scand. J. Clin. Lab. Invest. 25(Suppl. 113):82.

Eranko, O., Rechardt, L., Eranko, L., and Cunningham, A., 1970, Light and electron microscopic histochemical observations on cholinesterase-containing sympathetic nerve fibers in the pineal body of the rat, Histochem. J. 2:479-489.

Fellenberg, A. J., Phillipou, G., and Seamark, R. F., 1980, Specific quantitation of urinary 6-hydroxymelatonin sulphate by gas chromatography mass spectrometry, Biomed. Mass. Spec. 7:84-87.

Hanssen, T., Heyden, T., Sundberg, T., and Wetterberg, L., 1977, Effect of propanolol on serum-melatonin, Lancet 2:309-310.

Jimerson, D. C., Lynch, H. J., Post, R. M., Wurtman, R. J., and Bunney, W. E., 1977, Urinary melatonin rhythms during sleep deprivation in depressed patients and normals, Life Sci. 20:1501-1508.

Klein, D. C., and Parfitt, A., 1976, A protective role of nerve endings in stress-stimulated increase in pineal N-acetyltransferase activity, in: Catecholamines and Stress, E. Usdin, R. Kvetnanasky, and I. J. Kopin, eds., Pergamon Press, New York, pp. 119-128.

Kneisley, L. W., Moskowitz, M. H., and Lynch, H. J., 1978, Cervical spinal cord lesions disrupt the rhythm in human melatonin excretion, J. Neurol. Transm., Suppl. 13:311-323.

Lewy, A. J., and Markey, S. P., 1978, Analysis of melatonin in human plasma by gas chromatography negative chemical ionization mass spectrometry, Science 201:741-743.

Lewy, A. J., Tetsuo, M., Markey, S. P., Goodwin, F. K., and Kopin, I. J., 1980, Pinealectomy abolishes plasma melatonin in the rat, J. Clin. Endocrinol. Metab. 50:204-205.

Lewy, A. J., and Neuwelt, E. A., Disappearance of plasma melatonin after surgical removal of a neoplastic pineal gland, in preparation.[a]

Lewy, A. J., Siever, L., Uhde, T. W., Murphy, D. L., Post, R. A., Markey, S. P., and Goodwin, F. K., Clonidine reduces human melatonin secretion, in preparation.[b]

Lewy, A. J., Wehr, T. A., Gold, P. W., and Goodwin, F. K., Melatonin secretion in manic-depressive patients, in preparation.[c]

Machado, A. B. M., and Lemos, V. P. J., 1971, Histochemical evidence for a cholinergic sympathetic innervation of the rat pineal body, J. Neurovisc. Relat. 32:104-111.

Manocha, S. L., 1970, Histochemical distribution of acetylcholinesterase and simple esterases in the brain of squirrel monkey (Saimiri sciureus), Histochemie. 21:236-248.

Markey, S. P., and Buyell, P. E., 1982, Pinealectomy abolishes 6-hydroxymelatonin excretion by male rats, Endocrinology 111:425-426.

Mikuni, M., Saito, Y., Koyama, T., Yamashita, I., 1981, Circadian variation of cyclic AMP in the rat pineal gland. J. Neurochem. 36:1295-1297.

Moller, M., 1978, Presence of a pineal nerve (nervus pinealis) in the human fetus; a light and electron microscopical study of the innervation of the pineal gland, Brain Res. 154:1-12.

Nielsen, J. T., and Moller, M., 1975, Nervous connections between the brain and the pineal gland in the cat (Felis catus) and the monkey (Cercopithecus aethiops), Cell. Tiss. Res. 161:293-301.

Nijjar, M.S., Smith, T. L., and Hauser, G., 1980, Evidence against dopaminergic and further support for beta-adrenergic receptor involvement in the pineal phosphatidylinositol effect, J. Neurochem. 34:813-821.

Parfitt, A., and Klein, D. C., 1977, Increase caused by desmethylimipramine in the production of (^{3}H) melatonin by isolated pineal glands, Biochem. Pharmacol. 26:904-905.

Pelayo, F., Dubocovich, M. L., Langer, S. Z., 1977, Regulation of noradrenaline release in the rat pineal through a negative feedback mechanism mediated by presynaptic alphaadrenoceptors, Eur. J. Pharmacol. 45:317-318.

Pelham, R. W., 1975, A serum melatonin rhythm in chickens and its abolition by pinealectomy, Endocrinology 96:543-546.

Pelham, R. W., Ralph, C. L., and Campbell, I. M., 1972, Mass spectral identification of melatonin in blood, Biochem. Biophys. Res. Commun. 46:1236-1241.

Rollag, M.D., 1981, Methods for measuring pineal hormones, in: The Pineal Gland, Anatomy and Biochemistry, vol. 1, R. J. Reiter, ed., CRC Press, Boca Raton, Fla., pp. 273-302.

Romijn, H. J., 1973, Parasympathetic innervation of the rabbit pineal gland, Brain Res. 55:431-436.

Ronnekleiv, O. K., Kelly, M. J., and Wuttke, W., 1980, Single unit recordings in the rat pineal gland: evidence for habenulo-pineal neural connections, Exp. Brain Res. 39:187-192.

Schrier, B. K., and Klein, D. C., 1974, Absence of choline acetyltransferase in rat and rabbit pineal gland, Brain Res. 79:347-351.

Siever, L. J., Cohen, R. M., and Murphy, D. L., 1981, Antidepressants and alpha-adrenergic autoreceptor desensitization. Am. J. Psychiatry 138:681-682.

Snyder, A. H., Axelrod, J., Wurtman, R. J., and Fischer, J. E., 1965, Control of 5-hydroxytryptophan decarboxylase activity in the rat pineal gland by sympathetic nerves, J. Pharmacol Exp. Ther. 147:371-375.

Tetsuo, M., Markey, S. P., Kopin, I. J., 1980, Measurement of 6-hydroxymelatonin in human urine with its diurnal variation, Life Sci. 27:105-109.

Tetsuo, M., Perlow, M. J., Mishkin, M., and Markey, S. P., 1982, Light exposure reduces and pinealectomy virtually stops urinary excretion of 6-hydroxymelatonin by Rhesus monkeys, Endocrinology 110:997-1003.

Tetsuo, M., Polinsky, R. J., Markey, S. P., and Kopin, I. J., 1981, Urinary 6-hydroxymelatonin excretion in patients with orthostatic hypotension, J. Clin. Endocrinol. Metab. 53:607-610.

Trueman, T., and Herbert, J., 1970, The distribution of monoamines and acetylcholinesterase in the pineal gland and habenula of the ferret, J. Anat. 106:406.

Vaughan, G. M., McDonald, S. D., Bell, R., and Stevens, E. A., 1979, Melatonin, pituitary function and stress in humans, Psychoneuroendocrinology 4:351-362.

Vaughan, G. M., Pelham, R. W., Pang, S. F., Loughlin, L. L., Wilson, K. M., Sandock, K. L., Vaughan, M. K., Koslow, S. H., and Reiter, F. J., 1976, Nocturnal elevation of plasma melatonin and urinary 5-hydroxy-indoleacetic acid: attempts at modification by brief changes in environmental lighting and sleep and by autonomic drugs, J. Clin. Endocrinol. Metab. 42:752-754.

Wartman, S. A., Branch, B. J., George, R., and Taylor, A. N., 1969, Evidence for a cholinergic influence on pineal hydroxyindole-0-methyltransferase activity with changes in environmental lighting, Life Sci. 8:1263-1270.

Wetterberg, L., Beck-Friis, J., Aperia, B., and Pettersen, U., 1979, Melatonin/cortisol ratio in depression, Lancet 2:1361.

HUMAN MELATONIN SECRETION, ITS ENDOGENOUS CIRCADIAN PACEMAKER AND THE EFFECTS OF LIGHT

Alfred J. Lewy

Departments of Psychiatry, Pharmacology and Opthalmology
Oregon Health Sciences University
Portland, Oregon, U.S.A.

INTRODUCTION

Elucidation of pineal physiology must be based on an understanding of the circadian as well as adrenergic regulation of melatonin production. For more than 50 years it has been known that the endogenous circadian pacemaker is located in the brain (Richter, 1965; Richter, 1967). In many species of animals, this pacemaker is thought to be located in the suprachiasmatic nucleus (SCN) of the hypothalamus (Moore and Eichler, 1972; Stephan and Zucker, 1972; Ibuka and Kawamura, 1975; Rusak, 1977). There may be more than one pacemaker; however, with regard to melatonin production, there is increasing agreement that there is one pacemaker located in the SCN.

Lesioning of the SCN results in a decrease of N-acetyltransferase activity (Moore and Klein, 1974; Klein and Moore, 1979). It was for this reason that it was thought that the SCN "turns on" the sympathetic stimulation of the pineal. In other words, when the SCN is "on," the pineal is "on." Recent studies may have challenged this view, since stimulation of some SCN neurons causes a decrease in sympathetic activity (Nishino et al., 1976). Also, both inhibitory and excitatory responses have been recorded in response to light (Nishino et al., 1976; Groos and Mason, 1980). However, these neurophysiological studies are difficult to interpret and do not necessarily rule out that SCN activity results in an "on" signal. Studies of uptake of 2-deoxy-D-(C^{14})glucose show increased uptake during the day compared to night (Schwartz and Gainer, 1977; Schwartz et al., 1980). (Should the SCN prove to be "off" at night, then there must be a center of sympathetic activity caudal to the SCN, that is inhibited by the SCN during the day).

Future studies are necessary to determine with certainty whether the SCN is "on" or "off" at night, with respect to sympathetic stimulation of the pineal.

THREE BIOLOGICAL EFFECTS OF LIGHT

Light has three effects on melatonin production. One, light suppresses melatonin production. Two, the light-dark cycle entrains the endogenous pacemaker that drives the melatonin production rhythm. Three, the changing lengths of daylight and darkness regulate the annual (seasonal) patterns of melatonin production.

Suppression of Melatonin Production

The suppressant effect of light appears to be unique to melatonin production by the pineal gland. It is thought that the retinohypothalamic tract (RHT) mediates both the entrainment effect of light and the suppressant effect of light (Klein and Moore, 1979). Consequently, the suppressant effect may be a means of assessing the properties of light important for the entrainment effect.

In experimental animals very dim light is sufficient to suppress melatonin production. As little as 1-2 lux will suppress N-acetyltransferase activity by 50% in the rat (Minneman et al., 1974). Less than 500 lux will completely suppress N-acetyltransferase activity in the sheep and the monkey (Rollag and Niswender, 1976; Perlow et al., 1981). Suppression occurs immediately, and the decline in melatonin levels corresponds to its half-life in plasma or CSF (Reppert et al., 1979). Upon return to darkness, N-acetyltransferase and melatonin levels resume their pretreatment values within a few minutes. However, in the rat, melatonin production resumes slowly, and only if the light pulse occurs during the first half of the night (Deguchi and Aelrod, 1972; Illnerova and Vanecek, 1979). One minute of light is sufficient for suppression of melatonin production in the rat or the hamster (Illnerova and Vanecek, 1979; Hoffman et al., 1981).

Previous investigators have had difficulty in attempting to demonstrate effects of light in man (Vaughan et al., 1976; Jimerson et al., 1977; Lynch et al., 1977; Arendt, 1978; Weitzman et al., 1978; Wetterberg, 1978; Akerstedt et al., 1979; Vaughan et al., 1979) leading many to speculate that humans are quantitatively, and perhaps qualitatively, different in their response to light compared to all other warm-blooded species tested, including nonhuman primates (Perlow et al., 1980; Reppert and Klein, 1980). Recently, however, we have shown that light suppresses human melatonin secretion (Lewy et al., 1980); humans apparently require bright artificial light or sunlight for suppression of melatonin secretion that is accomplished with much lower intensities of light in other species.

First, sunlight was tested (Lewy et al., 1980). The idea that sunlight might be effective in suppressing human melatonin secretion (though ordinary room light was ineffective) came from an instance of self-experimentation: shortly after returning to Washington, D.C., from a two-week stay in Sydney, Australia, the author's morning plasma melatonin level was quite low. (It should have been at a higher level, based on the phase of the author's endogenous pacemaker.)

The sleep/wake cycles of two healthy volunteer subjects were then shifted in the laboratory; that is, they slept between 3 and 11 a.m. for several days. After seven days melatonin secretion was maximal during the late morning (between 9 and 10 a.m.). On the eighth day of the study, the volunteers were awakened at 7 a.m. and exposed to sunlight. Melatonin concentrations declined precipitously. It appeared that sunlight was indeed capable of suppressing melatonin secretion in man.

Why was sunlight (and not ordinary artificial light) effective? Since sunlight is more intense than ordinary room light (at all wave lengths), intensity was studied next. The intensity of sunlight on a sunny afternoon is 100,000 lux; ordinary room light is rarely above 500 lux and is usually 200-300 lux (Thorington, 1980). By increasing intensity from 500 to 2500 lux, the intensity at practically each wavelength will increase, despite a change in light spectra.

Six volunteers were studied under different light intensities during an interruption of sleep between 2 and 4 a.m. Five hundred lux fluorescent light had little or no effect in reducing melatonin concentrations, but 2500 lux incandescent light profoundly reduced melatonin secretion (Figure 1, left). (Fluorescent light [Vita-Lite] of this intensity is also effective, produces less glare and heat, and is currently being used in these studies.) Two of these subjects who were exposed to 1500 lux appeared to have a 50% reduction of melatonin secretion at this intensity, thus suggesting a dose-response relationship between light intensity and suppression of melatonin secretion (Figure 1, right). This relationship had been previously shown in the rat with respect to N-acetyltransferase activity (Minneman et al., 1974).

These findings have several implications. First, the human pineal appears to be regulated in the same way as in other species. (The SCN has been identified in man as well as in other primates

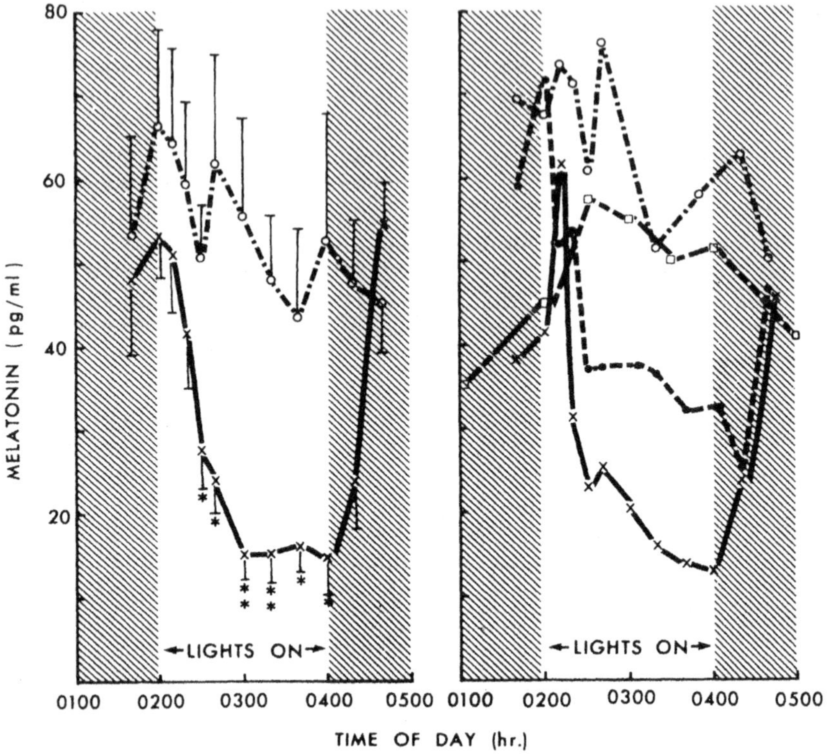

Fig. 1 (left). Effect of light on melatonin secretion. Each point represents the mean concentration of melatonin (± standard error) for six subjects. A paired t-test, comparing exposure to 500 lux with exposure to 2500 lux, was performed for each data point. A two-way analysis of variance with repeated measures and the Newman-Keuls statistic for the comparison of means showed significant differences between 2:30 a.m. and 4 a.m. (*, $P < .05$; **, $P < .01$)). Fig. 1 (right). Effect of different light intensities on melatonin secretion. The averaged values for two subjects are shown. Symbols: (○) 500 lux; (x) 2500 lux; (●) 1500 lux; and (□) asleep in the dark.
From Lewy et al., 1980. Copyright, 1980, AAAS.

(Moore, 1979; Lydic et al., 1980).) These results also suggest that the human pineal may function in a similar way as in other mammals.

The second implication is that (perhaps on the basis of intensity) humans have adapted to ordinary room light, yet may still remain sensitive to the natural (brighter) sunlight/dark cycle. Thus, the changing length of the natural photoperiod might affect humans in similar ways in which other animal species are affected. These results may stimulate more epidemiological studies of seasonal rhythms in humans.

The third implication of these findings is that a "bright light/dim light" paradigm may be a useful research strategy for investigating the effects of light in man. Most previous studies of the effects of light in man have resulted in negative or ambiguous findings (Hollwich, 1979). Previous studies that used insufficiently intense light should be repeated with bright artificial light or sunlight.

Why does man require brighter light? It is unlikely that it is evolutionary, because the discovery of fire was not such a long time ago. A more likely explanation is that short-term adaptation occurs in the retina, retinohypothalamic tract, or SCN. Recent evidence suggests that there may be an annual rhythm in light sensitivity (author's unpublished data): if a person is exposed to 100,000 lux during a sunny summer day and is tested that night with 500 lux, he will have a relatively subsensitive response than after having been exposed to 20,000 lux during a winter day. This type of short-term adaptation may be responsible for apparent species differences. To confirm this, laboratory animals should be tested who have been reared outdoors. Either a toxic effect of light or a type of short-term adaptation has recently been demonstrated in the rat: dim light permitted melatonin production, when it was alternated with bright light (Lynch et al., 1981; Rivest et al., 1981).

Entrainment by the Light-Dark Cycle

The second effect of light on melatonin production is entrainment of its circadian secretory rhythm. (Circadian means "about a day" or about 24 hours [Halberg, 1959]). The effect of the light-dark cycle on the circadian rhythm of melatonin production is so profound that at first it was thought that the light-dark cycle directly regulated melatonin production, i.e., that melatonin production passively followed environmental conditions of light and dark. Later, it was discovered that the rhythm persisted in blinded animals or animals kept in constant darkness (Klein and Weller, 1970; Klein et al., 1971; Ralph et al., 1971). Identification of the RHT and the SCN led scientists to conclude that the light-dark cycle regulates melatonin production through its effects on an endogenous hypothalamic circadian pacemaker.

The activity rhythm and the melatonin production rhythm are "hands of the clock" and are not actually the "clock" itself (which is presumedly the SCN). When an animal is no longer entrained to the light-dark cycle, it displays a circadian rhythm of approximately, but not precisely, 24 hours; this is termed "free-running."

In constant darkness the melatonin production rhythm free-runs with the activity/rest cycle (melatonin production occurs during the activity phase of nocturnal animals or during the rest phase of diurnal animals) (Ralph et al., 1971; Pohl and Gibbs, 1978; Tamarkin et al., 1978).

The period (τ, or tau) of the free-running endogenous pacemaker is a characteristic of an individual animal (and usually of a particular species). Some animals have endogenous rhythms that are greater than 24 hours ($\tau > 24$), some less than 24 hours ($\tau < 24$) (Aschoff, 1960; Aschoff, 1969). In general, the former is more common in diurnal animals, the latter in nocturnal animals (Aschoff, 1960; Aschoff, 1969). Why is the endogenous free-running period (τ) of the biological clock not precisely 24 hours? Perhaps because a biological clock can never be as precise as a physical clock. Perhaps being slightly different than 24 hours permits greater flexibility for accommodation to the changes in the times of dawn and dusk throughout the year. Pittendrigh has reasoned that internal phase relationships would be unstable if the period of the endogenous pacemaker were to be exactly 24 hours (Pittendrigh, 1981).

When endogenous rhythms are synchronized by an environmental time cue, such as the 24-hour light-dark cycle, they are "entrained." The period of the environmental cycle must be close to the endogenous period (τ) for entrainment to occur. The "range of entrainment" is the range in the periods of the environmental cycle that are capable of synchronizing the endogenous pacemaker.

The interaction of light as a zeitgeber (literally, time giver) with the endogenous pacemaker can be described and predicted by a phase response curve (PRC). The PRC was first discovered by exposing an animal free-running in constant darkness to a short (15-minute) pulse of light. The animal responded with an advance or a delay in phase, depending on when the light pulse occurred (Pittendrigh, 1981). Because the animal is in constant darkness, the term "subjective day" is used to indicate circadian time represented by behavior typical during the day (rest for nocturnal animals, activity for diurnal animals). Subjective night indicates circadian time represented by behavior typical during the night (rest for diurnal animals, activity for nocturnal animals). (Subjective dawn and dusk, therefore, refer to the transitions between activity and rest in constant dark.)

If the light pulse occurs during the animal's subjective night, the animal will respond with either an advance or a delay; if the light pulse occurs during the animal's subjective day, phase shifts will be minimal. (This is true for both diurnal and nocturnal animals.) During subjective night, the closer the pulse is to subjective dawn, the more likely the animal will respond with an advance in phase; the closer to dusk, the more likely there will be a delay in phase. In the middle of the night there is an inflection point where an advance is separated from a delay by only a few minutes. In general, the closer the pulse is to the middle of the night the greater the magnitude of the response -- in either direction. Thus, the middle of the night is a time when the maximum phase advance is separated from the maximum phase delay by only a few minutes (Figure 2). Since the animal is in constant darkness, it continues to free-run throughout this procedure, relative to the transient phase shift due to the light pulse. PRCs have recently been described

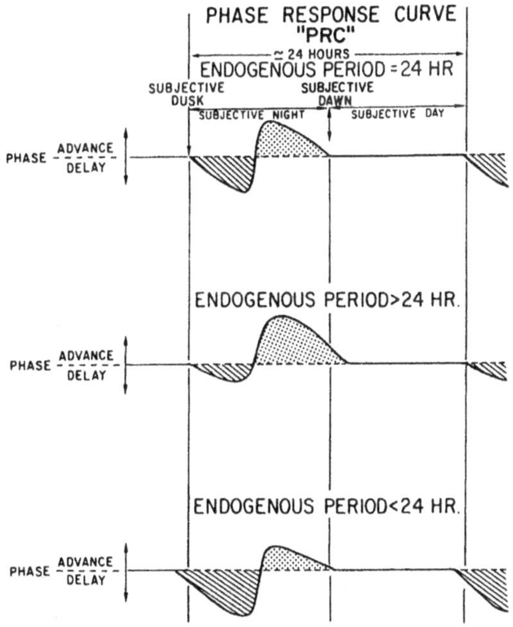

Figure 2. Schematicized phase response curves (PRCs) for animals with endogenous periods equal to (top), greater than (middle), and less than (bottom) 24 hours; these curves may help to explain how animals use light to maintain stable steady-state entrainment to the 24-hour day.

measuring chicken (Binkley et al., 1981) and rat (Illnerova and Vanecek, 1982) N-acetyltransferase activity. In the rat, a light pulse as short as one minute can shift the phase of the N-acetyltransferase activity rhythm (Vanecek and Illnerova, 1979). What happens under entrained conditions? Another experimental paradigm for evaluating the PRC is the skeleton photoperiod. This is accomplished by two 15-minute pulses of light, several hours apart. The pulse that occurs near the phase-advance portion of the PRC functions as dawn, the pulse that occurs close to the phase-delay part of the PRC functions as dusk. The relative magnitude of the phase advance compared to the phase delay produces the net effect.

In an animal with $\tau > 24$ hours, each day the light-dark cycle must advance the endogenous pacemaker one hour for stable steady-state entrainment to 24 hours. The net effect from the PRC must also include a phase advance to compensate for any phase delay that might result each day from the action of light on the phase-delay portion of the PRC (see Figure 2, middle). If $\tau < 24$ hours, then the animal must delay more than advance each day for entrainment to a 24-hour period (see Figure 2, bottom). A major factor in determining the shape of the PRC is whether an animal's endogenous period is greater than, or less than, 24 hours (see Figure 2).

In order to evaluate the entrainment effects of light in humans, we studied totally blind subjects. Previous work on blind subjects has produced suggestive, but not conclusive, data (Hollwich, 1979). This might have resulted because most studies were not done longitudinally. For example, in an initial study of melatonin patterns in ten blind subjects, subjects were studied for only one day (Lewy and Newsome, in preparation). "Phase dispersion" was found in this group. Melatonin onset occurred in these blind subjects at all times of the evening or early morning. (In sighted subjects, melatonin secretion begins regularly between 10 p.m. and 1 a.m.)

More information data was obtained, however, by studying two of these subjects longitudinally once a week for four weeks. Studied in this way these two subjects displayed markedly unusual circadian melatonin secretory rhythms (Lewy and Newsome, in preparation). One subject appeared to be entrained to a 24-hour period but was more than 120° phase-delayed from normal. The other subject was "free-running" with a rhythm of approximately 24.7 hours.

Although only two subjects were studied, these data suggest that light is important in the entrainment of the human melatonin circadian secretory rhythm. Many researchers have thought that, unlike other animals, light has little or no effect in the entrainment of human circadian rhythms. Perhaps past negative results were due to the use of insufficiently intense light. Bright artificial light or sunlight should be used in future experiments.

There are some past data that suggest that sunlight is more effective than ordinary-intensity artificial light, particularly in entraining human circadian rhythms.. Sunlight deprivation may increase the number of days necessary for reentrainment of circadian rhythms after air travel across several time zones (Klein and Wegmann, 1974). Field studies, compared to laboratory simulation, of travel across time zones have been noted to produce different results (Wever, 1980). Perhaps these differences are due to the fact that the laboratory simulations were conducted with ordinary room light. In a few subjects studed under isolated conditions, the light-dark cycle has been shown to be able to entrain the temperature rhythm, while the activity/rest rhythm continued to free-run (Aschoff and Wever, 1981). In other studies, the range of entrainment of the human temperature rhythm appears to be greater under the natural light of the Arctic summer (Lewis and Lobban, 1959) than under ordinary room light (Aschoff and Wever, 1981). Finally, the light-dark cycle appears to be capable of entraining at least part of the cortisol circadian secretory rhythm (Orth and Island, 1969).

Photoperiodism

There are a few epidemiological studies of seasonal rhythms in human physiology and disease (Aschoff, 1981). Nonetheless, human seasonal rhythms have generally not been attributed to changes in the photoperiod, probably because previous data on effects of light in humans has been unimpressive. It is hoped that recent melatonin research demonstrating suppression of secretion by sunlight and bright artificial light will stimulate more epidemiological studies of seasonal rhythms in humans, as well as stimulate more testing of possible photoperiodic responses. Again, the bright light/dim light paradigm may be useful here.

In a recent study of one individual, photoperiodic effects of light were evaluated (Lewy et al., 1982). A manic-depressive patient with a 13-year history of winter depressions and springtime remissions was exposed to bright artificial light between 6 and 9 a.m. and between 4 and 7 p.m. during the first week of December, 1980. After four days of exposure to a spring photoperiod using bright full spectrum light (Vita-Lite), he "switched" out of his typical winter depression (which typically happens to him in the spring).

MELATONIN AS A "BIOLOGICAL MARKER"

Melatonin secretion is unique in that its active phase occurs at night, completely confined between dawn and dusk. (The peaks and troughs of the circadian rhythms of cortisol and thyroid stimulating hormone secretion and core body temperature straddle the dawn and dusk transitions [Aschoff, 1979]). Melatonin secretion is also unique among circadian rhythms in that its active phase occurs at night in both diurnal or nocturnal species of animals. (Other circadian rhythms have a 180° phase-angle difference when diurnal animals are compared to nocturnal animals [Aschoff, 1979].) For these reasons, melatonin has been thought to be a highly useful marker for the phase and period of its endogenous pacemaker, and perhaps functions as part of the endogenous biological clock as well

One of the major methodological problems concerning circadian rhythm research has been that of "masking." For example, motor activity causes an increase in core body temperature; inactivity (and sleep) decrease temperature (Weitzman et al., 1979). Stress increases cortisol secretion (Czeisler et al., 1979). However, melatonin secretion appears to be free from these types of masking effects in that acute changes in the activity/rest cycle do not seem to affect the rhythm of melatonin secretion (Jimerson et al., 1977; Lynch et al., 1977); that is, sleep deprivation has no effect (Jimerson et al., 1977; Akerstedt et al., 1979). It is also likely that diet does not affect the rhythm of melatonin secretion (Arendt, 1979; Herbert and Reiter, 1981), nor does stress (Lynch et al., 1973; Illnerova, 1976; Klein and Parfitt, 1976; Parfitt and Klein, 1977; Vaughan et al., 1978[a]; Vaughan et al., 1979[b]). Melatonin secretion is not related to the stages of sleep (Vaughan et al., 1978[a]; Weinberg et al., 1979; author's unpublished data). Aside from specific drugs which can affect the amplitude of melatonin secretion, melatonin secretion is masked only by light. By reducing light intensity below threshold (in human research a fairly high intensity is permissible), this masking effect can be avoided.

The study of manic-depressive patients provides an opportunity to examine "abnormal" conditions of melatonin secretion. Many of these patients are also thought to have phase-advanced circadian rhythms in cortisol secretion, core body temperature, and other physiological variables (Wehr and Goodwin, 1981). The circadian rhythm of melatonin secretion appears to be consistent with the "phase-advance hypothesis," in that it appears to have an earlier timing in mania compared to depression and in depression compared to healthy subjects (Lewy et al., in preparation[a]).

Recently, light suppression of melatonin secretion was studied in a group of manic-depressive patients compared to a group of controls matched for age but not sex (Lewy et al., 1981). Patients were more sensitive to light than the control subjects; that is, melatonin secretion was suppressed to twice the extent in the patient group compared to the control group when exposed to 500 or 1500 lux. A group of euthymic (i.e., neither manic nor depressed) patients also appeared to be supersensitive to light, suggesting that this finding might possibly be a "trait marker," and might possibly explain the phase-advanced circadian rhythms observed in these patients (Lewy et al., in preparation[b]).

SUMMARY

Melatonin appears to be a useful biological marker, not only for adrenergic activity in the pineal, but also for the endogenous circadian pacemaker that regulates the nighttime increase in melatonin production and for the effects of light on this system. These physiologic processes are potentially important in human health and disease. The study of the effects of light on melatonin production has also increased our interest in the possible effects of light on human physiology and behavior and in seasonal rhythms in humans.

REFERENCES

Akerstedt, T., Froberg, J. E., Friberg, Y., and Wetterberg, L., 1979, Melatonin secretion, body temperature and subjective arousal during 64 hours of sleep deprivation, Psychoneuroendocrinology 4:219-225.

Arendt, J., 1979, Radioimmunoassayable melatonin: circulating patterns in man and sheep, in: The Pineal Gland of Vertebrates Including Man, Progress in Brain Research, vol. 52, J. Ariens Kappers, and P. Pevet, eds., Elsevier North-Holland Biomedical Press, New York, pp. 249-258.

Arendt, J., 1978, Melatonin assays in body fluids, J. Neural. Transm., Suppl.13:265-278.

Aschoff, J., 1960, Exogenous and endogenous components in circadian rhythms, Cold Spring Harbor Symp. Quant. Biol. 25:11-28.

Aschoff, J., 1969, Desynchronization and resynchronization of human circadian rhythms, Aerosp. Med. 40:844-849.

Aschoff, J., 1979, Circadian rhythms: general features and endocrinological aspects, in: Comprehensive Endocrinology, D. T. Krieger, ed., Raven Press, New York, pp. 1-61.

Aschoff, J., 1981, Annual rhythms in man, in: Handbook of Behavioral Neurobiology, Biological Rhythms, vol. 4, J. Aschoff, ed., Plenum Press, New York, pp. 475-487.

Aschoff, J., and Wever, R., 1981, The circadian system of man, in: Handbook of Behavioral Neurobiology, vol. 4, J. Aschoff, ed., Plenum Press, New York, pp. 311-331.

Binkley, S., Muller, G., and Hernandez, T., 1981, Circadian rhythm in pineal N-acetyltransferase activity: phase-shifting by light pulses (I), J. Neurochem. 37:798-800.

Czeisler, C. A., Moore-Ede, M. C., Regestein, Q. R., Kisch, E. S., Fang, V. S., and Erhlich, E. N., 1979, Episodi 24-hour cortisol secretory rhythm during cardiac surgery, J. Clin. Endocrinol. Metab. 42:273-283.

Deguchi, T., and Axelrod, J., 1972, Control of circadian change of serotonin N-acetyltransferase in the pineal organ by the beta-adrenergic receptor, Proc. Natl. Acad. Sci. USA 69:2547-2550.

Groos, G. A., and Mason, R., 1980, The visual properties of rat and cat suprachiasmatic neurones, J. Comp. Physiol. 135:349-356.

Halberg, F., 1959, Physiologic 24-hour periodicity; general and procedural considerations with reference to the adrenal cycle, Z. Vitamin Hormon Fermentforschung 10:225-296.

Herbert, D. C., and Reiter, R. J., 1981, Influence of protein-calorie malnutrition on the circadian rhythm of pineal melatonin in the rat, Soc. Exp. Biol. Med. 166:360-363.

Hoffmann, K., Illnerova, H., and Vanecek, J., 1981, Effect of photoperiod and of one minute light at night-time on the pineal rhythm on N-acetyltransferase activity in the Djungarian hamster Phodopus syngorus, Biol. Reprod. 24:551-556.

Hollwich, F., 1979, The Influence of Ocular Light Perception on Metabolism in Man and in Animal, Springer-Verlag, New York.

Ibuka, N., and Kawamura, H., 1975, Loss of circadian rhythm in sleep-wakefulness cycle in the rat by suprachiasmatic nucleus lesions, Brain Res. 96:76-81.

Illnerova, H., 1976, The effects of immobilization of the activity of serotonin N-acetyltransferase in the rat epiphysis, in: Catecholamines and Stress, E. Usdin, R. Kvetnansky, and I. J. Kopin, eds., Pergamon Press, New York, pp. 129-136.

Illnerova, H., and Vanecek, J., 1979, Response of rat pineal serotonin N-acetyltransferase to one min. light pulse at different night times, Brain Res. 167:431-434.

Jimerson, D. C., Lynch, H. J., Post, R. M., Wurtman, R. J., and Bunney, W. E., 1977, Urinary melatonin rhythms during sleep deprivation in depressed patients and normals, Life Sci. 20:1501-1508.

Klein, D. C., and Moore, R. Y., 1979, Pineal N-acetyltransferase and hydroxyindole-O-methyl transferase: control by the retinohypothalamic tract and the suprachiasmatic nucleus, Brain Res. 174:245-262.

Klein, D. C., and Parfitt, A., 1976, A protective role of nerve endings in stress-stimulated increase in pineal N-acetyltransferase activity, in: Catecholamines and Stress, E. Usdin, R. Kvetnansky, and I. J. Kopin, eds., Pergamon Press, New York, pp. 119-128.

Klein, D. C., Reiter, R. J., and Weller, J. L., 1971, Pineal N-acetyltransferase activity in blinded and anosmic rats, Endocrinology 89:1020-1023.

Klein, D. C., and Weller, J., 1970, Indole metabolism in the pineal gland: a circadian rhythm in N-acetyltransferase, Science 169:1093-1095.

Klein, K. E., and Wegmann, H.-M., 1974, The resynchronization of psychomotor performance circadian rhythm after transmeridian

flights as a result of flight direction and mode of activity, in: Chronobiology, L. E. Scheving, F. Halberg, and J. E. Pauly, eds., George Thieme Purl, Stuttgart, pp. 564-570.
Lewis, P. R., and Lobban, M. C., 1957, Dissociation of diurnal rhythms in human subjects living on abnormal time routines, Q. J. Exp. Physiol. 42:371-376.
Lewy, A. J., Wehr, T. A., Goodwin, F. K., Newsome, D. A., and Markey, S. P., 1980, Light suppresses melatonin secretion in humans, Science 210:1267-1269.
Lewy, A. J., Wehr, T. A., Goodwin, F. K., Newsome, D. A., and Rosenthal, N.E., 1981, Manic-depressive patients may be supersensitive to light, Lancet 1:383-384.
Lewy, A. J., Kern, H., Rosenthal, N. E., Wehr, T. A., Newsome, D. A., Gillin, J. C., and Goodwin, F. K., 1982, Bright artificial light treatment of a manic-depressive patient with a sasonal mood cycle, Amer. J. Psych. 139:1496-1498.
Lewy, A. J., and Neuwelt, E. A., Disappearance of plasma melatonin after surgical removal of a neoplastic pineal gland, in preparation.[a]
Lewy, A. J., and Newsome, D. A., Unusual melatonin secretion in some blind subjects, in preparation.
Lewy, A. J., Wehr, T. A., Gold, P. W., and Goodwin, F. K., Melatonin secretion in manic-depressive patients, in preparation.[b]
Lydic, R. Schoene, W. C., Czeisler, C. A., and Moore-Ede, M. C., 1980, Suprachiasmatic region of the human hypothalamus: homolog to the primate circadian pacemaker? Sleep 2:355-361.
Lynch, H. J., Eng, J. P., and Wurtman, R. J., 1973, Control of pineal indole biosynthesis by changes in sympathetic tone caused by factors other than environmental lighting, Proc. Natl. Acad. Sci. USA 70:1704-1708.
Lynch, H. J., Jimerson, D. C., Ozaki, Y., Post, R. M., Bunney, W. E., and Wurtman, R. J., 1977, Entrainment of rhythmic melatonin secretion in man to a 12-hour phase shift in the light dark cycle, Life Sci. 23:1557-1564.
Lynch, H. J., Rivest, R. W., Ronsheim, P. M., and Wurtman, R. J., 1981, Light intensity and the control of melatonin secretion in rats, Neuroendocrinology 33:181-185.
Minneman, K. P., Lynch, H., and Wurtman, R. J., 1974, Relationship between environmental light intensity and retina-mediated suppression of rat pineal serotonin-N-acetyltransferase, Life Sci. 15:1791-1796.
Moore, R. Y., and Eichler, V. B., 1972, Loss of circadian adrenal corticosterone rhythm following suprachiasmatic lesions in the rat, Brain Res. 42:201-206.
Moore, R. Y., and Klein, D. C., 1974, Visual pathways and the central neural control of a circadian rhythm in pineal serotonin N-acetyltransferase activity, Brain Res. 71:17-33.
Moore, R. Y., 1979, The anatomy of central neural mechanisms regulating endocrine rhythms, in: Comprehensive Endocrinology, D. T. Krieger, ed., Raven Press, New York, pp. 63-87.
Nishino, E., Koizumi, K., and Brooks, C. M., 1976, The role of the

suprachiasmatic nuclei of the hypothalamus in the production of circadian rhythms, Brain Res. 112:45-59.
Orth, D. P., and Island, D. P., 1969, Light synchronization of the circadian rhythm in plasma cortisol (17-OHCS) concentration in man, J. Clin. Endocrinol. Metab. 29:479-486.
Parfitt, A., and Klein, D. C., 1977, Increase caused by desmethylimipramine in the production of (^{3}H) melatonin by isolated pineal glands, Biochem. Pharmacol. 26:904-905.
Perlow, M. J., Reppert, S. M., Tamarkin, L., Wyatt, R. J., and Klein, D. C., 1980, Photic regulation of the melatonin rhythm: monkey and man are not the same, Brain Res. 182:211-216.
Perlow, M. J., Reppert, S. M., Boyar, R. M., and Klein, D. C., 1980, Daily rhythms in cortisol and melatonin in primate cerebrospinal fluid. Effects of constant light and dark, Neuroendocrinology 32:193-196.
Pittendrigh, C. S., 1981, Circadian systems: entrainment, in: Handbook of Behavioral Neurobiology Biological Rhythms, vol. 4, J. Aschoff, ed., Plenum Press, New York, pp. 95-124.
Pohl, C. R., and Gibbs, F. P., 1978, Circadian rhythms in blinded rats: correlation between pineal activity cycles, Am. J. Physiol. 234:110-114.
Ralph, C. L., Hull, D., Lynch, H. J., and Hedlund, L., 1971, A melatonin rhythm persists in rat pineals in darkness, Endocrinology 89:1361-1366.
Reppert, S. M., Perlow, M. J., Tamarkin, L., Klein, D. C., 1979, A diurnal melatonin rhythm in primate cerebrospinal fluid, Endocrinology 104:295-301.
Reppert, S. M., and Klein, D. C., 1980, Mammalian pineal gland: basic and clinical aspects, in: The Endocrine Functions of the Brain, M. Motta, ed., Raven Press, New York, pp. 327-371.
Richter, C. P., 1965, Biological Clocks in Medicine and Psychiatry, Charles C. Thomas, Publisher, Springfield, Ill.
Richter, C. P., 1967, Sleep and activity: their relation to the 24-hour clock, Proc. Assoc. Res. Ner. Ment. Dis. 45:8-27.
Rivest, R. W., Lynch, H. J., Ronsheim, P. M., and Wurtman, R. J., 1981, Effect of light intensity on regulation of melatonin secretion and drinking behavior in the albino rat, in: Melatonin -- Current Status and Perspectives, Advances in the Biosciences, vol. 29, N. Birau, and W. Schloot, eds., Pergamon Press, Oxford, pp. 119-121.
Rollag, M. D., and Niswender, G. D. 1976, Radioimmunoassay of serum concentrations of melatonin in sheep exposed to different light regimens, Endocrinology 98:482-489.
Rusak, B., 1977, The role of the suprachiasmatic nuclei in the generation of circadian rhythm in the golden hamster, Mesocricetus auratus, J. Comp. Physiol. 118:145-164.
Schwartz, W. J., and Gainer, H., 1977, Suprachiasmatic nucleus: use of ^{14}C-labeled deoxyglucose uptake as a functional marker, Science 197:1089-1091.

Schwartz, W. J., Davidsen, L. C., and Smith, C. B., 1980, In vivo metabolic activity of a putative circadian oscillator, the rat suprachiasmatic nucleus, J. Comp. Neurol. 189:157-167.

Stephan, F. K., and Zucker, I., 1972, Circadian rhythms in drinking behavior and locomotor activity of rats are eliminated by hypothalamic lesions, Proc. Natl. Acad. Sci. USA 69:1583-1586.

Tamarkin, L, Reppert, S., Anderson, A., Pratt, B., Goldman, B. D., and Klein, D. C., 1978, Regulation of pineal melatonin in the Syrian hamster, Pharmacologist 20:151, 1978.

Thorington, L., 1980, Actinic effects of light and biological implications, Photochem. Photobiol. 32:117-129.

Vanecek, J., and Illnerova, H., 1979, Changes of a rhythm in rat pineal serotonin N-acetyltransferase following a one-minute light pulse at night, in: The Pineal Gland of Vertebrates Including Man, Progress in Brain Research, J. Ariens Kappers, and P. Pevet, eds., Elsevier North-Holland Biomedical Press, New York, pp. 245-248.

Vaughan, G. M., Allen, J. P. Tullis, U., Silar-Khodr, T. J., De La Pena, A., and Sackman, J. W., 1978,[a] Overnight plasma profiles of melatonin and certain adenohypophysial hormones in men, J. Clin. Endocrinol. Metab. 47:566-571.

Vaughan, G. M., McDonald, S. D., Jordan, R. M., Allen, J. P., Bohm Falk, A. L., Abou-Samro, M., and Story, J. L., 1978,[b] Melatonin concentration in human blood and cerebrospinal fluid, J. Clin. Endocrinol. Metab. 47:220-223.

Vaughan, G. M., Bell, R., and De La Pena, A., 1979,[a] Nocturnal Plasma melatonin in humans: episodic pattern and influence of light, Neuroscience Letters, 14:81-84.

Vaughan, G. M., McDonald, S. D., Bell, R., and Stevens, E. A., 1979,[b] Melatonin, pituitary function and stress in humans, Psychoneuroendocrinology 4:351-362.

Wehr, T. A., and Goodwin, F. K., 1981, Biological rhythms and psychiatry, in: American Handbook of Psychiatry, vol. 7, 2nd ed., S. Arieti, and H. K. H. Brodie, eds., Basic Books, New York, pp. 46-74.

Weinberg, J., D'Eletto, R. D., Weitzman, E. D., Erhlich, S., and Hollander, C. S., 1979, Circulatory melatonin in man: episodic secretion throughout the dark-light cycle, J. Clin. Endocrinol. Metab. 48:114-118.

Weitzman, E. D., Weinberg, U., D'Eletto, R., Lynch, H. J., Wurtman, R. J., Czeisler, C. A., and Erlich, S., 1978, Studies of the 24 hour rhythm of melatonin in man, J. Neural Transm., Suppl. 13:325-337.

Weitzman, E. D., Czeisler, C. A., and Moore-Ede, M. C., 1979, Sleep-wake, neuroendocrine and body temperature circadian rhythms under entrained and non-entrained (free-running) conditions in man, in: Biological Rhythms and Their Central Mechanisms, M. Suda, O. Hayaishi, and H. Nakagawa, eds., Elsevier North-Holland, New York, pp. 199-227.

Wetterberg, L., 1978, Melatonin in humans: physiological and clinical studies, J. Neural. Transm. Suppl. 13:289-310.
Wever, R. A., 1980, Phase shifts of human circadian rhythms due to shifts of artificial Zeitgebers, Chronobiologia 7:303.

THE SECRETION AND EFFECTS OF MELATONIN IN HUMANS

Richard J. Wurtman, Franz Waldhauser** and Harris R. Lieberman*

Laboratory of Neuroendocrine Regulation
Department of Nutrition and Food Science
and *Department of Psychology
Massachusetts Institute of Technology
Cambridge, MA 02139

** Present address: Univ. - Kinderklinik
Waehringer Guertel 74-76
A-1090 Wien 9 Austria

I. INTRODUCTION

Although it has been well established for two decades that the human pineal gland produces and secretes melatonin, and although investigators began describing effects of exogenous melatonin on humans (principally on sleepiness and subjective vigor) soon after its discovery, relatively little information has been available until recently about the normal range of melatonin concentrations in human plasma and other body fluids, and even less about the effects of exogenous melatonin or its roles in human physiological processes.

In part this state of affairs can be ascribed to the lack of sensitive, specific, and reasonably convenient assays for melatonin until the mid-1970's; in part it probably also reflects the fact that, until recently, the cost of synthetic melatonin was so great as to preclude its administration in large doses to more than relatively few subjects. Good assays for melatonin do now exist based on radioimmunoassay and GCMS, and these are in routine use within a growing number of laboratories; moreover the cost of synthetic melatonin has decreased considerably. Hence it appears likely that information about melatonin's levels in, and effects on, humans will accumulate rapidly during the next few years.

Table 1. Blood Melatonin Values in Adult Humans

Sample	Daytime X ± SE pg/ml	Nighttime X ± SE pg/ml	Subject No.	Subject Sex	Reference	
BIOASSAY:						
PLASMA	ND** - 12‡	49.6 ± 10.5	5	M	Pelham et al.	1973
PLASMA	162 ± 41	258 ± 44	3 2	M F	Arendt et al.	1975
PLASMA	ND**	20 - 150‡	16	M	Vaughan et al.	1976
RADIOIMMUNOASSAY:						
PLASMA	132 ± 10	188 ± 15	14	M	Arendt et al.	1975
SERUM	47 - 481‡	---	5	F	Wetterberg et al.	1976
SERUM	63 ± 22† 100 ± 45†	--- ---	8 7	M F	Arendt et al.	1977a
PLASMA	<20	50 - 100‡	2	M	Kennaway et al.	1977
SERUM	20 ± 3	78 ± 13	5	M	Smith et al.	1977
SERUM	~ 12-14 spring & autumn ~ 20-24 winter & summer		5	M	Arendt et al.	1977b
PLASMA	14.6 ± 0.9	49.1 ± 3.8	47	M	Arendt & Wilkinson	1978
	13.9 ± 0.5	66.1 ± 5.3	50	F		
PLASMA	23 ± 7	97 ± 33	2	M	Lynch et al.	1978
PLASMA	32 ± 8	179 ± 26	5 Mult. sample	M	Vaughan et al.	1978
SERUM	19 ± 7†	74 ± 26†	7		Smith et al.	1979
PLASMA	---	12 PM 36±14 4 AM 51±11	7	M	Sizonenko et al.	1979
PLASMA	50 - 180***	150 - 450***	5	M	Weinberg et al.	1979
SERUM	11-21‡	---	18		Silman et al.	1979
SERUM	24±1.6	79±8.6	8 15	M F	Waldhauser et al.	1981

Table 1. (continued)

Sample	Daytime X ± SE pg/ml	Nighttime X ± SE pg/ml	Subject No.	Sex	Reference	
GCMS:						
PLASMA	25 - 40‡	125 - 440‡	3		Wilson et al.	1977
PLASMA	10 - 55‡		5			
PLASMA	1.5 - 4.9‡	19.4 - 42.6‡	2	M	Lewy & Markey	1978
			2	F		

** ND = nondetectable
† SD
*** range of mean values from 5 individuals obtained by sampling every 15 min over 24 h.
‡ range
PL plasma
SE serum

This article summarizes our present knowledge about melatonin levels in body fluids from normal humans and from people with particular diseases. It also describes the little information available about melatonin's physiologic roles and the consequences of its administration to humans.

II. MELATONIN LEVELS IN BODY FLUIDS

A. Presence and Levels in Blood, Urine and CSF

Melatonin levels apparently are similar in plasma and serum (Wetterberg, 1978), and values obtained using RIA are in good agreement with those based on bioassays or GC-MS, i.e., up to about 40 pg/ml during daytime and 40-150 pg/ml during nighttime (Table I). No sex difference was noted in plasma melatonin levels (Arendt et al., 1975). Measurements of melatonin levels in human urine have been difficult to make because of the abundance of unknown cross-reacting substances. However, available data (Table II) do indicate that it is present, that it exhibits a substantial daily rhythm, and that a very good correlation exists between blood and urinary melatonin levels at times when these levels are changing (Lynch et al., 1978b; Lang et al., 1981). At present, CSF melatonin levels have been measured mainly in patients with particular diseases, and only in samples obtained

Table 2. Urine Melatonin Values in Adult Humans

	Daytime X ± SE ng/4 h	Nighttime X ± SE ng/4 h	Subject No.	Subject Sex	Reference	
BIOASSAY:						
	ND - 1.3‡	0.55 - 13.4‡	2	M	Lynch et al.	1975b
			4	F		
RADIOIMMUNOASSAY:						
	1.3 ± 0.1	4.3 ± 0.4	4	M	Lynch et al.	1975a
	---	38.2 ± 7.8*	3	M	Wetterberg	1979
	---	31.7 ± 9.5*	3	F		
	10.1 ± 1.7	23.2 ± 2.4	4	M	Lang et al.	1981
			10	F		
	67.75 ± 29§		12		Lemaitre et al.	1981a

* on basis of first morning urine sample

§ on basis of a 24-h urine collection

‡ range

Table 3. Melatonin Content in Human CSF

(All samples except those indicated otherwise were obtained by lumbar puncture)

	Daytime X ± SE pg/ml	Nighttime X ± SE pg/ml	Subject No.	Subject Sex	Reference	
GC/MS:						
	55 - 80[‡]	---	2		Wilson et al.	1977
RADIOIMMUNOASSAY:						
	59 ± 33[†]	---	8	M	Arendt et al.	1977a
	57 ± 28[†]	---	7	F		
	<10 - 30[‡]	---	10	M	Vaughan et al.	1978b
	<10 - 23[‡]	---	5	F		
	1.7 - 68.7[‡]	---	12	M	Brown et al.	1979
	2.2 - 25.2[‡]	---	7	F		
	<2.5 - 41.8[‡,]	---	15	M		19
	1.6 - 35.7[‡,‡]	---	11	F		
	39 ± 15[†]	---	3		Smith et al.	1979
	64 ± 11.5[**]	---	4		Tan and Khoo	1981
	41 ± 4.6[‡‡]		4			

† SD

‡ range

** CSF obtained by lumbar puncture

‡ CSF obtained by cisternal puncture

‡‡ CSF obtained by ventricular callicles?

during the daytime (Table III). These levels are apparently similar to (Vaughan et al., 1978b) or slightly lower than (Arendt et al., 1977a; Tan and Khoo, 1981) those in blood. In an early study (Smith et al., 1976a) based on samples from children with leukemia, the opposite conclusion was drawn: serum melatonin levels averaged 14 pg/ml while melatonin concentrations in CSF samples drawn simultaneously averaged 98 pg/ml.

B. Rhythms in Melatonin Concentrations

1. Day-night rhythm: Day-night rhythms in melatonin levels of human blood (Table I) and urine (Table II) have been demonstrated by all investigators who have looked for them. Highest blood levels have been noted in samples taken between midnight and 2 AM (Arendt et al., 1977a; Smith et al., 1977), and lowest levels between noon and 2 PM (Vaughan et al., 1978a). Expressed in cosinor terminology, the acrophase of the melatonin rhythm in nine healthy subjects was −357° from mid-sleep, and its amplitude was 65% from the mesor (Scheving et al., 1981). Similar results were reported by Birkeland et al. (1980) and Fevre-Montange et al. (1981). The physiologic basis of the day-night rhythm in melatonin secretion was originally thought to be an endogenous oscillator, whose effects could be amplified or diminished by light and darkness but didn't completely depend on the lighting environment. Consistent with this hypothesis, Vaughan et al. (1976) found that a plasma melatonin rhythm persisted among subjects kept under continuous illumination (12 ft-candles) for 60 h. Jimerson et al. (1977) and Akerstedt et al. (1979) also described persistent urinary melatonin rhythms among subjects exposed to continuous illumination. Furthermore, some investigators have found a clear dissociation between the time of onset of darkness and the time of melatonin secretion: Arendt et al. (1977a) and others (Vaughan et al., 1978a) observed that melatonin secretion can precede dark onset, while Weinberg et al. (1979) observed a considerable delay between dark onset and the nocturnal rise in melatonin secretion. If light or darkness *per se* did not trigger the day-night changes in melatonin secretion, what then did? Apparently not sleep stage (Vaughan et al., 1978a), even though, in one very preliminary study (Sizonenko et al., 1979), a correlation was noted between the nocturnal plasma melatonin peak and the number of preceding REM sleep periods.

More recent observations have again focused on light as the phase-setter for the melatonin rhythm. Lynch et al. (1978b) found that an artificial phase shift of 180° in the light/dark regime did cause a corresponding shift in plasma and urinary melatonin rhythms after 5-7 days. A similar delay was observed after a natural phase shift, i.e., a transatlantic flight (Fevre-Montange et al., 1981; Wetterberg, 1978). Blindness was also found to affect melatonin secretion from the human pineal: among two of four blind subjects studied by Lynch et al. (1975a), highest urinary melatonin levels were found in the morning or at noon. Serum melatonin contents in four blind subjects sampled at nighttime (11 PM) and during the day (2 PM) were higher at daytime (Smith & O'Hara, 1981). This may indicate a free-running diurnal melatonin cycle in blind humans.

Most compelling, Lewy et al. (1980a) found that light of sufficient intensity could cause serum melatonin levels to fall rapidly to daytime levels, among normal volunteers awakened at 2 AM and then exposed to light for 2 hours. If subjects were then returned to darkness, melatonin levels again rose within 30 minutes. A light intensity of 500 lux had no effect on melatonin levels, while light intensity of 2500 lux was extremely effective (2500 lux is about 3 times the intensity normally present in an artificially lit office, but only 3-5% of the intensity present out of doors on a sunny day.)

Hence the only difference between humans and the laboratory rodents usually used as experimental animals, in terms of pineal responses, may be their different sensitivities to light. In retrospect it should not be surprising that humans -- who normally work outside in daylight -- are less sensitive to light than albino rats -- which normally spend daylight hours sleeping in a completely dark burrow.

2) <u>Menstrual rhythms</u>: Wetterberg et al. (1976) measured morning serum melatonin levels in five healthy women at 2-3 day intervals. Melatonin was lowest at the time of ovulation, increased during the following days, and -- after a small decrease during the late follicular phase -- exhibited a second peak at the time of menstruation. A 4.5-fold increase was found between ovulation and menstruation. Later, using a more specific antibody, Arendt (1978a) confirmed this relationship.

3) <u>Annual rhythms</u>: By estimating morning (8 AM) serum melatonin levels monthly for one year, Arendt et al. (1977b) found a significant seasonal variation in melatonin among five men, with peak values in winter and summer and lower values in spring and autumn. In a later study (Arendt et al., 1979) the same pattern was observed among females, using either midday or midnight melatonin levels. The latter finding was taken as evidence that the annual rhythm is real, and not simply due to a shift in the day-night rhythm. Wetterberg et al. (1981b) confirmed this interpretation by showing no seasonal alteration in the acrophase of the 24-hour urinary melatonin excretion among 13 normal volunteers.

4) Pulsatile secretion: Three groups of investigators (Vaughan et al., 1978a; Weitzman et al., 1978; Mullen et al., 1981) have sampled melatonin levels in blood with sufficient frequency (i.e., at intervals of 30 minutes or less) to look for - and affirm - pulsatile secretion of the hormone (like LH, prolactin, and ACTH). All of the groups observed evidence for such pulses and concluded that this pattern is superimposed on the day-night melatonin secretory rhythm. The frequency of pulses appears to be about one per hour (Weinberg et al., 1979), and their amplitude sufficient to raise plasma levels by as much as 200%. In addition to these hourly pulses, melatonin may also be secreted in even more frequent pulses: when Vaughan et al. (1979b, 1979a) measured melatonin levels in blood samples at intervals of 2.5 min, they observed nighttime secretory bursts with frequencies of 8.25 and 12.14 minutes, and amplitudes of 35 and 33% (nadir to peak). These amplitudes are not much greater than melatonin's usual interassay variance; thus the bursts, if real, aren't very robust. Arendt et al., 1982, measuring daytime serum levels every 60 minutes, did not find evidence for pulsatile secretion. This is in agreement with preliminary observations by the present authors who could not find evidence for melatonin pulses in 7 male volunteers sampled every 30 minutes between 7 AM and 11 AM.

C. Changes With Age

That age in general, and the stage of puberty in particular, are factors which influence pineal activity has been suggested since the first publications concerning the pineal (Marburg, 1909; Kitay & Altschule, 1954).

Several studies have addressed the question of changes in melatonin levels with puberty. Arendt (1978a) estimated serum midday and midnight levels in six prepubertal and three pubertal children: In all nine, noontime levels were less than 14 pg; in the postpubertal children midnight levels averaged 128 pg/ml, and in the prepubertal subjects, 162 pg/ml. Thus, in this small sample no significant differences in melatonin levels were noted. Lenko et al. (1982) recorded daytime plasma and 24-hour urinary melatonin values among children aged 7 years and more. Plasma levels (n=116) did not change with pubertal stage. Urine levels (n=43) averaged 42.1 ng/24 h/m^2 for prepubertal and 30.5 ng/24 h/m^2 for pubertal children (Tanner Stage III - V). These did not differ significantly. Similar results were obtained by Willig et al. (1979), who measured daytime and nighttime plasma melatonin levels.

Opposite results were recorded by other groups. Silman et al. (1979) measured daytime serum melatonin levels by GC/MS in a school class of 51 healthy boys and girls aged 11 1/2 -14 years. Melatonin concentrations in girls ranged from less than 5 to 280 pg/ml, and didn't change with puberty. However, levels in boys ranged from less than 10 to 2,300 pg/ml, and were significantly higher during Tanner Stage I than during other stages. Waldhauser et al. (1981) performed a cross-sectional study throughout childhood by measuring daytime and nighttime melatonin levels in 101 children and 25 adults. They found highest concentrations in the group aged 1-3 years and a tendency to decreasing values in both daytime and nighttime during childhood. Melatonin levels were lower in infancy than at ages 2-6. Lemaitre et al. (1981b), who estimated morning blood samples in 26 male infants, found undetectable or low values during the first 3 months of life and thereafter values up to 1508 pg/ml.

These data suggest that melatonin levels exhibit an inverse pattern to that of LH, which is high during the first months of life and then suppressed until the onset of puberty (Winter et al., 1975).

The view that melatonin production is elevated in children is supported by observations on urinary melatonin and 6 OH-melatonin, its main metabolite. Among 58 male subjects, Lemaitre et al. (1981a) found significantly higher melatonin excretions among children aged 4-15 years than in adults. Tetsuo et al. (1981), measuring 6 OH-melatonin by GC/MS in 101 children and 20 adults, observed that the total quantities of 6 OH-melatonin excreted per day were similar in children and adults. Unfortunately, results were not analyzed with regard to the body weights of the children. However, if small children excrete the same amount of 6 OH-melatonin as adults, it can be assumed that relative excretions per bodyweight are greater in the children.

The available results are conflicting. However, it should be noted that in all studies except one involving a large subject population studied throughout childhood, decreases of serum and urinary melatonin were observed during childhood. Even when different RIA and GC/MS methods were used a wide range of serum melatonin values was noted, from undetectable levels up to more than 2000 pg/ml.

The extent to which increasing pineal calcification during aging affects melatonin secretion is uncertain. Increasing age was not associated with changes in the activity of various pineal enzymes (Wurtman et al., 1964c), but was associated with reduced melatonin concentration in CSF (Brown et al., 1979), and blood (Touitou et al., 1981).

D. Effects of Various Treatments

1) Stimulation: It would be very useful to identify an innocuous treatment capable of stimulating melatonin secretion in humans. Such a treatment could serve as the basis of a pineal function test. Since the pineal is controlled by norepinephrine released from its sympathetic nerves and since administation of L-Dopa, a norepinephrine precursor, increases melatonin synthesis in rats (Deguchi & Axelrod, 1972; Lynch et al., 1973) L-Dopa seemed a reasonable candidate as a pineal stimulator. Unfortunately, L-Dopa was without effect on melatonin secretion in humans (Arendt, 1978a; Wetterberg, 1978; Moore et al., 1979; Vaughan et al., 1979c). Infusions of β-receptor agonists like isoproterenol (Vaughan et al., 1976) or orciprenaline (Moore et al., 1979) also failed to increase melatonin secretion. One possible explanation for this lack of action, i.e., an inhibitory effect of daylight or a diminished number of pineal-β-receptors during daytime, was excluded by the observation that L-Dopa and orciprenaline given in the evening also failed to increase blood melatonin levels (Moore et al., 1979).

Psychosocial stress (Akerstedt et al., 1979) and stress due to insulin administration (Vaughan et al., 1979c; Wetterberg, 1979), pneumoencephalography (Vaughan et al., 1978b), electroconvulsive therapy (Wetterberg, 1978) or sprinting (Vaughan et al., 1979c), also failed to elevate blood melatonin. However, Carr et al. (1981) did find a 100-200% increase in blood levels among seven healthy women after 1 h of exercise on a bicycle ergometer during daytime. If this preliminary report can be confirmed it might provide the basis for a pineal function test.

In very preliminary studies, such other neuroactive substances as scopolamine (Vaughan et al., 1976), amphetamine, TRH, LHRH and TRH, desaminocys-D-arg-vasopressin (Wetterberg, 1978; Wetterberg, 1979) also failed to affect melatonin secretion.

2) Suppression: As described above, bright light (2500 lux) can suppress nocturnal melatonin secretion in normal humans (Lewy et al., 1980a). The nocturnal release of melatonin can also be prevented by administration of propanolol, a β-receptor blocker (Vaughan et al., 1976; Wetterberg 1978; Moore et al., 1979; Lewy et al., 1981). This effect can be produced by as little as 40 mg propanolol, given p.o. at 8 PM (Moore et al., 1979). It is not clear why β-receptor blockers prevent nocturnal melatonin secretion while β-receptor agonists fail to elicit its secretion in the daytime. Perhaps a particular β-1 or β-2 receptor is involved.

Preliminary reports suggest that clonidine (Lewy et al., 1981) and dexamethasone (Wetterberg, 1979) can inhibit melatonin secretion.

III. MELATONIN LEVELS ASSOCIATED WITH DISEASE STATES

In the last few years much speculation has been offered suggesting relationships between melatonin secretion and various disease states. Unfortunately, very little supporting evidence seems to be available.

A. Neurological Diseases

To our knowledge, melatonin secretory patterns have been described in seven patients with pineal tumors, unfortunately most with inadequate clinical and histological data (Barber et al., 1978; Arendt, 1978b; Tapp, 1978; Kennaway et al., 1979; Vaughan et al., 1979c). Kennaway et al. (1979) described two males, one with a histologically unidentified pineal tumor, the other with a pineoblastoma. Both had undetectable serum melatonin levels on several occasions. Arendt (1978b) described two males, one with low and the other with unusually high melatonin levels. Interestingly, Barber et al. (1978) found elevated serum melatonin levels in one patient whose pinealoma showed much lower HIOMT activity than pineal tissue itself. A similar case with a pineal germinoma associated with high serum melatonin levels was reported by Tapp (1978). Thus it seems possible that high or low serum melatonin levels can occur in association with pineal tumors, and perhaps circumstances are more complex than Marburg (1909) and Kitay & Altschule (1954) supposed in suggesting that parenchymatous pineal tumors produce an excess of an antigonadal hormone, while pineal-destroying tumors cause a lack of the hormone.

Kneisley et al. (1978) demonstrated abolition of the day-night urinary melatonin rhythm among six patients with clinical evidence of transected cervical spinal cords. This constitutes evidence that similar neural pathways mediate the central control of pineal function in humans and rats.

B. Psychiatric Diseases

Since depression and perhaps other psychiatric diseases can be associated with alterations in neuroendocrine rhythmicity (Bunney et al., 1977) and since these disorders often have a seasonal pattern of exacerbations and remissions, it is not surprising that psychiatrists have become interested in examining disease-related rhythms in melatonin secretion.

Wetterberg et al. (1979) described a single depressed woman in whom nocturnal melatonin levels were lower and cortisol levels higher during depression than during remission. Later similar patterns were found in four out of twelve depressive patients (Wetterberg et al., 1981a), and in a number of schizophrenics

(Ferrier et al., 1982). Mendlewicz et al. (1979) described diminished circadian melatonin rhythms in four depressed patients, during both the depressive and recovery phases.

Lewy et al. (1979) described significantly higher melatonin levels in four manic patients than in healthy controls. However, possible effects of medication were apparently not ruled out and may have been involved: chlorpromazine has been reported to elevate melatonin levels in rats (Ozaki et al., 1976) and humans (Smith et al., 1979) (by impairing its hepatic metabolism), and certain antidepressive drugs have similar effects in rats (Wirz-Justice et al., 1980).

Very recently Lewy et al. (1981) proposed that depressive patients may be considerably more sensitive than normal controls to the photic inhibition of nocturnal melatonin secretion. Failure of other investigators to take light intensities into consideration may explain why, for example, Jimerson et al. (1977) found no difference in melatonin excretion between controls and depressive patients. At this point, perhaps the best lead relating depression to melatonin and pineal function seems to be the possibility of altered (i.e., enhanced) light sensitivity. It will be important to determine how general this alteration is, and whether or not its presence is correlated with such other neuroendocrine disturbances as abnormal dexamethasone-suppression tests.

C. Carcinoma

In several animal models, melatonin administration reportedly diminished the growth of tumors and their metastases (Lapin, 1979; Karmali et al., 1978). Cohen et al. (1978b) proposed that diminished melatonin production might be a factor in the etiology of breast cancer. And the same group (Tamarkin et al., 1982) subsequently reported lower nocturnal melatonin levels among 10 women with estrogen-receptor-positive breast cancers than among control subjects, or women whose cancers lacked estrogen receptors. Tapp et al. (1980) found no differences in serum melatonin levels among patients with benign or malignant tumors.

D. Other Disorders

Birau & Schloot (1979) reported plasma melatonin levels to be elevated in spina bifida and sarcoidosis and diminished in psoriasis vulgaris, Turner's Syndrome and Klinefelter's Syndrome.

IV. EFFECTS OF EXOGENOUS MELATONIN ON HUMANS

A. Physiological Disposition

The fate of small doses of exogenous melatonin has been studied in rats (Wurtman et al., 1968). The indole enters all tissues, especially the pineal and certain endocrine organs (Wurtman et al., 1964b); it disappears rapidly from the circulation because of its metabolism within the liver (by 6-hydroxylation followed by conjugation); its major urinary metabolites are thus 6-hydroxymelatonin sulfate and glucuronide (Kopin et al., 1961).

Few if any data are available on the fate of exogenous melatonin in humans, and, to our knowledge, none of the disposition of large doses such as might be useful to produce pharmacological and behavioral effects. We have thus initiated studies on the levels of melatonin in serum before and after subjects receive single large (80 mg) oral doses of melatonin, by capsule, at 11:00 AM. Among the three young male subjects described in Fig. 1, daytime serum melatonin levels rose after one hour to levels 500 to 8000-fold higher then peak nocturnal values in untreated subjects. The hormone subsequently disappeared from the circulation with an apparent half-life of 45-60 minutes. Among individual subjects, those with highest peak serum melatonin levels also had greatest urinary secretions. In all cases less than 0.5% of the administered melatonin was present unchanged in urine samples collected during the next 18 hours. Among our three subjects receiving the same melatonin dose, we observed at least 20-fold differences in peak serum melatonin levels. The sources of this variability await characterization. It will be interesting to determine whether people with highest nocturnal serum melatonin levels do or do not also attain highest peak levels after melatonin administration.

B. Pharmacological Effects

Published reports describe approximately 100 subjects who have received melatonin (Lerner & Nordlund, 1978). Its only consistent effect (among those sought) in the doses given has been to cause mild sedation and analgesia, probably by acting on the CNS (Cramer et al., 1974; Wetterberg, 1977; Vollrath et al., 1981). No effect, or in some cases an exacerbation of clinical findings, has been observed after its administration to people with schizophrenia (Lerner & Nordlund, 1978), parkinsonism (Papavasilion et al., 1972), and depression (Carman et al., 1976). Its influence on epilipsy is not clear, but justifies further exploration (Anton-Tay et al., 1971).

Very few studies have been performed on melatonin's gonadal effects in humans. Apparently melatonin does not suppress gonado-

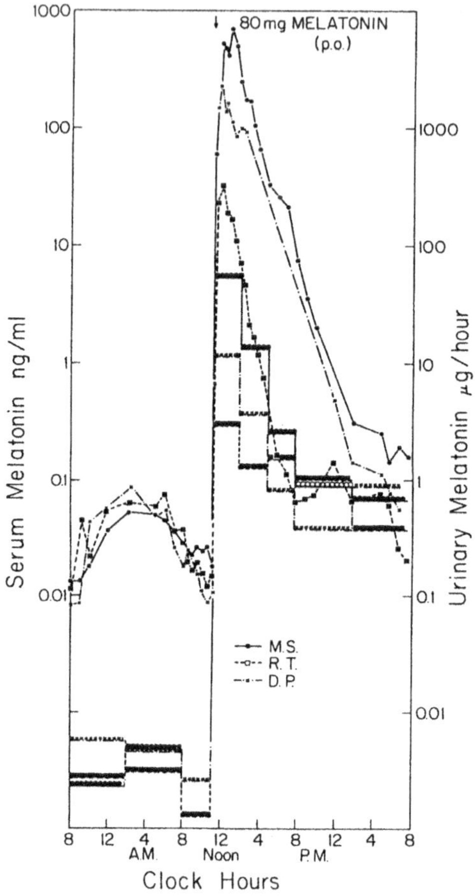

Fig. 1. Effect of exogenous melatonin on serum and urinary melatonin levels. Three young male subjects provided serum and urine samples for 15 hours before, and 22 hours after, receiving melatonin-containing capsules (80 mg) orally at 11:00 AM. Subjects were exposed to darkness between 10 PM and 7 AM. Horizontal bars represent total mealtonin present in urine samples collected during a 3 or 6 hour interval, divided by the number of hours in the collection period.

tropin secretion caused by giving LHRH (Fideloff et al., 1976; Weinberg et al., 1980); gonadotropin values in people receiving only melatonin were reported to be unaffected (Fideloff et al., 1976) or blunted (Nordlund & Lerner, 1977). Acute melatonin administration reportedly reduces the rise in plasma growth hormone after insulin (Smythe & Lazarus, 1974) or L-tryptophan (Kolu & Lammintausta, 1979) administration. Melatonin did not modify the increase in plasma growth hormone after apomorphine or L-Dopa (Kolu & Lammintausta, 1979).

No side effects of melatonin have been described except for mild headache or abdominal cramps, even when as much as 6.6 g were administered orally each day for 35 days (Lerner & Nordlund, 1978). Melatonin is thus probably a highly non-toxic substance, and deserves considerable further study in humans. In mice even 800 mg/kg failed to cause death; the LD50 could not be estimated because solubility limitations precluded testing higher doses (Barchas et al., 1967).

C. Behavioral Effects of Exogenous Melatonin

Several published studies suggest that melatonin's secretion may have significant behavioral sequence: In man, melatonin reportedly acts as a potent sleep-inducing agent. Thus, Cramer et al. (1974) found that administration of 50 mg (I.V.) of melatonin in the daytime induced sleep within 15-40 minutes, which lasted 26-60 minutes, reachng stages 3 and 4. When administered at night, 50 mg of melatonin (I.V.) decreased sleep onset time significantly, but did not alter total sleep time nor time spent in particular sleep stages. They also reported that melatonin administration may have decreased anxiety in their more anxious patients. In a more recent study, Vollrath et al. (1981) observed similar effects of a small intranasal dose of melatonin (1.7 mg) on healthy volunteers. Most of their subjects (70%) fell asleep within 40-60 minutes, and others reported feelings of mild sedation or mild tiredness. All subjects but one, who described depression, reported feelings of well-being and felt emotionally well-balanced after melatonin-induced sleep. The potent effect of such a small dose of melatonin was attributed to its intranasal route of administration. Weinberg et al. (1980) found that small amounts of slowly-infused melatonin (30 μg/min, total dose of 7.2 mg) had no readily observed behavioral effect. This study should not be considered as contradicting the Cramer et al. (1974) and Vollrath et al. (1981) studies in light of its lower doses and very slow rate of infusion.

Related investigations carried out on rats suggest that melatonin can reduce certain behavioral concomitants of emotional stress. For example, melatonin administration facilitates extincticn in a passive shock avoidance paradigm, and decreases defecation induced by this aversive situation (Datta and King, 1977). Golus et al. (1979) reported that melatonin significantly attenuated the normal fear responses of rats given a novel drinking solution. Melatonin also significantly increased open field activity in rats (Golus and King, 1981). These observations are consistent with recently-obtained biochemical evidence suggesting that melatonin itself, or one of its brain metabolites, may be an endogenous ligand for benzodiazepine receptors (Marangos et al., 1981).

Since melatonin release occurs primarily at night, its behavioral effects should be considered with respect to this fact: perhaps the pineal, by secreting melatonin, helps to synchronize certain diurnal behavioral rhythms to changes in the external lighting environment. It can be speculated that cyclically-secreted melatonin may act on the human brain to modulate behaviors that vary with 24-hour cycles. Since the nocturnal release of melatonin in humans is immediately suppresed by exposing them to bright light (Lewy et al., 1980), it is clear that environmental lighting also controls melatonin secretion in humans.

We are examining the hypothesis that melatonin may synchronize central human behavioral rhythms to diurnal changes in lighting, specifically by reducing levels of arousal, general responsiveness, and sensitivity to the environment, at night. The behavioral studies cited above, which indicate that melatonin can induce sleep and reduce anxiety, would tend to support this hypothesized role. We are thus studying the effects of exogenous melatonin on numerous indices of behavioral and perceptual responsiveness, many of which are known to vary diurnally in man (Colquhoun, 1971).

We administer oral melatonin (80 mg in 1 or 3 doses) using a double-blind placebo-controlled crossover design, to healthy males aged 18-36. The tests we have selected measure sensation, motor performance, mood and memory. To evaluate sensory sensitivity, visual critical flicker fusion threshold (CFF) is measured. CFF is known to vary diurnally (Musumeci and Misiak, 1974). Performance is assessed using a variety of different tasks. Wilkinson 4-choice simple reaction time (RT) and simple auditory RT are not difficult tasks but require sustained attention for optimal performance. The Digit Symbol Substitution Test (DSST) is more intellectually demanding, but lasts for only 90 seconds. Finally, the Grooved Pegboard Test requires sustained performance and considerable motor coordination.

Mood is assessed using 3 self-reporting questionnaires. The Profile of Mood States (POMS) yields 6 factor analytically-derived scales: tension, depression, anger, vigor, fatigue and confusion. The Visual Analog Mood Scale (VAMS) yields 3 scales: vigor, calm, and sad. The Stanford Sleepiness Scale (SSS) quantifies the progressive steps in sleepiness. Finally, memory, which also varies diurnally (Colquhoun, 1971) is measured via a category list recall test. In addition, we are monitoring core temperature and collecting blood and urine so that we can relate melatonin levels to any effects we may observe.

REFERENCES

Akerstedt, T., Froberg, J. E., Friberg, Y., and Wetterberg, L., 1979, Melatonin excretion, body temperature and subjective arousal during 64 hours of sleep deprivation, Psychoneuroendrocrinology, 4:219-225.

Anton-Tay, F., Diaz, J. L., and Fernandez-Guardiola, A., 1971, On the effect of melatonin upon human brain: Its possible therapeutic implications, Life Sci., 10:841-850.

Arendt, J., 1978a, Melatonin assays in body fluids, J. Neural. Transm., Suppl., 13:265-278.

Arendt, J., 1978b, Melatonin as a tumour marker in a patient with pineal tumor, Br. Med. J. II, 635-636.

Arendt, J., and Wilkinson, M., 1978, Melatonin, in: "Methods of Hormone Radioimmunnoassay," Second Edition, B.M. Jaffe and M.R. Behrman, eds., Academic Press, New York (1978).

Arendt, J., Paunier, L., and Sizonenko, P. C., 1975, Melatonin radioimmunoassay, J. Clin. Endocrinol. Metab., 40:347-350.

Arendt, J., Wetterberg, L., Heyden, T., Sizonenko, P. C., and Paunier, L., 1977a, Radioimmunoassay of melatonin: human serum and cerebrospinal fluid, Horm. Res., 8:65-75.

Arendt, J., Wirz-Justice, A., and Bradtke, J., 1977b, Annual rhythm of serum melatonin in man, Neurosci. Lett., 7:327-330.

Arendt, J., Hampton, S., English, J., Kwasowkis, P., and Marks, V., 1982, 24-hour profiles of melatonin, cortisol, insulin, c-peptide, and GIP following a meal and subsequent fasting. Clin. Endocrinol., 16:89-95.

Barber, S. G., Smith, J. A., and Hughes, R. C., 1978, Melatonin as a tumour marker in a patient with pineal tumour, Br. Med. J. I, 328.

Barchas, J. D., Da Costa, F., and Spector, S., 1967, Acute pharmacology of melatonin, Nature, 214:919-920.

Birau, N., and Schloot, W., 1979, Pathological nyctohumeral rhythm of melatonin secretion in psoriesis, IRCS Medical Science, 7:400.

Birkelund, A. J., and Halberg, F., 1980, Individual circadian assessment of human blood melatonin and comparative physiologic integration, Chronobiologia, 7:277.

Brown, G. M., Young, S. N., Gauthier, S., Tsui, H., and Grota, L. J., 1979, Melatonin in human cerebrospinal fluid in daytime: its origin and variation with age, Life Sci., 25:929-936.

Bunney, W. E., Wehr, T. R., Gillin, J. C., Post, R., Goodwin, F. K., and van Kammen, D. P., 1977, The switch process in manic-depressive psychosis, Ann. Intern. Med., 87:319-335.

Carman, J. S., Post, R. M., Buswell, R., and Goodwin, F. K., 1976, Negative effects of melatonin on depression, Am. J. Psychiatry, 133:1181-1186.

Carr, D. B., Reppert, S. M., Bullen, B., Skrinar, G., Beitins, I., Arnold, M., Rosenblatt, M., Martin, J. B., and McArthur, J. W., 1981, Plasma melatonin increases during exercise in women, J. Clin. Endocrinol. Metab., 53:224-225.

Cohen, M., Lippman, M., and Chabner, B., 1978b, Role of pineal gland in aetiology and treatment of breast cancer, Lancet II, 814-816.

Colquhon, W. P., 1971, Circadian variations in mental efficiency, in: "Biological Rhythms and Human Performance," W. P. Colquhon, ed., Academic Press, London.

Cramer, M., Rudolph, J., Consbruch, U., and Kendel, K., 1979, On the effects of melatonin on sleep and behavior in man, Adv. Biochem. Psychopharmacol. 11:187-191.

Datta, P. C., and King, M. G., 1977, Effects of melanocyte-stimulating hormone (MSH) and melatonin on passive avoidance and on an emotional response, Pharmacol. Biochem. Behav., 6:449-452.

Deguchi, T., and Axelrod, J., 1972, Induction and superinduction of serotonin N-acetyl-transferase by adrenergic drugs and denervation in rat pineal organ, Proc. Natl. Acad. Sci. USA, 69:2208-2211.

Ferrier, I. N., Johnstone, E. C., Crow, T. J., Arendt, J., 1982, Melatonin/cortisol ratio in psychiatric illness, Lancet i, 1070.

Fevre, M. Segel, T., Marks, J. F., and Boyar, R. M., 1978, LH and melatonin secretion patterns in pubertal boys, J. Clin. Endocrinol. Metab., 47:1383-1386.

Fevre-Montange, M., van Cauter, E., Refetoff, S., Desir, D., Tourniaire, J., and Copinschi, G., 1981, Effects of "jet lag" on hormonal patterns. II. Adaption of melatonin circadian periodicity, J. Clin. Endocrinol. Metab., 52:642-649.

Fideloff, M., Aparico, N. J., Guitelman, A., Debeljuk, L., Mancini, A., and Cramer, C., 1976, Effects of melatonin on the basal and stimulated gonadotropin levels in normal man and postmenopausal women, J. Clin. Endocrinol. Metab., 42:1014-1017.

Golus, P., McGee, R., and King, M. G., 1979, Attenuation of saccharin neophobia by melatonin, Pharmacol. Biochem. Behav., 11:367-369.

Golus, P., and King, M., 1981, The effects of melatonin on open field behavior, Pharmacol. Biochem. Behav., 15:883-885.

Jimerson, D. C., Lynch, H. J., Post, R. M., Wurtman, R. J., and Bunney, W. E., 1977, Urinary melatonin rhythms during sleep deprivation in depressed patients and normals, Life Sci., 23:1501-1508.

Karmali, R. A., Morrobin, D. F., and Ghayur, T., 1978, Role of pineal gland in aetiology and treatment of breast cancer, Lancet ii, 1002.

Kennaway, D. J., Frith, R. G., Phillipou, G., Matthews, C. D., and Seamark, R. F., 1977, A specific radioimmunoassay for melatonin in biological tissue and fluids and its validation by gas chromatography-mass spectrometry, Endocrinology, 101:119-127.

Kennaway, D. J., McCulloch, G., Matthews, C. D., and Seamark, R. F., 1979, Plasma melatonin, luteinizing hormone, follicle-stimulating hormone, prolactin and corticoids in two patients with pinealoma, J. Clin. Endocrinol. Metab., 49:144-145.

Kitay, J. I., and Altschule, M. D., 1954, "The Pineal Gland. A Review of the Physiologic Literature," Harvard University Press, Cambridge, Mass.

Kneisley, L. W., Moskowitz, M. A., and Lynch, H. J., 1978, Cervical spinal cord lesions disrupt the rhythm in human melatonin excretion, J. Neural. Transm. Suppl. 13, 311-324.

Kolu, M., and Lammintausta, R., 1979, Effect of melatonin on L-Tryptophan- and Apemorphine-stimulated GH secretion in man, J. Clin. Endocrinol. Metab., 49:70-72.

Kopin, I. J., Pierce, C. M., Axelrod, J., and Weissbach, H., 1961, The fate of melatonin in animals, J. Biol. Chem., 236:3072-3075.

Lang, U., Kornemark, M., Aubert, M. L., Paunier, L., and Sizonenko, P. C., 1981, Radioimmunological determination of urinary melatonin in humans: Correlation with plasma levels and typical 24-hour rhythmicity, J. Clin. Endocrinol. Metab., 53:645-650.

Lapin, V., 1979, Pineal influence on tumor, Prog. Brain Res., 52:523-533.

Lemaitre, B. J., Bouillie, J., and Hartmann, L., 1981a, Variations of urinary melatonin excretion in humans during the first 30 years of life, Clin. Chim. Acta, 110:77-84.

Lemaitre, B. J., Roger, M., Gendrel, D., Chaussain, J. L., and Hartmann, L., 1981b, Plasma and urinary melatonin in male infants during the first twelve months of life, in: "Proceedings of the First Joint Meeting of LWPS and ESPE," Geneva, p. 67.

Lenko, H. L., Lang, U., Aubert, M. L., Paunier, L., and Sizonenko, P. C., 1981, Hormonal changes in puberty. VII. Lack of variation of daytime plasma melatonin, J. Clin. Endocrinol. Metab., 54:1056-1058.

Lerner, A. B., and Nordlund, J. J., 1978, Melatonin: clincial pharmacology, J. Neural. Transm. Suppl. 13, 339-347.

Lewy, A. J., and Markey, S. P., 1978, Analysis of melatonin in human plasma by gas chromatography: negative chemical ionization mass spectrometry, Science, 201:741-743.

Lewy, A. J., Wehr, T. A., Gold, P. W., and Goodwin, F. K., 1979, Plasma melatonin in manic-depressive illness, in: "Catecholamines: Basic and Clinical Frontiers", Vol. II, Usdin, Kopin and Barchas, eds., Pergamon Press, NY, Oxford.

Lewy, A. J., Wehr, T. A., Goodwin, F. K., Newsome, D. A., and Markey, S. P., 1980a, Light suppresses melatonin secretion in humans, Science, 210:1267-1269.

Lewy, A. J., Wehr, T. A., Goodwin, F. K., Newsome, D. A., and Rosenthal, N. E., 1981, Manic-depressive patients may be supersensitive to light, Lancet i, 383-384.

Lewy, A. J., 1981, Regulation of human melatonin secretion by environmental light via a central pacemaker and peripheral sympathetic stimulation, in: Proceedings of the 11th Annual Meeting of Society for Neuroscience,", ---Publisher?---, Los Angeles, California.

Lynch, H. J., Ozaki, Y., Shakal, D., and Wurtman, R. J., 1975a, Melatonin excretion of man and rats: effect of time of day, sleep, pinealectomy and food consumption, Int. J. Biometerol., 19:267-279.

Lynch, H. J., Wurtman, R. J., Moskowitz, M. A., Archer, M. C., and Ho, M. H., 1975b, Daily rhythm in human melatonin, Science, 17:169-171.

Lynch, H. J., Jimerson, D. C., Ozaki, Y., Post, R. M., Bunney, W. E., and Wurtman, R. J., 1978, Entrainment of rhythmic melatonin secretion in man to a 12-hour phase shift in the light/dark cycle, Life Sci., 23:1557-1564.

Marangos, P. J., Patel, J., Hirata, F., Sondheim, D., Paul, S. M., Skolnich, P., and Goodwin, F.K., 1981, Inhibition of diazepam binding by tryptophan derivatives including melatonin and its' brain metabolite N-acetyl-5 methoxy-kynuramine, Life Sci., 29:259-267.

Marburg, O., 1909, Zur Kenntnis der normalen und pathologischen Histologie der Zirbeldruese, Arb. Neur. Inst. Wien., 12:217-279.

Mendlewicz, J., Linkowski, P., Branchey, L., Weinberg, U., Weitzman, E. D., and Branchey, M., 1979, Abnormal 24-hour pattern of melatonin secretion in depression, Lancet, ii, 1362.

Moore, D. C., Paunier, L., and Sizonenko, P. C., 1979, Effect of adrenergic stimulation and blockade on melatonin secretion in the human, Prog. Brain Res., 52:517-521.

Mullen, P. E., Linsell, G. R., Leone, R. M., Silman, R. E., Smith, I., Hooper, R. J. L., Finnie, M., and Parrot, J., 1981, Melatonin and 5-methoxytryptophol, the 24-hour pattern of secretion in man, Advances in the Biosciences, 29:337.

Musumeci, M., Misiak, H., 1979, Arcadian variation of critical flicker frequency among children, Perceptual on Motor Skills, 38:751-754.

Nordlund, J. J., and Lerner, A. B., 1977, The effects of oral melatonin on skin color and on the release of pituitary hormones, J. Clin. Endocrinol. Metab., 45:768-774.

Ozaki, Y., Lynch, H. J., and Wurtman, R. J., 1976, Melatonin in rat pineal, plasma, and urine: 24-hour rhythmicity and effect of chlorpromazine, Endocrinology, 98:1418-1424.

Papavasiliou, P. S., Cotzias, G. C., Duby, S. E., Steck, A. J., Bell, M., and Lawrence, W. H., 1972, Melatonin and Parkinsonism, J. Am. Med. Assoc., 221:88.

Pelham, R. W., Vaughan, G. M., Sandock, K. L., and Vaughan, M. K., 1973, Twenty-four-hour cycle of a melatonin-like substance in the plasma of human males, J. Clin. Endocrinol. Metab., 37:341-344.

Scheving, L. E., Wetterberg, L., Kanabrocki, E. L., and Halberg, F., 1981, Individualized circadian assessment of blood melotonin of chemically healthy men in the U.S.A., Chronobiologica, 8:188.

Silman, R. E., Leone, R. M., Hooper, R. J. L., and Preece, M. A., 1979, Melatonin, the pineal gland and huyman puberty, Nature, 282:301-303.

Sizonenko, P. C., Moore, D. C., Paunier, L., Beaumanoir, A., and Nahory, A., 1979, Melatonin secretion in relation to sleep in epileptics, Prog. Brain Res., 52:549-551.

Smith, J. A., and O'Hara, J., 1981, Altered diurnal serum melatonin rhythms in blind men, Lancet ii, 933.

Smith, J. A., Mee, T. J. X., Barnes, N. D., Thornburn, R. J., and Barnes, J.L.C., 1976a, Lancet ii, 425.

Smith, J. A., Padwick, D., Mee, T. J. X., Minneman, K. P., and Bird, E. D., 1977, Synchronous nyctohemeral rhythms in human blood melatonin and in human post-mortem pineal enzyme, Clin. Endocrinol., 6:219-225.

Smith, J. A., Barnes, J. L., and Mee, T. J., 1979, The effect of neuroleptic drugs on serum and cerebrospinal fluid melatonin concentrations in psychiatric subjects, J. Pharm. Pharmacol., 31:246-248.

Smythe, G. A., and Lazurus, L., 1974, Supression of GH secretion by melatonin and cyroheptadine, J. Clin. Invest., 54:116-121.

Tamarkin, L., Danforth, D., Lichter, A., De Moss, E, Cohen, M., Chabner, B. Lippman, M., 1982, Decreased nocturnal plasma melatonin peak in patients with estrogen receptor positive breast cancer, Science, 216:1003-1005.

Tan, C. H., Khoo, J. C. M., 1981, Melatonin concentrations in human serum, ventricular and lumbar cerebrospinal fluids as an index of the secretory pathway of the pineal gland, Hormone Res., 16:224-233.

Tetsuo, M., Poth, M., and Markey, S. P., 1981, Melatonin metabolite excretion during childhood and puberty, (in press).

Touitou, Y., Fevre, M., Lagoguey, M., Carayon, A., Bogdan, A., Reinberg, A., Beck, H., Cesselin, F., and Touitou, C., 1981, Age- and mental health related circadian rhythms of plasma levels of melatonin, prolactin, luteinizing hormone and follicle-stimulating hormone in man, J. Endocrinol, 91:467-475.

Vaughan, G. M., Pelham, R. W., Pang, S. F., Loughlin, L. L., Wilson, K. M., Sandock, K. L., Vaughan, M. K., and Koslow, S. H., 1976, Nocturnal elevation of plasma melatonin and urinary 5-hydroxyindoleacetic acid in young men: attempts at modification by brief changes in environmental lighting and sleep and by autonomic drugs, J. Clin. Endocrinol. Metab., 42:752-764.

Vaughan, G. M., Allen, J. P., Tullis, W., Siler-Khodr, T. M., de la Pena, A., and Sackman, J. W., 1978a, Overnight plasma profiles of melatonin and certain adenolypo-physeal hormones in men, J. Clin Endocrinol. Metab., 47:566-571.

Vaughan, G. M., McDonald, S. D., Jordan, R. M., Allen, J. P., Bohmfalk, G. L., Abou-Samra, M., and Story, J. L., 1978b, Melatonin concentration in human blood and cerebrospinal fluid: relationship to stress, J. Clin. Endocrinol. Metab., 47:220.

Vaughan, G. M., Allen, J. P., and de la Pena, A., 1979a, Rapid melatonin transients, Waking and Sleeping, 3:169-173.

Vaughan, G. M., Bell, R., de la Pena, A., 1979b, Nocturnal plasma melatonin in humans: episodic pattern and influence of light, Neurosci. Lett., 14:81-84.

Vaughan, G. M., McDonald, S. D., Jordan, R. M., Allen, J. P., Bell, R., and Stevens, E. A., 1979c, Melatonin, pituitary function and stress in humans, Psychoneuroendocrinology, 4:351-362.

Vollrath, L., Semm, P., and Gammel, G., 1981, Sleep induction by intranasal application of melatonin, Advances in the Biosciences, 29:327-329.

Waldhauser, F., Frisch, M., Weissenbacher, G., Zeitlhuber, U., and Toifl, K., 1981, Day-and-night-time serum melatonin in children and adults, Proceedings of the 1st Joint Meeting of Lawson, Wilkinson Pediatric Endocrine Society and the European Society for Pediatric Endocrinology, Geneva.

Weinberg, U., D'Eletto, R. D., Weitzman, E. D., Erligh, S., and Hollander, C., 1979, Circulating melatonin in man: episodic secretion throughout the light-dark cycle, J. Clin. Endocrinol. Metab., 48:114-118.

Weinberg, U., Weitzman, E. D., Fukushima, D. K., Cancel, G. F., and Rosenfeld, R.S., 1980, Melatonin does not suppress the pituitary LH-response to LMRM in man, J. Clin. Endocrinol. Metab., 51:161-162.

Weitzman, E. D., Weinberg, U., D'Eletto, R., Lynch, H., Wurtman, R. J., Czeisler, C., and Erlich, S., 1978, Studies of the 24-hour rhythm of melatonin in man, J. Neural. Transm. Suppl., 13:325-337.

Wetterberg, L., 1977, Melatonin in serum, Nature, 269:696.

Wetterberg, L., 1978, Melatonin in humans: phsiological and clinical studies, J. Neural. Transm. Suppl., 13:289-310.

Wetterberg, L., 1979, Clinical importance of melatonin, Prog. Brain Res., 52:523-533.

Wetterberg, L., Arendt, J., Paunier, L., Sizonenko, P. C., van Donselaar, W., and Heyden, T., 1976, Human serum melatonin changes during the menstrual cycle, J. Clin. Endocrinol. Metab., 42:185-188.

Wetterberg, L., Beck-Friis, J., Aperia, B., and Petterson, U., 1979, Melatonin/cortisol ratio in depression, Lancet ii, 1361.
Wetterberg, L., Aperia, B., Beck-Friis, J., Kjellman, B. F., Ljunggren, J. G., Petterson, U., Sjolin, A., Tham, A., and Unden, F., 1981a, Pineal-hypothalamic-pituitary function in patients with depressive illness, in: "Steroid Hormone Regulation of the Brain," K. Fuxe, J. A. Gustafson, and L. Wetterberg, eds., Pergamon Press, Oxford.
Wetterberg, L., Halberg, F., Haus, E., Kawasaki, T., Uezono, K., Ueno, M., and Oma, T., 1981b, Circadian rhythmic urinary melatonin excretion in four seasons by clinically healthy Japanese subjects in Kyushu, Chronobiologia, 8:188-189.
Wetterberg, L., Aperia, B., Beck-Friss, J., Kjellman, B. F., Ljunggren, J.-G., Petterson, U., Sjolin, A., Tham, A., and Unden, F., 1981, Hormone changes in effective diseases, in: "Brain Peptides and Hormones," Collu, R., Ducharme, J. C., Polis, G., and Barbeau, A., eds. Proceedings from XII Congress of the International Society of Psychoneuroendocrinology, Raven Press, New York.
Wetterberg, L., Halberg, F., Tarquini, B., Cagnoni, M., Haus, E., Griffith, K., Kawasaki, T., Wallach, L.-A., Ueno, M., Vezo, K., Matsuoka, M., Kuzel, M., Malberg, E., and Oma, T., 1979, Circadian variation in urinary melatonin in clinically healthy women in Japan and the United States of America, Experientia, 35:416-479.
Willig, R. P., and Schroder, C., 1979, Melatonin plasma levels and circadian rhythms in children, Pediatr. Res., 13:1184.
Wilson, B. W., Snedden, W., Silman, R. E., Smith, I., and Mullen, P., 1977, A gas chromatography-mass spectrometry method for the quantitative analysis of melatonin in plasma and cerebrospinal fluid, Anal. Biochem., 81:283-291.
Winter, J. S. D., Faiman, C. H., Hobson, W. C., Prasad, A. V., and Reyes, F. I., 1975, Pituitary-gonadal relations in infancy. I. Patterns of serum gonadotrophin concentrations from birth to four years of age in man and chimpanzee, J. Clin. Endocrinol. Metab., 40:545-551.
Wirz-Justice, A., Arendt, J., and Marston, A., 1980, Antidepressant drugs elevate rat pineal and plasma melatonin, Experientia, 36:442-444.
Wurtman, R. J., Axelrod, J., and Kelly, D. E., 1968, "The Pineal Gland," Academic Press, New York.
Wurtman, R. J., Axelrod, J., and Potter, L. T., 1964b, The uptake of H^3-melatonin in endocrine and nervous tissues and the effects of constant light exposure, J. Pharmacol. Exp. Ther., 143:314-318.
Wurtman, R. J., Axelrod, J., and Barchas, J. D. F., 1964c, Age and enzyme activity in the human pineal, J. Clin. Endocrinol. Metab., 24:299-301.

MELATONIN AS A CHRONOBIOLOGICAL MARKER IN HEALTH AND DISEASE

Lennart Wetterberg

Karolinska Institute, Department of Psychiatry
St Göran's Hospital, Box 125 00
S-112 81 Stockholm, Sweden

INTRODUCTION

The psychiatric literature has reported possible relationship between mental illness and time of day, season of birth and even lunar cycle. Many psychiatric patients e.g. with affective disorders cycle with more or less regularity. The modern life style has led scientists to study body rhythms in shift-workers, air-travellers across time zones and astronauts. The influence of rapid shifts in light-dark cycles and magnetic fields, social isolations and even weightlessness has been examined. In order to understand the effects of these changes there is a need for new markers for endogenous rhythms. Hormonal variations may be associated with boredom, withdrawl and irritability in healthy persons. Desynchronization, upsetting the circadian, circaannual and other rhythms may also cause fatigue, decreased performance, insomnia, anxiety, depression and other psychiatric and somatic symptoms.

The objective of this review is to summarize the results of the recent and ongoing studies in which the pineal gland hormone melatonin has been used as a chronobiological marker in health and disease in man. Melatonin (5-methoxy-N-acetyl-tryptamine) is synthesized from tryptophan via 5-hydroxytryptophan, serotonin and acetylserotonin and this pathway is illustrated in Fig. 1.

Measurement of melatonin

The current status of assay methods of melatonin was recently summarized by Arendt (1981). In her review tadpole bioassay, gas-chromatography-mass spectrometry and radioimmunoassay (RIA) were

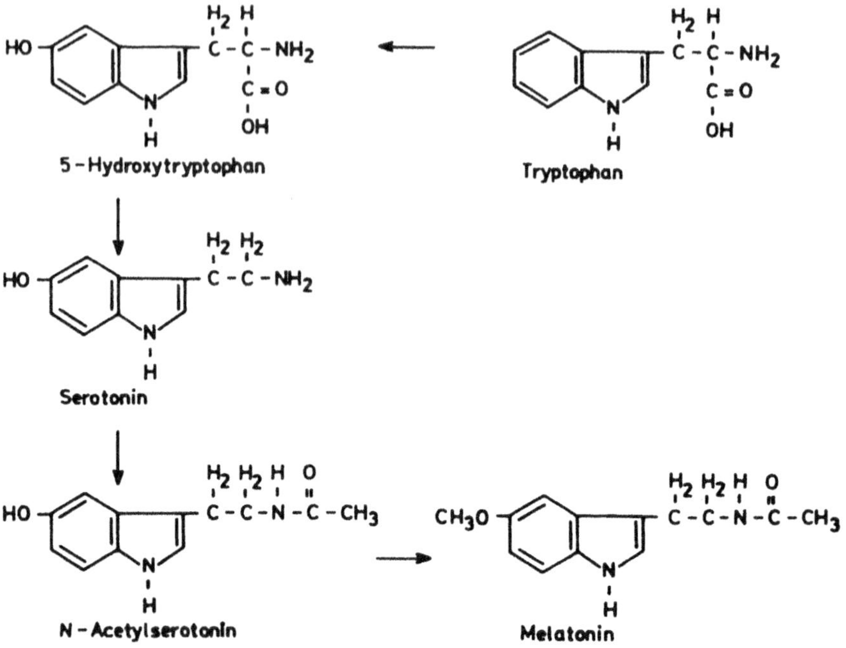

Fig. 1. Metabolic pathway of melatonin from tryptophan

discussed. Several RIA-methods were compared in an international collaborative study (Wetterberg and Eriksson, 1981). Most of the 12 participating laboratories reported comparable results. The most sensitive and specific melatonin antibody reported so far has been used in more than 30 studies and is now commercially available in a RIA melatonin kit (WHB, Box 19018, S-16142 Bromma, Sweden). Using this assay melatonin has so far been found in amniotic fluid cerebrospinal fluid, serum, plasma, lymph, saliva, tears, urine and pineal gland tissue (Wetterberg, 1979).

Melatonin as a marker for adrenergic β-receptor blockade and stimulation

Melatonin in serum or plasma and urine from healthy controls on a regular sleep-dark cycle displays a diurnal rhythm with the lowest concentrations during the day and peak levels during the night. This nocturnal peak of melatonin occurs in most people around 02 hr when they are sleeping in the dark (Fig. 2).

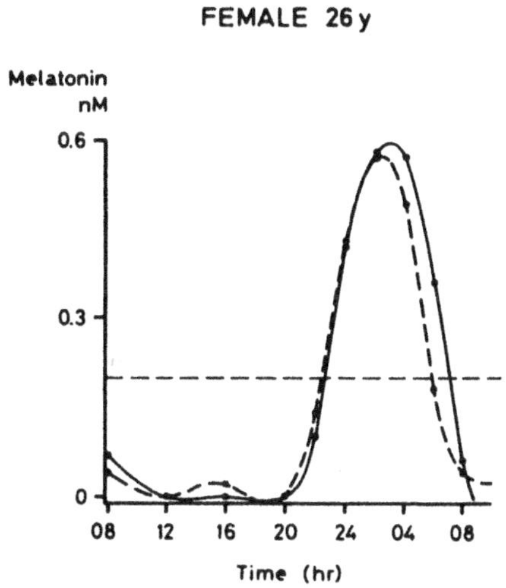

Fig. 2. Serum melatonin levels in a person with normal night/day pattern studied twice, one month apart. The woman was sleeping in the dark between 23 and 07 hr. The dotted horizontal line indicates the low normal range of nocturnal peak level of serum melatonin.

The human pineal is thought to be innervated by sympathetic neurons. Melatonin in serum decreases following treatment with adrenergic β-receptor blocking agents (Vaughan et al., 1976; Hanssen et al., 1980; Moore et al., 1979). This blockade occurs even at moderate doses of 60 mg propranolol per 24 hr which indicate that melatonin production in humans may also be under noradrenergic control.

However, administration of adrenergic β-receptor-stimulating agents have not been effective in stimulating melatonin production (Vaughan et al., 1976; Moore et al., 1979). A shortcoming in the human studies of the pineal function is that there is, as yet no reliable methodology for a stimulatory test of the melatonin production.

Effect of light on melatonin in healthy humans

In an early report by Wetterberg (1978) bright light from day light tubes suppressed blood melatonin at night (Fig. 3), although short-time (30 min) exposure of a 40 W light bulb did not cause a significant decrease.

Urinary excretion of melatonin was measured during 64 hours of sleep deprivation in 12 healthy males when the subjects were exposed to artificial light in the laboratory. The light was kept

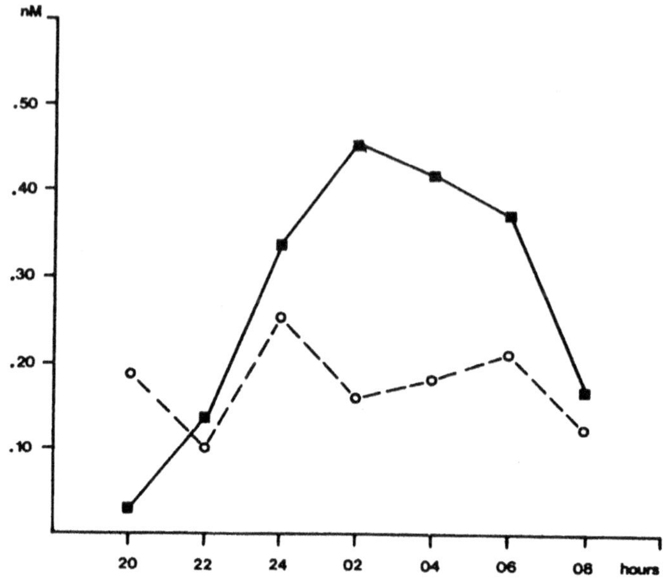

Fig. 3. Plasma melatonin concentration in a 25 year old healthy male during the night sleeping in darkness (■——■) and awake in bright light (○ - - - ○). (From Wetterberg, 1978).

constant and consisted of five light bulbs in the range 40-100 W (Åkerstedt et al., 1979). In this experiment the urinary excretion of melatonin followed a circadian pattern even though the subjects were exposed to light of the same intensity which rapidly suppressed melatonin in rat serum and pineal gland (Illnerova et al., 1978). The practical conclusion of these studies for blood sampling of melatonin at night is that dim light does not suppress melatonin in humans but bright light should be avoided when the physiological status is examined.

Humans are however under certain conditions sensitive to alteration of the light/dark cycle. In four healthy subjects exposed to an extended dark period of four hours in the morning for ten days the melatonin rhythm was shifted with peak levels later in the night (Wetterberg, 1978). This result was interpreted "that the rhythm does not seem to be regulated only by the time of onset of darkness and sleep but rather dependent on the total length of the period of darkness".

In a series of experiments Lewy et al. (1980) further studied the effects of bright light and confirmed its suppressive effect in healthy humans. Although the qualitative effect of light on serum melatonin in human is similar to that of other mammals, including monkeys, the quantitative difference is marked (Reppert et al., 1981). The results of these experiments indicate that the differences between man and other mammals may be due to the fact that man has been exposed to artifical light at night for many hundred thousand years through this use of fire. The important difference in sensitivity to light between man and animal however, has been revealed using melatonin as a marker for circadian rhythms.

Melatonin as a marker for circadian rhythm disorder in affective disease.

Halberg (1968) proposed that manic-depressive cycles could result from desynchronization of two biological clock systems. Experimental shifts in the timing of sleep cause dysphoric mood and poor performance (Kripke et al., 1970). A circadian covariation of fatigue and urinary melatonin has been reported by Åkerstedt et al. (1982). In 12 healthy male subjects, vigilance performance and self-rated sleepiness showed a correlation to melatonin. High levels of the hormone was associated with reduced performance and increased sleepiness. The relation between melatonin and fatigue was comparable with the suggestion that the pineal hormone might be involved in the physiological sleep-wake regulation.

The occurrence of mood change and disturbed 24 hr rhythm in depressed patients (Sachar et al., 1973) has prompted studies of melatonin secretion in affective disorders. Both the hypothesis of a low β-adrenergic activity and estimations of time shift for the evaluation of the phase advance hypothesis in depression (Wehr et al., 1982) may be tested using melatonin as a marker for amplitude and rhythm.

A possible relationship between high cortisol levels found in many patients with depression (Gibbons, 1964; Sachar et al., 1970) and melatonin has recently been reported (Beck-Friis et al., 1981; Wetterberg et al., 1979a; 1981a; 1982a,b,c,d). The hypothesis that the dexamethasone non-suppressors may have a dysfunction in the pineal gland was also investigated. These studies comprise so far more than 50 patients with affective disorder and 30 healthy control subjects.

Melatonin and cortisol was assayed in blood samples drawn every four hr between 08 - 20 hr and every two hr during the night. The patients were sleeping in the dark between 23 and 07 hr. A dexamethasone suppression test (DST) was included in the

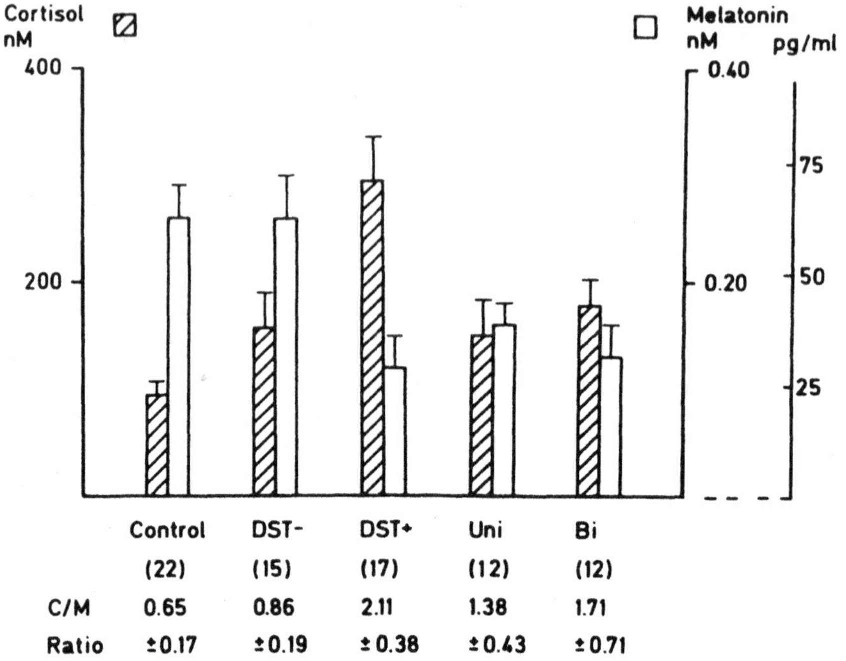

Fig. 4. Cortisol (C) and melatonin (M) in serum at 02 hr (mean ± SEM) in apparently healthy controls and patients with acute major depressive disorder. DST - = normal and DST + = abnormal dexamethasone suppression test. Uni = unipolar and Bi = bipolar affective disorder. For details see Wetterberg et al., 1982d.

test protocol using one mg of oral dexamethasone according to Carroll (1982). The main results of the studies from our laboratory are summarized in Fig. 4 adapted from a recent report by Wetterberg et al. (1982d). The 32 patients with acute major depressive disorder have been divided into two groups according to their response to DST. It is seen that the patients with an abnormal DST had significantly higher cortisol to melatonin ratio than did the patients with normal DST. The 24 patients with with unipolar and bipolar affective disorder in Fig. 4 all had a severe disease history but were studied in a state free from acute clinical manifestations. The patients had significantly lower melatonin levels than the 22 healthy controls. Results showing significantly higher hereditary factors for major depressive disorders among probands with low melatonin levels in serum compared to probands with higher levels points to the possibility that low melatonin production may be associated with the vulnerability of depression (Wetterberg et al., 1982d). Melatonin may thus be a genetic trait marker of the gene for affective disorder. The identification of persons at risk for depression may be helpful in prophylaxis and choice of therapy.

Lewy et al. (1981) reported that manic-depressive patients might be supersensitive to light and Kripke (1981) used bright early morning light to palliate depressive symptom. Weller and Jauhar (1981) found that changing of time zone by two hr or more cause a significant excess of affectively ill persons compared to passengers flying within the same time zone. No such correlation was found for schizophrenic break-downs. It will be important to further study the melatonin levels in persons at risk for depression. The rapid change of circadian light ratio between day/night in the spring and fall in several regions have been associated with depressive disorders (see Parker and Walter, 1982). These authors have also discussed a hypothetical link between depression and a pineal gland dysfunction.

Urinary melatonin rhythms have been studied in depressed patients and normals during sleep deprivation which is used as treatment for acute depressive episodes (Jimerson et al., 1977). Lynch et al. (1978) used melatonin measurement as marker for entrainment of rhythmic secretion in man in an experiment when the light-dark cycle was shifted 12 hr in phase.

Melatonin and pituitary and pineal tumours

A relationship between melatonin and pituitary insufficiencies has been reported by Werner et al. (1981) who found that six patients with pituitary tumours had low melatonin levels in serum. In ten patients with Cushing's disease six had low melatonin at 02 hr. This could possibly be interpreted as indicating a subgroup of patients with Cushing's disease who have a suprapituitary defect as part of their dysfunction. Brismar et al. (1982) have proposed that Cushing's disease, in some cases, might possibly be associated with a defect in the pineal gland. Such a defect has been hypothesized to be a decrease of a corticotropin-releasing factor-inhibiting factor (CRF-IF) which in its turn may be related to melatonin (Wetterberg et al., 1981a).

Young (1981) has reported abnormally high melatonin levels in three of four patients with Cushing's disease. Obviously, more studies with larger samples of patients are needed to clarify the possible relationship between the pineal and pituitary glands in Cushing's disease.

Melatonin may be a marker for melatonin-producing tumours in the pineal gland but it is also clear that some pinealoma do not have excess production of melatonin (Arendt, 1978; Barbet et al., 1978; Tapp, 1978, Vaughan et al., 1978; Wetterberg, 1979).

Melatonin and breast cancer

A possible relationship between breast cancer and melatonin has been discussed by Cohen et al. (1978) and Lapin (1978). Pineal extracts reportedly counteracts malignancy and pinealectomy in rats is associated with accelerated tumour growth. In a study by Wetterberg et al. (1979b) testing the possible association between melatonin levels and risk for breast cancer it was found that Japanese women had lower melatonin levels than American women, although Japanese women are presumably at lower risk than white Americans. The melatonin levels have to be followed in longitudinal studies to test the hypothesis that the brake to carcinogenesis is diagnosed as a case of chronopathology, a rhythm alteration.

Some indications that chronopathology indeed may occur in breast cancer has been provided by Bartsch et al. (1981) who measured urinary melatonin in ten Indian women suffering from advanced stages of breast cancer and in nine controls. The 24 hr urinary melatonin excretion in the cancer patients was on the average 31 % decreased and the cancer patients showed a dyssynchronization of their melatonin rhythm compared to controls.

A recent report by Tamarkin et al. (1982) showed decreased plasma melatonin in patients with breast cancer. The authors suggest that "low nocturnal melatonin concentrations may indicate the presence of estrogen receptor positive breast cancer and could conceivably have etiologic significance".

Monthly and yearly rhythms of melatonin

A melatonin rhythm over the menstrual cycle has been reported (Wetterberg et al., 1976; Birau et al., 1981). However, Matthews et al. (1981) did not confirm these changes using 6-hydroxymelatonin sulphate measurement of urinary excretion during a normal menstrual cycle. More studies of both the phase and amplitude of melatonin secretion are needed in relation to the menstrual cycle in women.

Seasonal variations in the melatonin rhythm was reported by Arendt et al. (1977) in healthy persons from middle Europe. The circadian rhythmic urinary melatonin excretion over four seasons in Kyushu was also studied in clinically healthy Japanese subjects (Wetterberg et al., 1981b). In this report the average urine melatonin exhibited remarkably stable acrophase in a group assessment throughout the seasons.

Concluding comments

Melatonin as a marker in relation to reproduction has been reviewed at this meeting by Reiter. The reader is also referred to each original paper for methodological details and to earlier reviews about the clinical aspects of the pineal gland (eg. Wetterberg, 1981a).

Although the normal function of melatonin is not yet known, the hormone has, during the last few years, found its place as a useful chronobiological marker. One reason for this is the pronounced rhythmic day/night fashion in which it is produced. Another is the close relationship between melatonin and the regulation of the pituitary-adrenal axis, as well as its correlation to behavioural states in man. Melatonin is an additional tool in chronobiology which allows us to look into the neuropsychoendocrine window for causes of suprapituitary desynchronization, a major disturbance in many psychiatric conditions.

Acknowledgements

This review is based upon studies supported by grants from the Swedish Medical Research Council no 3371, Karolinska Institute and St. Göransfonden för psykiatrisk forskning. The secretarial assistance of Elsa Rylander is acknowledged.

References

Åkerstedt, T., Fröberg, J.E., Friberg, Y., and Wetterberg, L., 1979, Melatonin excretion, body temperature and subjective arousal during 64 hours of sleep deprivation, Psychoneuroendocrinology, 4:219.

Åkerstedt, T., Gillberg, M., and Wetterberg, L., 1982, The circadian covariation of fatigue and urinary melatonin, Biol. Psychiatry, 17:547.

Arendt, J., 1978, Melatonin as a tumour marker in a patient with pineal tumour, Brit.med.J., 2:635.

Arendt, J., 1981, Current status of assay methods of melatonin, in: Melatonin - current status and perspectives, N. Birau, W. Schloot, eds., p 3, Pergamon Press, Oxford.

Arendt, J., Wirz-Justice, A., and Bradtke, J., 1977, Annual rhythm of serum melatonin in man, Neurosci.Letters, 7:327.

Barber, S.G., Smith, J.A., Cove, D.H., Smith, S.C.H., and London, D.R., 1978, Marker for pineal tumours, Lancet, 2:372.

Bartsch, C., Bartsch, H., Jain, A.K., Laumas, K.R., and Wetterberg, L., 1981, Urinary melatonin levels in human breast cancer patients, J.Neural.Transm. 52: 281.

Beck-Friis, J., Aperia, B., Kjellman, B.F., Ljunggren, J.-G., Nilsonne, Å., Petterson, U., Tham, A., Undén, F., and Wetterberg, L., 1981, Hormonal changes in acute depression, in: Biological Psychiatry, C.Perris, G. Struwe, B. Jansson, eds., p. 1244, Elsevier/North-Holland, Biomedical Press, Amsterdam.

Birau, N., Birau, M., and Schloot, W., 1981, Melatonin rhythms in human serum, in: Melatonin - curent status and perspectives, N. Birau, W. Schloot, eds. p. 287, Pergamon Press, Oxford.

Brismar, K., Werner, S., and Wetterberg, L., 1982, Melatonin and cortico-steroid response to Metyrapone in patient's with pituitary disease, in: The pineal and its hormones, R.J. Reiter, ed. p. 000, Alan R. Liss, New York.

Carroll, B.J., 1982, The dexamethasone suppression test for melancholia, Br.J.Psychiatry, 140: 292.

Cohen, M., Lippman, M., and Chagner, B., 1978, Role of pineal gland in aetiology and treatment of breast cancer, Lancet, 2:814.

Gibbons, J.L.,1964, Cortisol secretion rate in depressive illness, Arch.Gen.Psychiatry, 10:572.

Halberg, F., 1968, Physiologic considerations underlying rhythmometry with special reference to emotional illness, in: Cycles Biologiques et Psychiatrie, J. deAjuriaguerra, ed., p. 73, Masson et Cie., Paris.

Hanssen, T., Heyden, T., Sundberg, I., Alfredsson, G., Nybäck, H., and Wetterberg, L., 1980, Propranolol in schizophrenia: clinical, metabolic and pharmacological findings. Arch.Gen. Psychiatry, 35: 685.

Illnerova, H., Bäckström, M., Sääf, J., Wetterberg, L., and Vangbo, B., 1978, Melatonin in rat pineal gland and serum; rapid parallel decline after light exposure at night, Neurosci. Letters, 9: 189.

Jimerson, D.C., Lynch, H.J., Post, R.M., Wurtman, R.J., and Bunney, W.E., 1977, Urinary melatonin rhythms during sleep deprivation in depressed patients and normals, Life Sci, 20: 1501.

Kripke, D.F., 1981, Photoperiodic mechanisms for depression and its treatment, in: Biological Psychiatry, C. Perris, G. Struwe, B. Jansson, eds., p. 1249, Elsevier/North-Holland. Biomedical Press, Amsterdam.

Kripke, D.F., Cook, B., and Lewis, O.F., 1970, Sleep in night workers: EEG recordings, Psychophysiology, 7: 377.

Lapin, V., 1978, Effects of reserpine on the incidence of 9,10-dimethyl-1,2-benzanthracene-induced tumors in pinealectomised and thymectomised rats, Oncology, 35: 132.

Lewy, A.J., Wehr, T.A., Goodwin, F.K., Newsome, D.A., and

Markey, S.P., 1980, Light suppressess melatonin secretion in humans, Science, 20: 1267.
Lewy, A.J., Wehr, T.A., Goodwin, F.K., Newsome, D.A., and Rosenthal, N.E., 1981, Manic-depressive patients may be supersensitive to light, Lancet, i, 383.
Lynch, H.J., Jimerson, D.C., Ozaki, Y., Post, R.M., Bunney, W.E., and Wurtman, R.J., 1978, Entrainment of rhythmic melatonin secretion in man to a 12-hour phase shift in light-dark cycle, Life Sci, 23: 1557.
Matthews, C.D., Kennaway, D.J., Fellenberg, A.J.G., Phillipou, G., Cox, L.W., and Seamark, R.F., 1981, Melatonin in man, in: Melatonin - current status and perspectives, N. Birau, W. Schloot, eds, p. 371, Pergamon Press, Oxford.
Moore, D.C., Paunier, L., and Sizonenko, P.C., 1979, Effect of adrenergic stimulation and blockade on melatonin secretion in the human, in: The Pineal gland of Vertebrates including Man, J. Ariëns Kappers, P. Pévet, eds. p. 517, Elsevier/ North-Holland, Biomedical Press, Amsterdam.
Parker, G., and Walter, S., 1982, Seasonal variation in depressive disorders and suicidal deaths in New South Wales. Br.J.Psychiatry, 140: 626.
Reppert, S.M., Perlow, M.J., Tamarkin, L., Orloff, D., and Klein, D.C., 1981, The effects of environmental lighting on the daily melatonin rhythm in primate cerebrospinal fluid, Brain Res., 223: 313.
Sachar, E.J., Hellman, L., Fukushima, D., and Gallagher, T., 1970, Cortisol production in depressive illness: A clinical and biochemical clarification. Arch. Gen.Psychiatry, 23: 289.
Sachar, E.J., Hellman, L., Roffwarg, H., Halpern, F., Fukushima, D., and Gallagher, T., 1973, Disrupted 24-hours pattern of cortisol secretion in psychotic depression, Arch. Gen. Psychiatry, 28: 19.
Tamarkin, L., Danforth, D., Lichter, A., DeMoss, E., Cohen, M., Chabner, B., and Lippman, M., 1982, Decreased nocturnal plasma melatonin peak in patients with estrogen receptor positive breast cancer, Science, 216: 1003.
Tapp, E., 1978, Melatonin as a tumour marker in a patient with pineal tumour. Brit. med. J., 2:635.
Vaughan, G.M., McDonald, S.D., Jordan, R.M., Allen, J.P., Bohmfalk, G.L., Abousamra, M., and Story, J.L., 1978, Melatonin concentration in human blood and cerebrospinal fluid - relationship to stress, J.clin.Endocr.Metab., 47: 220.
Vaughan, G.M., Pelham, R.W., Pang, S.F., Loughlin, L.L.,Wilson, K.M., Sandock, K.L., Vaughan, M.K., Koslow, S.H., and Reiter, R.J., 1976, Nocturnal elevation of plasma melatonin and urinary 5-hydroxyindoleacetic acid in young men: Attempts at modification by brief changes in environmental lighting and sleep and by autonomic drugs. J. clin. Endocr. Metab., 42:752.

Wehr, T.A., Lewy, A.J., Wirz-Justice, A., Craig, C., and Tamarkin, L., 1982, Antidepressants and a circadian rhythm phase-advance hypothesis of depression, in: Brain Peptides and Hormones, R. Collu, R.J. Ducharme, A.Barbeau, G. Tolis, eds, p. 263, Raven Press, New York.

Weller, M.O.I., and Jauhar, P., 1981, Travel induces disturbances in circadian rhythms as precipitants of affective illness, in: Biological Psychiatry, C. Perris, G. Struwe, B. Jansson, eds., p. 1253, Elsevier/North-Holland, Biomedical Press, Amsterdam.

Werner, S., Brismar, K., Wetterberg, L., and Eneroth, P., 1981, Circadian rhythms of melatonin, prolactin growth hormone and cortisol in patients with pituitary adenomas, empty sella syndrome and Cushing's syndrome due to adrenal tumours, in: Melatonin: Current status and perspectives, N. Birau, W. Schloot, eds., p. 357, Pergamon Press Ltd, Oxford.

Wetterberg, L., 1978, Melatonin in humans; physiological and clinical studies, J.Neural. Transm., Suppl. 13: 289.

Wetterberg, L., 1979, Clinical importance of melatonin, in: The Pineal gland of Vertebrates including man, J. Ariens Kappers, P. Pévet, eds., p. 539, Elsevier/North-Holland, Biomedical Press, Amsterdam.

Wetterberg, L., Aperia, B., Beck-Friis, J., Kjellman, B.F., Ljunggren, J.-G., Petterson, U., Sjölin, Å., Tham, A., and Undén. F., 1981a,Pineal-hypothalamic-pituitary function in patients with depressive illness, in: Steroid Hormone Regulation of the Brain, K. Fuxe, J.-A. Gustafsson, L. Wetterberg, eds., p. 397, Pergamon Press Ltd, Oxford.

Wetterberg, L., Aperia, B., Beck.Friis, J., Kjellman, B.F., Ljunggren, J.-G., Petterson, U., Sjölin, Å., Tham, A., and Undén, F., 1982a, Hormonal changes in affective disorders, in: Brain Peptides and Hormones, R. Collu, J.R. Ducharme, A. Barbeau, G. Tolis, eds., p. 257, Raven Press, New York.

Wetterberg, L., Aperia, B., Beck-Friis, J., Kjellman, B.F., Ljunggren, J.-G., Nilsonne, Å., Petterson, U., Tham, A., and Undén, F., 1982b, Melatonin and cortisol levels in pschiatric illness, Lancet, July 10, 100.

Wetterberg, L., Aperia, B., Beck-Friis, J., Kjellman, B.F., Ljunggren, J.-G., Nilsonne, Å., Petterson, U., Tham, A., and Undén, F., 1982c, The relationship between cortisol and melatonin serum levels in patients with affective disorders and in healthy subjects, in: Proceedings of XV Int.Conf. of the Int. Soc. Chronobiol., Minneapolis, Minn.

Wetterberg, L., Beck-Friis, J., Kjellman, B.F., and Ljunggren, J.-G., 1982d, Circadian rhythms in melatonin and cortisol secretion in depression, in: Frontiers in biochemical and pharmacological research in depression, F. Sjöqvist, M. Åsberg, E. Usdin, eds., p. 000, Raven Press, New York.

Wetterberg, L., Beck-Friis, J., Aperia, B., and Petterson, U., 1979a, Melatonin/cortisol ratio in depression, Lancet, ii, 1361.

Wetterberg, L., and Eriksson, O., 1981, Melatonin in human serum - A collaborative study of current radioimmunoassays, in: Melatonin: Current status and perspectives, N. Birau, W. Schloot, eds., p. 15, Pergamon Press Ltd, Oxford.

Wetterberg, L., Halberg, F., Haus, E., Kawasaki, T., Uezono, K., Ueno, M., and Omae, T., 1981b, Circadian rhythmic urinary melatonin excretion in four seasons by clinically healthy Japanese subjects in Kyushu, Chronobiologia, VIII: 188.

Wetterberg, L., Halberg, F., Tarquini, B., Cagnoni, M., Haus, E., Griffith, K., Kawasaki, T., Wallach, L.-A., Ueno, M., Uezo, K., Matsuoka, M., Kuzel, M., Halberg, E., and Omae,T., 1979b, Circadian variation in urinary melatonin in clinically healthy women in Japan and the United States of America, Experientia, 35: 416.

Young, I.M., 1981, The pineal indole hormones in Cushing's disease and acromegaly, in: Pineal Function, C.D. Matthews, R.F. Seamark, eds., p. 7, Elsevier/North-Holland, Biomedical Press, Amsterdam.

INDEX

ODE TO THE ERICE MEETING

Erice was the venue which NATO did pick
We avoided the water, so as not to get sick.
Our "clocks" were confused, when first we met
This caused many to sleep, directly on the set.

The speakers were good, the students were great
Discussion was ample, and we never ran late
All matters were discussed, well into the night
And we did it all, without a fist fight.

Axelrod began, the meeting with skill
While Wurtman stayed only, briefly on the hill
Oksche from Giessen gave a fine talk
He rarely used slides, and never used chalk.

Vollrath and Kappers considered the gland
Many items were mentioned, even brain sand
The biochemistry was investigated by Cardinali's group
When he finished his talk, it was alphabet soup.

Bob Moore described retinohypothalamic way
By which light tells the SCN exactly what to say
Lewy and Poth, looked at the human in depth
Urine and blood were collected while the subjects slept.

Wetterberg and Brown, they did it with class
While the Frenchman Pevet, was with his lass
Semm from Frankfurt, had a new thought
I expected some doubt but no one fought.

From Milano, Calabrino did not come
But Demaine and Ebels, provided the fun
Mess and Quay, each spoke twice
While the discussants themselves, got their slice.

On a visit to Marsala to see a ship
A pig in the sea was the hit of the trip
On the beach in San Vito, we all turned red
While to Segesta and Selinunte we were kindly led.

To Fraschini and Velo, we bid you adieu
We wish success, to you and your crew
Reiter is concerned about the short days of fall
As a result he says, arrivederci to all.

Russel J Reiter

GPSR Compliance

The European Union's (EU) General Product Safety Regulation (GPSR) is a set of rules that requires consumer products to be safe and our obligations to ensure this.

If you have any concerns about our products, you can contact us on ProductSafety@springernature.com

In case Publisher is established outside the EU, the EU authorized representative is:

Springer Nature Customer Service Center GmbH
Europaplatz 3
69115 Heidelberg, Germany

Batch number: 10761588

Printed by Printforce, the Netherlands